NUTRITION AND HEALTH

Adrianne Bendich, PhD, FACN, SERIES EDITOR

For further volumes:
http://www.springer.com/series/7659

Ronald Ross Watson · Victor R. Preedy · Sherma Zibadi
Editors

Magnesium in Human Health and Disease

Humana Press

Editors
Ronald Ross Watson
Mel and Enid Zuckerman College
of Public Health
University of Arizona
Tucson, AZ, USA

Victor R. Preedy
Department of Nutrition and Dietetics
King's College London
London, UK

Sherma Zibadi
Division of Health Promotion Sciences
Mel and Enid Zuckerman
University of Arizona
Tucson, AZ, USA

ISBN 978-1-4939-6290-7 ISBN 978-1-62703-044-1 (eBook)
DOI 10.1007/978-1-62703-044-1
Springer New York Heidelberg Dordrecht London

Printed on acid-free paper

Humana Press is a brand of Springer
Springer is part of Springer Science+Business Media (www.springer.com)

Series Editor Page

The great success of the Nutrition and Health Series is the result of the consistent overriding mission of providing health professionals with texts that are essential because each includes: 1) a synthesis of the state of the science, 2) timely, in-depth reviews by the leading researchers in their respective fields, 3) extensive, up-to-date fully annotated reference lists, 4) a detailed index, 5) relevant tables and figures, 6) identification of paradigm shifts and the consequences, 7) virtually no overlap of information between chapters, but targeted, inter-chapter referrals, 8) suggestions of areas for future research and 9) balanced, data-driven answers to patient as well as health professionals questions which are based upon the totality of evidence rather than the findings of any single study.

The Series volumes are not the outcome of a symposium. Rather, each editor has the potential to examine a chosen area with a broad perspective, both in subject matter as well as in the choice of chapter authors. The editor(s), whose training(s) is (are) both research and practice oriented, have the opportunity to develop a primary objective for their book, define the scope and focus, and then invite the leading authorities to be part of their initiative. The authors are encouraged to provide an overview of the field, discuss their own research and relate the research findings to potential human health consequences. Because each book is developed de novo, the chapters are coordinated so that the resulting volume imparts greater knowledge than the sum of the information contained in the individual chapters.

"Magnesium and Health", edited by Professor Ronald Ross Watson PhD and Professor Victor R. Preedy PhD DSc FRIPH, is a welcome addition to the Nutrition and Health Series. Magnesium, as an essential mineral and the 4th most abundant mineral in the body and 2nd most abundant mineral within cells, is critical to the formation of energy within each cell of the body and is also involved in muscle contraction, neurological and cardiovascular functions and bone metabolism as examples. Over the past decade, there has been an increased interest in the importance of magnesium in the area of cellular biology as well as clinical nutrition. Thus, it is timely that the first volume solely dedicated to objective reviews of the latest research on magnesium and human health and disease be developed. This is the first volume on magnesium for the health professional which is neither the result of a scientific conference nor a single chapter that attempts to review an entire field of research.

The 20 chapters in this comprehensive volume examine the biological as well as clinical consequences of magnesium deficiency and review the data related to the beneficial effects of optimal magnesium status. The book is logically organized into five sections and begins with an overview section that includes informative chapters on the assessment of magnesium status, dietary sources of this essential nutrient, and reviews the essential functions of magnesium in inflammation, endothelial function and cytokine regulation using the clinically relevant data on the role of magnesium in cardiovascular disease as well as placental and neonatal neurological function. The extensively referenced first chapter includes over 500 citations as well as excellent tables of clinical importance.

The second section contains four chapters that describe the importance of magnesium status in certain clinical conditions and chronic diseases. The first chapter examines the negative effects of magnesium deficiency that has been seen in asthmatics. Evidence suggests that magnesium can directly influence lung function by regulating smooth muscle contractility and neuromuscular excitability, immune function, inflammation and oxidative stress. Research indicates that adults with mild to moderate asthma may have inadequate magnesium intakes and may benefit from taking magnesium supplements. Another critical role of magnesium is electrolyte homeostasis. The chapter on magnesium and kidney function examines the importance of magnesium-dependent enzymes in controlling the electrolyte pumps in the glomerulus. This chapter clearly demonstrates the importance of the kidney to optimal cardiovascular function and their interactions during renal disease. Since the kidney controls magnesium blood levels, any alteration in kidney function will affect systemic magnesium levels. Adequate magnesium intake is also important in maintaining glucose and insulin homeostasis. Magnesium is an essential mineral needed for activation of over 300 enzymes, glucose transportation between membranes, glucose oxidation, all reactions involving phosphorylation, energy exchange, and for the proper activity of insulin. The next two chapters examine the association of magnesium status, genetic factors involved with magnesium metabolism and the risk of development of type 2 diabetes. The chapter by Song et al. includes informative tables and figures that include prospective studies linking magnesium status and type 2 diabetes and an extensive list of potential candidate genes involved in magnesium metabolism.

The third section contains unique chapters that examine the potential for magnesium supplementation to beneficially affect certain disease conditions. Type 2 diabetes is reviewed in two unique chapters; the first examines the potential for supplementation to affect insulin actions in the metabolic syndrome, hypertension and type 2 diabetes. The second chapter looks at the data from Asian versus non-Asian populations with regard to magnesium intake and risk of type 2 diabetes and considers the potential role of ethnically-specific genetic factors in differences in epidemiological findings linking magnesium to type 2 diabetes. As mentioned above, magnesium is important in bone formation and is a major component of the outer surface of bone. Magnesium also is involved in the regulation of parathyroid secretion as well as its actions and also affects vitamin D metabolism. The data on magnesium status and osteoporosis as well as magnesium supplementation and bone density are reviewed in detail. The last chapter in this section describes the adverse effects of certain cancer chemotherapy drugs including cisplatin, 5- fluorouracil and leucovorin. Due to the loss of absorptive surface of the gastrointestinal tract following chemotherapy, magnesium levels may be significantly reduced and some of the serious adverse effects associated with these drugs, including significant neuropathy, may be due in part to lowered magnesium status. Tables included in this chapter outline the clinical studies where magnesium supplementation has been provided to cancer patients treated with the above-mentioned drugs and the resultant reduction in adverse effects.

The fourth section contains four chapters that specifically examine the critical role of magnesium in hypertension and cardiovascular disease. Magnesium's role as a natural calcium channel blocker theoretically should affect blood pressure. The chapter objectively examines the epidemiological data as well as mechanisms of action of magnesium and posits that higher than recommended intakes of magnesium may lower blood pressure especially if combined with reduced sodium intake. Magnesium addition in vitro to heart cells prevents intracellular depletion of magnesium, potassium and high-energy phosphates. Laboratory animal studies report that magnesium supplementation improves myocardial metabolism, prevents intra-mitochondrial calcium accumulation and reduces vulnerability to oxygen-derived free radicals. Magnesium affects vascular tone, platelet aggregation, endothelial function, infarct size, lipid metabolism, cardiac arrhythmias, myocardial infarction and heart failure. The next three chapters provide detailed review of the studies that show the beneficial impact of magnesium on cardiovascular tissues and resultant data linking low magnesium status with increased risk of cardiovascular diseases and their consequences. Data on the use of intravenous magnesium in patients with arrhythmias are also reviewed objectively. Although intravenous magnesium has been used for a variety of ventricular arrhythmias, the evidence to support its anti-arrhythmic effects is strongest in the prevention of atrial fibrillation after cardiac surgery, reduction of ventricular rate in acute-onset atrial fibrillation, and prevention and treatment of certain types of tachycardia.

The final section on magnesium and neurological function contains five comprehensive, clinically relevant chapters. The chapters include investigations that span the entire lifetime from prenatal development to loss of neurological function associated with aging. The chapters include chronic diseases as well as in-depth discussions of the mechanisms by which acute injuries such as thermal or electrical burns, head or musculoskeletal trauma, subarachnoid hemorrhage and/or intracerebral bleeds can result in loss of magnesium or increased requirement. Acute stressor states involve systemic inflammatory responses accompanied by neurohormonal activation. The adrenergic nervous and renin-angiotensin-aldosterone systems and effector hormones are integral to stressor responses that require optimal levels of magnesium for reduction of adverse effects. Intravenous magnesium is also used in the treatment of acute brain injuries and the clinical studies are tabulated for the reader. The final chapter reviews the multitude of effects of acute and chronic alcohol exposure. Two effects are to decrease total diet quality including magnesium intake and at the same time, increase magnesium excretion. Low magnesium status has been implicated in the neurological dysfunctions seen with excess alcohol intake.

The logical sequences of the Sections as well as the chapters within each Section enhance the understanding of the latest information on the current standards of practice with regard to the physiological and pharmacological uses of magnesium. The volume is of value to clinicians, related health professionals including dieticians, nurses, pharmacists, physical therapists, and others involved in the successful treatment of hypomagnesemia. This comprehensive volume also has great value for academicians involved in the education of graduate students and post-doctoral fellows, medical students and allied health professionals who plan to interact with patients with relevant disorders.

The volume contains over 55 detailed tables and figures that assist the reader in comprehending the complexities of the metabolism as well as the potential benefits and risks of magnesium on human health. The over-riding goal of this volume is to provide the health professional with balanced documentation and awareness of the newest research and therapeutic approaches including an appreciation of the complexity of the effects magnesium can have on virtually every organ system within the body. Hallmarks of the 20 chapters include key words and bulleted key points at the beginning of each chapter, complete definitions of terms with the abbreviations fully defined for the reader and consistent use of terms between chapters. There are over 2,400 up-to-date references; all chapters include a conclusion to highlight major findings. The volume also contains a highly annotated index.

This unique text provides practical, data-driven resources based upon the totality of the evidence to help the reader understand the basics, treatments and preventive strategies that are involved in the understanding of how magnesium may affect healthy individuals as well as those with acute injuries and/or chronic diseases. Of equal importance, critical issues that involve patient concerns, such as malnourishment, potential effects on mental, cardiovascular and immune functions are included in well-referenced, informative chapters. The overarching goal of the editors is to provide fully referenced information to health professionals so they may have a balanced perspective on the value of various preventive and treatment options that are available today as well as in the foreseeable future.

In conclusion, "Magnesium and Health", edited by Ronald Ross Watson, PhD and Victor R. Preedy, PhD DSc FRIPH FRSH FIBiol FRCPath provides health professionals in many areas of research and practice with the most up-to-date, well referenced and comprehensive volume on the current state of the science concerning magnesium. This volume will serve the reader as the most authoritative resource in the field to date and is a very welcome addition to the Nutrition and Health Series.

Adrianne Bendich, Ph.D., FACN, FASN

Preface

Magnesium is an essential mineral which is required for growth and survival of humans. Since magnesium is a mineral and not synthesizable, it must be obtained through dietary foods and/or supplements. Magnesium is found in many sources, primarily whole grains, green leafy vegetables, nuts, and legumes. Even with many dietary sources, only about one-third of Americans maintain the appropriate dietary intake of magnesium. A very small number of people have drug-induced severe magnesium deficiency. So major issues reviewed are the benefits of magnesium supplementation to reach (a) recommended intakes or (b) *above*-recommended intakes to promote health or treat various diseases and risk factors.

While two-thirds of people have intakes of magnesium below recommended amounts, only a small group is frankly deficient. Symptoms of magnesium deficiency include excitability, weak muscles, and fatigue. Deficiency of magnesium can cause low serum potassium and calcium levels, retention of sodium, and low circulating levels of regulatory hormones. These changes in nutrients cause neurological and muscular symptoms, such as tremor and muscle spasms. Further, magnesium deficiency may cause loss of appetite, nausea, vomiting, personality changes, and death from heart failure. Just as high magnesium intakes improve insulin resistance and diabetics' health, low serum levels play an important role in carbohydrate metabolism and worsen insulin resistance. Causes of magnesium deficiency include alcohol abuse, poorly controlled diabetes, excessive or chronic vomiting, and/or diarrhea. Thus, the effects of inadequate or deficient intakes of magnesium are critical to health and are be reviewed by experts in this book.

Hypermagnesemia is a *rare* electrolyte disturbance caused due to very high serum levels of magnesium. Normally, the kidney is very effective in excreting excess magnesium. Hypermagnesemia occurs due to excessive intakes of antacids or laxatives which contain magnesium salts. Often, very high serum potassium and low calcium are also major causes. These may result in muscle weakness, cardiac arrhythmia, or sudden death. Certain drugs can also deplete magnesium levels, such as osmotic diuretics, some anticancer drugs, cyclosporine, amphetamines, and proton pump inhibitors. Magnesium is absorbed orally at about 30 % bioavailability from any water-soluble salt. Magnesium citrate is a common oral magnesium salt available in 100- and 200-mg magnesium supplements typically per capsule. Some multinutrient supplements sold in developed countries contain magnesium. Insoluble magnesium salts, such as milk of magnesia (magnesium hydroxide) and magnesium oxide, are released by the stomach acid for neutralization before they can be absorbed and, thus, offer poor oral magnesium sources. Severe low serum magnesium levels are treated medically with intravenous or intramuscular magnesium sulfate solutions which are bioavailable and effective. As magnesium excess occurs rarely and usually due to drug use, it will be reviewed but not as a major focus of the book.

Since magnesium is a mineral and not synthesizable, it must be obtained through the dietary foods and/or supplements. Magnesium is found in many sources primarily whole grains, green leafy vegetables, nuts, and legumes. Even with many dietary sources, only about one-third of Americans maintain the appropriate dietary intake of magnesium. Interestingly, higher intakes of magnesium positively affect insulin resistance in type 2 diabetics, suggesting that the optimum intake for people with this disease may be higher than the recommended daily intake. In addition, hypertension, cholesterol levels, and cardiovascular disease are all modified positively for health promotion by high intakes of magnesium. Recent research found that magnesium supplementation in overweight individuals decreased insulin markers and led to changes in genes related to metabolism and inflammation. The benefits of dietary supplements to produce *high* levels and/or treat deficiency are be reviewed by several authors. New research is suggesting more roles of magnesium supplementation as a therapy to reach intakes above the recommended ones. For example, magnesium supplementation are reviewed as a modifier of diseases of old age and for treatment of preeclampsia, asthma, ocular health, etc. Thus, the *primary goal* of this book are to get expert reviews of the potential benefits, or lack thereof, of normal and high magnesium supplementation. Animal model research and early human trials are reviewed to document other disease states that would benefit from increased magnesium intake.

Tucson, AZ, USA — Ronald Ross Watson
London, UK — Victor R. Preedy
Tucson, AZ, USA — Sherma Zibadi

Acknowledgments

The work of our editorial assistant, Bethany L. Stevens, and Kevin M. Wright of Humana Press in communicating with authors and working with the manuscripts and the publisher was critical to the successful completion of the book and is much appreciated. Their daily responses to queries and collection of manuscripts and documents were extremely helpful. Support for Ms. Stevens' work was graciously provided by the Natural Health Research Institute, as part of its mission to communicate to scientists about bioactive foods and dietary supplements was vital (http://www.naturalhealthresearch.org). This was part of its efforts to educate scientists and the lay public on the health and economic benefits of nutrients in the diet as well as supplements. Mari Stoddard of the Arizona Health Sciences Library was instrumental in finding the authors and their addresses in the early stages of the book's preparation. The support of Humana Press staff as well as the input by the series editor, Adrianne Bendich, is greatly appreciated for the improved organization of this book.

Biography

Ronald R. Watson, Ph.D. attended the University of Idaho but graduated from Brigham Young University in Provo, Utah, with a degree in chemistry in 1966. He earned his Ph.D. in biochemistry from Michigan State University in 1971. His postdoctoral schooling in nutrition and microbiology was completed at the Harvard School of Public Health, where he gained 2 years of postdoctoral research experience in immunology and nutrition. From 1973 to 1974 Dr. Watson was assistant professor of immunology and performed research at the University of Mississippi Medical Center in Jackson. He was assistant professor of microbiology and immunology at the Indiana University Medical School from 1974 to 1978 and associate professor at Purdue University in the Department of Food and Nutrition from 1978 to 1982. In 1982 Dr. Watson joined the faculty at the University of Arizona Health Sciences Center in the Department of Family and Community Medicine of the School of Medicine. He is currently professor of health promotion sciences in the Mel and Enid Zuckerman Arizona College of Public Health. Dr. Watson is a member of several national and international nutrition, immunology, cancer, and alcoholism research societies. Among his patents he has one on a dietary supplement; passion fruit peel extract with more pending. He had done DHEA research on its effects on mouse AIDS and immune function for 20 years. He edited a previous book on melatonin (Watson RR. *Health Promotion and Aging: The Role of Dehydroepiandrosterone (DHEA).* Harwood Academic Publishers, 1999, 164 pages). For 30 years he was funded by Wallace Research Foundation to study dietary supplements in health promotion. Dr. Watson has edited more than 100 books on nutrition, dietary supplements and over-the-counter agents, and drugs of abuse as scientific reference books. He has published more than 500 research and review articles.

Victor R. Preedy, BSc, DSc, FSB, FRCPath, FRSPH is currently Professor of Nutritional Biochemistry in the Department of Nutrition and Dietetics, King's College London, and Honorary Professor of Clinical Biochemistry in the Department of Clinical Biochemistry, King's College Hospital. He is also Director of the Genomics Centre, King's College London, and a member of the School of Medicine, King's College London. King's College London is one of the world's leading universities. Professor Preedy gained his Ph.D. in 1981, and in 1992, he received his Membership of the Royal College of Pathologists (MRCPath), based on his published works. He was elected a Fellow of the Royal College of Pathologists (FRCPath) in 2000. In 1993, he gained his second doctorial degree (D.Sc.) for his outstanding contribution to protein metabolism. In 2004, Professor Preedy was elected a Fellow to both the Royal Society for the Promotion of Health (FRSH) and the Royal Institute of Public Health (FRIPHH). In 2009, he was elected a Fellow of the Royal Society for Public Health (RSPH). He is also a Fellow of the Society of Biology (FSB). Professor Preedy has written or edited over 550 articles, which include over 160 peer-reviewed manuscripts based on original research, 85 reviews, and 30 books. His interests pertain to matters concerning nutrition and health at the individual and societal levels.

Dr. Sherma Zibadi, MD, Ph.D., received her Ph.D. in nutrition from the University of Arizona and is a graduate of the Mashhad University of Medical Sciences, where she earned her M.D. She has recently completed her postdoctoral research fellowship awarded by the American Heart Association. Dr. Zibadi engages in the research field of cardiology and complementary medicine. Her main research interests include maladaptive cardiac remodeling and heart failure, studying the underlying mechanisms and potential mediators of the remodeling process, which helps to identify new targets for treatment of heart failure. Dr. Zibadi's research interest also extends into alternative medicine, exploring the preventive and therapeutic effects of natural dietary supplements on heart failure and their major risk factors in both basic animal and clinical studies, translating lab research findings into clinical practice. Dr. Zibadi is an author of multiple research papers published in peer-reviewed journals and books as well as coeditor of several books.

Contents

Contributors

Alaa Amash, Ph.D. Department of Microbiology and Immunology, Ben-Gurion University of the Negev, Beer Sheva, Israel

Tayze T. Antunes, Ph.D. Kidney Research Center, Ottawa Hospital Research Institute, University of Ottawa, Ottawa, Canada

Hasan Aydın, M.D. Department of Internal Medicine, Section of Endocrinology and Metabolism, Yeditepe University Hospital, Istanbul, Turkey

Syamal K. Bhattacharya, Ph.D. Division of Cardiovascular Diseases, University of Tennessee Health Science Center, Memphis, TN, USA

Lubomir Bodnar, M.D., Ph.D. Department of Oncology, Military Institute of Medicine, Warsaw, Poland

Irina Burd, M.D., Ph.D. Department of Gynecology and Obstetrics, Division of Maternal Fetal Medicine, Johns Hopkins University, Baltimore, MD, USA

Neuroscience Laboratory, Kennedy Krieger Institute, Baltimore, MD, USA

Glaucia Callera, Ph.D. Kidney Research Center, Ottawa Hospital Research Institute, University of Ottawa, Ottawa, Canada

Sara A. Chacko, Ph.D., M.P.H. UCLA Department of Epidemiology, Center for Metabolic Disease Prevention, Los Angeles, CA, USA

Dharam Paul Chaudhary, Ph.D. Department of Biochemistry, Directorate of Maize Research, New Delhi, India

Qi Dai, M.D., Ph.D. Vanderbilt Epidemiology Center, Institute for Medicine and Public Health, Nashville, TN, USA

Dwight A. Dishmon, M.D. Division of Cardiovascular Diseases, University of Tennessee Health Science Center, Memphis, TN, USA

Abigail E. Duffine, M.S. Department of Nutrition Sciences, Drexel University, Philadelphia, PA, USA

Tamar Eshkoli, M.D. Obstetrics and Gynecology, Soroka University Medical Center, Beer Sheva, Israel

Valeria Feinshtein, M.Sc. Department of Clinical Pharmacology, Ben-Gurion University of the Negev, Beer Sheva, Israel

Ernest Graham, M.D. Department of Gynecology and Obstetrics, Division of Maternal Fetal Medicine, Johns Hopkins University, Baltimore, MD, USA

Elżbieta Grochowska-Niedworok, Ph.D. Department of Internal Medicine and Department of Public Health, Medical University of Silesia, Bytom, Poland

Kwok M. Ho, M.P.H., Ph.D., FRCP, FANZCA, FCICM Department of Intensive Care Medicine, Royal Perth Hospital, Perth, Australia

School of Population Health, University of Western Australia, Perth, Australia

Gershon Holcberg, M.D. Department of Obstetrics and Gynecology, Soroka University Medical Center, Ben-Gurion University of the Negev, Beer Sheva, Israel

Jack Hou, M.D. Zhang Neuroscience Research Laboratories, Loma Linda University Medical Center, Loma Linda, CA, USA

Mark Houston, M.D., M.S., FACP, FAHA, FASH Department of Medicine, Vanderbilt University School of Medicine, Nashville, TN, USA

Department of Medicine, Hypertension Institute, Saint Thomas Hospital, Nashville, TN, USA

Mahmoud Huleihel, Ph.D. Department of Microbiology and Immunology, Ben-Gurion University of the Negev, Beer Sheva, Israel

Adel A.A. Ismail, B.Pharm. (Hon.), Ph.D., FRCPath Retired consultant in Clinical Biochemistry and Chemical Endocrinology, Mid-Yorkshire and Leeds Teaching Hospitals Trust, Wakefield, West Yorkshire, UK

Yasmin Ismail, B.Sc., M.D., MRCP Bristol Heart Institute, Bristol Royal Infirmary, Bristol, UK

Abbas A. Ismail, B.Sc., M.Sc., M.D., FRCP Stepping Hill Hospital, Poplar Grove; Hazel Grove, Stockport, Cheshire, UK

Michael V. Johnston, M.D. Department of Gynecology and Obstetrics, Division of Maternal Fetal Medicine, Johns Hopkins University, Baltimore, MD, USA

Alexandra Kazaks, Ph.D. Department of Nutrition and Exercise Science, Bastyr University, Kenmore, WA, USA

M. Usman Khan, M.D. Division of Cardiovascular Diseases, University of Tennessee Health Science Center, Memphis, TN, USA

Rami N. Khouzam, M.D. Division of Cardiovascular Diseases, University of Tennessee Health Science Center, Memphis, TN, USA

Teresa Kokot, M.D., Ph.D. Department of Internal Medicine and Department of Human Nutrition, Medical University of Silesia, Bytom, Poland

Babatunde O. Komolafe, M.D. Division of Cardiovascular Diseases, University of Tennessee Health Science Center, Memphis, TN, USA

Simin Liu, M.D., Sc.D., M.P.H. UCLA Departments of Epidemiology, Medicine, and Obstetrics & Gynecology, Center for Metabolic Disease Prevention, Los Angeles, CA, USA

Jesse E. McGee, M.D. Division of Cardiovascular Diseases, University of Tennessee Health Science Center and Veterans Affairs Medical Center, Memphis, TN, USA

Tetsuya Mizoue, M.D., Ph.D. Department of Epidemiology and Prevention, Clinical Research Center, National Center for Global Health and Medicine, Tokyo, Japan

Małgorzata Muc-Wierzgoń, Prof. Department of Internal Medicine and Department of Human Nutrition, Medical University of Silesia, Bytom, Poland

Akiko Nanri, M.D. Department of Epidemiology and Prevention, Clinical Research Center, National Center for Global Health and Medicine, Tokyo, Japan

Kevin P. Newman, M.D. Division of Cardiovascular Diseases, University of Tennessee Health Science Center, Memphis, TN, USA

Elisabeth Nigrini, M.D. Department of Gynecology and Obstetrics, Division of Maternal Fetal Medicine, Johns Hopkins University, Baltimore, MD, USA

Ewa Nowakowska-Zajdel, M.D., Ph.D. Department of Internal Medicine and Department of Human Nutrition, Medical University of Silesia, Bytom, Poland

Dimitrios G. Oreopoulos, M.D., Ph.D., FRCPC, FACP Department of Nephrology, University Health Network and University of Toronto, Toronto, ON, Canada

Alon Shechter, M.D. Department of Medicine, Tel Aviv University, Tel Hashomer, Israel

Michael Shechter, M.D., M.A. Department of Medicine, Tel Aviv University, Tel Hashomer, Israel

Clinical Research Unit, Leviev Heart Center, Chaim Sheba Medical Center, Tel Hashomer, Israel

Eyal Sheiner, M.D., Ph.D. Department of Obstetrics and Gynecology, Soroka University Medical Center, Ben-Gurion University of the Negev, Beer Sheva, Israel

Yiqing Song, M.D., Sc.D. Division of Preventive Medicine, Brigham and Women's Hospital, Harvard Medical School, Boston, MA, USA

Rhian M. Touyz, M.D., Ph.D. Kidney Research Center, Ottawa Hospital Research Institute, University of Ottawa, Ottawa, Canada

Institute of Cardiovascular and Medical Sciences, BHF Glasgow Cardiovascular Research Centre, University of Glasgow, Glasgow, Scotland, United Kingdom

Renée J. Turner, Ph.D. Adelaide Centre for Neuroscience Research, School of Medical Sciences, University of Adelaide, Adelaide, SA, Australia

Ioannis P. Tzanakis, M.D., Ph.D. Department of Nephrology, General Hospital of Chania, Chania, Crete, Greece

Nephrological Department, Gen. Hospital of Chania, Mournies, Greece

Robert Vink Adelaide Centre for Neuroscience Research, School of Medical Sciences, University of Adelaide, Adelaide, SA, Australia

Stella Lucia Volpe, Ph.D., RD, LDN, FACSM Department of Nutrition Science, Drexel University, Philadelphia, PA, USA

Lu Wang, M.D., Ph.D. Division of Preventive Medicine, Brigham and Women's Hospital, Harvard Medical School, Boston, MA, USA

Christopher Wayock, M.D. Department of Gynecology and Obstetrics, Division of Maternal Fetal Medicine, Johns Hopkins University, Baltimore, MD, USA

Gabriel Wcislo, M.D., Ph.D. Department of Oncology, Military Institute of Medicine, Warsaw, Poland

Karl T. Weber, M.D. Division of Cardiovascular Diseases, University of Tennessee Health Science Center, Memphis, TN, USA

Cuilin Zhang, M.D., Ph.D. Epidemiology Branch, Division of Epidemiology, Statistics, and Prevention Research, Eunice Kennedy Shriver National Institute of Child Health and Human Development, National Institutes of Health, Bethesda, MD, USA

John H. Zhang, M.D., Ph.D. Zhang Neuroscience Research Laboratories, Loma Linda University Medical Center, Loma Linda, CA, USA

Section A
Introduction and Mechanism of Action

Chapter 1
Clinical Assessment of Magnesium Status in the Adult: An Overview

Adel A.A. Ismail, Yasmin Ismail, and Abbas A. Ismail

Key Points

- Serum magnesium measurement is potentially flawed, can be normal despite deficiency.
- Magnesium deficiency is common but under diagnosed; assessment of the patient's lifestyle can help diagnosis.
- Hypermagnesaemia is easy to diagnose analytically in serum but not clinically.
- Treatment of magnesium deficiency and hypermagnesaemia is straightforward and beneficial.

Keywords Magnesium • Hypomagnesaemia • Hypermagnesaemia • Symptoms • Clinical and laboratory assessment • Diagnosis • Treatment

Introduction

The relationship between magnesium and health has been recognized some 400 years ago and well before magnesium was even identified as an element. The English summer in 1618 was exceptionally hot and dry. A farmer by the name of Henry Wicker in Epsom, Surrey, dug out a few wells in his farm to get water for his herd of cows. He noticed that his thirsty animals refused to drink this water because it had a tart and bitter taste. However, he noted that this water has the ability to rapidly heal scratches, sores and rashes both in animals and humans. Tried by others, the fame of this water spread by the word of mouth. Londoners flocked to Epsom which became a spa town, surpassing other more fashionable ones at the time such as Tunbridge wells in Kent for its water and salt. A physician (also a botanist) with extensive practice in London by the name of Nehemiah Grew (Fig. 1.1) noted that the salt in this water had a laxative effect. This "mind-boggling" discovery was patented as a purging salt, and a

A.A.A. Ismail, B. Pharm. (Hon.), Ph.D., FRCPath (✉)
Retired Consultant in Clinical Biochemistry and Chemical Endocrinology, Mid-Yorkshire and Leeds Teaching Hospitals Trust, 4 Chevet Lane, Sandal, Wakefield, West Yorkshire, UK
e-mail: Adelaaismail@aol.com

Y. Ismail, B.Sc., M.D., MRCP
Bristol Heart Institute, Bristol Royal Infirmary, Marlborough street, Bristol BS2 8HW, UK

A.A. Ismail, B.Sc., M.Sc., M.D., FRCP
Stepping Hill Hospital, Poplar Grove; Hazel Grove, Stockport, Cheshire SK2 7JE, UK

R.R. Watson et al. (eds.), *Magnesium in Human Health and Disease*, Nutrition and Health,
DOI 10.1007/978-1-62703-044-1_1, © Springer Science+Business Media New York 2013

Fig. 1.1 *Nehemiah Grew*, English physician and microscopist, 1641–1712 (Reprinted from Makers of British Botany, Cambridge University Press, 1913)

factory in London was established for worldwide marketing. In England this salt was (and still is) known as "Epsom salt" and in continental Europe as "salt anglicum".

Late in the seventeenth century and thereafter, Epsom salt was one of the most popular medicinal drugs. The people who used it did not know exactly why it was so beneficial, but they did understand that in some way, it was good for health and promoted longevity. Even now, it is surprising to know that there is an "Epsom Salt Council" in the UK whose members are wild about the goodness of "Epsom salt". Currently, 13 wonderful ways have been described for the use of "Epsom salt" by this council.

Epsom salt is hydrated magnesium sulphate ($MgSO_4.7H_2O$). In 1755, the Scottish chemist Joseph Black in Edinburgh identified magnesium as an element, and the English chemist Sir Humphry Davy was the first to isolate magnesium by electrolysis in 1808 and suggested the name "magnium" (using the suffix noun "ium" as in sodium, potassium, etc.). The name magnesium was subsequently used, derived from the Greek district of "Magnesio" in Thessaly in which magnesium carbonate (magnesio alba) was abundant. Permutation of that name is used worldwide apart from Slavic countries in which magnesium in Cyrillic is known as "horcik".

The nineteenth century was the age of chemistry of magnesium; its biology, however, became clearer during the twentieth century. Magnesium is the fourth most abundant mineral in the body after calcium, potassium and sodium. Approximately 40% of magnesium is intracellular and some 60% in bone and teeth with 1% or less present in the circulation [1].

Biochemically, magnesium is regarded with justification as a "chronic regulator" and physiologically as a "forgotten electrolyte". An adequate magnesium store is necessary for the function of hundreds of widely distributed kinases, a group of magnesium-dependent enzymes that catalyse the transfer of a phosphate group and attach it to the recipient molecule, i.e. phosphorylation. The underlying mechanism seems to be the same for all known kinases and necessitates the presence of magnesium. Kinases can only bind "ATP-Mg" molecules, cleave the γ phosphate group which is subsequently transferred to the recipient molecule. Phosphorylation is an ion-radical, electron-spin selective process [2] which transforms (switches on) an inactive molecule into an active or "functional" one, which can then perform a specific biological/biochemical task (or vice versa). In addition to the phosphorylation of small organic molecules, up to 30% of functional body proteins are activated by magnesium-dependent kinases.

Magnesium-dependent kinases [3–10] are paramount in regulating the cell cycle and growth, as well as apoptosis. It has also a vital role in signal transduction and the production and actions of second messengers such as c-AMP, diacylglycerol, calmodulin and c-GMP. Central to all these intracellular functions is that each protein must be at the right place and work at the right time. Individual kinases regulate and control a particular subset of proteins in these highly complex systems within each cell.

Magnesium plays an important role in electrolyte homeostasis, being necessary for the activation of ATP/ATPase pumps such as Na^+/K^+ pump, Na^+/Ca^{++}, Na^+/Mg^{++} and Mg^{++}/Ca^{++} pumps which, if deficient, causes impairment and reduction in their efficacy and activities. Chronic magnesium deficiency with time may eventually lead to overt pathology and electrolyte disturbances such as "refractory" hypokalaemia and/or hypocalcaemia. Neither the former nor the latter can be corrected by potassium or calcium treatment alone, and magnesium replacement becomes essential for restitution. It is therefore paramount to note that magnesium itself is an electrolyte which plays a major role in the homeostasis of other major electrolytes, namely, Na^+, K^+ and Ca^{++}. Furthermore, magnesium plays an important role in bioenergetics, regulating oxidative energy metabolism of protein, carbohydrate and fat metabolism, energy transfer, storage and use. It is also necessary for bone mineral density. About 150 magnesium-dependent kinases are linked to a wide variety of diseases; it is not therefore surprising that magnesium deficiency can potentially cause/exacerbate a wide range of disorders [11–17].

Considering the many vital roles of magnesium, there was surprisingly a lack of information regarding its homeostasis. Only in the last decade, two ion channels have been suggested as magnesium transporters which appear to play a pivotal role in its homeostasis through the dual processes of its absorption from the gut and reabsorption by the kidneys. Ion channels conduct a particular ion after which it is named whilst excluding others, e.g. Na^+, K^+ and Ca^{++} channels. Ion hydration energy (water shell surrounding each ion) and the charges at the binding sites by the ligand make the internal milieu within each channel favourable for conducting only a specific ion. The two dedicated ion channels specifically aimed at transporting Mg^{++} belong to the transient receptor potential melastatin (TRPM), a subfamily of the transient receptor potential protein superfamily involved in transporting other cellular cations such as calcium by TPRM3. Recently, TRPM6 and TRPM7 have been suggested as unique transporters for Mg^{++} termed chanzymes because they possess a channel and a kinase domain. These two chanzymes may therefore represent molecular mechanism aimed at regulating magnesium homeostasis at cellular level [18–24]. They are differentially expressed, with TRPM6 being found primarily in colon and renal distal tubules. Up-regulation of TRPM6 occurs in response to reduction in intracellular magnesium; this in turn enhances magnesium absorption from the gut and its reabsorption by the kidneys and can therefore alter whole-body magnesium homeostasis. TRPM7 is ubiquitous, occurring in numerous organs (e.g. lung). These two chanzymes may therefore represent a molecular mechanism specifically aimed at regulating body magnesium balance [18–24].

Clinical Conditions Associated with Magnesium Deficiency in Adults

Magnesium deficiency is common in the general population as well as in hospitalized patients and can occur in individuals with an apparently healthy lifestyle. Latent magnesium deficiency may be present in >10% of population, more common in the elderly, probably exacerbated by oestrogen which decline in women and men with age. Oestrogen influences body magnesium balance through its effect on TRPM6 which may help in explaining the hypermagnesuria in the elderly in general and postmenopausal in particular. Magnesium deficiency is a clinically under-diagnosed condition yet surprisingly easy to treat [25–29].

Table 1.1 Conditions associated with magnesium deficiency (The numbers between brackets are additional references published from 1990 to April 2011 for each entity)

Electrolytes(1–24): Hypocalcaemia, Hypokalaemia
CVS (25–162): Ventricular arrhythmias esp. torsades de pointes
Cardiac conduction abnormalities: SVTs
Abnormal vascular tone, congestive cardiac failure
Ischaemic heart disease/myocardial infarction
Hypertension (163–215): Pre-eclampsia/eclampsia, primary hypertension
Endocrine (216–285): Type II diabetes mellitus
Metabolic (286–294): The metabolic syndrome
Bone (295–335): BMD and osteoporosis
Muscular (336–353): Muscle weakness, fatigue, numbness, tingling, spasms/cramps/tetany and fibromyalgia
Neurological (354–387): Irritability, depression, migraines and vertical and horizontal nystagmus
Cancer (388–398): Colorectal
Alcoholics (399–420): Exhibiting any of the above manifestations
Respiratory (421–472): Asthma

We have researched peer-reviewed articles on magnesium published in English between 1990 and April 2011 in MEDLINE and EMBASE using database keywords "magnesium, deficiency, diagnosis, treatment and hypomagnesaemia". Bibliographies of retrieved articles have been searched and followed. We have also carried out a manual search of each individual issue of major clinical and biochemical journals in which most of these reports have appeared.

Clinically, magnesium deficiency may present acutely or with chronic latent manifestations. Clinical presentation of chronic/latent magnesium deficiency may vary from vague and non-specific symptoms to causing and/or exacerbating the progression of a wide range of diseases such as cardiovascular pathology (CVS), primary hypertension and Type II Diabetes Mellitus.

Magnesium is a physiological calcium antagonist and natural calcium channel blocker in skeletal and smooth muscle, promoting relaxation, whilst calcium stimulates contraction. A high calcium/magnesium ratio caused by magnesium deficiency and/or high calcium intake may affect this finely regulated homeostatic balance and may be a factor in the increased risk of cardiovascular events in patients receiving calcium supplementation [30, 31]. Magnesium deficiency is implicated/present in almost all patients with hypokalaemia and those with magnesium-dependent hypocalcaemia [32–38].

A growing body of literature has demonstrated a wide pathological role for magnesium deficiency. In 201 peer-reviewed studies published from 1990 to April 2011, magnesium deficiency was associated with increased risk and prevalence in the 11 conditions listed in Table 1.1 (irrespective of the nature, design, parameters, size and statistical approach of these studies). Such an inverse relationship was also demonstrable irrespective of the wide range of methods used to assess magnesium body stores (see references in Table 1.1 for each of these conditions).

Similarly, in 72 studies over the same period, magnesium deficiency was found to predict adverse events, and a reduced risk of pathology was noted when supplementation/treatment was instituted. In a recent study [39], a direct aetiological link between magnesium deficiency, impaired glucose tolerance and CVS pathology was demonstrated. In this study, 13 postmenopausal American women (12 Caucasian and 1 African-American) volunteered to reduce their dietary magnesium intake to ~ one-third of the recommended daily requirement (average 101 mg/day). In less than 3 months, five subjects had cardiac rhythm abnormalities, and three exhibited atrial fibrillation/flutter that responded quickly to magnesium supplementation [39]. Impaired glucose homeostasis was found in ten volunteers who underwent intravenous glucose tolerance test (IV GTT). The clinical manifestation was reflected in reduced levels in red-cell membranes; however, serum levels remained within reference range. This study, though small, is consistent with epidemiological surveys, supplementation trials and animal studies [40, 41] (see Table 1.1 for more references).

"Modus Vivendi" and Potential Magnesium Deficiency

Potential causes of magnesium deficiency are outlined in Table 1.2. It may not be difficult to surmise potential magnesium deficiency from an individual's lifestyle as body stores are dependent on the balance between daily intake and renal loss [23, 42–44]. Approximately 30–70% of dietary magnesium intake is absorbed by a healthy gut with negative magnesium store and high gastric acidity enhancing absorption [23, 42–47]. The commonly recommended daily intake for adults is 320–400 mg/day (or 6 mg/kg/body weight for both genders) [48] and increases during pregnancy, lactation and regular strenuous exercise [49–51] which increases magnesium losses in urine and sweat. An average healthy daily diet supplies ~250 mg of magnesium (120 mg/1,000 calories), with green vegetables, cereals, fish and nuts being a rich source (see Table 1.3). Refined grains and white flour are generally low in magnesium. Unrefined sea salt is very rich in magnesium occurring at ~ 12% of sodium mass; however, because this makes raw sea salt bitter, magnesium (and calcium) is removed, making purified table salt essentially ~99% sodium chloride.

Another important source is water [52, 53], with some (but not all) hard tap water containing more magnesium than soft water. Local water supplier can provide information regarding magnesium concentration in tap water to each location (e.g. postcode area in the UK). The bioavailability of magnesium in water is generally good at ~ 60%; however, its absorption from water significantly declines with age [54, 55].

The magnesium content in tap and/or bottled water varies greatly. Hardness of water is caused by dissolved calcium and magnesium and is usually expressed as the equivalent quantity of calcium carbonate in mg/l (e.g. a hardness of 100 mg/l would contain 40 mg/l of elemental Ca and/or Mg and 60 mg as carbonate). Water containing > 200 mg/l equivalent calcium carbonate is considered hard; medium hardness is between 100 and 200 mg/l, moderately soft < 100 mg/l and soft < 50 mg/l calcium carbonate equivalent. Hardness above 200 mg/l results in scale deposition on heating if large amount of calcium carbonate is present because it is less soluble in hot water.

Table 1.2 Factors contributing to chronic/latent magnesium deficiency

Age: Elderly absorb less and lose more magnesium
Daily diet low in magnesium
Soft drinking water, bottled or hard water low in magnesium
Refined salt for cooking and in food
Pregnancy, lactation and regular strenuous exercise
Regular alcohol intake esp. spirits
Malabsorption (also short-bowel syndrome/intestinal surgery)
Drugs such as diuretics

Table 1.3 Magnesium content in food

Magnesium-rich food contains > 100 mg per measure. A measure is a cup of vegetables, grains, legumes or 2 oz (or 56 g) of nuts and seeds
Vegetables: Green and leafy, e.g. spinach, seaweed and artichoke
Fish: Halibut (4 oz)
Grains: Barley, wheat, oat and bran (whole grain bread)
Legumes: Soybean, *adzuki* and black bean
Nuts: Almond, Brazil, cashews, pine and peanuts (peanut butter)
Seeds: (Dried) pumpkin, sunflower and watermelon
Chocolate: Dark (2 oz)
Intermediate values of magnesium are present in other vegetables, fruits, meats, dairy products and fish

It may be important to point out that the ratio of calcium to magnesium in hard water varies. Hard water may, in some cases, have predominantly high concentration of calcium but low in magnesium or vice versa. Furthermore, the type of anion in the calcium salt is important. For example, hard water which is rich in calcium carbonate is usually regarded as "temporarily hard" because on heating, calcium carbonate precipitates. In other forms of hard waters, magnesium and/or calcium may combine with anions other than carbonate, such as sulphate, and in this case, water is referred to as "permanently hard" because these elements are not affected by heating. All naturally occurring magnesium salts unlike those of calcium are relatively more soluble in both cold and heated water, including $MgCO_3$. Although hard water is a general term which encompasses wide ratios of calcium to magnesium, the magnesium contents in most hard water (but not all) are 5–20 times more than in soft water and can potentially provide up to 30% of daily requirement.

The term soft water is straightforward because it is used to describe types of water that contain few calcium or magnesium ions. Soft water usually comes from peat or igneous rock (volcanic rocks which make 95% of the earth's crust after the cooling of magma); other sources are granite and sandstone. All such sedimentary rocks are usually low in calcium and magnesium. The magnesium content of soft drinking water is between 2 and 20 mg/l, average ~ 6 mg/l. The content of magnesium in bottled water varies from 0 to 126 mg/l [56], whilst carbonated tonic and soda water contain little or no magnesium. One gram of instant coffee granules releases ~ 5 mg of magnesium in hot water; the corresponding figure for tea is ~ 0.6 mg [57].

Significant magnesium deficiency has been reported in both elderly self-caring in the community and in hospitalized Norwegians [58]. In a consensus survey involving 37,000 Americans, 39% were found to ingest less than 70% of the recommended daily magnesium intake and 10% of women over the age of 70 years consume less than 42% of the recommended dietary requirement [59–61]. When dietary magnesium intake is poor, the kidney can compensate by increasing fractional reabsorption from the filtered load, mainly in the loop of Henle with further reabsorption in the distal tubule. Normally, plasma magnesium is filtered at the glomeruli apart from the fraction bound to albumin. Reabsorption of the filtered load can vary depending on the body store, being lowest when body stores are adequate to maximum in deficiencies. Prolonged periods of poor dietary intake, however, would eventually lead to a decline in intracellular magnesium concentration.

Excessive renal loss is, however, a common cause of negative magnesium stores. Alcohol is a known cause, being magnesium diuretic as even moderate amounts produce magnesiuresis. Alcohol increases urinary magnesium loss above baseline by an average of 167% (range 90–357%), and its effect is rapid [62–67] and occurs even in individuals with an already negative magnesium balance. Alcohol consumption has increased with availability and cheaper cost [66, 68] and in moderate amounts, is considered socially and culturally acceptable (taken as 2–4 units, i.e. 16–32 g of alcohol a day, though there is no standard definition). It may be of interest to point out that spirits such as gin, rum, brandy, cognac, vodka and whisky contain little or no magnesium; fermented apple ciders have 10–50 mg/l of magnesium whilst beer and wine have levels ranging from ~ 30–250 mg/l. Although drinks such as some ciders, beer and wine may be considered "magnesium-rich", they cannot be recommended as a reliable source. Furthermore, large consumption of magnesium-rich beer and wine can have a laxative or even diarrhoeic effect, potentially impeding bioavailability and absorption.

It appears reasonable therefore to suggest that a lifestyle associated with low dietary magnesium intake in food and drinking water, purified table salt for cooking and in food, and regular and strenuous exercise coupled with moderate and regular consumption of alcoholic drinks which cause a net renal magnesium loss can additively lead to negative balance over time. Magnesium deficiency can be further compounded with malabsorption and those receiving medications [69–73] such as diuretics (loop and thiazide); proton pump inhibitors (e.g. omeprazole); tacrolimus; chemotherapeutic agents such as cisplatin, cyclosporine and cetuximab; and some phosphate-based drugs.

In summary, modus vivendi when carefully examined can determine the potential of latent magnesium deficiency which may be associated with a wide range of major pathologies. It is, however, a common practice for clinicians to rely more on laboratory tests in the diagnosis of magnesium deficiency.

Laboratory Tests and Assessment of Magnesium Deficiency

Assessment of magnesium status is biochemical. Serum magnesium is the most commonly requested test and is informative when magnesium is reduced, indicating hypomagnesaemia. However, normal serum magnesium (commonly reported between ~0.75 and <1.2 mmol/l) remained problematical because in patients suspected with magnesium deficiency, serum concentration can be normal despite whole-body deficiency [74–77]. This is not surprising due to the fact that magnesium in the circulation does not represent total body magnesium, being only 1% or less of total body content. A fraction of bone magnesium appears to be on a surface-limited pool, present either within the hydration shell or else on the crystal lattice. Based largely on animal studies, it has been speculated that this form of bone surface magnesium may represent a limited buffering capacity [38].

Magnesium in serum is subdivided into three heterogeneous fractions, namely, magnesium-bound to albumin (~ 30%), a fraction loosely complexed with anions such as phosphate, citrate and bicarbonate (~20%) and a free ionized fraction. Although serum ionized magnesium which represents ~ 50% of total serum magnesium correlates with total magnesium concentrations [78], it is mistakenly regarded by some to be the biologically active moiety, i.e. analogous to ionized calcium. Unlike calcium, however, the bulk of magnesium is intracellular, bound to numerous subcellular components, and these are the moieties which account for its biological role. In other words, it is the intracellular bound magnesium which expresses its primary biological role, and normal serum magnesium, total or ionized, must be interpreted with caution [77]. However, ionized magnesium measurement has an advantage because it can be made on whole blood thus avoiding the laborious and time-consuming step of separating serum from cells [78, 79].

Deficiency of other major minerals such as calcium, sodium and potassium are commonly reflected in their serum concentrations; similarly, low serum magnesium (with normal albumin) in a fasting or random sample also confirms significant deficiency warranting supplementation. However, some cases with latent magnesium deficiency may not be associated with low serum levels. For this reason, the practicable, inexpensive and commonly used serum magnesium must be regarded as potentially flawed test, capable of identifying magnesium deficiency in some but not all patients with deficiency and negative body stores.

To exclude with confidence latent/chronic magnesium deficiency in cases with high index of suspicion albeit normal serum magnesium, a dynamic study, namely, magnesium loading test, would be appropriate if renal function is normal. This procedure is probably the best physiological "gold standard test" within the capability of all routine hospital laboratories. It involves the administration of elemental magnesium load (as sulphate or chloride) intravenously followed by assessment of the amount of elemental magnesium excreted in the urine in the following 24 h [80–84]. A large fraction of the given magnesium load is retained, and a smaller amount of the given dose appears in the urine in patients with latent magnesium deficiency. Such a procedure in the experience of one of us was valuable, accurate and informative; however, it is time-consuming and (understandably) not commonly used in clinical practice. It is also contraindicated in individuals with renal impairment.

Magnesium Loading Test

The loading test measures the body's retention of magnesium and therefore reflects the degree of deficiency [80–84]. Attention to details is, however, paramount for valid interpretation of data. Patients should empty their bladder immediately before the test. The test involves intravenous administration of 30 mmol of elemental magnesium (1 mmol = 24 mg) in 500 ml 5% dextrose over a period of 8–12 h. A slow-rate infusion is important because plasma magnesium concentration affects the renal reabsorption threshold, and abrupt elevation of plasma concentration above the normal range would reduce magnesium retention and increase urinary excretion with its potential misinterpretation. Urine collection begins with the onset of magnesium infusion and continues over the next 24-h period, including a last void at the end of this period.

Patients with adequate body magnesium stores retain less than 10% of the infused elemental magnesium load. Latent magnesium deficiency is considered present if less than 25 mmol of elemental magnesium is excreted in the 24-h collection. Repeat of magnesium loading test to check repletion can also be informative because average difference between repeats is ~2%. Magnesium body stores are considered repleted when >90% of the elemental magnesium load is excreted in the following 24-h urine.

Magnesium loading test is contraindicated in patients with renal impediment, salt-losing nephropathy, respiratory failure and medications which affect renal tubular function such as diuretics, cisplatin, cyclosporine... etc.

A number of studies attempted to simplify the magnesium loading test [85] by reducing the infused magnesium load to 0.1 mmol of elemental magnesium per kilogram body weight, reducing the infusion time to 1–2 h and collecting urine over a shorter period of 12 h. Oral magnesium loading test was also described. However, although these modifications are simpler, their usage was limited and the 8–12-h infusion of 30 mmol remained the standard test.

Cellular Magnesium Concentration In Vitro

Since ~99% of magnesium is intracellular, it would be reasonable to assess body magnesium status by measuring cellular magnesium concentration rather than serum magnesium levels. Fractionating cellular components in blood and tissues can be time-consuming, and its handling can be inaccurate. Cellular zone separation commonly relies on centrifugal force (sedimentation velocity) which in blood allows sample to separate into RBC-rich bottom layer, an intermediate "buffy-coat" layer of WBCs and an upper plasma portion.

Magnesium concentration in RBC's membrane has been used for assessing body magnesium status [39]. After isolating RBCs, they are lysed by suspension into a hypotonic medium. Their uptake of water by osmosis causes the cells to explode, leaving behind an empty membrane sack (commonly referred to as "ghost") which can be subsequently separated as a pellet after centrifugation at 12,000 g. However, extra care is necessary because pellet can still be contaminated by tangles of fibrin, white cells, platelets and unlysed RBCs. Intra-erythrocyte, mononuclear blood cells and granulocytes magnesium levels are also used; however, levels showed poor correlation with other clinical and biochemical parameters [86–88].

Methods for Assessing Intracellular Magnesium In Vivo

Non-invasive methods are research tools of limited pragmatic clinical applications [89–95]. Example is the use of nanorod potentiometric selective sensor electrodes to measure intracellular magnesium. ZnO nanorods were functionalized for selectivity of Mg^{++} by applying a coat of polymeric membrane with Mg^{++}-selective ionophores which exhibited an Mg^{++}-dependent electrochemical potential difference versus Ag/AgCl reference microelectrode [89]. Others used fluorescent probes or magnetic resonance spectroscopy to assess intracellular magnesium.

Another non-invasive technique for quantitating cellular magnesium uses sublingual cells and energy-dispersive X-ray microanalysis (EXA). Sublingual cells have advantages such as accessibility, turnover of <3 days, non-cornified with long shelf-life exhibiting 99% viability. Excitation of cellular atoms displaces inner orbital electrons which are replaced by electrons from high-energy cells, releasing fluoresced X-ray which allows quantitation. Magnesium contents in sublingual cells correlated favourably with cardiac and muscle biopsies taken during bypass surgery [93] ($r=0.68$; $p<0.002$).

In summary, serum measurement is simple, pragmatic and informative if results are low. However, normal result is of limited value because it does not exclude latent/chronic magnesium deficiency. Magnesium loading test though laborious is analytically within the domain of all clinical laboratories and if carried properly is highly reliable and informative. Cellular magnesium in RBCs and WBCs is neither practicable nor accurate, able to detect only 60% of those found to be deficient by the magnesium loading test. Non-invasive methods, though innovative, remained research tools outside the scope of clinical laboratories.

Biochemical Monitoring of Magnesium Therapy in Patients with Magnesium Deficiency

Magnesium has low toxicity in people with normal renal function. Deficiency, however, may not be corrected through nutritional supplementation only. The most common therapeutic modalities are intravenous infusion in patients with depletion manifesting as significant hypomagnesaemia and orally (occasionally subcutaneously [96]) for individuals requiring long-term supplementation. Aerosolized magnesium sulphate was also used in patients with acute asthma [97, 98].

Intravenous magnesium (up to ~30 mmol of elemental magnesium; 1 mmol = 24 mg) is given over a period of hours. A slow-rate infusion is important because plasma magnesium concentration affects the renal reabsorption threshold, and abrupt elevation of plasma concentration above the normal range can reduce magnesium retention. Magnesium body stores are considered repleted when >90% of the elemental magnesium load is excreted in the following 24-h urine (see magnesium loading test). On the other hand, persistent elevation in serum magnesium in samples taken longer than 24 h after treatment would be indicative of over-treatment. Other analytes which may be associated with magnesium deficiency are calcium, potassium, phosphate and vitamin D [99].

Common oral magnesium supplement exists in two forms: chelated and non-chelated. In the chelated forms, magnesium is attached to organic radicals; in the non-chelated forms, magnesium is in the form of sulphate, chloride or oxide. Magnesium attached to organic/amino acid radicals appears to be better tolerated with superior bioavailability [100, 101] than the commonly available magnesium oxide. Generally, over-treatment leading to significant hypermagnesaemia is unlikely to occur in patients on the recommended oral magnesium supplement. This is because when the intake exceeds daily requirement, absorption of magnesium from the gut is reduced, and its excretion can exceed 100% of the filtered load caused by active renal secretion in the urine.

It may be of interest to point out that net magnesium absorption rises with increasing intake; however, fractional absorption falls as magnesium intake increases (e.g. from 65% at 40-mg intake to 11% at 960 mg). Magnesium absorption from the gut is slow with ~ 80% of oral magnesium being absorbed within 6–7 h [102]. Note also that calcium and magnesium competes for absorption; thus, too much calcium in diet/medication can impede magnesium absorption. A ratio for calcium to magnesium of ~ 2:1 would allow adequate absorption of magnesium. However, high oral calcium intake or consumption of large amounts of calcium-rich products such as dairy foods which have a ratio of calcium to magnesium of ~ 10:1 can sufficiently alter the balance, potentially reducing magnesium absorption. Dosage regimen of oral magnesium should therefore take into account the degree of patients' magnesium deficiency in the first place, the basic chemical composition of oral magnesium supplement and its bioavailability plus other concurrent medications which can increase magnesium loss (e.g. diuretics, regular intake of spirits) and/or impede its absorption, e.g. GI disorders.

Claims that Epsom salt can be absorbed through the skin are widespread throughout the internet with numerous products which can be added to bath water, in oil, gel or lotions to be directly messaged to skin for "extra-relaxation, detoxification and exfoliation". However, no peer-reviewed systematic or controlled studies could be found on this subject. In a widely quoted but rather limited, not peer-reviewed study involving 19 subjects (Waring R; School of Bioscience, Birmingham University, UK) who for 1 week bathed in water containing 1% Epsom salt at temperature of ~ 50°C for ~ 15 mins, 16 have increased their baseline serum magnesium concentration before the test by up to 40% and doubled their magnesium content in urine. However, in view of the dearth of scientifically peer-reviewed studies, transdermal absorption of magnesium should remain speculative.

Hypermagnesaemia and Magnesium Toxicity

Unlike magnesium deficiency which can be associated with normal serum concentration, magnesium excess and toxicity is manifested as identifiable hypermagnesaemia in serum (concentration > 1.2 mmol/l). Hypermagnesaemia is not common in patients with normal renal function because of the efficacy of kidneys in excreting excess magnesium in urine. Hypermagnesaemia per se inhibits the unidirectional reabsorption of magnesium in the renal distal tubule (DCT) through a complex mechanism involving the activation of the calcium-sensing receptor (CaSR), a member of the family of G-protein–coupled receptors, causing a decrease in protein kinase A and TRPM6, the gatekeeper of magnesium reabsorption [103].

Hypermagnesaemia is commonly encountered in two settings, namely, in individuals with renal impediment and in those repeatedly given large magnesium load, intravenously, orally or as an enema [104–108]. Iatrogenic examples are excessive magnesium administration in the treatment of patients with eclampsia (pregnancy-induced hypertension), total parenteral nutrition and overzealous replacement of magnesium deficiency. Other less-known but insidious causes of iatrogenic overload are over-the-counter (OTC) magnesium-rich medications [107–111] which may be clinically overlooked because of lack of awareness of their chemical formulation and patients' perception as innocuous products, e.g. antacids, laxative, analgesic (magnesium salicylate), cathartics containing magnesium and repeated use of Epsom salt as a gargle for halitosis. When OTC medications are taken by elderly patients with intestinal hypomotility from any cause such as narcotics, anticholinergic drugs or obstruction [111], magnesium absorption can be significantly increased, causing profound hypermagnesaemia [109, 111] regardless of renal function. Some 310 cases of adverse events were associated with administration of OTC magnesium-containing products in the USA and reported to the FDA between 1968 and 1994. Of these, 45 were classified as serious including 14 deaths.

Neuromuscular toxicity is a common complication of hypermagnesaemia. Excessive extracellular magnesium can block neurosynaptic transmission by interfering with the release of acetylcholine, causing fixed, dilated pupils, producing a curare-like effect. Hypermagnesaemia can also modulate and interfere with the release of catecholamines from the adrenal medulla including its response to stress.

Mild hypermagnesaemia may be encountered in some medical conditions such as Addison's disease, hypothyroidism, hypoparathyroidism, hyperparathyroidism, diabetic ketoacidosis, rhabdomyolysis, lithium intoxication and rapid mobilization of magnesium from soft tissues following trauma, burns or tumour lysis syndrome and severe haemolysis. Hypermagnesaemia delays thrombin formation and platelet clumping.

Symptoms of mild hypermagnesaemia are non-specific and may include nausea, vomiting, flushing, lethargy, weakness and dizziness. Assessing the rate of rise in serum magnesium levels is paramount because a rapid increase in serum magnesium is clinically more detrimental than a slower progressive elevation in blood levels. For this reason, the severity of symptoms of hypermagnesaemia per se does not always correlate with serum concentrations.

A rough clinical guide of symptoms of early and mild hypermagnesaemia is the appearance of muscle weakness and a disappearance of deep tendon reflexes which usually occurs at serum concentrations of ~ 2.0 mmol/l. Magnesium is an effective calcium channel blocker both extracellularly and intracellularly. A further increase in intracellular magnesium profoundly blocks several cardiac potassium channels which can combine to impair cardiovascular function. Bradycardia, vasodilatation and hypotension [111, 112] begin to appear at serum magnesium concentration of ~ 2.5 mmol/l. A variety of ECG changes can be seen at concentrations of 2.5–5 mmol/l, including prolongation of the P-R interval, an increase in QRS duration and an increase in Q-T interval. Flaccid skeletal muscle and cardiac manifestations [113, 114] such as arrhythmias, including atrial fibrillation and intraventricular conduction delay, occurs at higher concentration. At serum concentration of ~ 5.0 mmol/l or more, ventilatory failure [113], widened QRS complex and delays in intraventricular conduction occur. Complete heart block and cardiac arrest may occur at a serum magnesium concentration of ~ 7.5 mmol/l.

Signs and symptoms of hypermagnesaemia per se may be difficult to diagnose clinically. For example, in a large survey involving more than 1,000 hospitalized patients in whom serum magnesium was elevated, only 13% were suspected on clinical grounds [115]. Furthermore, hypermagnesaemia can mimic signs and symptoms of many other CVS/neurological illnesses, even masquerading in severe cases as pseudo-coma and brainstem stroke [111, 116]. Serum magnesium measurement can be important in differential diagnosis because it can accurately establish hypermagnesaemia, if present. It may be therefore prudent to consider serum magnesium as an additional test [115] to the commonly requested U&E profile; this would marginally increase the actual cost of this automated analysis. The high morbidity of hypermagnesaemia as well as its reversibility makes it important to identify these cases regardless of the normality of renal function. Magnesium suppression of PTH or suppressing non-PTH–mediated renal tubular calcium reabsorption may lead to aberration of other electrolytes, namely, hyperkalaemia, hypercalcaemia or hypocalcaemia. Other complementary tests such as TFT, cortisol, CK and/or urine myoglobin and glucose may be also considered as appropriate.

Treatment of mild to moderate hypermagnesaemia is by diuresis; loop diuretics (frusemide) have a greater effect on magnesium wasting because of their site of action on reabsorption in the cortical aspect of the thick ascending limb of Henle where ~ 3% of filtered magnesium load is normally excreted. Frusemide diminishes loop reabsorption, thereby allowing more magnesium to be excreted in the urine. Intravenous calcium gluconate or chloride antagonizes the cardiac and neuromuscular effect of excess extracellular magnesium. Dialysis to augment elimination of magnesium removal may be reserved for patients with morbid life-threatening symptoms of hypermagnesaemia. Patients do well upon the restoration of magnesium concentration to normality.

Summary and Conclusions

Both significant magnesium deficiency and magnesium excess are conditions associated with consequential morbidity and mortality especially in patients with other co-morbidities. Both conditions are reversible once identified; however, suspecting these two conditions on clinical ground is poor, and appropriate biochemical tests play an important role in establishing diagnosis and monitoring treatment as appropriate (see summary in Table 1.4).

Magnesium deficiency is an undervalued multifactorial disorder, common particularly in the elderly. Serum magnesium is a useful test because low serum concentration indicates significant deficiency, warranting replacement. However, normal magnesium concentration must not be used to exclude negative body stores. Modus vivendi has an important role in identifying at-risk patients, such as adults living in areas with soft drinking water or hard water with low magnesium contents plus other factors listed in Table 1.2, notably diet and diuretics. The most informative laboratory investigation is magnesium loading test.

Table 1.4 Take-home message

Magnesium deficiency	
	Magnesium deficiency is common
	It is under-diagnosed
	It has clinical consequences
	The commonly used serum magnesium is potentially flawed
	Low serum magnesium indicates deficiency, but normal concentration must not be used to exclude deficiency and negative body store
	Modus vivendi and medications can help in identifying individuals at risk
	Magnesium loading test, though laborious, is reliable and informative
	Treatment is straightforward and clinically beneficial
Hypermagnesaemia	
	Clinically difficult to diagnose but biochemically easy to identify
	Addition of magnesium to a U&E profile marginally increase cost of automated analysis
	Neurological and CVS are common presentations
	OTC magnesium-rich medications can be insidious causes of hypermagnesaemia even in individuals with normal renal function
	Reversible with favourable outcome upon restoration to normal magnesium concentration

Magnesium deficiency should be considered in cases such as electrolyte disturbances (hypocalcaemia and/or hypokalaemia); arrhythmia, esp. torsades de pointes; regular/excessive alcohol intake; and muscular spasms/cramps in both normocalcaemic and hypocalcaemic patients. In other conditions listed in Table 1.1, however, it is important that patients at risk in each category are also identified. The limitations of serum magnesium, though well known among laboratorians, are not widely disseminated nor emphasized to clinicians. The perception that "normal" serum magnesium excludes deficiency has therefore contributed to the under-diagnosis of latent/chronic magnesium deficiency. Based on literature in the last two decades, magnesium deficiency remained common and undervalued, warranting a proactive approach because restoration of magnesium stores is simple, tolerable and inexpensive and can be clinically beneficial.

Hypermagnesaemia on the other hand is known to occur in patients with renal impediment as well as during therapy with large doses of magnesium. Monitoring serum magnesium concentration in such cases would be prudent. However, magnesium overload can be insidious and more difficult to suspect in patients using OTC magnesium-rich medications. Significant overload by these drugs can cause serious hypermagnesaemia regardless of renal function. Clinical diagnosis of hypermagnesaemia per se is generally poor, and symptoms can mimic and masquerade as neurological/CVS disorders of various severity. Unlike patients with magnesium deficiency, serum magnesium measurement is a pragmatic and inexpensive test in identifying patients with hypermagnesaemia. Adding serum magnesium to test repertoire such as U&E would marginally increase cost of analysis but can quickly identify unexpected magnesium excess if present. Hypermagnesaemia is reversible, and patients do well upon the restoration of magnesium concentration to normality.

Acknowledgement Wish to thank Dr David Ramsey for his helpful comments and constructive advice. We are also grateful for Mr. Martin Saunders for thorough literature searching.

References

1. Swaminathan R. Magnesium metabolism and its disorders. Clin Biochem Rev. 2003;24(2):47–66.
2. Buchachenko A, Kouznetsov DA, Orlova MA, Markarian AA. Magnetic isotope effect of magnesium in phosphoglycerate kinase phosphorylation. Proc Natl Acad Sci. 2005;102:10793–6.
3. Grubbs RD, Maguire ME. Magnesium as a regulatory cation: criteria and evaluation. Magnesium. 1987;6:113–27.

4. Laires MJ, Monteiro CP, Bicho M. Role of cellular magnesium in health and human disease. Front Biosci. 2004;9:262–76.
5. Rubin H. The logic of the membrane, magnesium, mitosis (MMM) model for the regulation of animal cell proliferation. Arch Biochem Biophys. 2007;458:16–23.
6. Harms KL, Chen X. The C terminus of p53 family proteins is a cell fate determinant. Mol Cell Biol. 2005;25:2014–30.
7. Ventura A, Kirsch DG, McLaughlin ME, Tuveson DA, Grimm J, Lintault L, et al. Restoration of p53 function leads to tumour regression in vivo. Nature. 2007;445:661–5.
8. Killilea DW, Ames BN. Magnesium deficiency accelerates cellular senescence in cultured human fibroblasts. Proc Natl Acad Sci. 2008;105:5768–73.
9. Cheung PCF, Campbell DG, Nebreda A, Cohen P. Feedback control of the protein kinase TAK1 by SAPK2a/p38. EMBO J. 2003;22:5793–805. doi:10,1093/emboj/cdg 552.
10. Gabibov AG, Kochetkov SN, Sashchenko LP, Smirnov I, Severin ES. Studies on the mechanism of action of histone kinase dependent on adenosine 3′,5′-monophosphate. Eur J Biochem. 2005;115:297–301.
11. Alexander RT, Hoenderop JG, Bindels R. Molecular determinants of magnesium homeostasis: insights from human disease. J Am Soc Nephrol. 2008. doi:10.1681 ASN. 2008010098.
12. Rude RK. Magnesium deficiency: a cause of heterogeneous disease in humans. J Bone Miner Res. 1998;13:749–58.
13. Altura BM, Altura BT. Cardiovascular risk factors and magnesium: relationships to atherosclerosis, ischemic heart disease and hypertension. Magnes Trace Elem. 1992;10:182–92.
14. Al-Delaimy WK, Rimm EB, Willett WC, Stampfer MJ, Hu FB. Magnesium intake and risk of coronary heart disease among men. J Am Coll Nutr. 2004;23:63–70.
15. Touyz RM. Transient receptor potential melastatin 6 and 7 channels magnesium transport and vascular biology: implications in hypertension. Am J Physiol Heart Circ Physiol. 2008;294:1103–18.
16. Larsson SC, Bergkvist L, Wolk A. Magnesium intake in relation to risk of colorectal cancer in women. JAMA. 2005;293:86–9.
17. Ismail AA, Thurston A. How magnesium deficiency affects bone health. Osteoporos Rev. 2010;8:9–12.
18. Murphy E. Mysteries of magnesium homeostasis. Circulation. 2000;86:245.
19. Abed E, Moreau R. Importance of melastatin-like transient receptor potential 7 and cations magnesium and calcium in human osteoblast. Cell Prolif. 2007;40:849–65.
20. Moomaw AS, Maguire ME. The unique nature of mg2+ channels. Physiology. 2008;23:275–85.
21. Kim BJ, Lim HH, Yang DK, Jun JY, Chang IY, Park CS, et al. Melastatin-type transient receptor channel 7 in the intestine. Gastroenterology. 2005;129:1504–17.
22. Wolf FI. Channeling the future of cellular magnesium homeostasis. Sci STKE. 2004;233:23.
23. Bindels RJ. Minerals in motion: from new ion transporters to new concepts. J Am Soc Nephrol. 2010;21:1263–9.
24. Glaudemans B, Knoers NV, Hoenderop JG, Bindels RJ. New molecular players facilitating Mg(2+) reabsorption in the distal convoluted tubule. Kidney Int. 2010;77:17–22.
25. Killilea DW, Maier JA. A connection between magnesium deficiency and aging: new insights from cellular studies. Magnes Res. 2008;21:77–82.
26. Barbagallo M, Belvedere M, Dominguez LJ. Magnesium homeostasis and aging. Magnes Res. 2009;22:235–46.
27. Musso CG. Magnesium metabolism in health and disease. Int Urol Nephrol. 2009;41:357–62.
28. Arinzon Z, Peisakh A, Schrire S, Berner YN. Prevalence of hypomagnesemia (HM) in a geriatric long-term care (LTC) setting. Arch Gerontol Geriatr. 2010;51:36–40.
29. Barbagallo M, Dominguez LJ. Magnesium and aging. Curr Pharm Des. 2010;16:832–9.
30. Bolland MJ, Barber PA, Doughty RN, Mason B, Horne A, Ames R, et al. Vascular events in healthy older women receiving calcium supplementation: randomised controlled trial. BMJ. 2008;336(7638):262–6.
31. Rowe WJ. Calcium-magnesium-ratio and cardiovascular risk. Am J Cardiol. 2006;98:140–2.
32. Hermans C, Lefebvre CH, Devogelaer JP, Lambert M. Hypocalcaemia and chronic alcohol intoxication: transient hypoparathyroidism secondary to magnesium deficiency. Clin Rheumatol. 1996;15:193–6.
33. Loughrey C. Serum magnesium must also be known in profound hypokalaemia. BMJ. 2002;324:1039.
34. Wang R, Flink EB, Dyckner T. Magnesium depletion as a cause of refractory potassium repletion. Arch Intern Med. 1985;145:1686.
35. Whang R, Whang DD, Ryan MP. Refractory potassium depletion: a consequence of magnesium depletion. Arch Intern Med. 1992;152:40–5.
36. Jones BJ, Twomey PJ. Comparison of reflective and reflex testing for hypomagnesaemia in severe hypokalaemia. J Clin Pathol. 2009;62:816–9.
37. Srivastava R, Bartlett WA, Kennedy IM, Hiney A, Fletcher C, Murphy MJ. Reflex and reflective resting: efficiency and effectiveness of adding on laboratory tests. Ann Clin Biochem. 2010;47:223–7.
38. Ismail AAA. On the efficiency and effectiveness of added-on serum magnesium in patients with hypokalaemia and hypocalcaemia. Ann Clin Biochem. 2010;47:492–3.
39. Forrest H, Nielson DB, Milne LM, Klevay SG, LuAnn J. Dietary magnesium deficiency induces heart rhythm changes, impairs glucose tolerance, and decreases serum cholesterol in post menopausal women. J Am Coll Nutr. 2007;26:121–32.

40. Fung TT, Manson JE, Solomon CG, Liu S, Willett WC, Hu FB. The association between magnesium intake and fasting insulin concentration in healthy middle-aged women. J Am Coll Nutr. 2003;22:533–8.
41. Rumawas ME, McKeown NM, Rogers G, Meigs JB, Wilson PWF, Jacques PF. Magnesium intake is related to improved insulin homeostasis in the Framingham offspring cohort. J Am Coll Nutr. 2006;25:486–92.
42. Ladefoged K, Hessov I, Jarnum S. Nutrition in short-bowel syndrome. Scand J Gastroenterol Suppl. 1996;216:122–31.
43. Rude KR. Magnesium metabolism and deficiency. Endocrinol Metab Clin North Am. 1993;22:377–95.
44. Quamme GA. Recent developments in intestinal magnesium absorption. Curr Opin Gastroenterol. 2008;24:230–5.
45. Saris NE, Mervaala E, Karppanen H, Khawaja JA, Lewenstam A. Magnesium: an update on physiological, clinical, and analytical aspects. Clin Chim Acta. 2000;294:1–26.
46. Ford ES, Mokdad AH. Dietary magnesium intake in a national sample of U.S. adults. J Nutr. 2003;133:2879–82.
47. Bialostosky K, Wright JD, Kennedy-Stephenson J, McDowell M, Johnson CL. Dietary intake of macronutrients, micronutrients and other dietary constituents: United States 1988–94. Vital Health Stat. 2002;11(245):168, National Center for Health Statistics.
48. Rayssiguier Y, Durlach J, Boirie Y. Apports Nutritionnels Conseille´s pour la population francaise. A Martin coordonnateur. Paris: TEC DOC Lavoisier; 2000.
49. Bohl CH, Volpe SL. Magnesium and exercise. Crit Rev Food Sci Nutr. 2002;42:533–63.
50. Nielsen FH, Lukaski HC. Update on the relationship between magnesium and exercise. Magnes Res. 2006;19:180–9.
51. Laires MJ, Monteiro C. Exercise, magnesium and immune function. Magnes Res. 2008;21:92–6.
52. Rubenowitz E, Axelsson G, Rylander R. Magnesium in drinking water and body magnesium status measured by an oral loading test. Scand J Clin Lab Invest. 1998;58:423–8.
53. Monarca S, Donato F, Zerbini I, Calderon R, Craun GF. Review of epidemiological studies on drinking water hardness and cardiovascular disease. Eur J Cardiovasc Prev Rehabil. 2006;13:495–506.
54. Verhas M, de La Gueronniere V, Grognet J-M, Paternot J, Hermanne A, Van den Winkel P, et al. Magnesium bioavailability from mineral water. A study in adult men. Eur J Clin Nutr. 2002;56:442–7.
55. Durlach J, Durlach V, Bac P, Rayssiguier Y, Bara M, Guiet-Bara A. Magnesium and ageing. II. Clinical data: aetiological mechanism and pathophysiological consequences of magnesium deficit in the elderly. Magnes Res. 1993;6:379–94.
56. Garzon P, Mark J, Eisenberg MJ. Variation in the mineral content of commercially available bottled waters: implications for health and disease. Am J Med. 1998;105:125–30.
57. Gillies ME, Birbeck JA. Tea and coffee as source of some minerals in the New Zealand diet. Am J Clin Nutr. 1983;38:936–42.
58. Gullestad L, Nes M, Ronneberg R, Midtvedt K, Falch D, Kjekshus J. Magnesium status in healthy free-living elderly Norwegians. J Am Coll Nutr. 1994;13:45–50.
59. Marier JR. Magnesium content of the food supply in the modern-day world. Magnesium. 1986;5:1–8.
60. Costello RB, Moser-Veillon PB. A review of magnesium intake in the elderly. A cause for concern? Magnes Res. 1992;5:61–7.
61. Byrd Jr RP, Roy TM. Magnesium: its proven and potential clinical significance. South Med J. 2003;96:104.
62. Martin HE, McCuskey Jr C, Tupikova N. Electrolyte disturbance in acute alcoholism (with particular reference to magnesium). Am J Clin Nutr. 1959;7:191–6.
63. Rylander R, Megevand Y, Lasserre B, Granbom AS. Moderate alcohol consumption and urinary excretion of magnesium and calcium. Scand J Clin Lab Invest. 2001;61:401–5.
64. Rivlin RS. Magnesium deficiency and alcohol intake: mechanisms, clinical significance and possible relation to cancer development. J Am Coll Nutr. 1994;13:416–23.
65. Kalbfleisch JM, Linderman RD, Ginn HE, Smith WO. Effects of ethanol administration on urinary excretion of magnesium and other electrolytes in alcoholics and normal subjects. J Clin Invest. 1963;42:1471–5.
66. Peel S. Utilizing culture and behaviour in epidemiological models of alcohol consumption and consequences for western nations. Alcohol Alcohol. 1997;32:51–64.
67. Poikolainen K, Alho H. Magnesium treatment in alcoholics: a randomized clinical trial. Subst Abuse Treat Prev Policy. 2008;3:1.
68. Sheron N. Alcohol in Europe: the EU alcohol forum. Clin Med. 2007;7:323–4.
69. Cundy T, MacKay J. Proton pump inhibitors and severe hypomagnesaemia. Curr Opin Gastroenterol. 2011;27:180–5.
70. Cao Y, Liao C, Tan A, Liu L, Gao F. Meta-analysis of incidence and risk of Hypomagnesemia with Cetuximab for advanced cancer. Chemotherapy. 2010;56:459–65.
71. Joo Suk O. Paradoxical hypomagnesemia caused by excessive ingestion of magnesium hydroxide. Am J Emerg Med. 2008;26:837. e1-2.
72. Cundy T, Dissanayake A. Severe hypomagnesaemia in long-term users of proton-pump inhibitors. Clin Endocrinol. 2008;69:338–41.

73. Shabajee N, Lamb EJ, Sturgess I, Sumathipala RW. Omeprazole and refractory hypomagnesaemia. BMJ. 2008;337:a425. 0959-535X;1468-5833 (2008).
74. Liebscher D-H, Liebscher D-E. About the misdiagnosis of magnesium deficiency. J Am Coll Nutr. 2004;23:730S–1.
75. Franz KB. A functional biomarker is needed for diagnosing magnesium deficiency. J Am Coll Nutr. 2004;23:738S–41.
76. Arnaud MJ. Update on the assessment of magnesium status. J Nutr. 2008;99 Suppl 3:S24–36.
77. Ismail Y, Ismail AA, Ismail AAA. The underestimated problem of using serum magnesium measurements to exclude magnesium deficiency in adults; a health warning is needed for "normal" results. Clin Chem Lab Med. 2010;48:323–7.
78. Longstreet D, Vink R. Correlation between total and ionic magnesium concentration in human serum samples is independent of ethnicity or diabetic state. Magnes Res. 2009;22:32–6.
79. Ben Rayana MC, Burnett RW, Covington AK, D'Orazio P, Fogh-Andersen N, Jacobs E, et al. International Federation of Clinical Chemistry and Laboratory Medicine (IFCC); IFCC Scientific Division Committee on Point of Care Testing. IFCC guideline for sampling, measuring and reporting ionized magnesium in plasma. Clin Chem Lab Med. 2008;46:21–6.
80. Ismail AAA. Disorders of parathyroid hormone, calcitonin and vitamin D metabolism. In: Biochemical investigation in endocrinology. London: Academic; 1983. p. 67.
81. Rasmussen HS, Nair P, Goransson L, Balslov S, Larsen OG, Aurup P. Magnesium deficiency in patients with ischaemic heart disease with and without acute myocardial infarction uncovered by an intravenous loading test. Arch Intern Med. 1988;148:329–32.
82. Goto K, Yasue H, Okumura K, Matsuyama K, Kugiyama K, Miyagi H, et al. Magnesium deficiency detected by iv loading test in variant angina pectoris. Am J Cardiol. 1990;65:709–12.
83. Gullestad L, Midtvedt K, Dolva LO, Narseth J, Kjekshus J. The magnesium loading test: reference value in healthy subjects. Scand J Clin Lab Invest. 1994;54:23–31.
84. Gullestad L, Dolva LO, Waage A, Falch D, Fagerthun H, Kjekshus J. Magnesium deficiency diagnosed by an intravenous loading test. Scand J Clin Lab Invest. 1992;52:245–53.
85. Seyfert T, Dick K, Renner F, Rob PM. Simplification of the magnesium loading test for use in outpatients. Trace Met Electrolyte. 1998;15:120–6.
86. Ulger Z, Ariogul S, Cankurtaran M, Halil M, Yavuz BB, Orhan B, et al. Intra-erythrocyte magnesium levels and their clinical implications. J Nutr Health Aging. 2010;14:810–4.
87. Elin RJ, Hosseini JM. Magnesium content of mononuclear blood cells. Clin Chem. 1985;31:377–80.
88. Hosseini JM, Yang XY, Elin RJ. Determination of magnesium in granulocytes. Clin Chem. 1989;35:1404–7.
89. Asif MH, Ali SMU, Nur O, Willander M, Englund U, Elinder F. Functionalized ZnO nanorod-based selective magnesium ion sensor for intracellular measurements. Biosens Bioelectron. 2010;26:1118–23.
90. Trapani V, Farruggia G, Marraccini C, Lotti S, Cittadini A, Wolf FI. Intracellular magnesium detection: imaging a brighter future. Analyst. 2010;135:1855–66.
91. Wong KC, Yeung DTW, Ahuja AT, King A, Lam WKC, Chan MTV, et al. Intracellular free magnesium of brain and cerebral phosphorus-containing metabolites after subarachnoid haemorrhage and hypermagnesaemic treatment. A 31P-Magnetic resonance spectroscopy study. J Neurosurg. 2010;113:763–9.
92. Silver BB. Development of cellular magnesium nano-analysis in treatment of clinical magnesium deficiency. J Am Coll Nutr. 2004;23:732S–7.
93. Haigney MCP, Silver B, Tanglao E, Silverman HS, Hill JD, Shapiro E, et al. Non-invasive measurement of tissue magnesium and correlation with cardiac levels. Circulation. 1995;92:2190–7.
94. Iotti S, Malucelli E. In vivo assessment of Mg2+ in human brain and skeletal muscle by 31P-MRS. Magnes Res. 2008;21:157–62.
95. Aslam A, Pejovic-Milic A, McNeill FE, Byun SH, Prestwich WV, Chettle DR. In vivo assessment of magnesium status in human body using accelerator-based neutron activation measurement of hands: a pilot study. Med Phys. 2008;35:608–16.
96. Martinez-Riquelme A, Rawlings J, Morley S, Kindall J, Hosking D, Allison S. Self-administered subcutaneous fluid infusion at home in the management of fluid depletion and hypomagnesaemia in gastro-intestinal disease. Clin Nutr. 2005;24:158–63.
97. Blitz M, Hughes R, Diner B, Beasley R, Knopp J, Rowe BH. Aerosolized magnesium sulphate for acute asthma: a systematic review. Chest. 2005;128:337–44.
98. Villeneuve EJ, Zed PJ. Nebulized magnesium sulphate in the management of acute exacerbations of asthma. Ann Pharmacother. 2006;40:1118–24.
99. Bringhurst FR, Demay MB, Krane SM, Kronenberg HM. Bone and mineral metabolism in health and disease. In: Fauci AS, Braunwald E, Kasper DL, Hauser SL, Longo DL, Jameson JL, Loscalzo J, editors. Harrison's principles of internal medicine. 17th ed. New York: McGraw Hill Medical; 2008. p. 2372–3.

100. Firoz M, Graber M. Bioavailability of US commercial magnesium preparation. Magnes Res. 2001;14:257–62.
101. Walker AF, Marakis G, Christie-byng M. Magnesium citrate found more bioavailability than other magnesium preparations in a randomized double-blind study. Mag Res. 2003;16:183–91.
102. Graham LA, Caesae JJ, Burgen ASV. Gastrointestinal absorption and excretion of magnesium in man. Metabolism. 1960;9:646–9.
103. Joost QX, Hoenderop GJ, Bindels RMJ. Regulation of magnesium reabsorption in DCT. Eur J Physiol. 2009;458:89–98.
104. Schelling JR. Fatal hypermagnesaemia. Clin Nephrol. 2000;53:61–5.
105. Onishi S, Yoshino S. Cathartic-induced fatal hypermagnesaemia in the elderly. Intern Med. 2006;45:207–10.
106. Smilkstein MJ, Smolinske SC, Kulig KW, Rumack BH. Severe hypermagnesaemia due to multiple-dose cathartic therapy. West J Med. 1988;148:208–11.
107. Castelbaum AR, Donotrio PD, Walker FO, Troost BT. Laxative abuse causing hypermagnesaemia, quadriparesis and neuromuscular junction defect. Neurology. 1989;39:746–7.
108. Jaing TH, Hung IJ, Chung HT, Lai CH, Liu WM, Chang KW. Acute hypermagnesaemia: a rare complication of antacid administration after bone marrow transplantation. Clin Chim Acta. 2002;326:201–3.
109. McLaughlin SA, McKinney PE. Antacid-induced hypermagnesaemia in a patient with normal renal function and bowel obstruction. Ann Pharmacother. 1998;32:312–5.
110. Van Hook JW. Endocrine crises. Hypermagnesaemia. Crit Care Clin. 1991;1:215–23.
111. Fung MC, Weintraub M, Bowen DL. Hypermagnesaemia; elderly over-the-counter drug users at risk. Arch Fam Med. 1995;4:718–23.
112. Zwerling H. Hypermagnesaemia-induced hypotension and hypoventilation. JAMA. 1991;266:2374–5.
113. Ferdinandus J, Pederson JA, Whang R. Hypermagnesaemia as a cause of refractory hypotension, respiratory depression and coma. Arch Intern Med. 1981;141:669–70.
114. So M, Ito H, Sobue K, Tsuda T, Katsuya H. Circulatory collapse caused by unnoticed hypermagnesaemia in a hospitalized patient. J Anaesth. 2007;21:273–6.
115. Whang R, Ryder KW. Frequency of hypomagnesaemia and hypermagnesaemia: requested vs routine. JAMA. 1990;263:3063–4.
116. Rizzo MA, Fisher M, Lock JP. Hypermagnesaemic pseudocoma. Arch Intern Med. 1993;153:1130–2.

Supplemental Data[1,2]

Abbott LG, Rude RK, Tong GM. Magnesium deficiency in critical illness. J Intensive Care Med. 2005;20:3–17.
Abbott LG, Rude RK. Clinical manifestations of magnesium deficiency. Mineral and Electrol Metab. 1993;19:314–22.
Agus ZS. Hypomagnesaemia. J Am Soc Nephrol. 1999;10:1616–22.
Bringhurst FR, Demay MB, Krane SM, Kronenberg HM. Bone and mineral metabolism in health and disease. In: Fauci AS, Braunwald E, Kasper DL, Hauser SL, Longo DL, Jameson JL, Loscalzo J, editors. Harrison's principles of internal medicine. 17th ed. New York: Publ McGraw Hill Medical; 2008. p. 2372–3.
Cohn JN, Kowey PR, Whelton PK, Prisant M. New guidelines for potassium replacement in clinical practice. Arch Intern Med. 2000;160:2429–36.
Dorup I. Magnesium and potassium deficiency. Acta Physiol Scand. 1994;150(Suppl):1–55.
Efstratopoulos AD, Voyaki SM, Meikopoulos M, Nourgos LN, Baltas AA, Lekakis V, et al. Alterations of serum magnesium under chronic therapy with diuretics and/or angiotensin-converting enzymes inhibitor in hypertensive patients. Am J Hypertens. 2003;16:105–8.
Elisaf M, Milionis H, Siamopoulos KC. Hypomagnesemic hypokalemia and hypocalcemia: clinical and laboratory characteristics. Miner Electrol Metab. 1997;23:105–12.
Gibb MA, Wolfson AB, Tayal VS. Electrolyte disturbances. In: Rosen P, Barkin RM, editors. Emergency medicine concepts and clinical practice, vol. 3. St Louis: Mosby; 1998. p. 2445–6.
Hermans C, Lefebvre C, Devogelaer J-P, Lambert M. Hypocalcemia and chronic alcohol intoxication: transient hypoparathyroidism secondary to magnesium deficiency. Clin Rheumatol. 1996;15:193–6.
Hoenderop JG, Bindels RJ. Epithelial calcium and magnesium channels in health and disease. J Am Soc Nephrol. 2005;16:15–26.

[1] Additional References from 1990 to April 2011 for Each Category Listed in Table 1.1
[2] Electrolytes

Kaye P, O'Sullivan I. The role of magnesium in the emergency department. Emerg Med J. 2002;19:288–91.
Kelepouris E, Agus ZS. Hypomagnesemia: renal magnesium handling. Semin Nephrol. 1998;18:58–73.
Konrad M, Schlingmann KP, Gudermann T. Insight into the molecular nature of magnesium homeostasis. Am J Physiol Renal Physiol. 2004;286:F599–605.
Leicht E, Biro G. Mechanisms of hypocalcaemia in the clinical form of severe magnesium deficit in the human. Magnes Res. 1992;5:37–44.
Levine BS, Coburn JW. Magnesium, the mimic/antagonist of calcium. N Engl J Med. 1984;310:1255–9.
Liebscher D-H, Liebscher D-E. About the misdiagnosis of magnesium deficiency. J Am Coll Nutr. 2004;23:730S–1S.
Matz R, Whang R. Refractory potassium repletion due to magnesium deficiency. Arch Intern Med. 1992;152:2346–7.
Myerson RM. Magnesium – a neglected element? J Pharm Med. 1991;1:89–97.
Nijenhuis T, Vallon V, van der Kemp AW. Enhanced passive calcium reabsorption and reduced magnesium channel abundance explains thiazide-induced hypocalcuria and hypomagnesaemia. J Clin Invest. 2005;115:1651–8.
Rude RK. Physiology of magnesium metabolism and the important role of magnesium in potassium deficiency. Am J Cardiol. 1989;63:31G–4G.
Tong GM, Rude RK. Magnesium deficiency in critical illness. J Intensive Care Med. 2005;20:3–17.
Touitou T, Godard JP, Ferment O, Chastang C, Proust J, Bogdan A, et al. Prevalence of magnesium and potassium deficiencies in the elderly. Clin Chem. 1987;33:518–23.
Whang R, Whang DD, Ryan MP. Refractory potassium repletion: a consequence of magnesium deficiency. Arch Intern Med. 1992;152:40–5.

Cardiovascular: arrhythmia, myocardial infarction, coronary heart disease (CHD)

Adamopoulos C, Pitt B, Sui X, Love TE, Zannad F, Ahmed A. Low serum magnesium and cardiovascular mortality in chronic heart failure: a propensity-matched study. Int J Cardiol. 2009;136:270–7.
Almoznino-Sarafian D, Sarafian G, Berman S, Shteinshnaider M, Tzur I, Cohen N, et al. Magnesium administration may improve heart rate variability in patients with heart failure. Nutr Metab Cardiovasc Dis. 2009;19:641–5.
Alon I, Gorelik O, Almoznino-Sarafian D, Shteinshaider M, Weissgarten J, Modai D, et al. Intracellular magnesium in elderly patients with heart failure: effects of diabetes and renal dysfunction. J Trace Elem Med Biol. 2006;20:221–6.
Alper CM, Mattes RD. Peanut consumption improves indices of cardiovascular disease risk in healthy adults. J Am Coll Nutr. 2003;22:133–41.
Altura BM, Altura BT. Cardiovascular risk factors and magnesium: relationships to atherosclerosis, ischemic heart disease and hypertension. Magnes Trace Elem. 1991;10:182–92.
Altura BM, Altura BT. Magnesium and cardiovascular biology: an important link between cardiovascular risk factors and atherogenesis. Cell Mol Biol Res. 1995;41:347–59.
Altura BM, Shah NC, Jiang XC, Li Z, Perez-Albela JL, Altura BT. Short-term magnesium deficiency results in decreased levels of serum sphingomylin, lipid peroxidation and apoptosis in cardiovascular tissues. Am J Physiol Heart Circ Physiol. 2009;297:H86–92. doi:10.1152/ajpheart.01154, 2008.
Field JM, Hazinski MP, Sayre MR, Chameides L, Schexnader SM, et al. American Heart Association guidelines for cardiopulmonary resuscitation and emergency cardiovascular care: life threatening electrolyte abnormalities. Circulation 2005;112:IV121–5.
Angomachalelis NJ, Tiyopoulos HS, Tsoungas MG, Gavrielides A. Red cell magnesium concentration in cor pulmonale. Correlation with cardiopulmonary findings. Chest. 1993;103:751–5.
Antman EM. Magnesium in acute MI: timing is critical. Circulation. 1995;92:2367–72.
Arsenian MA. Magnesium and cardiovascular disease. Progr Cardiovasc Dis. 1993;35:271–310.
Bert AA, Reinert SE, Singh AK. A b blocker, not magnesium, is effective prophylaxis for atrial tachyarrhythmias after coronary artery bypass graft surgery. J Cardiothorac Vasc Anesth. 2001;15:204–9.
Bharucha DB, Kowey PR. Management and prevention of atrial fibrillation after cardiovascular surgery. Am J Cardiol. 2004;85:20–4.
Booth JV, Phillips-Bute B, McCants CB, Podgoreanu MV, Smith PK, Mathew JP, et al. Low serum magnesium predicts major adverse cardiac events after coronary artery bypass graft surgery. Am Heart J. 2003;145:1108–13.
Brugada P weditorialx. Magnesium: an antiarrhythmic drug, but only against very specific arrhythmias. Eur Heart J 2000;21:1116.
Bupesh K, Solchenberger J. Magnesium deficiency and perioperative cardiac dysrhythmia. South Med J. 1996;89:S11–4.
Burgess DC, Kilborn MJ, Keech AC. Interventions for prevention of post-operative atrial fibrillation and its complications after cardiac surgery: a meta-analysis. Eur Heart J. 2006;27:2846–57.

Byrd Jr RP, Roy TM. Magnesium: its proven and potential clinical significance. Southern Med J. 2003;96:104.
Casthely PA, Yoganathan T, Komer C, Kelly M. Magnesium and arrhythmias after coronary artery bypass surgery. J Cardiothorac Vasc Anesth. 1994;8:188–9.
Catling LA, Abubakar I, Lake IR, Swift L, Hunter PR. A systematic review of analytical observational studies investigating the association between cardiovascular disease and drinking water hardness. J Water Health. 2008;6:433–42.
Cheng TO. Mitral valve prolapse and hypomagnesemia: how are they casually related? Am J Cardiol. 1997;80:976–8.
Chiuve SE, Korngold EC, Januzzi Jr JL, Gantzer ML, Albert CM. Plasma and dietary magnesium and risk of sudden cardiac death in women. Am J Clin Nutr. 2011;93:253–60.
Christensen CW, Rieder MA, Silverstein EL, Gencheff NE. Magnesium sulphate reduces myocardial infarction size when administered before but not after coronary reperfusion in a canine model. Circulation. 1995;92:2617–21.
Colquhoun IW, Berg GA, el-Fiky M, Hurle A, Fell GS, Wheatley DJ. Arrhythmia prophylaxis after coronary artery surgery. A randomized controlled trial of intravenous magnesium chloride. Eur J Cardio-Thorac Surg. 1993;7:520–3.
Cooper HA, Dries DL, Davis CE, Shen YL, Domansk MJ. Diuretics and risk of arrhythmic death in patients with left ventricular dysfunction. Circulation. 1999;100:1311–5.
Costello RB, Moser-Veillon PB, DiBianco R. Magnesium supplementation in patients with congestive heart failure. J Am Coll Nutr. 1997;16:22–31.
Crystal E, Connolly SJ, Sleik K, Ginger TJ, Yusuf S. Interventions on prevention of postoperative atrial fibrillation in patients undergoing heart surgery: a meta-analysis. Circulation. 2002;106:75–80.
Duning J, Treasure T, Versteegh M, Nashef SA. Guidelines on the prevention and management of de novo atrial fibrillation after cardiac and thoracic surgery. Eur J Cardiothorac Surg. 2006;30:852–72.
Douban S, Brodsky MA, Whang DD, Whang R. Significance of magnesium in congestive heart failure. Am Heart J. 1996;132:664–71.
Dunning J, McKeown P. Are the American college of chest physicians guidelines for the prevention and management of atrial fibrillation after cardiac surgery already obsolete? Chest. 2006;129:1112–3.
Dyckner T. Relation of cardiovascular disease to potassium and magnesium deficiencies. Am J Cardiol. 1990;65:44K–6K.
Eisenberg MJ. Magnesium deficiency and sudden death. Am Heart J. 1992;124:544–9.
Elwood PC, Fehily AM, Sweetnam PM, Yarnell JW. Dietary magnesium and prediction of heart disease. Lancet. 1992;340:483.
England MR, Gordon G, Salem M, Chernow B. Magnesium administration and dysrhythmia after cardiac surgery. A placebo-controlled, double blind, randomised trial. J Am Med Assoc. 1992;268:2395–402.
Eray O, Akca S, Pekdemir M, Eray E, Cete Y, Oktay C. Magnesium efficacy in magnesium deficient and no deficient patients with rapid ventricular response. Eur J Emerg Med. 2001;7:287–90.
Fangzi L, Folsom AR, Brancadi FL. Is low magnesium concentration a risk factor for coronary heart disease? The Atherosclerosis risk in communities (ARIC). Am Heart J. 1998;136:480–90.
Fanning W, Thomas C, Roach A, Tomichek R, Alford W, Stoney W. Prophylaxis of atrial fibrillation with magnesium sulfate after coronary artery bypass grafting. Ann Thorac Surg. 1991;52:529–33.
Faulk EA, McCully JD, Hadlow NC, Tsukube T, Krukenkamp IB, Federman M, et al. Magnesium cardioplegia enhances mRNA levels and the maximal velocity of cytochrome oxidase I in the senescent myocardium during global ischemia. Circulation. 1995;92:405–12.
Feldstedt M, Boesgaard S, Bouchelouche P, Svenningsen A, Brooks L, Lech Y, et al. Magnesium substitution in acute ischaemic heart syndrome. Eur Heart J. 1991;12:1215–8.
Finkenberg P, Merasto S, Louhelainen M, Vapaatalo H, Muller D, Luft F, et al. Magnesium supplementation prevents angiotensin II – induced myocardial damage and CTGE over expression. J Hypertens. 2005;23:375–80.
Forlani S, De Paulis R, de Notaris S, Nardi P, Tomai F, Proietti I, et al. Combination of sotalol and magnesium prevents atrial fibrillation after coronary artery bypass grafting. Ann Thorac Surg. 2002;74:720–5.
Fox C, Ramsoomair D, Carter C. Magnesium: its proven and potential clinical significance. South Med J. 2001;94:1195–201.
Fox CH, Mahoney MC, Ramsoomair D, Carter CA. Magnesium deficiency in African-Americans: does it contribute to increased cardiovascular risk factors? J Nat Med Assoc. 2003;95:257–62.
Frick M, Darpo B, Ostergren J, Rosenqvist M. The effect of oral magnesium alone or as an adjunct to sotalol, after cardio version in patients with persistent atrial fibrillation. Eur Heart J. 2000;21:1177–85.
Fuentes JC, Salmon AA, Silver MA. Acute and chronic oral magnesium supplementation: effects on endothelial function, exercise capacity, and quality of life in patients with symptomatic heart failure. Congest Heart Fail. 2006;12:9–13.
Fujioka Y, Yokoyama M. Magnesium, cardiovascular risk factors and atherosclerosis. Clin Calcium. 2005;15:221–5.
Geertman H, van derStarre P, Sie H, Beukema W, van Rooyen-Butjin M. Magnesium in addition to sotalol does not influence the incidence of postoperative atrial tachyarrhythmias after coronary artery bypass surgery. J Cardiothorac Vasc Anesth. 2004;18:309–12.

Gums JG. Magnesium in cardiovascular and other disorders. Am J Health Syst Pharm. 2004;61:1569–76.
Gyamlani G, Parikh C, Kulkarni AG. Benefits of magnesium in acute myocardial infarction: timing is crucial. Am Heart J. 2000;139:703.
Hazelrigg SR, Boley TM, Cetindag IB, Moulton KP, Trammell GL, Polancic JE, et al. The efficacy of supplemental magnesium in reducing atrial fibrillation after coronary artery bypass grafting. Ann Thorac Surg. 2004;77:824–30.
Heesch CM, Eichhorn EJ. Magnesium in acute myocardial infarction. Ann Emerg Med. 1994;24:1154–60.
Henyan NN, Gillespie EL, White CM, Coleman CI. Impact of intravenous magnesium on post-cardiothoracic surgery atrial fibrillation and length of hospital stay: a meta-analysis. Ann Thorac Surg. 2005;80:2402–6.
Herzog WR, Schlossberg ML, MacMurdy KS, Edenbaum LR, Gerber MJ, Vogel RA, et al. Timing of magnesium therapy affects experimental infarct size. Circulation. 1995;92:2622–6.
Hiroki T, Kato M, Yamagata T, Matsuura H, Goro K. The preventative effect of magnesium on coronary spasm in patients with vasospastic angina. Chest. 2000;118:1690–5.
Horner SM. Efficacy of intravenous magnesium in acute myocardial infarction in reducing arrythmias and mortality. Metaanalysis of magnesium in acute myocardial infarction. Circulation. 1992;86:774–9.
Kaplan M, Kut MS, Icer UA, Demirtas MM. Intravenous magnesium sulfate prophylaxis for atrial fibrillation after coronary artery bypass surgery. J Thorac Cardiovasc Surg. 2003;125:344–52.
Kartha CC, Eapen JT, Radhakumary C, Kutty VR, Ramani K, Lal AV. Pattern of cardiac fibrosis in rabbits periodically fed a magnesium-restricted diet and administered rare earth chloride through drinking water. J Cardiovasc Risk. 2000;7:31–5.
Kasaoka S, Tsuruta R, Nakashima K, Soejima Y, Miura T, Sadamitsu D, et al. Effect of intravenous magnesium sulfate on cardiac arrhythmias in critically ill patients with low serum ionized magnesium. Jpn Circ J. 1996;60:871–5.
Keller KB, Lemberg L. The importance of magnesium in cardiovascular disease. Am J Crit Care. 1993;2:348–50.
Keller PK, Aronson RS. The role of magnesium in cardiac arrhythmias. Prog Cardiovasc Dis. 1990;32:433–48.
Khan AM, Sullivan L, McCabe E, Levy D, Vasan RS, Wang TJ. Lack of association between serum magnesium and the risks of hypertension and cardiovascular disease. Am Heart J. 2010;160:715–20.
King DE, Mainous III AG, Geesey ME, Woolson RF. Dietary magnesium and C-reactive protein levels. J Am Coll Nutr. 2005;24:166–71.
Kisters K, Al-Tayar H, Nguyen MQ, Liebscher H, Wessels F, Buntzel J, et al. Magnesium metabolism and cardiovascular diseases. Trace Elem Electrol. 2011;28:70–3.
Kisters K, Liebscher D, Hausberg M. Magnesium metabolism and cardiovascular diseases. Trace ElemElectrol. 2010;27:152–3.
Kitlinski M, Konduracka E, Piwowarska W, Stepniewski M, Nessler J, Mroczek-Czernecka D, et al. Magnesium deficiency in mitral valve prolapse syndrome. Med Sci Monit. 1999;5:904–7.
Klevay LM, Milne DB. Low dietary magnesium increases supraventricular ectopy. Am J Clin Nutr. 2002;75:550–4.
Kurita T. Antiarrhythmic effect of parenteral magnesium on ventricular tachycardia associated with long QT syndrome. Magnes Res. 1994;7:155–7.
Laires MJ, Monteiro CP, Bicho M. Role of cellular magnesium in health and human disease. Front Biosci. 2004;9:262–76.
Lasserre B, Spoerri M, Moullet V, Theubet MP. Should magnesium therapy be considered for the treatment of coronary heart disease? II. Epidemiological evidence in outpatients with and without coronary heart disease. Magnes Res. 1994;7:145–53.
Lichodziejewska B, KI-os J, Rezler J, Grudzka K, DI-uzzniewska M, Buday A, et al. Clinical symptoms of mitral valve proplapse are related to hypomagnesemia and attenuated by magnesium supplementation. Am J Cardiol. 1997;79:768–72.
Maier JA, Malpuech-Brugere C, Zimowska W, Rayssiguier Y, Mazur A. Low magnesium promotes endothelial cell dysfunction: implications for atherosclerosis, inflammation and thrombosis. Biochim Biophys Acta. 2004;1689:13–21.
Marcus A, Chanut E, Laurant P, Gaume V, Berthelot A. A long-term moderate magnesium-deficient diet aggravates cardiovascular risks associated with aging and increases mortality in rats. J Hypertens. 2008;26:44–52.
Mathers TW. Beckstrand RL Oral magnesium supplementation in adults with coronary heart disease or coronary heart disease risk. J Am Acad Nurse Prac. 2009;21:651–7.
Matsusaka T, Hasebe N, Jin Y-T, Kawabe J, Kikuchi K. Magnesium reduces myocardial infarct size via enhancement of adenosine mechanism in rabbits. Cardiovasc Res. 2002;54:568–75.
Miyagi H, Yasue H, Okumura K, Ogawa H, Goto K, Oshima S. Effect of magnesium on anginal attack induced by hyperventilation in patients with variant angina. Circulation. 1989;79:597–602.
Mori H, Narita H, Hoshi Y, Kanazawa T, Onodera K, Metoki H, et al. The relationship between coronary artery diseases and magnesium deficiency, with regard to coronary artery spasms. Ann New York Acad Sci. 1993;676:334–7.
Nadler JL, Rude RK. Disorders of magnesium metabolism. Endocrinol Metab Clin North Am. 1995;24:623–41.
Novo S, Abrignani MG, Novo G, Nardi E, Dominguez LJ, Strano A, et al. Effect of drug therapy on cardiac arrhythmias and ischemia in hypertensives. Am J Hypertens. 2001;14:637–43.

Nurozler F, Tokgozoglu L, Pasaoglu I, Boke E, Ersoy U, Bozer AY. Atrial fibrillation after coronary artery bypass surgery: predictors and the role of MgSO4 replacement. J Card Surg. 1996;11:421–7.

Ohtsuka S, Yamaguchi I. Magnesium in congestive heart failure. Clin Calcium. 2005;15:181–6.

Onalan O, Crystal E, Daoulah A, Lau C, Crystal A, Lashevsky I. Meta-analysis of magnesium therapy for acute management of rapid atrial fibrillation. Am J Cardiol. 2007;99:1726–32.

Parikka H, Toivonen L, Naukkarinen V, Tierala I, Pohjola-Sintonen S, Heikkila J, et al. Decrease by magnesium of QT dispersion and ventricular arrhythmias in patients with acute myocardial infarction. Eur Heart J. 1999;20: 111–20.

Parikka H, Toivonen T, Pellinen K, Verkkala K, Jarvinen A, Nieminen MS. The Influence of intravenous magnesium sulphate on the occurrence of atrial fibrillation after coronary artery bypass operation. Eur Heart J. 1993;14:251–8.

Peacock JM, Ohira T, Post W, Sotoodehnia N, Rosamond W, Folsom A. Serum magnesium and risk of sudden cardiac death in the atherosclerosis risk in communities (ARIC) study. Am Heart J. 2010;160:464–70.

Prabha A, Shetty M, Thomas J, Nityanada C, Prabhu MV. Chelation therapy for coronary heart disease. Am J Heart. 2002;144(5):e-10.

Purvis JR, Cummings DM, Landsman P, Carroll R, Barakat H, Bray J, et al. Effect of oral magnesium supplementation on selected cardiovascular risk factors in non-insulin-dependent diabetics. Arch Fam Med. 1994;3:503–8.

Purvis JR, Movahed A. Magnesium disorders and cardiovascular diseases. Clin Cardiol. 1992;15:556–68.

Rasmussen HS, McNair P, Goransson L, Belslor S, Larsen OG, Aurup P. Magnesium deficiency in patients with ischaemic heart disease with and without acute myocardial infarction uncovered by an intravenous magnesium loading test. Arch Int Med. 1988;148:329–32.

Rasmussen SH. Justification for magnesium therapy in acute ischaemic heart disease. Clinical and experimental studies. Dan Med Bull. 1993;40:84–99.

Redwood SR, Bashir Y, Huang J, Leatham EW, Kaski J-C, Camm AJ. Effect of magnesium sulphate in patients with unstable angina: a double blind, randomized, placebo-controlled study. Eur Heart J. 1997;18:1269–77.

Reffelmann T, Dorr M, Ittermann T, Schwahn C, Volzke H, Ruppert J, et al. Low serum magnesium concentrations predict increase in left ventricular mass over 5 years independently of common cardiovascular risk factors. Atherosclerosis. 2010;213:563–9.

Reinhart RA. Clinical correlates of the molecular and cellular actions of magnesium on the cardiovascular system. Am Heart J. 1991;121:1513–21.

Roffe C, Fletcher S, Woods KL. Investigation of the effects of intravenous magnesium sulphate on cardiac rhythm in acute myocardial infarction. Br Heart J. 1994;71:141–5.

Rostron A, Sanni A, Dunning J. Does magnesium prophylaxis reduce the incidence of atrial fibrillation following coronary bypass surgery? Interact Cardiovasc Thorac Surg. 2005;4:52–8.

Roth A, Eshchar Y, Keren G, Kerbel S, Harsat A, Villa Y, et al. Effect of magnesium on restenosis after percutaneous transluminal coronary angioplasty: a clinical and angiographic evaluation in a randomized patient population. A pilot study. Eur Heart J. 1994;15:1164–73.

Roth A, Kornowski R, Agmon Y, Vardinon N, Sheps D, Graph E, et al. High-dose intravenous magnesium attenuates complement consumption after acute myocardial infarction treated by streptokinase. Eur Heart J. 1996;17:709–14.

Rubenowitz E, Axelsson G, Rylander R. Magnesium and calcium in drinking water and death from acute myocardial infarction in women. Epidemiology. 1999;10:31–6.

Rylander R. Environmental magnesium deficiency as a cardiovascular risk factor. J Cardiovasc Risk. 1996;3:4–10.

Sasaki S, Oshima T, Matsura H. Abnormal magnesium status in patients with cardiovascular diseases. Clin Sci (Colch). 2000;98:175–81.

Satake K, Lee J-D, Shimizu H, Ueda T. Relation between severity of magnesium deficiency and frequency of anginal attacks in men with variant angina. J Am Coll Cardiol. 1996;28:897–902.

Satur CM, Anderson JR, Jennings A, Newton K, Martin PG, Nair U, et al. Magnesium flux caused by coronary artery bypass operation: three patterns of deficiency. Ann Thorac Surg. 1994;58:1674–8.

Satur CM. Magnesium and cardiac surgery. Ann R Coll Surg Engl. 1997;79:349–54.

Seelig MS. Consequences of magnesium deficiency on the enhancement of stress reactions; preventive and therapeutic implications (a review). J Am Coll Nutr. 1994;13:429–46.

Seelig MS. Interrelationship of magnesium and estrogen in cardiovascular and bone disorders, eclampsia, migraine and premenstrual syndrome. J Am Coll Nutr. 1993;12:442–58.

Shechter M, Bairey Merz CN, Stuehlinger H-G, Slany J, Pachinger O, Robinowitz L. Effects of oral magnesium therapy on exercise tolerance, exercise-induced chest pain, and quality of life in patients with coronary artery disease. Am J Cardiol. 2003;91:517–21.

Shechter M, Merz CN, Rude RK, Paul Labrador MJ, Meisel SR, Shah PK, et al. Low intracellular magnesium levels promote platelet-dependent thrombosis in patients with coronary artery disease. Am Heart J. 2000;140:212–8.

Shechter M, Sharir M, Maura J, Labrador P, Forrester J, Silva B, et al. Oral magnesium therapy improves endothelial function in patients with coronary artery disease. Circulation. 2000;102:2353–8.

Shiga T, Wajima Z, Inoue T, Ogawa R. Magnesium prophylaxis for arrhythmia after cardiac surgery: a meta-analysis of randomized controlled trials. Am J Med. 2004;117:325–33.
Solomon AJ, Berger AK, Trivedi KK, Hannan RL, Katz NM. The combination of propranolol and magnesium does not prevent postoperative atrial fibrillation. Ann Thorac Surg. 2000;69:126–9.
Spasov AA, Iezhitsa IN, Kharitonova MV, Gurova NA. Arrhythmogenic threshold of myocardium under conditions of magnesium deficiency. Bull Exp Biol Med. 2008;146:63–5.
Speziale G, Ruvolo G, Fattouch K, Macrina F, Tonelli E, Donnetti M, et al. Arrhythmia prophylaxis after coronary artery bypass grafting: regimens of magnesium sulfate administration. J Thorac Cardiovasc Surg. 2000;48:22–6.
Stepura OB, Martynow AI. Magnesium ortate in severe congestive heart failure. Int J Cardiol. 2007;131:293–5. doi:10.1016/ijcard 2007.11.022.
Storm W, Zimmerman JJ. Magnesium deficiency and cardiogenic shock after cardiopulmonary bypass. Ann Thorac Surg. 1997;64:572–7.
Sueda S, Fukuda H, Watanabe K, Suzuki J, Saeki H, Ohtani T, et al. Magnesium deficiency in patients with recent myocardial infarction and provoked coronary artery spasm. Jpn Circ J. 2001;65:643–8.
Sueta CA, Clarke SW, Dunlap SH, Jensen L, Blauwet MB, Koch G, et al. Effect of acute magnesium administration on the frequency of ventricular arrhythmia in patients with heart failure. Circulation. 1994;89:660–6.
Tanabe K, Noda K, Kamegai M, Miyake F, Mikawa T, Murayama M, et al. Variant angina due to deficiency of intracellular magnesium. Clin Cardiol. 1990;13:663–5.
Tanabe K, Noda K, Ozasa A, Mikawa T, Murayama M, Sugai J. The relation of physical and mental stress to magnesium deficiency in patients with variant angina. J Cardiol. 1992;22:349–55.
Tejero-Taldo MI, Kramer JH, Mak IT, Komarov AM, Weglicki WB. The nerve-heart connection in the pro-oxidant response to magnesium-deficiency. Heart Fail Rev. 2006;11:35–44.
Thel MC, O'Connor CM. Magnesium in acute myocardial infarction. Coron Artery Dis. 1995;6:831–7.
Toraman F, Karabulut EH, Alhan HC, Dagdelen S, Tarcan S. Magnesium infusion dramatically decreases the incidence of atrial fibrillation after coronary artery bypass grafting. Ann Thorac Surg. 2001;72:1256–62.
Touyz RM. Magnesium in clinical medicine. Front Biosci. 2004;9:1278–93.
Tsuji A, Araki K, Maeyama K, Hashimoto K. Effectiveness of oral magnesium in a patient with ventricular tachycardic due to hypomagnesaemia. J Cardiovasc Pharmacol Ther. 2005;10:205–8.
Tsutsui M, Shmokawa H, Yoshihara S, Sobashima A, Hayashida K, Higuchi S, et al. Intracellular magnesium deficiency in acute myocardial infarction. Jpn Heart J. 1993;34:391–401.
Turlapaty PDMV, Altura BM. Magnesium deficiency produces spasms of coronary arteries: relationship to etiology of sudden death ischemic heart disease. Science. 1980;208:198–200.
Tzivani D, Banai S, Schuger C, Benhorin J, Keren A, Gottlieb S, et al. Treatment of Torsades de Pointes with magnesium sulphate. Circulation. 1988;77:392–7.
Ueshima K. Magnesium and ischemic heart disease. Clin Calcium. 2005;15:175–80.
Viskin S, Belhassen B, Laniado S. Deterioration of ventricular tachycardia to ventricular fibrillation after rapid intravenous administration of magnesium sulphate. Chest. 1992;101:1445–7.
Vitale JJ. Magnesium deficiency and cardiovascular disease. Lancet. 1992;340:1224–5.
Weiss M, Lasserre B. Should magnesium therapy be considered for the treatment of coronary heart disease? A critical appraisal of current facts and hypotheses. Magnes Res. 1994;7:135–44.
Whang R, Hampton EM, Whang DD. Magnesium homeostasis and clinical disorders of magnesium deficiency. Ann Pharmacother. 1994;28:220–6.
Wilkes NJ, Mallett SV, Peachey T, Di Salvo C, Walesby R. Correction of ionized plasma magnesium during cardiopulmonary bypass reduces the risk of cardiac arrhythmia. Anesth Analg. 2002;95:828–34.
Wolf FI, Trapani V, Simonacci M, Ferre S, Maier JA. Magnesium deficiency and endothelial dysfunction: is oxidative stress involved? Magnes Res. 2008;21:58–64.
Wolf MA. President's address: mother was right: the health benefits of milk of magnesia. Trans Am Clin Climatol Assoc. 2006;117:1–11.
Woods KL, Fletcher S. Long-term outcome after intravenous magnesium sulphate in suspected acute myocardial infarction: the second Leicester Intravenous Magnesium Intervention Trial(LIMIT-2). Lancet. 1994;343:816–9.
Yamori Y, Mizushima S. A review of the link between dietary magnesium and cardiovascular risk. Biol Trace Elem Res. 1998;63:19–30.
Yellon DM, Hausenloy DJ. Myocardial reperfusion injury. N Engl J Med. 2007;357:1121–35.
Zipes DP, Camm J, Borggrefe M, Buxton AE, Chaitman B, Framer M, et al. A report of the American college of cardiology, American heart association task force and the European society of cardiology committee for practice. Guidelines for the management of patients with ventricular arrhythmias and the prevention of sudden cardiac death. Circulation. 2006;114:1088–132.
Zuccala G, Pahor M, Lattanzio F, Vagnoni S, Rodola F, De Sole P, et al. Detection of arrhythmogenic cellular magnesium depletion in hip surgery patients. Br J Anaesth. 1997;79:776–81.

Hypertension (preeclampsia/eclampsia, primary hypertension)

Adam B, Malatyalioglu E, Alvur M, Talu C. Magnesium, zinc and iron levels in pre-eclampsia. J Matern-Fetal Med. 2001;10:246–50.

Altman D, Carroli G, Duley L, Farrell B, Moodley J, Neilson J, et al. Magpie Trial Collaboration Group. Do women with pre-eclampsia, and their babies, benefit from magnesium sulphate? The Magpie Trial: a randomised placebo-controlled trial. Lancet. 2002;359:1877–90.

Ascherio A, Hennekens C, Willett WC, Sacks F, Rosner B, Manson J, et al. Prospective study of nutritional factors, blood pressure and hypertension among US women. Hypertension. 1996;27:1065–72.

Ascherio A, Rimm EB, Hernan A, Giovannucci EL, Kawachi I, Stampfer MJ, et al. Intake of potassium, magnesium, calcium and fiber and risk of stroke among US men. Circulation. 1998;98:1198–204.

Baker WL, Kluger J, White CM, Dale KM, Silver BB, Coleman CI. Effect of magnesium L-lactate on blood pressure in patients with implantable cardioverter defibrillator. Ann Pharmacother. 2009;43:569–76.

Barbagallo M, Dominguez LJ, Resnick LM. Magnesium metabolism in hypertension and type 2 diabetes mellitus. Am J Ther. 2007;14:375–85.

Barbieri RL, Repke JT. Medical disorders during pregnancy; preeclampsia. In: Fauci AS, Braunwald E, Kasper DL, Hauser SL, Longo DL, Jameson JL, Loscalzo J, editors. Harrison's principles of internal medicine. 17th ed. New York: Publ McGraw Hill Medical; 2008. p. 44.

Beyer FR, Dickinson HO, Nicolson DJ, Ford GA, Mason J. Combined calcium, magnesium and potassium supplementation for the management of primary hypertension in adults. Cochrane Database Syst Rev. 2006;3:CD004805.

Bukoski RD. Reactive oxygen species: the missing link between magnesium deficiency and hypertension? J Hypertens. 2002;20:2141–3.

Cappuccio FP, Markandu ND, Beynon GW, Shore AC, Sampson B, MacGregor GA. Lack of effect of oral magnesium on high blood pressure: a double blind study. Br Med J. 1985;291:235–8.

Carlin M, Franz KB. Magnesium deficiency during pregnancy in rats increases systolic blood pressure and plasma nitrite. Am J Hypertens. 2002;15:1081–6.

Champagne CM. Magnesium in hypertension, cardiovascular disease, metabolic syndrome, and other conditions: a review. Nutr Clin Pract. 2008;23:142–51.

Crowther CA, Hiller JE, Doyle LW. Magnesium sulphate for preventing preterm birth in threatened preterm labour. Cochrane Database Syst Rev. 2002;4:CD001060.

Dickinson HO. Magnesium supplementation for the management of essential hypertension in adults. Cochrane Database Syst Rev. 2006;3:CD004640.

Ferrara LA, Iannuzzi R, Castaldo A, Iannuzzi A, Dello Russo A, Mancini M. Long-term magnesium supplementation in essential hypertension. Cardiology. 1992;81:25–33.

Fujita T, Ito Y, Ando K, Noda H, Ogata E. Attenuated vasodilator responses to magnesium in young patients with borderline hypertension. Circulation. 1990;82:384–93.

Geleinjoise JM, Witteman JC, den Breeijen JH, Hofman A, de Jong PT, Pols HA, et al. Dietary electrolyte intake and blood pressure in older subjects: the Rotterdam study. J Hypertens. 1996;14:737–41.

Greene MF. Magnesium sulphate for preeclampsia. N Engl J Med. 2003;348:275–6.

Guerrero-Romero F, Rodriguez-Moran M. The effect of lowering blood pressure by magnesium supplementation in diabetic hypertensive adults with low serum magnesium levels: a randomized, double-blind, placebo-controlled clinical trial. J Human Hypertens. 2009;23:245–51.

Itoh K, Kawasaka T, Nakamura M. The effects of high oral magnesium supplementation on blood pressure, serum lipids and related variables in apparently healthy Japanese subjects. Br J Nutr. 1997;78:737–50.

Jee SH, Miller III ER, Guallar E, Singh VK, Appel LJ, Klag MJ. The effect of magnesium supplementation on blood pressure: a meta-analysis of randomized clinical trials. Am J Hypertens. 2002;15:691–6.

Joffres MR, Reed DM, Yano K. Relationship of magnesium intake and other dietary factors to blood pressure: the Honolulu Heart Study. Am J Clin Nutr. 1987;45:469–75.

Katz A, Rosenthal T, Maoz C, Peleg E, Zeidenstein R, Levi Y. Effect of a mineral salt diet on 24 hr blood pressure monitoring in elderly hypertensive patients. J Hum Hypertens. 1999;13:777–80.

Kawasaki T, Itoh K, Kawasaki M. Reduction in blood pressure with a sodium-reduced potassium- and magnesium-enriched salt in subjects with mild essential hypertension. Hypertens Rev. 1998;21:235–43.

Kisters K, Barenbrock M, Louwen F, Hausberg M, Rahn KH, Kosch M, et al. Membrane, intracellular, and plasma magnesium and calcium concentrations in preeclampsia. Am J Hypertens. 2000;13:765–9.

Kisters K, Hoffmann O, Gremmler B, Hausberg M. Plasm magnesium deficiency is correlated to pulse pressure values in essential hypertension – influence of a metabolic syndrome. Am J Hypertens. 2005;18:215–9.

Kisters K, Tomak F, Hausberg M. Role of the Naq/Mg2q exchanger in hypertensions. Am J Hypertens. 2003;16:95–6.

Kisters K, Wessels F, Kuper H, Tokmak F, Krefting R, Gremmler B, et al. Incresed calcium and decreased magnesium concentrations and an increase calcium/magnesium ratio in spontaneously hypertensive rats versus Wister-Kyoto rats: relation to atherosclerosis. Am J Hypertens. 2004;17:59–62.

Konrad M, Schlingmann KP, Gudermann T. Insights into the molecular nature of magnesium homeostasis. Am J Physiol Renal Physiol. 2004;286:F599–605.

Kosch M, Housberg M, Westermann G, Koneke J, Matzkies F, Rahn KH, et al. Alterations in calcium and magnesium content of red cell membranes in patients with primary hypertension. Am J Hypertens. 2001;14:254–8.

Lind L, Lithell H, Pollare T, Ljunghall S. Blood pressure response during long-term treatment with magnesium is dependent on magnesium status. A double-blind, placebo-controlled study in essential hypertension and in subjects with high-normal blood pressure. Am J Hypertens. 1991;4:674–9.

Lucas MJ, Leveno KJ, Cunningham FG. A comparison of magnesium sulfate with phenytoin for the prevention of eclampsia. N Engl J Med. 1995;333:201–5.

Ma J, Folsom AR, Melnick SL, Eckfeldt JH, Sharrett AR, Nabulsi AA, et al. Associations of serum and dietary magnesium with cardiovascular disease, hypertension, diabetes, insulin, and carotid arterial wall thickness: the ARIC study: atherosclerosis risk in communities study. J Clin Epidemiol. 1995;48:927–40.

Meltzer JI. A clinical specialist in hypertension critiques. Am J Hypertens. 2005;18:894–8.

Musso CG. Magnesium metabolism in health and disease. Int J Urol Nephrol. 2009;41:357–62.

Newman JC, Amarasingham JL. The pathogenesis of eclampsia: the 'magnesium ischaemia' hypothesis. Med Hypotheses. 1993;40:250–6.

Ozono R, Oshima T, Matsuura H, Higashi Y, Ishida T, Watanabe M, et al. Systemic magnesium deficiency disclosed by magnesium loading test in patients with essential hypertension. Hypertens Res–Clin E. 1995;18:39–42.

Paolisso G, Barbagallo M. Hypertension, diabetes mellitus, and insulin resistance. Am J Hypertens. 1997;10:346–55.

Pascal L, Touyz RM. Physiological and pathophysiological role of magnesium in cardiovascular system: implications in hypertension. J Hypertens. 2000;18:1177–91.

Patil VP, Choudhari NA. A study of serum magnesium in preeclampsia and eclampsia. Indian J Clin Biochem. 1991;6:69–72.

Peebles DM, Marlow N, Brocklehurst P. Antenatal magnesium sulphate reduces risk of cerebral palsy if given before preterm delivery. BMJ. 2011;342:1095–6.

Resnick LM, Bardicef O, Altura BT, Alderman MH, Altura BM. Serum ionized magnesium: relation to blood pressure and racial factors. Am J Hypertens. 1997;10:1420–4.

Roberts JM. Magnesium for preeclampsia and eclampsia. N Engl J Med. 1995;333:250–1.

Schiffrin EL, Touyz RM. Calcium, magnesium and oxidative stress in hyperaldosteronism. Circulation. 2005;111:830–1.

Seelig C. Magnesium deficiency in two hypertensive patients groups. South Med J. 1990;83:739–42.

Sontia B, Touyz RM. Role of magnesium in hypertension. Arch Biochem Biophys. 2007;458:33–9.

Touyz RM, Milne FJ. Magnesium supplementation attenuates, but does not prevent, the development of hypertension in spontaneously hypertensive rats. Am J Hypertens. 1999;12:757–65.

Touyz RM. Magnesium and hypertension. Curr Opin Nephrol Hy. 2006;15:141–4.

Touyz RM. Transient receptor potential melastatin 6 and 7 channels magnesium transport and vascular biology: implications in hypertension. Am J Physiol Heart Circ Physiol. 2008;294:1103–18.

Van Leer EM, Seidell JC, Kromhout D. Dietary calcium, potassium, magnesium and blood pressure in the Netherlands. Int J Epidemiol. 1995;24:1117–23.

Wells IC, Agrawal DK, Anderson RJ. Abnormal magnesium metabolism in etiology of salt-sensitive hypertension and type 2 diabetes mellitus. Biol Trace Elem Res. 2004;98:97–108.

Widman L, Wester PO, Stegmayr BK, Wirell M. The dose dependent reduction in blood pressure through administration of magnesium. A double blind placebo controlled cross-over study. Am J Hypertens. 1993;6:41–5.

Yang C-Y, Chiu H-F. Calcium and magnesium in drinking water and the risk of death from hypertension. Am J Hypertens. 1999;12:894–9.

Diabetes (diabetes type 2; insulin resistance)

Agrawal P, Arora S, Singh B, Manamalli A, Dolia PB. Association of macrovascular complications of type 2 diabetes mellitus with serum magnesium levels. Diab Metab Syndr Clin Res Rev. 2011;5:41–4.

Altura BT, Brust M, Bloom S, Barbour RL, Stempak JG, Altura BM. Magnesium dietary intake modulates blood lipid levels and atherogenesis. Proc Natl Acad Sci. 1990;87:1840–4.

Balon TW, Gu JL, Tokuyama Y, Jasman AP, Nadler JL. Magnesium supplementation reduces development of diabetes in a rat model of spontaneous NIDDM. Am J Physiol. 1995;269:E745–52.

Barbagallo M, Dominguez LJ, Galioto A, Ferlisi A, Cani C, Malfa L, et al. Role of magnesium in insulin action, diabetes and cardio-metabolic syndrome X. Mol Aspects Med. 2003;24:39–52.

Barragan-Rodriguez L, Rodriguez-Moran M, Guerrero-Romero F. Efficacy and safety of oral magnesium supplementation in the treatment of depression in the elderly with type 2 diabetes: a randomized equivalent trial. Magnes Res. 2008;21:218–23.

Bloomgarden ZT. American Diabetes Association scientific sessions, 1995. Magnesium deficiency, atherosclerosis, and health care. Diabetes Care. 1995;18:1623–7.

Bo S. Dietary magnesium and fiber intake, inflammatory and metabolic parameters in middle-aged subjects from a population- based cohort. Am J Clin Nutr. 2006;84:1062–9.

Chambers EC, Heshka S, Gallagher D, Wang J, Pi-Sunyer X, Pierson Jr RN. Serum magnesium and type-2 diabetes in African American and Hispanics: a New York cohort. J Am Coll Nutr. 2006;25:509–13.

Chaudhary DP, Sharma R, Bansal DD. Implications of magnesium deficiency in type 2 diabetes: a review. Biol Trace Elem Res. 2010;134:119–29.

Corica F, Corsonella A, Ientile R, Cucinotta D, Benedetto AD, Perticone F, et al. Serum ionized magnesium levels in relation to metabolic syndrome in type 2 diabetic patients. J Am Coll Nutr. 2006;25:210–5.

Curiel-Garcia JA, Rodriguez-Moran M, Guerrero-Romero F. Hypomagnesemia and mortality in patients with type 2 diabetes. Magnes Res. 2008;21:163–6.

De Lourdes Lima M, Lima MD, Cruz T, Pousada JC, Rodrigues LE, Barbosa K, et al. The effect of magnesium supplementation in increasing doses on the control of type 2 diabetes. Diabetes Care. 1998;21:682–6.

De Valk HW. Magnesium in diabetes mellitus. Neth J Med. 1999;54:139–46.

De Valk HW, Verkaaik R, van Rigin HJM. Oral magnesium supplementation in insulin-requiring type 2 diabetic patients. Diab Med. 1998;15:503–7.

Eibl NL, Kopp HP, Nowak HR, Schnack CJ, Hopmeier PG, Schernthaner G. Hypomagnesemia in type II diabetes: effect of a 3-month replacement therapy. Diabetes Care. 1995;18:188–92.

Fox CH, Ramsoomair D, Mahoney MC, Carter C, Young B, Graham R. An investigation of hypomagnesemia among ambulatory urban African Americans. J Fam Pract. 1999;48:636–9.

Fung TT, Manson JE, Solomon CG, Liu S, Willett WC, Hu FB. The association between magnesium intake and fasting insulin concentration in healthy middle-aged women. J Am Coll Nutr. 2003;22:533–8.

Grafton G, Baxter MA. The role of magnesium in diabetes mellitus. A possible mechanism for the development of diabetic complications. J Diabetes Complicat. 1992;6:143–9.

Guerrero-Romero F, Rascon-Pacheco RA, Rodriguez-Moran M, de la Pena JE, Wacher N. Hypomagnesaemia and risk of metabolic glucose disorder: a 10-year follow-up study. Eur J Clin Invest. 2008;38:389–96.

Guerrero-Romero F, Rodriguez-Moran M. Hypomagnesemia is linked to low serum HDL-cholesterol irrespective of serum glucose values. J Diabetes Complications. 2000;14:272–6.

Guerrero-Romero F, Tamez-Perez HE, Gonzalez-Gonzalez G, Salinas-Martinez AM, Montes-Villarreal J, Trevino-Ortiz JH, et al. Oral magnesium supplementation improves insulin sensitivity in non-diabetic subjects with insulin resistance: a double- blind placebo-controlled randomized trial. Diabetes Metab. 2004;30:253–8.

Huerta MG, Roemmich JN, Kington ML, Bovbjerg VE, Weltman AL, Holmes VF, et al. Magnesium deficiency is associated with insulin resistance in obese children. Diabetes Care. 2005;28:1175–81.

Humphries S, Kushner H, Falkner B. Low dietary magnesium is associated with insulin resistance in a sample of young, nondiabetic black Americans. Am J Hypertens. 1999;12:747–56.

Joffres MR, Reed DM, Yano K. Relationship of magnesium intake and other dietary factors to blood pressure: the Honolulu Heart Study. Am J Clin Nutr. 1987;45:469–75.

Kandeel FR, Balon E, Scott S, Nadler JL. Magnesium deficiency and glucose metabolism in rat adipocytes. Metab Clin Exp. 1996;45:838–43.

Kao WH, Folsom AR, Nieto FJ, Mo JP, Watson RL, Brancati FL. Serum and dietary magnesium and the risk for type 2 diabetes mellitus: the atherosclerosis risk in communities study. Arch Intern Med. 1999;159:2151–9.

Laughlin MR, Thompson D. The regulatory role for magnesium in glycolytic flux of the human erythrocyte. J Biol Chem. 1996;271:977–83.

Longstrret DA, Heath DL, Panaretto KS, Vink R. Correlation suggests low magnesium may lead to higher rate of type 2 diabetes in indegenous Australians. Rural Remote Health. 2007;7:843.

Lopez-Ridaura R, Willett WC, Rimm EB, Liu S, Stampfer MJ, Manson JE, et al. Magnesium intake and risk of type 2 diabetes in men and women. Diabetes Care. 2004;27:134–40.

Maier JA, Malpuech-Brugere C, Zimowska W, Rayssiguier Y, Mazur A. Low magnesium promotes endothelial cell dysfunction: implications for atherosclerosis, inflammation and thrombosis. Biochim Biophys Acta. 2004;1689:13–21.

Mann JI, De Leeuw I, Hermansen K, Karamanos B, Karlstrom B, Katsilambros N, et al. Diabetes and nutrition study group (DNSG) of the European association for the study of diabetes (EASD). Nutr Metab Cardiovasc Dis. 2004;14:373–94.

Mann JI, Riccardi G. Evidence-based European guidelines on diet and diabetes. Nutr Metab Cardiovasc Dis. 2004;14:332–3.

Montagnana M, Lippi G, Targher G, Salvagno GL, Guidi GC. Relationship between hypomagnesemia and glucose homeostasis. Clin Lab. 2008;54:5–6.

Nadler JL, Buchanan T, Natarajan R, Antonipillai I, Bergman R, Rude R. Magnesium deficiency produces insulin resistance and increased thromboxane synthesis. Hypertension. 1993;21:1024–9.

Nadler JL, Malayan S, Luong H, Shaw S, Natarajan RD, Rude RK. Intracellular free magnesium deficiency plays a key role in increased platelet reactivity in type II diabetes mellitus. Diabetes Care. 1992;15:835–41.

Orchard TJ. Magnesium and type 2 diabetes mellitus. Arch Intern Med. 1999;159:2119–20.

Paolisso G, Scheen A, D'Onofrio F, Lefebvre P. Magnesium and glucose homeostasis. Diabetologia. 1990;33:511–4.

Paolisso G, Sgambato S, Gambardella A, Pizza G, Tesauro P, Varricchio M, et al. Daily magnesium supplements improve glucose handling in elderly subjects. Am J Clin Nutr. 1992;55:1161–7.

Paolisso G, Sgambato S, Pizza G, Passariello N, Varricchio M, D'Onofrio F. Improved insulin response and action by chronic magnesium administration in aged NIDDM subjects. Diabetes Care. 1989;12:262–9.

Paolisso G. Changes in glucose turnover parameters and improvement of glucose oxidation after 4-week magnesium administration in elderly non-insulin-dependent (type II) diabetic patients. J Clin Endocrinol Metab. 1994;78:1510–4.

Pham PC, Pham PM, Pham PA, Pham SV, Pham HV, Miller JM, et al. Lower serum magnesium levels are associated with more rapid decline of renal function in patients with diabetes mellitus type 2. Clin Nephrol. 2005;63:429–33.

Pham PC, Pham PM, Pham SV, Miller JM, Pham PT. Hypomagnesaemia in patients with type 2 diabetes. Clin J Am Soc Nephrol. 2007;2:366–73.

Pham PC, Pham PM, Pham PT, Pham SV, Pham PA, Pham PT. The link between lower serum magnesium and kidney function in patients with diabetes mellitus Type 2 deserves a closer look. Clin Nephrol. 2009;71:375–9.

Rumawas ME, McKeown NM, Rogers G, Meigs JB, Wilson PW, Jacques PF. Magnesium intake in relation to improved insulin homeostasis in the Fragmingham offspring cohort. J Am Coll Nutr. 2006;25:486–92.

Reis MA, Latorraca MQ, Carneiro EM, Boschero AC, Saad MJ, Velloso LA, et al. Magnesium deficiency improves glucose homeostasis in the rat: studies in vivo and in isolated islets in vitro. Br J Nutr. 2001;85:549–52.

Rodriguez-Hernandez H, Gonzalez JL, Rodriguez-Moran M, Guerro-Romero F. Hypomagnesemia, insulin resistance, and non-alcoholic steatohepatitis in obese subjects. Arch Med Res. 2005;36:362–6.

Rodriguez-Moran M, Guerrero-Romero F. Oral magnesium supplementation improves insulin sensitivity and metabolic control in type 2 diabetic subjects: a randomized double-blind controlled trial. Diabetes Care. 2003;26:1147–52.

Romero JR, Ferreira A, Ricupero DA, Rivera A. Dysregulation of cellular magnesium homeostasis by glucose in type II diabetic patients. Am J Hypertens. 2005;18:166–70.

Rosolova H, Mayer Jr O, Reaven GM. Insulin-mediated glucose disposal is decreased in normal subjects with relatively low plasma magnesium concentrations. Metabolism. 2000;49:418–20.

Rude RK. Magnesium deficiency and diabetes mellitus. Causes and effects. Postgrad Med. 1992;92:217–9.

Sales CH, Pedrosa Lde F. Magnesium and diabetes mellitus: their relation. Clin Nutr. 2006;25:554–62.

Schnack C, Bauer I, Pregant P, Hopmeier P, Schernthaner G. Hypomagnesaemia in type 2 (non-insulin-dependent) diabetes mellitus is not corrected by improvement of long-term metabolic control. Diabetologia. 1992;35:77–9.

Schulze MB, Schulze M, Heidemann C, Schienkiewitz A, Hoffman K, Boeing H. Fiber and magnesium intake and incidence of type 2 diabetes. A prospective study and meta-analysis. Arch Intern Med. 2007;167:956–65.

Seyoum B, Siraj ES, Saenz C, Abdulkadir J. Hypomagnesaemia in Ethiopians with diabetes mellitus. Ethn Dis. 2008;18:147–51.

Seyoum B, Siraj ES, Saenz C, Abdulkadir J. Low levels of magnesium place diabetics at increases risk. Ethn Dis. 2008;18:238–9.

Sharma A, Dabla S, Agrawal RP, Barjatya H, Kochar DK, Kothari RP. Serum magnesium: an early predictor of course and complications of diabetes mellitus. J Indian Med Assoc. 2007;105:16–20.

Sheehan JP. Magnesium deficiency and diabetes mellitus. Magnes Trace Elem. 1991;10:215–9.

Simmons D, Joshi S, Shaw J. Hypomagnesaemia is associated with diabetes: not pre-diabetes, obesity or the metabolic syndrome. Diabetes Res Clin Pract. 2010;87:261–6.

Singh RB, Rastogi SS, Sharma VK, Saharia RB, Kulshretha SK. Can dietary magnesium modulate lipoprotein metabolism? Magnes Trace Elem. 1990;9:255–64.

Song Y, Ka H, Levitan EB, Manson JE, Liu S. Effect of oral magnesium supplementation on glycaemic control in type 2 diabetes: a meta-analysis of randomized double blind controlled trials. Diabetes Care. 2006;23:1050–6.

Song Y, Manson JE, Buring JE, Liu S. Dietary magnesium intake in relation to plasma insulin levels and risk of type 2 diabetes in women. Diabetes Care. 2004;27:59–65.

Song Y, Ridker PM, Manson JE, Cook NR, Buring JE, Liu S. Magnesium intake, C-reactive protein, and the prevalence of metabolic syndrome in middle-aged and older U.S. women. Diabetes Care. 2005;28:1438–44.

Song Y, Hsu YH, Niu T, Manson JE, Buring JE, Liu S. Common genetic variants of the ion channel transient receptor potential membrane melastatin 6 and 7 (TRPM6 and TRPM7), magnesium intake, and risk of type 2 diabetes in women. BMC Med Genet. 2009;10:4–10.

Tosiello L. Hypomagnesemia and diabetes mellitus. A review of clinical implications. Arch Intern Med. 1998;156:1143–8.

Volpe SL. Magnesium, the metabolic syndrome, insulin resistance, and type 2 diabetes mellitus. Crit Rev Food Sci. 2008;48:293–300.

Walti MK, Zimmermann MB, Spinas GA, Hurrell RF. Low plasma magnesium in type 2 diabetes. Swiss Med Wkly. 2003;133:289–92.

White Jr JR, Campbell RK. Magnesium and diabetes: a review. Ann Pharmacother. 1993;27:775–80.
Wolf FI, Trapani V, Simonacci M, Ferre S, Maier JA. Magnesium deficiency and endothelial dysfunction: is oxidative stress involved? Magnes Res. 2008;21:58–64.
Yang CY, Chiu HF, Cheng MF, Tsai SS, Hung CF, Tseng YT. Magnesium in drinking water and the risk of death from diabetes mellitus. Magnes Res. 1999;12:131–7.
Yokota K. Diabetes mellitus and magnesium. Clin Calcium. 2005;15:203–12.

The metabolic syndrome

Evangelopoulos AA, Vallianou NG, Panagiotakos DB, Georgiou A, Zacharias GA, Alevra AN, et al. An inverse relationship between cumulating components of the metabolic syndrome and serum magnesium levels. Nutr Res. 2008;28:659–63.
Ford ES, Li C, McGuire LC, Mokdad AH, Liu S. Intake of dietary magnesium and the prevalence of the metabolic syndrome among US adults. Obesity. 2007;15:1139–46.
Ghasemi A, Zahediasl S, Syedmoradi L, Azizi F. Low serum magnesium levels in elderly subjects with metabolic syndrome. Biol Trace Elem Res. 2010;136:18–25.
Guerrero-Romero F, Rodriguez-Moran M. Low serum magnesium levels and metabolic syndrome. Acta Diabetol. 2002;39:209–13.
Hoerzer S, Wascher TC, Lipp RW, Haas J, Klein WW, Zweiker R. Effect of oral magnesium and potassium therapy on thallium perfusion scan defects, quality of life and pulse wave analysis in hypertensive patients with syndrome X. Am J Hypertens. 2004;17:71–5.
Ka H, Liu K, Daviglus ML, Morris SJ, Loria CM, Van Horn L, et al. Magnesium intake and evidence of metabolic syndrome among young adults. Circulation. 2006;113:1675–82.
Ka H, Song Y, Belin RJ, Chen Y. Magnesium intake and the metabolic syndrome: epidemiologic evidence to date. J Cardiometab Syndr. 2007;1:351–5.
Kumeda Y, Inaba M. Metabolic syndrome and magnesium. Clin Calcium. 2005;15:97–104.
Lima MD, Cruz T, Rodrigues LE, Bomfim O, Melo J, Correia R, et al. Serum and intracellular magnesium deficiency in patients with metabolic syndrome. Evidence for its relationship to insulin resistence. Diabetes Res Clin Pract. 2009;83:257–62.

Osteoporosis and bone mineral density (BMD)

Abed E, Moreau R. Importance of melastatin-like transient receptor potential 7 and magnesium in the stimulation of osteoblast proliferation and migration of PDGF. Am J Physiol Cell Physiol. 2009;297:C360–8. doi:10.1152/ajpcell, 00614, 2008.
Abraham G, Grewal H. A total dietary program emphasizing magnesium instead of calcium: effect on mineral density of cancellous bone in postmenopausal women on hormone therapy. J Reprod Med. 1990;35:503–7.
Abraham GE. The importance of magnesium in the management of primary postmenopausal osteoporosis. J Nut Med. 1991;2:165–78.
Angus RM, Sambrook PN, Pocock NA, Eisman JA. Dietary intake and bone mineral density. Bone Miner. 1988;4:265–77.
Boskey AL, Rimnac CM, Bansal M, Federman M, Lian J, Boyan BD. Effect of short-term hypomagnesemia on the chemical and mechanical properties of rat bone. J Orthop Res. 1992;10:774–83.
Carpenter TO, Barton CN, Park YK. Usual dietary magnesium intake in NHANES III is associated with femoral bone mass. J Bone Miner Res. 2000;15 Suppl 1:S292.
Carpenter TO, DeLucia MC, Zhang JH, Bejnerowicz G, Tartamella L, Dziura J, et al. A randomized controlled study of the effects of dietary magnesium oxide supplementation on bone mineral density in healthy girls. J Clin Endocrinol Metab. 2006;91:4866–72.
Carpenter TO, Mackowiak SJ, Troiano N, Gundberg CM. Osteocalcin and its message: relationship to bone histology in magnesium-deprived rats. Am J Physiol. 1992;263:E107–14.
Carpenter TO. Disturbances of vitamin D metabolism and action during clinical and experimental magnesium deficiency. Magnes Res. 1988;1:131–9.
Cohen L, Laor A, Kitzes R. Magnesium malabsorption in postmenopausal osteoporosis. Magnesium. 1983;2:139–43.
Cohen L. Recent data on magnesium and osteoporosis. Magnes Res. 1988;1:85–7.

Creedon A, Flynn A, Cashman K. The effect of moderately and severely restricted dietary magnesium intakes on bone composition and bone metabolism in the rat. Br J Nutr. 1999;82:63–71.

Gruber HE, Rude PK, Wei L, Frausta A, Mills BG, Norton HJ. Magnesium deficiency: effect on bone mineral density in the mouse appendicular skeleton. BMC Musculoskelet Disord. 2003;4:7–10.

Houtkooper LB, Ritenbaugh C, Aickin M, Lohman TG, Going SB, Weber JL, et al. Nutrients, body composition and exercise are related to change in bone mineral density in premenopausal women. J Nutr. 1994;125:1229–37.

Jackson RD. The impact of magnesium intake on fractures: results from the women's health initiative observational study (WHI-OS). In Am Soc for Bone and Mineral Research; ASBMR, 2003. p. 31.

Kenney MA, McCoy H, Williams L. Effects of magnesium deficiency on strength, mass and composition of rat femur. Calcif Tissue Int. 1994;54:44–9.

Launius BK, Brown PA, Cush EM, Mancini MC. Osteoporosis: the dynamic relationship between magnesium and bone mineral density in the heart transplant patients. Crit Care Nurs Q. 2004;27:96–100.

Nakamura K, Ueno K, Nishiwaki T, Saito T, Tsuchiya Y, Yamamoto M. Magnesium intake and bone mineral density in young adults women. Magnes Res. 2007;20:250–3.

New SA, Bolton-Smith C, Grubb DA, Reid DM. Nutritional influences on bone mineral density: a cross-sectional study in premenopausal women. Am J Clin Nutr. 1997;65:1831–9.

Odabasi E, Turan M, Aydin A, Akay C, Kutlu M. Magnesium, zinc, copper, manganese, and selenium levels in postmenopausal women with osteoporosis. Can magnesium play a key role in osteoporosis? Ann Acad Med (Singapore). 2008;37:564–7.

Reginster JY, Strause L, Deroisy R, Lecart MP, Saltman P, Franchimont P. Preliminary report of decreased serum magnesium in postmenopausal osteoporosis. Magnesium. 1989;8:106–9.

Rude PK, Gruber HE. Magnesium deficiency and osteoporosis; animal and human observations. J Nutr Biochem. 2004;15:710–6.

Rude RK, Gruber HE, Norton J, Wei L, Frausto A, Kilburri J. Dietary magnesium reduction to 25% of nutrient requirement disrupts bone and mineral metabolism. Bone. 2005;37:211–9.

Rude RK, Olerich M. Magnesium deficiency: possible role in osteoporosis associated with gluten-sensitive enteropathy. Osteoporos Int. 1996;6:453–61.

Rude RK. Magnesium deficiency: a cause of heterogeneous disease in humans. J Bone Miner Res. 1998;13:749–58.

Ryder KM, Shorr RI, Bush AJ, Kritchevsky SB, Harris T, Stone K, et al. Magnesium intake from food and supplements is associated with bone mineral density in healthy older white subjects. Am J Geriatr Soc. 2005;53: 1875–80.

Ryschon TW, Rosenstein DL, Rubinow DR, Niemela JE, Elin RJ, Balaban RS. Relationship between skeletal muscle intracellular ionized magnesium and measurements of blood magnesium. J Lab Clin Med. 1996;127:207–13.

Saito N, Tabata N, Saito S, Andou Y, Onaga Y, Iwamitsu A, et al. Magnesium deficiency affects the bones and could be a cause of osteoporosis. J Am Coll Nutr. 2004;23:701–3.

Seeling MS. Increased need for magnesium with the use of combined oestrogen and calcium for osteoporosis treatment. Magnes Res. 1990;3:197–215.

Sojka JE. Magnesium supplementation and osteoporosis. Nutr Rev. 1995;53:71–80.

Song CH, Barrett-Conner E, Kim SH, Kim KS. Association of calcium and magnesium in serum and hair with bone mineral density in premenopausal women. Biol Trace Elem Res. 2007;118:1–9.

Stendig-Lindberg G, Koeller W, Bauer A, Rob PM. Experimentally induced prolonged magnesium deficiency causes osteoporosis in the rat. Eur J Intern Med. 2004;15:97–107.

Stendig-Lindberg G, Koeller W, Bauer A, Rob PM. Prolonged magnesium deficiency causes osteoporosis in the rat. J Am Coll Nutr. 2004;23:704S–11S.

Stendig-Lindberg G, Tepper R, Leichter I. Trabecular bone density in a two year controlled trial of peroral magnesium in osteoporosis. Magnes Res. 1993;6:155–63.

Struijs A, Mulder H. Treatment of postmenopausal osteoporosis and low serum magnesium with intermittent cyclical EHDP and magnesium. Neth J Med. 1996;48:76–8.

Toba Y, Kajita Y, Masuyama R, Takada Y, Suzuki K, Aoe S. Dietary magnesium supplementation affects bone metabolism and dynamic strength of bone in ovariectomized rats. J Nutr. 2000;130:216–20.

Tranquilli AL, Lucino E, Garzetti GG, Romanini C. Calcium, phosphorus, and magnesium intakes correlate with bone mineral content in postmenopausal women. Gynecol Endocrinol. 1994;8:55–8.

Tucker KL, Hannan MT, Chen H, Cupples LA, Wilson PW, Kiel DP. Potassium, magnesium, and fruit and vegetable intakes are associated with greater bone mineral density in elderly men and women. Am J Clin Nutr. 1999;69:727–37.

Vormann J. Magnesium: nutrition and metabolism. Mol Aspects Med. 2003;24:27–37.

Wang MC, Moore EC, Crawford PB, Hudes M, Sabry ZI, Marcus R, et al. Influence of pre-adolescent diet on quantitative ultrasound measurements of the calcaneus in young adult women. Osteoporos Int. 1999;9:532–5.

Whelan AM. Natural health products in the prevention and treatment of osteoporosis: systematic review of randomised controlled trials. Ann Pharmacother. 2006;40:836–49.

Musculoskeletal (hyperexcitability, spasms, cramps, tetany)

Bazzichi L, Giannaccini G, Betti L, Fabbrini L, Schmid L, Palego L, et al. ATP, calcium and magnesium levels in platelets of patients with primary fibromyalgia. Clin Biochem. 2008;41:1084–90.
Bilbey DL, Prabhakaran VM. Muscle cramps and magnesium deficiency: case reports. Can Fam Physician. 1996;42:1348–51.
Clauw DJ, Ward K, Wilson B, Katz P, Rajan SS. Magnesium deficiency in the eosinophilia-myalgia syndrome: report of clinical and biochemical improvement with repletion. Arthritis Rheum. 1994;37:1331–4.
Dahle LO, Berg G, Hammar M, Hurtig M, Larsson L. The effect of oral magnesium substitution on pregnancy-induced leg cramps. Am J Obstet Gynecol. 1995;173:175–80.
Dominguez LJ, Barbagallo M, Lauretani F, Bandinelli S, Bos A, Corsi AM, et al. Magnesium and muscle performance in older persons: the InCHIANTI study. Am J Clin Nutr. 2006;84:419–26.
Hantoushzadeh S, Jafarabadi M, Khazardoust S. Serum magnesium levels, muscle cramps, and preterm labor. Int J Gynaecol Obstet. 2007;98:153–4.
Iannello S, Spina M, Leotta P, Prestipino M, Spina S, Ricciardi N, et al. Hypomagnesemia and smooth muscle contractility: diffuse esophageal spasm in an old female patient. Miner Electrol Metab. 1998;24:348–56.
Keen CL, Lowney P, Gershwin ME, Hurley LS, Stern JS. Dietary magnesium intake influence exercise capacity and haematologic parameters in rats. Metabolism. 1987;36:788–93.
Mahajan S, Engel WK. Assessment: symptomatic treatment for muscle cramps (an evidence-based review): report of the Therapeutics and Technology Assessment Subcommittee of the American Academy of Neurology. Neurology. 2010;75:1397–8. author reply 1398–9.
Morikawa H, Yoshida S. Toxemie of pregnancy and magnesium. Clin Calcium. 2005;15:213–9.
Rogers R, Fairfield JE. Latent hypomagnesaemic tetany. Lancet. 1994;343:63.
Romano TJ, Stiller JW. Magnesium deficiency in fibromyalgia syndrome. J Nutr Med. 1994;4:165–7.
Romano TJ. Magnesium deficiency in systemic lupus erythematosus. J Nutr Environ Med. 1997;7:107–11.
Ross RM, Baker T. An effect of magnesium on neuromuscular function in parturient. J Clin Anesth. 1996;8:202–4.
Saris NE, Mervaala E, Karppanen H, Khawaja JA, Lewenstam A. Magnesium: an update on physiological, clinical and analytical aspects. Clin Chim Acta. 2000;294:1–26.
Sendur OF, Tastaban E, Turan Y, Ullman C. The relationship between serum trace element levels and clinical parameters in patients with fibromyalgia. Rheumatol Int. 2008;28:1117–21.
Smets YF, Bokani N, De Meijer PH, Meinders AE. Tetany in excessive use of alcohol: a possible magnesium deficiency. Ned Tijdschr Genees. 2004;148:641–4.
Smith WO, Clark RM, Mohr J, Whang R. Vertical and horizontal nystagmus in magnesium deficiency. South Med J. 1980;73:269–72.

Neurological; migraines; Depression

Aloisi P, Marrelli A, Porto C, Tozzi E, Cerone G. Visual evoked potentials and serum magnesium levels in juvenile migraine patients. Headache. 1997;37:383–5.
Barragan-Rodriguez L, Rodriguez-Moran M, Guerrero-Romero F. Efficacy and safety of oral magnesium supplementation in the treatment of depression in the elderly with type 2 diabetes: a randomized, equivalent trial. Magnes Res. 2008;21:218–23.
Diener HC, Kaube H, Limmroth V. Antimigraine drugs. J Neurol. 1999;246:515–9.
Eby GA, Eby KL. Magnesium for tretment-resistant depression: a review and hypotheses. Med Hypotheses. 2010;74:649–60.
Faccchinetti F, Sances G, Borella P. Magnesium prophylaxis of menstrual migraine: effect on intraceullar magnesium. Headache. 1991;31:298–301.
Gallai V, Sarchielli P, Morucci P. Magnesium content of mononuclear blood cells in migraine patients. Headache. 1994;34:160–5.
Gallai V, Sarchielli P, Morucci P. Red blood cell magnesium levels in migraine patients. Cephalalgia. 1993;13:94–8.
Gallai V, Sarchielli P, Coata G. Serum and salivary magnesium in migraine. Results in a group of juvenile patients. Headache. 1992;32:132–5.
Gupta VK. Magnesium therapy for migraine: do we need more trials or more reflection? Headache. 2004;44:445–6.
Jacka FN, Overland S, Stewart R, Tell GS, Bjelland I, Mykletun A. Association between magnesium intake and depression and anxiety in community-dwelling adults: the Hordaland Health Study. Aust N Z J Psychiatry. 2009;43:45–52.

Jain AC, Sethi NC, Babbar PK. A clinical electroencephalographic and trace elements study with special reference to zinc, copper and magnesium in serum and cerebrospinal fluid (CSF) in cases of migraine. J Neurol. 1985;232(Suppl):161–5.

Jung KI, Ock SM, Chung JH, Song CH. Associations of serum Ca and Mg levels with mental health in adult women without psychiatric disorders. Biol Trace Elem Res. 2010;133:153–61.

Lodi R, Kemp GJ, Montagna P, Pierangeli G, Cortelli P, Iotti S, et al. Quantitative analysis of skeletal muscle bioenergetics and proton efflux in migraine and other cluster headache. J Neurol Sci. 1997;146:73–80.

Mauskop A, Altura BM. Role of magnesium in the pathogenesis and treatment of migraines. Clin Neurosci. 1998;5:24–7.

Mauskop A, Altura BT, Alyura BM. Serum ionized magnesium levels and serum ionized calcium/ionized magnesium ratios in women with menstrual migraine. Headache. 2002;42:242–8.

Mauskop A, Altura BT, Cracco RQ, Altura BM. Deficiency in serum ionized magnesium but not total magnesium in patients with migraines. Possible role of ICa2q/IMg2q ratio. Headache. 1993;33:135–8.

Mauskop A. Evidence linking magnesium deficiency to migraines. Cephalalgia. 1999;19:766–7.

Mazzotta G, Sarchielli P, Alberti A, Gallai V. Intracellular magnesium concentration and electromyographical ischaemic test in juvenile headache. Cephalalgia. 1999;19:802–9.

Mazzotta G, Sarchielli P, Alberti A. Electromyographical ischemic test and intracellular and extracellular magnesium concentration in migraine and tension-type headache patients. Headache. 1996;36:357–61.

Mazzotta G, Sarchielli P, Alberti A. Intracellular magnesium concentration and electromyographical ischemic test in juvenile headache. Cephalalgia. 1999;19:802–9.

Nechifor M. Interactions between magnesium and psychotropic drugs. Magnes Res. 2008;21:97–100.

Nechifor M. Magnesium in major depression. Magnes Res. 2009;22:163S–6S.

Ohira T, Peacock JM, Iso H, Chambless LE, Rosamond WD, Folsom AR. Serum and dietary magnesium and risk of ischemic stroke: the atherosclerosis risk in communities study. Am J Epidemiol. 2009;169:1437–44.

Peikert A, Wilmzig C, Kohnel-Volland R. Prophylaxis of migraine with oral magnesium: results from a prospective, multi-center, placebo-controlled and double-blind randomized study. Cephalalgia. 1996;16:257–63.

Ramadan NM, Halvorson H, Vande-Linde A. Low brain magnesium in migraine. Headache. 1989;29:590–3.

Rouse DJ, Hirtz DG, Thom E, Varner MW, Spong CY, Mercer BM, et al. A randomized, controlled trial of magnesium sulphate for the prevention of cerebral palsy. N Engl J Med. 2008;359:895–905.

Sarchielli P, Bach FW, Sarchielli P, Bach FW. Magnesium in blood and spinal fluid in migraines. In: Olesen J, Goadsby PJ, editors. The headache. 3rd ed. Philadelphia: Lippincott Williams & Wilkins; 2006. p. 324.

Sarchielli P, Coata G, Firenze C, Morucci P, Abbritti G, Gallai V. Serum and salivary magnesium in migraine and tension type headache. Results in group of adult patients. Cephalalgia. 1992;12:21–7.

Schoenen J, Sianard-Gainko J, Lenaerts M. Blood magnesium levels in migraine. Cephalalgia. 1991;11:97–9.

Stanley FJ, Crowther C. Antenatal magnesium sulphate for neuroprotection before preterm birth. N Engl J Med. 2008;359:962–4.

Sun-Edelstein C, Mauskop A. Role of magnesium in the pathogenesis and treatment of migraine. Review of Neurotherapeutics. 2009;9:369–79.

Thomas J, Millot JM, Sebille S, Delabroise AM, Thomas E, Manfait M, et al. Free and total magnesium in lymphocytes of migraine patients – effect of magnesium-rich mineral water intake. Clin Chim Acta. 2000;295:63–75.

Thomas J, Thomas E, Tomb E. Serum and erythrocyte magnesium concentrations and migraine. Magnes Res. 1992;5:127–30.

Virtanen MJ, Mars M, Mannisto S, Pietinen P, Albanes D, Virtamo J. Magnesium, calcium, potassium and sodium intake and risk of stroke in male smokers. Arch Int Med. 2008;168:459–65.

Welch KM, Ramadan NM. Mitochondria, magnesium and migraine. J Neurol Sci. 1995;134:9–14.

Young LT, Robb JC, Levitt AJ, Cooke RG, Joffe RT. Serum magnesium and calcium/magnesium ratio in major depression disorder. Neuropsychobiology. 1996;34:26–8.

Cancer: colorectal

Dai Q, Shrubsok J, Ness RM, Schlundt D, Cai Q, Smalley WE, et al. The relation of magnesium and calcium intake and a genetic polymorphism in the magnesium transporter to colorectal neoplasia risk. Am J Clin Nutr. 2007;86:743–51.

Dai Q, Motley SS, Smith jr JA, Concepcion R, Barocas D, Byerly S, et al. Blood magnesium, and the interaction with calcium, on the risk of high-grade prostate cancer. PLoS One. 2011;6:1932–35.

Folsom AR, Hong CP. Magnesium intake and reduced risk of colon cancer in a prospective study in women. Am J Epidemiol. 2006;163:232–5.

Hartwig A. Role of magnesium in genomic stability. Mutat Res. 2001;475:113–21.
Killilea DW, Ames BN. Magnesium deficiency accelerates cellular senescence in cultured human fibroblasts. Proc Nat Acad Sci. 2008;105:5768–73.
Larsson SC, Bergkvist L, Wolk A. Magnesium intake in relation to risk of colorectal cancer in women. J Am Med Assoc. 2005;293:86–9.
Lin J, Cook NR, Lee I-M, Manson JA, Buring J, Zhang SM. Total magnesium intake and colorectal cancer incidence in women. Cancer Epidemiol Biomark. 2006;15:2006–9.
Rubin H. The logic of the membrane, mitosis (MMM) model for the regulation of animal cell proliferation. Arch Biochem Biophys. 2007;458:16–23.
Saif MW. Management of hypomagnesaemia in cancer patients receiving chemotherapy. J Support Oncol. 2008;6:243–8.
Wolf FI, Trapani V, Simonacci M, Boninsegna A, Mazur A, Maier JA. Magnesium deficiency affects mammary epithelial cell proliferation: involvement of oxidative stress. Nutr Cancer. 2009;61:131–6.
Yang CY, Chiu HF. Calcium and magnesium in drinking waterand risk of death from rectal cancer. Int J Cancer. 1998;77:528–32.

Alcohol

Abbott L, Nadler J, Rude RK. Magnesium deficiency in alcoholism: possible contribution to osteoporosis and cardiovascular disease in alcoholics. Alcohol Clin Exp Res. 1994;18:1076–82.
Babu AN, Cheng TP, Zhang A, Altura BT, Altura BM. Low concentrations of ethanol deplete type-2 astrocytes of intracellular free magnesium. Brain Res Bull. 1999;50:59–62.
Bohmer T, Mathiesen B. Magnesium deficiency in chronic alcoholic patients uncovered by an intravenous loading test. Scand J Clin Lab Invest. 1982;42:633–6.
Cohen L, Laor A, Kitzes R. Lymphocyte and bone magnesium in alcohol-associated osteoporosis. Magnesium. 1985;4:48–52.
Elisaf M, Bairaktari E, Kalaitzidis R, Siamopoulos KC. Hypomagnesemia in alcoholic patients. Alcohol Clin Exp Res. 1998;22:134–9.
Elisaf M, Merkouropoulos M, Tsianos EV, Siamopoulos KC. Pathogenetic mechanisms of hypomagnesemia in alcoholic patients. J Trace Elem Med Biol. 1995;9:210–4.
Embry CK, Lippmann S. Use of magnesium sulfate in alcohol withdrawal. Am Fam Physician. 1987;35:167–70.
Fankushen D, Raskin D, Dimich A, Wallsch S. The significance of hypomagnesemia in alcoholic patients. Am J Med. 1964;37:802–12.
Flink EB. Magnesium deficiency in alcoholism. Alcohol Clin Exp Res. 1986;10:590–4.
Gonzalez MM, Cavalcanti TC, Vianna CB, Timerman S. Hypomagnesaemia causing QT interval prolongation and torsades de pointes in an alcoholic patient. Resuscitation. 2006;70:346–7.
Gullestad L, Dolva LO, Soyland E, Manger AT, Falch D, Kjekshus J. Oral magnesium supplementation improves metabolic variables and muscle strength in alcoholics. Alcohol Clin Exp Res. 1992;16:986–90.
Kisters K, Schodjaian K, Tokmak F, Kosch M, Rahn KH. Effect of ethanol on blood pressure: role of magnesium. Am J Hypertens. 2004;17:455–6.
Lim P, Jacob E. Magnesium status of alcoholic patients. Metabolism. 1972;21:1045–51.
Miwa K, Igawa A, Miyagi Y, Fujita M. Importance of magnesium deficiency in alcohol-induced variant angina. Am J Cardiol. 1994;73:813–6.
Pall HS, Williams AC, Heath DA, Sheppard M, Wilson R. Hypomagnesaemia causing myopathy and hypocalcaemia in an alcoholic. Postgrad Med J. 1987;63:665–7.
Poikolainen K, Alho H. Magnesium treatment in alcoholics: a randomized clinical trial. Subst Abuse Treat Prev Policy. 2008;3:1747–59.
Rivlin RS. Magnesium deficiency and alcohol intake: mechanisms, clinical significance and possible relation to cancer development (a review). J Am Coll Nutr. 1994;13:416–23.
Romani AM. Magnesium homeostasis and alcohol consumption. Magnes Res. 2008;21:197–204.
Rylander R, Megevand Y, Lasserre B, Granbom AS. Moderate alcohol consumption and urinary excretion of magnesium and calcium. Scand J Clin Lab Invest. 2001;61:401–5.
Seelig MS. Electrographic patterns of magnesium depletion appearing in alcoholic heart disease. Ann New York Acad Sci. 1969;162:906–17.
Shane SR, Flink EB. Magnesium deficiency in alcohol addiction and withdrawal. Magnes Trace Elem. 1992;10:263–7.
Sullivan JF, Lankford HG, Swartz MJ, Farrell C. Magnesium metabolism in alcoholism. Am J Clin Nutr. 1963;13:297–303.

Asthma

Aggarwal P, Sharad S, Handa R, Dwiwedi SN, Irshad M. Comparison of nebulised magnesium sulphate and salbutamol combined with salbutamol alone in the treatment of acute bronchial asthma: a randomized study. Emerg Med J. 2006;23:358–62.

Alter HJ, Koepsell TD, Hilty WM. Intravenous magnesium as an adjuvant in acute bronchospasm: a meta-analysis. Ann Emerg Med. 2000;36:191–7.

Barnes PJ. Asthma. In: Fauci AS, Braunwald E, Kasper DL, Hauser SL, Longo DL, Jameson JL, Loscalzo J, Fauci AS, Braunwald E, Kasper DL, Hauser SL, Longo DL, Jameson JL, Loscalzo J, editors. Harrison's principles of internal medicine. 17th ed. New York: Publ McGraw Hill Medical; 2008. p. 1605.

Bernstein WK, Khastgir T, Khastgir A. Lack of effectiveness of magnesium in chronic stable asthma. A prospective, randomized, double-blind, placebo-controlled, crossover trial in normal subjects and in patients with chronic stable asthma. Arch Intern Med. 1995;155:271–6.

Bessmertny O, DiGregorio RV, Cohen H, Becker E, Looney D, Golden J, et al. A randomized clinical trial of nebulized magnesium sulfate in addition to albuterol in the treatment of acute mild-to-moderate asthma exacerbations in adults. Ann Emerg Med. 2002;39:585–91.

Bhatt SP, Khandelwal P, Nanda S, Stoltzfus JC, Fioravanti GT. Serum magnesium is an independent predictor of frequent readmissions due to acute exacerbation of chronic obstructive pulmonary disease. Respir Med. 2008;102:999–1003.

Bichara MD, Goldman RD. Magnesium for treatment of asthma in children. Can Fam Physician. 2009;55:887–9.

Blitz M, Blitz S, Beasely R, Diner BM, Hughes R, Knopp JA, et al. Inhaled magnesium sulfate in the treatment of acute asthma. Cochrane Database Syst Rev. 2005(4). doi:DOI:10.1002/14651858.CD003898.pub4.

Blitz M, Blitz S, Hughes R, Diner B, Beasley R, Knopp J, et al. Aerosolized magnesium sulfate for acute asthma: a systematic review. Chest. 2005;128:337–44.

Bloch H, Silverman R, Mancherje N, Grant S, Jagminas L, Scharf SM. Intravenous magnesium sulfate as an adjunct in the treatment of acute asthma. Chest. 1995;107:1576–81.

Britton J, Pavord I, Richards K, Wisniewski A, Knox A, Lewis S. Dietary magnesium, lung function, wheezing, and airway hyperreactivity in a random adult population sample. Lancet. 1994;344:357–62.

Cheuk DK, Chau TC, Lee SL. A meta-analysis on intravenous magnesium sulphate for treating acute asthma. Arch Dis Child. 2005;90:74–7.

Chipps B, Murphy K. Assessment and treatment of acute asthma in children. J Pediatr. 2005;147:288–94.

Ciarallo L, Sauer A, Shannon MW. Intravenous magnesium therapy for moderate to severe pediatric asthma: results of a randomized, placebo-controlled trial. J Pediatr. 1996;129:809–14.

de Valk HW, Kok PT, Struyvenberg A, van Rijn HJ, Haalboom JR, Kreukniet J, et al. Extracellular and intracellular magnesium concentrations in asthmatic patients. Eur Respir J. 1993;6:1122–5.

do Amaral AF, Rodrigues-Junior AL, Rodrigues-Junior AL, Terra Filho J, Vannucchi H, Martinez JA. Effects of acute magnesium loading on pulmonary function of stable COPD patients. Med Sci Monit. 2008;14:524–9.

Durlach J. Commentary on recent clinical advances: magnesium depletion, magnesium deficiency and asthma. Magnes Res. 1995;8:403–5.

Emelyanov A, Fedoseev G, Barnes PJ. Reduced intracellular magnesium concentrations in asthmatic patients. Eur Respir J. 1999;13:38–40.

Fantidis P, Cacho JR, Marin M, Jarabo RN, Solera J, Herrero E. Intracellular (polymorphonuclear) magnesium content in patients with brochial asthma between attacks. J Roy Soc Med. 1995;88:441–5.

Gilliland FD, Berhane KT, Li X-F, Kim DH, Margolis HG. Dietary magnesium, potassium, sodium and children's lung function. Am J Epidemiol. 2002;155:125–31.

Glover ML, Machado C, Totapally BR. Magnesium sulfate administered via continuous intravenous infusion in pediatric patients with refractory wheezing. J Crit Care. 2002;17:255–8.

Green SM, Rothrock SG. Intravenous magnesium for acute asthma: failure to decrease emergency treatment duration or need for hospitalization. Ann Emerg Med. 1992;21:260–5.

Hashimoto Y, Nishimura Y, Maeda H, Yokoyama M. Assessment of magnesium status in patients with bronchial asthma. J Asthma. 2000;37:489–96.

Hill J, Micklewright A, Lewis S, et al. Investigation of the effect of short-term change in dietary magnesium intake in asthma. Eur Respir J. 1997;10:2225–9.

Hill J, Lewis S, Britton J. Studies of the effects of inhaled magnesium on airway reactivity to histamine and adenosine monophosphate in asthmatic subjects. Clin Exp Allergy. 1997;27:546–51.

Hill JM, Britton J. Effect of intravenous magnesium sulphate on airway calibre and airway reactivity to histamine in asthmatic subjects. Br J Clin Pharmacol. 1996;42:629–31.

Johnson D, Gallagher C, Cavanaugh M, Yip R, Mayers I. The lack of effect of routine magnesium administration on respiratory function in mechanically ventilated patients. Chest. 1993;104:536–41.

Koepsell T, Alter H, Hilty W. Intravenous magnesium as an adjuvant in acute bronchospasm: a meta-analysis. Ann Emerg Med. 1999;36:191–7.

Kowal A, Panaszeka B, Barg W, Obojski A. The use of magnesium in bronchial asthma: a new approach to an old problem. Arch Immunol Ther Ex. 2007;55:35–9.

Landon R, Young E. Role of magnesium in regulation of lung function. J Am Diet Assoc. 1993;93:674–7.

Mangat HS, D'Souza GA, Jacob MS. Nebulized magnesium sulphate versus nebulized salbutamol in acute bronchial asthma: a clinical trial. Eur Respir J. 1998;12:341–4.

McLean RM. Magnesium and its therapeutic uses: a review. Am J Med. 1994;96:71.

Mohammed S, Goodacre S. Intravenous and nubulised magnesium sulphate for acute asthma: systematic review and meta-analysis. Emerg Med J. 2007;24:823–30.

Nannini LJ, Pendino JC, Corna RA, Mannarino S, Quispe R. Magnesium sulphate as a vehicle for nebulized salbutamol in acute asthma. Am J Med. 2000;108:193–7.

National Institutes of Health. Global strategy for asthma management and prevention. Bethesda: National Institutes of Health National Heart Lung and Blood Institute; 2002.

Noppen M wEditorialx. Magnesium treatment for asthma: where do we stand? Chest 2002;122:396–8.

Okayama H, Aikawa T, Okayama M, Sasaki H, Mue S, Takishima T. Bronchodilating effect of intravenous magnesium sulfate in bronchial asthma. J Am Med Assoc. 1987;257:1076–8.

Porter RS, Nester BA, Braitman LE, Geary U, Dalsey WC. Intravenous magnesium is ineffective in adult asthma, a randomised trial. Eur J Emerg Med. 2001;8:9–15.

Rodrigo GJ wletterx. There is no evidence to support the use of aerosolized magnesium for acute asthma. Chest 2006;130:304–6.

Rodrigo GJ, Rodrigo C, Noppen M, Silverman R. IV magnesium in the treatment of acute severe asthma? Chest. 2003;123:1314–6.

Rowe BH, Bretzlaff JA, Bourdon C, Bota GW, Camargo CA Jr. Magnesium sulfate for treating exacerbations of acute asthma in the emergency department. Cochrane Database Syst Rev. 1999, Issue 4. Art. No.:CD001490. DOI:10.1002/14651858.CD001490.

Rowe BH, Bretzlaff JA, Bourdon C, Bota GW, Camargo Jr CA. Intravenous magnesium sulfate treatment for acute asthma in the emergency department: a systematic review of the literature. Ann Emerg Med. 2000;36:181–90.

Schreck DM. Asthma pathophysiology and evidence-based treatment of severe exacerbations. Am J Health Syst Pharm. 2006;63:S5–13.

Silverman RA, Osborn H, Bunge J, Gallagher EJ, Chiarg W, Feldman J, et al. l IV magnesium sulfate in the treatment of acute severe asthma: a multicenter randomized controlled trial. Chest. 2002;122:489–97.

Skobeloff EM, Spivey WH, McNamara RM, Greenspon L. Intravenous magnesium sulfate for the treatment of acute asthma in the emergency department. J Am Med Assoc. 1989;262:1210–3.

Skorodin MS, Freebeck PC, Yetter B, Nelson JE, Van de Graaf WB, Walsh JM. Magnesium sulphate potentiates several cardiovascular and metabolic actions of terbutaline. Chest. 1994;105:701–5.

Spivey WH, Skobeloff EM, Levin RM. Effect of magnesium chloride on rabbit bronchial smooth muscle. Ann Emerg Med. 1990;19:1107–12.

Surendra K, Sharma K, Bhargava A, Pande JN. Effect of parenteral magnesium sulphate on pulmonary functions in bronchial asthma. J Asthma. 1994;31:109–15.

Tiffany BR, Berk WA, Todd IK, White SR. Magnesium bolus or infusion fails to improve expiratory flow in acute asthma exacerbations. Chest. 1993;104:831–4.

Villeneuve EJ, Zed PJ. Nebulized magnesium sulfate in the management of acute exacerbations of asthma. Ann Pharmacother. 2006;40:1118–24.

Wears RL. Dueling meta-analyses weditorialx. Ann Emerg Med 2000;36:234–6.

Zervas E, Papatheodorou G, Psathakis K, Panagou P, Georgatou N, Loukides S. Reduced intracellular Mg concentrations in patients with acute asthma. Chest. 2003;123:113–8.

Chapter 2
Dietary Mg Intake and Biomarkers of Inflammation and Endothelial Dysfunction

Simin Liu and Sara A. Chacko

Key Points

- In the past decade, systemic inflammation has garnered increasing attention as a fundamental process underlying metabolic disorders such as T2D, metabolic syndrome, and CVD.
- Although a growing body of experimental and observational research supports the link between dietary Mg intake and inflammatory processes, few high-quality randomized studies have tested the clinical efficacy of Mg supplementation for the prevention of inflammation and related metabolic disorders.
- Data from existing trials are too limited in scope to be conclusive and suggest the need for rigorous testing of Mg in diverse settings.
- Oral Mg supplementation is a well-tolerated, safe intervention that could be easily implemented in high-risk populations. Testing effects of dietary Mg is also achievable, requiring addition or substitution of foods to achieve higher dietary Mg intakes.
- Given the ease and practicality of Mg administration and potential for wide-ranging health benefits, future trials should test the potential impact of varying dosages of oral Mg supplementation and Mg from dietary sources on underlying chronic inflammation and endothelial dysfunction to explore potential therapeutic approaches for the prevention of metabolic disease in high-risk populations.

Keywords Dietary magnesium • Inflammation • C-reactive protein • Endothelial dysfunction • Diabetes

S. Liu, M.D., Sc.D., M.P.H. (✉)
UCLA Departments of Epidemiology, Medicine, and Obstetrics & Gynecology, Center for Metabolic Disease Prevention, 650 Charles E. Young Drive South, Los Angeles, CA 90095, USA
e-mail: siminliu@ucla.edu

S.A. Chacko, Ph.D., M.P.H.
UCLA Department of Epidemiology, Center for Metabolic Disease Prevention, 650 Charles E. Young Drive South, Los Angeles, CA 90095, USA
e-mail: sarachacko@ucla.edu

R.R. Watson et al. (eds.), *Magnesium in Human Health and Disease*, Nutrition and Health, DOI 10.1007/978-1-62703-044-1_2, © Springer Science+Business Media New York 2013

Introduction

Magnesium (Mg) is an essential mineral found abundantly in whole grains, leafy green vegetables, legumes, and nuts that plays a central role in hundreds of physiological processes in the human body. According to national survey data from National Health and Nutrition Examination Survey (NHANES) 1999–2000, a large proportion of the US population consumes inadequate dietary Mg [1]. Given the fundamental role of Mg in diverse cellular reactions, this is not without consequence. Low dietary Mg intake has been linked to a range of adverse health outcomes including those related to metabolic and inflammatory processes such as hypertension [2, 3], type 2 diabetes (T2D) [4], and metabolic syndrome [2] in both experimental and observational settings. The pathophysiologic mechanisms underlying these relations are not well understood; however, a maturing body of evidence suggests that suboptimal dietary Mg intake status may affect metabolism and inflammation pathways ultimately leading to the clinical manifestation of T2D, metabolic syndrome, and cardiovascular disease (CVD).

This chapter will review the available experimental and observational data linking low dietary Mg intake to complex metabolic diseases with a primary focus on evidence for the role of biomarkers of inflammation and endothelial dysfunction as an intermediate factor in this relation. Potential mechanistic explanations underlying the link between low dietary Mg intake and biomarkers of inflammation and endothelial dysfunction will also be discussed in detail with attention paid to both direct and indirect impacts of Mg status on inflammatory processes. Finally, future research directions and public health implications will be considered.

Overview of Mg Biology

Mg is a biologically active mineral involved in diverse cellular processes. Because Mg is the central ion in the chlorophyll molecule, which is critical for absorption of sunlight and photosynthetic production of energy in plants, dietary Mg exists naturally in high quantities in plant-based foods such as green leafy vegetables, legumes, nuts, and unrefined whole grains. Refined grains stripped of bran and germ in processing are generally low in essential vitamins and minerals including Mg. Historically, Mg was discovered in mineral-rich waters in England in the seventeenth century and used as Epsom salts for healing minor scratches and rashes [5]. Mg in the form of Mg sulfate was used medicinally as early as 1906 for the prevention of eclamptic seizures related to preeclampsia among pregnant women with hypertension [6]. High dosages of Mg are still used today to treat preeclampsia and as a laxative.

Over 300 enzymes in the human body require Mg for their action, especially those involved in ATP synthesis and production. Thus, Mg is critical for energy-dependent transport systems, glycolysis, oxidative energy metabolism, intracellular signaling systems, insulin receptor activity, and phosphorylation and dephosphorylation reactions [6]; severe Mg deficiency can result in muscular cramps, irritability, hypertension, and coronary and cerebral vasospasms [6]. Total body Mg is distributed across bodily compartments and tissue pools with approximately 60% stored in bone, 39% in the intracellular space, and 1% in the extracellular space [7]. Mg is the second most abundant intracellular cation in the body with 10% found in free ionized form and 90% bound to proteins, nucleic acids, ATP, and negatively charged phospholipids [5]. Absorption of Mg takes place primarily in the intestines through both passive and active transport mechanisms [8]. After intestinal absorption, Mg is transported to the tissues and taken up into cells to fulfill normal intracellular ionized Mg needs.

Biomarkers of Mg

Currently, there is no simple and established method to adequately measure total body stores of Mg [10]. This stems in part from the wide distribution and compartmentalization of Mg in the body and the slow exchange between body compartments and tissue pools; over half of total-body Mg is stored in long-term storage pools in bone and less than 1% of total body Mg is present in blood. Exchange of Mg between body compartments and tissue pools occurs slowly such that determining Mg concentration in one pool may not accurately reflect Mg status in another [9]. Although serum Mg is used frequently in research settings, only 0.01% of total body Mg is stored in serum, and therefore, serum Mg cannot accurately reflect total body stores. Further, serum concentrations are homeostatically regulated within narrow limits by the balance between intestinal absorption and renal excretion and therefore are not considered a sensitive marker of dietary intake except in cases of severe deficiency [10]. Intracellular concentrations in muscle, erythrocytes, and lymphocytes may more accurately reflect total body status; however, these measures are rarely feasible in large population studies [11].

Dietary Mg Intake in the United States

The recommended daily allowance (RDA) of dietary Mg for men is 400–420 mg/day and 310–320 mg/day for women, with higher requirements recommended among older age groups [12]. According to NHANES 1999–2000, the median of dietary Mg intake was below the RDA in all ethnic groups in a population-based sample of over 4,000 US adults [1]. However, Caucasians generally reported a higher level of Mg intake than other ethnic groups, and older age groups reported lower intakes. It is likely that the high prevalence of inadequate Mg intake in the US population is due to the predominance of highly processed foods in the Western diet stripped of natural sources of dietary Mg in the refining process [7].

Dietary Mg Intake and Complex Metabolic Diseases

A large body of observational evidence suggests a critical role for Mg in glucose metabolism, inflammation, insulin homeostasis, and CVD. Cross-sectional and prospective studies in diverse human populations indicate that higher dietary intake of Mg may be favorably associated with reduced risk of T2D [4, 13, 14], metabolic syndrome [15], and CVD [16, 17]. Among approximately 7,700 men and women aged ≥20 years in NHANES III (1984–1994), the estimated relative risk (RR) of metabolic syndrome defined using the criteria of the National Cholesterol Education Program (NCEP) was 0.56 (95% CI 0.34, 0.92) when comparing the highest quintile of dietary Mg intake to the lowest quintile [18] with an inverse trend observed across increasing quintiles of intake (p for trend = 0.03). Prospective inverse associations between dietary Mg intake and the risk of incident T2D were reported in several large cohort studies including the Nurses' Health Study (NHS) [19], the Health Professionals Follow-up Study (HPFS) [14], the Iowa Women's Health Study [20], and the Women's Health study (WHS) [4]. A meta-analysis of eight independent cohorts with 286,668 participants and 10,912 diabetes cases reported an overall RR per 100 mg/day Mg intake of 0.85 (CI, 0.79–0.92) with similar results reported for dietary and supplemental Mg. The apparent inverse relation between Mg intake and T2D risk was more pronounced in overweight women [4, 13, 14, 20]. Dietary intake of Mg has also been observed to be inversely associated with fasting insulin levels [4, 21], insulin resistance [22], and dyslipidemia [2, 15, 23].

Several small clinical trials report that Mg supplementation can directly improve insulin resistance among nondiabetic participants (i.e., lower fasting insulin and glucose concentrations) [24, 25]. In a randomized controlled study of 60 non-diabetic subjects with decreased serum Mg levels (≤ 0.74 mmol/l), Guerrero-Romero et al. reported that daily oral supplementation of 2.5 g/day Mg chloride for 3 months reduced fasting insulin levels (103.2 ± 56.4 to 70.2 ± 29.6 mmol/L) and insulin resistance as measured by the homeostasis model analysis for insulin resistance (HOMA-IR) (4.6 ±2.8 to 2.6 ± 1.1, $p < 0.0001$) [24]. Similarly, in a study of 12 elderly participants with insulin resistance, daily Mg supplementation with 4.5 g Mg pidolate/day for 4 weeks improved insulin action and total body and oxidative glucose metabolism [25]. A meta-analysis of oral Mg supplementation on glycemic control in 370 T2D patients reported that oral Mg supplementation for 4–16 significantly lowered fasting glucose in the treatment group [−0.56 mmol/l (95% CI, −1.10 to −0.01); P for heterogeneity = 0.02] and raised HDL-C [0.08 mmol/l (95% CI, 0.03–0.14); P for heterogeneity = 0.36] [26]. More recently, findings from a randomized crossover trial of overweight men and women reported decreases in fasting insulin and C-peptide concentrations after 4 weeks of Mg supplementation [27].

The link to impaired metabolic functioning is supported by several lines of biological evidence suggesting that intracellular Mg balance is important in maintaining peripheral glucose utilization [28, 29]. Mg is required for tyrosine kinase activity critical for insulin signaling [30] and may also exert indirect effects on insulin secretion in pancreatic β-cells through effects on cellular calcium homeostasis and oxidative stress [31, 32]. Systemic inflammation as an intermediate pathway has also been proposed as a plausible mechanism and has earned increasing research attention in recent years.

Dietary Mg and Inflammation and Endothelial Dysfunction

T2D and metabolic syndrome are inherently inflammatory disorders, and inadequate dietary Mg may contribute to their pathogenesis by exerting direct and indirect effects on underlying inflammatory processes and vascular functioning (Fig. 2.1). Although the exact molecular mechanisms remain elusive, experiments in animals suggest that Mg deficiency may contribute by inducing an acute state of inflammation characterized by leukocyte and macrophage activation paired with proatherogenic changes [33]. Both cross-sectional and prospective studies have also linked dietary Mg intake to biomarkers of systemic inflammation including high-sensitivity C-reactive protein (hsCRP) and interleukin-6 (IL-6) in large diverse human populations. The following section will provide an overview of systemic inflammation and related vascular changes associated with endothelial dysfunction, review available evidence for the potential link to low Mg status, and critically examine proposed biological mechanisms for this link.

Systemic Inflammation Overview

Systemic low-grade inflammation is a well-established intermediate pathogenic state for several degenerative chronic diseases including obesity, insulin resistance [34–36], T2D [37–45], hypertension [46], stroke [47], CVD [48, 49], and the metabolic syndrome [50]. Systemic inflammation is a complex biological response of the vascular tissues triggered by environmental factors including viral infections, persistent microbial infections, exposure to toxic agents, autoimmune disease, and metabolic imbalances [51]. The response is characterized by infiltration of the vascular and adipose tissues with mononuclear cells including macrophages and lymphocytes and widespread tissue destruction and repair involving new vessel proliferation and fibrosis [51]. Chronic inflammation differs from acute inflammation in that the response is persistent and characterized by the sustained release of

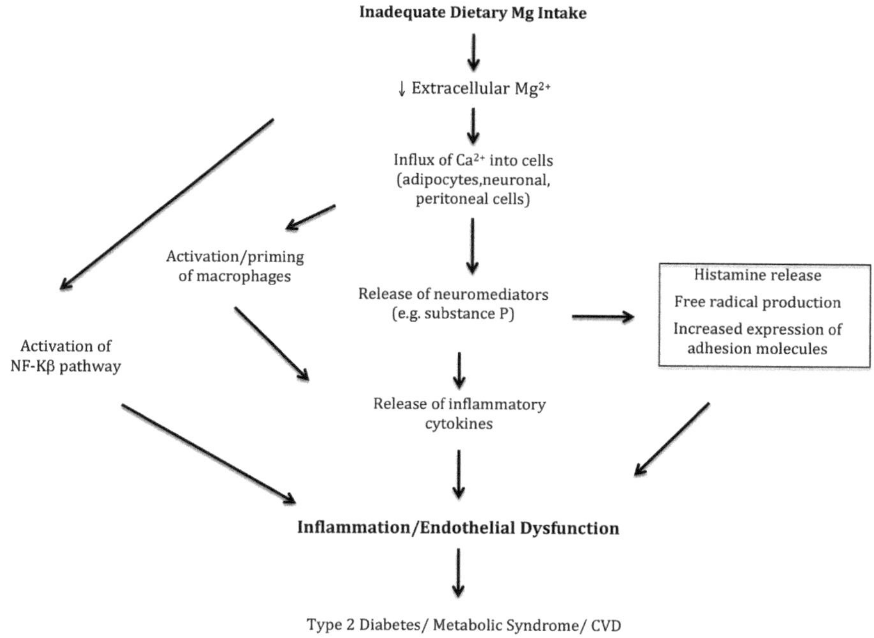

Fig. 2.1 Proposed pathways linking dietary Mg intake to systemic inflammation and endothelial dysfunction

proinflammatory cytokines from leukocytes (including macrophages and T-cells) and adipocytes. Major proinflammatory cytokines including interleukin-6 (IL-6) and tumor necrosis factor alpha (TNF-α) are released into the bloodstream and act as signaling molecules to promote the release of C-reactive protein (CRP) from the liver as part of the acute phase response, which further prolongs the inflammatory response in the body [51].

Low-grade chronic inflammation, as reflected by elevated levels of circulating inflammatory cytokines, has been linked to obesity, T2D, and overall impaired metabolic functioning. The overexpression of TNF-α in adipose tissue of obese mice [52] was the first experimental evidence linking inflammation to obesity. It is now well accepted that inflammation plays an important role in the development of metabolic disorders [37, 53]. Low-grade elevated concentrations of circulating inflammatory markers including CRP, IL-6, and TNF-α have been associated with the risk of obesity [36], metabolic syndrome [54], CVD [50], and T2D [37, 55, 56] in diverse settings. Although screening for systemic inflammation is not yet included in routine clinical practice, it is estimated that approximately 25% of the population has CRP concentrations > 3 mg/L, the cut point for increased risk [57].

Endothelial Dysfunction Overview

Endothelial dysfunction is characterized by a pathological state of endothelial cells lining the inner walls of blood vessels and is closely linked to inflammation. The response is marked by reduced nitric oxide generation, impaired vasodilation, increased oxidative stress, a proinflammatory state, and upregulation of cellular adhesion molecules [58]. Vascular endothelial cells located at the interface between blood and vessels are sensitive to circulating signaling proteins including proinflammatory cytokines. In response to increased concentrations of cytokines in the bloodstream, adhesion molecules on the surface of endothelial cells are expressed and facilitate the recruitment and binding of leukocytes across the endothelial surface into tissues to mount an inflammatory response [59]. E-selectin is an example of an adhesion molecule found primarily on the surface of stimulated endothelial cells which mediates the initial rolling of leukocytes along the endothelium. Intercellular adhesion molecule-1 (ICAM-1) and vascular adhesion molecule-1 (VCAM-1), two cellular adhesion molecules belonging to the immunoglobulin family, are primarily involved in the attachment and transendothelial migration of leukocytes.

Endothelial dysfunction, as reflected by elevated concentrations of adhesion molecules, has been associated with characteristics of the metabolic syndrome including hypertension [60], obesity [60, 61], dyslipidemia [62, 63], insulin resistance [64], as well as risk of T2D [53, 65]. Although several methods are currently available to measure endothelial dysfunction including direct measurement of coronary artery relaxation after acetylcholine infusions, assessment of brachial reactivity to vasodilator or vasoconstrictive agents, and flow-mediation vasodilation techniques [66], circulating concentrations of soluble forms of adhesion molecules in plasma can be easily measured in large populations and reasonably capture changes in endothelial functioning [66, 67]. Measurable levels of these molecules especially ICAM-1 were shown to be correlated with the direct assessment of endothelial function including flow-mediated brachial artery dilation [68, 69]. Elevated E-selectin may also be a sensitive measure of endothelial dysfunction [53] because it is expressed exclusively by endothelial cells, while ICAM-1 and VCAM-1 are expressed on other cells as well [67].

Experimental Data

Mg Deficiency and Inflammation

Several lines of experimental evidence in animal models indicate that diet-induced Mg deficiency leads to an impaired immune response [70, 71]. A well-documented link between Mg deficiency and an acute inflammatory response in rodent models characterized by leukocyte and macrophage activation, release of inflammatory cytokines and acute phase proteins, and excessive production of free radicals has been reported in experimental settings [33, 72]. Rats fed a Mg-deficient diet for 8 days clearly exhibited allergic-like inflammatory symptoms including peripheral vasodilation with hyperemia of the ears and marked increases in total circulating leukocytes including monocytes, eosinophils, and basophils, with a predominant increase observed among the neutrophil fraction. Resident macrophages collected from the peritoneal cavity were greater in number and appeared to be activated at higher levels among Mg-deficient rats as indicated by increased cell volume and an increased number of vacuoles. Release of inflammatory cytokines including IL-6, a known mediator of the acute phase response, were also increased in Mg-deficient rats compared with controls [72]. Similarly, dramatic increases in serum levels of inflammatory cytokines including interleukin 1 (IL-1), IL-6, and TNF-α were observed in rodents after 3 weeks on an Mg-deficient diet [73].

Plasma concentrations of substance P, a neuronal tachykinin released from neural tissues known to stimulate the production of cytokines and a potent vasodilator, were also elevated in rodents during the first week of Mg deficiency. Concomitant increases in histamine levels in response to Mg deficiency may have resulted from stimulation and degranulation of mast cells by substance P. These findings suggest that the release of substance P may be the earliest pathophysiological event leading to stimulation of the inflammatory cytokines [73]. Accompanying the release of proinflammatory cytokines, an increase in positive acute phase proteins including alpha2-macroglobulin, alpha1-acide glycoprotein, complement component C3, fetoprotein, haptoglobin, and fibrinogen and a decrease in negative acute phase proteins apolipoprotein E and retinol binding protein (RBP) expression and plasma levels have been reported in rodent models of Mg deficiency [72, 74]. The acute phase protein haptoglobin is of particular interest because diabetic individuals homozygous for the haptoglobin 2 allele (Hp 2-2) are at high risk for CVD events [75] and may potentially be targeted in dietary interventions, especially those concerning antioxidants including vitamin E [76].

Major immunological organs may also be affected by Mg deficiency. A reduction in thymus weight and histological changes including enlarged glands and a predominance of large lymphocytes in the thymus tissue were reported in rodents after several weeks of Mg deficiency [77]. Similar observations were reported after 8 days of Mg deficiency [78]; the mean thymus weight of Mg-deficient rats was lower than that of controls and concentrations of inflammatory cytokines including IL-6 were increased in the thymus tissue in the early stages of deficiency. These changes appeared to be linked to enhanced apoptosis in the thymus based on enumeration of apoptotic cells on the basis of morphological criteria and intranucleosomal degradation of genomic DNA [78]. Enlargement of the spleen in response to Mg deficiency has also been observed experimentally [74, 79–81]. After 8 days on an Mg-deficient diet, enlargement of the spleen was accompanied by neutrophilic, eosinophilic, and basophilic granular cell accumulation in the spleen [81]. Both red and white spleen pulps of deficient rats displayed an increased incidence of macrophage activity [79].

Mg Deficiency and Endothelial Dysfunction

Experimental data suggest Mg deficiency may directly influence endothelial functioning through enhanced expression of adhesion molecules and promotion of the thrombotic potential of the vessel wall, further exacerbating inflammation [33]. In vitro, low extracellular Mg induced the synthesis of vascular cell adhesion molecule (VCAM) [82] responsible for binding of antigens expressed by monocytes and lymphocytes. Adhesion of leukocytes to the vascular wall for transport to tissues is crucial to inflammatory processes. Low Mg concentrations were also shown to inhibit endothelial proliferation in cultured endothelial cells, correlating with a marked upregulation of interleukin-1 (IL-1), VCAM-1, and plasminogen activator inhibitor (PAI)-1 after Mg deficiency [83]. Low Mg acts to influence the synthesis of nitric oxide [82, 84] and may contribute to lipid peroxidation and activation of NF-κB as indicated in primary cultured cerebral vascular smooth muscle cells [85]. In rats, dietary Mg deficiency impaired lipoprotein profiles increasing atherogenic very low-density lipoprotein (VDLD) and decreasing antiatherogenic high-density lipoprotein (HDL) [86].

Observational and Randomized Studies in Humans

Dietary Mg Intake and Biomarkers of Inflammation

Dietary patterns high in Mg-rich foods have also been linked to lower circulating levels of inflammatory cytokines [87–90]. Inverse associations between dietary Mg intake and biomarkers of systemic inflammation including high-sensitivity C-reactive protein (hsCRP), interleukin-6 (IL-6), and TNF-α have consistently been reported in both cross-sectional and prospective cohort studies of diverse human populations [2, 87, 91–95] (Table 2.1). The inverse association between dietary Mg intake and hsCRP was first observed among 11,686 apparently healthy women enrolled in the WHS [2]. In multivariable analyses, hsCRP concentrations were 12% lower in the highest quintile of dietary Mg intake than in the lowest with mean hsCRP concentrations decreasing monotonically across increasing quintiles of Mg intake: 1.50, 1.39, 1.35, 1.34, and 1.31 mg/L (p for trend = 0.0003) [2]. In a nationally representative sample of US adults aged ≥20 years in NHANES 1999–2000, those who consumed less than the RDA of Mg were 1.48–1.75 times more likely to have elevated CRP (≥3.0 mg/L) than adults consuming the RDA amount or more (OR for intake <50 RDA = 1.75, 95% CI 1.08–2.87) [93].

Table 2.1 Observational studies examining association between dietary Mg intake and biomarkers of systemic inflammation and endothelial dysfunction in adult human populations

Author, year	Study population	Design	Mg measure	Outcomes	Adjustment	Results
Song et al. 2005	n = 11,686 women in the Women's Health Study free of CVD and cancer	Cross-sectional	Self-report semi-quantitative FFQ	hsCRP	Age, BMI, smoking, exercise, alcohol, energy intake, history of metabolic risk factors, dietary factors	CRP 12% lower and MetSyn risk 27% lower in highest quintile of Mg compared to lowest
King et al. 2005	n = 5,021 adults in NHANES 1999–2000	Cross-sectional	Self-report 24-h recall	hsCRP	Age, ethnicity, gender, BMI, smoking, income, alcohol, exercise, energy intake, metabolic risk factors	Adults consuming < RDA of Mg were 1.48–1.75× more likely to have CRP ≥ 3.0 mg/L
Bo et al. 2006	n = 1,653 middle-aged healthy participants	Cross-sectional	Self-report semi-quantitative FFQ	hsCRP	Age, gender, smoking, alcohol, BMI, physical activity, dietary factors	OR for CRP ≥ 3 mg/L: 2.05 (1.30, 3.25) comparing highest to lowest tertile of Mg intake
Song et al. 2007	n = 657 women enrolled in Nurses' Health Study aged 43-69 years free of CVD, cancer and diabetes	Cross-sectional	Self-report FFQ	hsCRP, IL-6, TNF-α-R2, sICAM-1, sVCAM-1, E-selectin	Age, physical activity, smoking, alcohol, hormone use, BMI	CRP 24% lower and E-selectin 14% lower in highest quintile of Mg compared with lowest
Chacko et al. 2010	n = 3,713 postmenopausal women enrolled in Women's Health Initiative	Cross-sectional	Self-report FFQ	hsCRP, IL-6, TNF-α-R2, sICAM-1, sVCAM-1, E-selectin	Age, ethnicity, matching factors, smoking, alcohol, physical activity, energy intake, BMI, diabetes status, dietary factors	Increase of 100 mg/day Mg inversely associated with hsCRP, Il-6, TNF-α-R2, and sVCAM-1
Kim et al. 2010	n = 4,497 African-American and Caucasian, aged 18–30 years, followed for 20 years (CARDIA study)	Prospective cohort	Interview-administered diet history questionnaire	hsCRP, IL-6 Fibrinogen	Age, ethnicity, gender, education, smoking, alcohol, physical activity, family history of T2D, BMI, blood pressure, energy intake, dietary factors	Inverse association with hsCRP, IL-6, and fibrinogen
De Oliveira Otto et al. 2011	n = 1,581 participants aged 45–85 years free of CVD/T2D enrolled in multi-ethnic population based study (MESA)	Cross-sectional	Self-report FFQ	CRP, IL-6	Age, gender, ethnicity, energy intake, education, physical activity, alcohol, smoking, supplement use	No association with CRP, IL-6

FFQ food frequency questionnaire, *hsCRP* high-sensitivity C-reactive protein, *IL-6* interleukin-6, *TNF-α-R2* tumor necrosis factor alpha receptor 2, *sICAM-1* soluble intercellular adhesion molecule-1 (ICAM-1), *sVCAM-1* soluble vascular adhesion molecule-1 (VCAM-1)

Similarly, among a population-based sample of middle-aged healthy subjects (n = 1,653), participants in the lowest tertile of Mg intake were twice as likely to have elevated hsCRP levels $\geq$ 3 mg/L (OR = 2.05, 95% CI: 1.30, 3.25) even after adjustment for related nutrients including dietary fiber [87]. More recently, dietary Mg intake was inversely associated with plasma concentrations of hsCRP but not IL-6 or TNF-α among 657 women aged 43–69 years enrolled in the NHS cohort (multivariable adjusted means of hsCRP: 1.70 $\pm$ 0.18 in highest quintile of dietary Mg intake compared with 1.30 $\pm$ 0.10 mg/dL in the lowest quintile; p for trend = 0.03) [92]. In a large, ethnically diverse cohort of postmenopausal women (n = 3,713) enrolled in the WHI, dietary Mg intake was inversely associated with hsCRP and IL-6 after adjustment for potential confounders including fruit and vegetable, folate, saturated and trans fat intake; multivariable adjusted geometric means across increasing quintiles of dietary Mg were 3.08, 2.63, 2.31, and 2.16 mg/L for hsCRP (p for linear trend = 0.005) and 2.91, 2.63, 2.45, 2.27, and 2.26 pg/mL for IL-6 (p for linear trend = 0.0005). Inverse correlations were more pronounced among overweight and obese women than among normal weight women in this population [91]. Findings from the younger CARDIA population of 4,497 Americans aged 18–30 years support the inverse association between dietary Mg intake and hsCRP and IL-6 [94]. Several cross-sectional and prospective studies have also reported a link between serum Mg concentrations and CRP [94, 96, 97] and TNF-α [98], although less research has explored this link.

Few randomized studies have assessed possible effects of dietary Mg on underlying states of chronic inflammation (Table 2.2). Among 35 patients with chronic systolic heart failure, Mg supplementation (300 mg/day in the form of Mg citrate) for 5 weeks appeared to reduce inflammation as reflected by a decrease in CRP concentrations not observed in the control group; (Decrease in log CRP in treated: 1.4 $\pm$ 0.4 to 0.8 $\pm$ 0.3, p < 0.001 vs. Decrease in control group: 1.5 $\pm$ 0.7 to 1.2 $\pm$ 0.6, p = 0.3) [99]. However, in a randomized crossover trial of 4 weeks of Mg supplementation (360 mg/day elemental Mg in the form of Mg citrate) among 14 overweight men and women [27], no consistent changes were observed in hsCRP, IL-6, and TNF-α concentrations after treatment, although a slight increase was observed for IL-6. After a 7-week intervention of 320 mg/day of oral Mg citrate among 100 adults with poor sleep quality, CRP concentrations decreased significantly after treatment, but only among those with CRP >3.0 mg/L [100]. Large, high-quality randomized studies are needed to rigorously test whether Mg supplementation may be a useful anti-inflammatory therapy for use in primary prevention of chronic metabolic disorders, especially among overweight populations at-risk for metabolic abnormalities.

Dietary Mg Intake and Biomarkers of Endothelial Dysfunction

Little research has explored the relation of dietary Mg intake to biomarkers of endothelial dysfunction in humans (Table 2.1). Although the general trend of available data is consistent with animal studies and suggests a protective role of dietary Mg intake, findings regarding specific biomarkers of endothelial dysfunction have been inconsistent. Among a homogeneous population of women enrolled in the NHS, dietary intake of Mg was inversely associated with concentrations of biomarkers of endothelial dysfunction including E-selectin, but not soluble ICAM-1 or VCAM-1 [92]. Multivariable-adjusted geometric means for women in the highest quintile of dietary Mg intake were 14% lower for E-selectin (48.5 $\pm$ 1.84 compared with 41.9 $\pm$ 1.58 ng/mL; p for trend = 0.01) than those women in the lowest quintile. Because E-selectin is expressed exclusively by activated endothelial cells in a membrane-bound form, it may be a more specific surrogate marker for endothelial functioning than sICAM-1 or sVCAM-1 which are found on various cells types including leukocytes and vascular smooth muscle cells [92]. In the same cohort, a Western dietary pattern low in Mg was reported to be positively associated with these same markers of endothelial dysfunction as well as E-selectin [88]. An inverse association with markers of endothelial dysfunction including sVCAM-1 and E-selectin was also reported among 3,713 postmenopausal women enrolled in the WHI after adjustment for known risk factors for

Table 2.2 Randomized controlled trials investigating effects of Mg supplementation on biomarkers of systemic inflammation and endothelial dysfunction in humans

Author, year	Study population	Treatment	Duration	Outcome	Results
Almoznino-Sarafian et al. 2007	N = 35 patients with chronic systolic heart failure	Oral Mg citrate (300 mg/d)	5 weeks	CRP	CRP decreased in Mg treatment group (Baseline log CRP: 1.4 ± 0.4; Post-treatment: 0.8 ± 0.3; p < 0.001); no change in control
Chacko et al. 2010	N = 14 overweight men and women aged	Oral Mg citrate (360 mg/d elemental Mg in the form of Mg citrate)	4 weeks	hsCRP, IL-6, TNF-α-R2, sICAM-1, sVCAM-1, E-selectin	Slight increase in IL-6 after Mg treatment (0.23 pg/mL compared to –0.37 after placebo); no significant changes in other markers
Nielsen et al. 2010	N = 100 adults with poor sleep quality aged 51–85 years	Oral Mg citrate (320 mg/day)	7 weeks	CRP	Decrease in CRP among those with CRP >3.0 mg/L
Rodriguez-Hernandez et al. 2010	N = 38 nonhypertensive obese women aged 30–65 years	5% solution of MgCl (450 mg elemental Mg)	4 months	hsCRP	Nonsignificant reduction in hsCRP concentrations after treatment

hsCRP high-sensitivity C-reactive protein, *IL-6* interleukin-6, *TNF-α-R2* tumor necrosis factor alpha receptor 2, *sICAM-1* soluble intercellular adhesion molecule-1 (ICAM-1), *sVCAM-1* soluble vascular adhesion molecule-1 (VCAM-1) metabolic outcomes. However, the association with E-selectin was attenuated after accounting for additional dietary factors associated with inflammation [91]. Although several randomized studies have explored effects of Mg supplementation on endothelial functioning [101, 102], few have examined the effects on biomarkers of endothelial dysfunction [27].

Potential Biological Mechanisms

Several biological mechanisms have been proposed to explain the link of low Mg to systemic inflammation and endothelial dysfunction. Potential pathways include intracellular calcium influx due to low Mg and subsequent phagocytic cell priming, release of neuromediators including substance P, and activation of the nuclear factor light chain enhancer of activated b-cells (NFκB) pathway involved in regulation of immune and inflammatory response [33]. Proposed biological pathways are shown in Fig. 2.1.

Subclinical Mg deficiency promotes cellular entry of calcium and subsequent release of inflammatory signaling molecules [103] by influencing the balance of intra- and extracellular Mg. Because calcium channels are controlled by regulatory gates with binding sites for Mg that block calcium influx when Mg binds, decreases in extracellular Mg inhibit blocking of the gate and allow for calcium influx into the cell [103, 104]. In a metabolic trial of 15 postmenopausal Caucasian women, moderate Mg deficiency achieved through diet alone (Mg intake 107 mg/day for 72 days) decreased urinary calcium excretion and increased calcium balance without affecting plasma calcium concentration, suggesting intracellular calcium retention [105]. Increases in intracellular calcium resulting from decreased extracellular Mg may enhance activation and priming of phagocytic cells such as macrophages which require second messengers and signal transduction pathways, further promoting the inflammatory response [33]. Several lines of experimental data in rats fed an Mg-deficient diet support this hypothesis including an enhanced Ca^{+2} response after in vitro stimulation of peritoneal cells of rats on a deficient diet [106] and reductions in the inflammatory response when deficient rates are fed a low-calcium diet [107]. Mg depletion also depresses both cellular and extracellular potassium [108, 109], further exerting effects on calcium metabolism [110]. Thus, imbalances in potassium as well as its ratio to Mg may also contribute to inflammation, although less research has directly examined this link.

Reductions in extracellular Mg may also promote release of neuromediators by lowering the threshold of excitatory amino acids (e.g. glutamate) required to activate neuronal receptors involved in the influx of calcium into the cell, specifically the ligand gated N-methyl-D-aspartate receptor (NMDA) [103]. Excessive calcium influx into neuronal tissues via NMDA or voltage-dependent calcium channels promotes the release of neuromediators believed to initiate the inflammatory response [111]. Substance P is one such neuropeptide hypothesized to be the initial trigger stimulating the inflammatory response including macrophage production of inflammatory cytokines, release of histamine by mast cells, and production of free radicals [73]. Among rats fed a Mg-deficient diet, elevations in substance P were observed after only 5 days of Mg deficiency and preceded changes in inflammatory cytokine profiles [73]. Substance P may also play a early role in promoting endothelial dysfunction and has been shown to induce expression of adhesion molecules in vitro [112].

Activation of NF-κB, a major pathway involved in regulating the inflammatory and immune response, may also be involved in the proinflammatory response observed in states of Mg deficiency [33]. NF-κB consists of a group of inducible transcription factors that increase the expression of specific cellular genes including cytokines [interleukin 1β (IL-1β) and TNF-α] and chemokines, the major histocompatibility complex (MHC), and receptors required for neutrophil adhesion and migration. NF-κB also regulates B-lymphocyte function and cytokine-induced proliferation of T lymphocytes [113]. Evidence from in vitro studies suggest low extracellular Mg ions induced lipid peroxidation and activation of the NFkB pathway in cultured canine cerebral vascular smooth muscle, suggesting these biochemical pathways may be signaling events in the proinflammatory actions of low Mg status [85].

Future Directions and Public Health Implications

Adequate dietary Mg intake (ranging from 310 to 420 mg/day for healthy adults) is fundamental for Mg homeostasis and optimal health [6], and current US dietary guidelines emphasize that individuals obtain adequate dietary Mg from food sources such as whole grains, green leafy vegetables, and nuts and minimize intake of processed foods [114]. Yet recent national surveys indicate a large proportion of the US population has suboptimal dietary Mg intake below the RDA [1], a statistic likely related to the widespread consumption of the typical Western diet low in Mg-rich foods.

In the past decade, systemic inflammation has garnered increasing attention as a fundamental process underlying metabolic disorders such as T2D, metabolic syndrome, and CVD. Although a growing body of experimental and observational research supports the link between dietary Mg intake and inflammatory processes, few high-quality randomized studies have tested the clinical efficacy of Mg supplementation for the prevention of inflammation and related metabolic disorders. Data from existing trials are too limited in scope to be conclusive and suggest the need for rigorous testing of Mg in diverse settings. Oral Mg supplementation is a well-tolerated, safe intervention that could be easily implemented in high-risk populations. Testing effects of dietary Mg is also achievable, requiring addition or substitution of foods to achieve higher dietary Mg intakes. Given the ease and practicality of Mg administration and potential for wide-ranging health benefits, future trials should test the potential impact of varying dosages of oral Mg supplementation and Mg from dietary sources on underlying chronic inflammation and endothelial dysfunction to explore potential therapeutic approaches for the prevention of metabolic disease in high-risk populations.

References

1. Ford ES, Mokdad AH. Dietary magnesium intake in a national sample of US adults. J Nutr. 2003;133(9): 2879–82.
2. Song Y, Ridker PM, Manson JE, Cook NR, Buring JE, Liu S. Magnesium intake, C-reactive protein, and the prevalence of metabolic syndrome in middle-aged and older U.S. women. Diabetes Care. 2005;28(6):1438–44.
3. Song Y, Sesso HD, Manson JE, Cook NR, Buring JE, Liu S. Dietary magnesium intake and risk of incident hypertension among middle-aged and older US women in a 10-year follow-up study. Am J Cardiol. 2006;98(12):1616–21.
4. Song Y, Manson JE, Buring JE, Liu S. Dietary magnesium intake in relation to plasma insulin levels and risk of type 2 diabetes in women. Diabetes Care. 2004;27(1):59–65.
5. Vormann J. Magnesium: nutrition and metabolism. Mol Aspects Med. 2003;24(1–3):27–37.
6. Saris NE, Mervaala E, Karppanen H, Khawaja JA, Lewenstam A. Magnesium. An update on physiological, clinical and analytical aspects. Clin Chim Acta. 2000;294(1–2):1–26.
7. Chaudhary DP, Sharma R, Bansal DD. Implications of magnesium deficiency in type 2 diabetes: a review. Biol Trace Elem Res. 2010;134(2):119–29. Epub 2009 Jul 24.
8. Wolf FI, Torsello A, Fasanella S, Cittadini A. Cell physiology of magnesium. Mol Aspects Med. 2003;24(1–3):11–26.
9. Arnaud MJ. Update on the assessment of magnesium status. Br J Nutr. 2008;99 Suppl 3:S24–36.
10. Elin RJ. Laboratory tests for the assessment of magnesium status in humans. Magnes Trace Elem. 1991;10(2–4): 172–81.
11. Stipanuk MH. Biochemical and physiological aspects of human nutrition. Philadelphia: W.B. Saunders; 2000.
12. Academy N. Dietary reference intakes for calcium, phosphorous, magnesium, vitamin D, and flouride. Washington: National Academy Press; 1997.
13. Colditz GA, Manson JE, Stampfer MJ, Rosner B, Willett WC, Speizer FE. Diet and risk of clinical diabetes in women. Am J Clin Nutr. 1992;55(5):1018–23.

14. Salmeron J, Ascherio A, Rimm EB, Colditz GA, Spiegelman D, Jenkins DJ, et al. Dietary fiber, glycemic load, and risk of NIDDM in men. Diabetes Care. 1997;20(4):545–50.
15. He K, Liu K, Daviglus ML, Morris SJ, Loria CM, Van Horn L, et al. Magnesium intake and incidence of metabolic syndrome among young adults. Circulation. 2006;113(13):1675–82.
16. Abbott RD, Ando F, Masaki KH, Tung KH, Rodriguez BL, Petrovitch H, et al. Dietary magnesium intake and the future risk of coronary heart disease (the Honolulu Heart Program). Am J Cardiol. 2003;92(6):665–9.
17. Al-Delaimy WK, Rimm EB, Willett WC, Stampfer MJ, Hu FB. Magnesium intake and risk of coronary heart disease among men. J Am Coll Nutr. 2004;23(1):63–70.
18. Ford ES, Li C, McGuire LC, Mokdad AH, Liu S. Intake of dietary magnesium and the prevalence of the metabolic syndrome among U.S. adults. Obesity (Silver Spring). 2007;15(5):1139–46.
19. Salmeron J, Manson JE, Stampfer MJ, Colditz GA, Wing AL, Willett WC. Dietary fiber, glycemic load, and risk of non-insulin-dependent diabetes mellitus in women. JAMA. 1997;277(6):472–7.
20. Meyer KA, Kushi LH, Jacobs Jr DR, Slavin J, Sellers TA, Folsom AR. Carbohydrates, dietary fiber, and incident type 2 diabetes in older women. Am J Clin Nutr. 2000;71(4):921–30.
21. Fung TT, Manson JE, Solomon CG, Liu S, Willett WC, Hu FB. The association between magnesium intake and fasting insulin concentration in healthy middle-aged women. J Am Coll Nutr. 2003;22(6):533–8.
22. Paolisso G, Barbagallo M. Hypertension, diabetes mellitus, and insulin resistance: the role of intracellular magnesium. Am J Hypertens. 1997;10(3):346–55.
23. Ma J, Folsom AR, Melnick SL, Eckfeldt JH, Sharrett AR, Nabulsi AA, et al. Associations of serum and dietary magnesium with cardiovascular disease, hypertension, diabetes, insulin, and carotid arterial wall thickness: the ARIC study. Atherosclerosis risk in communities study. J Clin Epidemiol. 1995;48(7):927–40.
24. Guerrero-Romero F, Tamez-Perez HE, Gonzalez-Gonzalez G, Salinas-Martinez AM, Montes-Villarreal J, Trevino-Ortiz JH, et al. Oral magnesium supplementation improves insulin sensitivity in non-diabetic subjects with insulin resistance. A double-blind placebo-controlled randomized trial. Diabetes Metab. 2004;30(3):253–8.
25. Paolisso G, Sgambato S, Gambardella A, Pizza G, Tesauro P, Varricchio M, et al. Daily magnesium supplements improve glucose handling in elderly subjects. Am J Clin Nutr. 1992;55(6):1161–7.
26. Song Y, He K, Levitan EB, Manson JE, Liu S. Effects of oral magnesium supplementation on glycaemic control in type 2 diabetes: a meta-analysis of randomized double-blind controlled trials. Diabet Med. 2006;23(10):1050–6.
27. Chacko SA, Sul J, Song Y, Li X, LeBlanc J, You Y, Butch A, Liu S. Magnesium supplementation, metabolic and inflammatory markers, and global genomic and proteomic profiling: a randomized, double-blind, controlled, crossover trial in overweight individuals. Am J Clin Nutr. 2011;93(2):463–73.
28. Kandeel FR, Balon E, Scott S, Nadler JL. Magnesium deficiency and glucose metabolism in rat adipocytes. Metabolism. 1996;45(7):838–43.
29. Hall S, Keo L, Yu KT, Gould MK. Effect of ionophore A23187 on basal and insulin-stimulated sugar transport by rat soleus muscle. Diabetes. 1982;31(10):846–50.
30. Suarez A, Pulido N, Casla A, Casanova B, Arrieta FJ, Rovira A. Impaired tyrosine-kinase activity of muscle insulin receptors from hypomagnesaemic rats. Diabetologia. 1995;38(11):1262–70.
31. Giugliano D, Ceriello A, Paolisso G. Oxidative stress and diabetic vascular complications. Diabetes Care. 1996;19(3):257–67.
32. Barbagallo M, Dominguez LJ, Galioto A, Ferlisi A, Cani C, Malfa L, et al. Role of magnesium in insulin action, diabetes and cardio-metabolic syndrome X. Mol Aspects Med. 2003;24(1–3):39–52.
33. Mazur A, Maier JA, Rock E, Gueux E, Nowacki W, Rayssiguier Y. Magnesium and the inflammatory response: potential physiopathological implications. Arch Biochem Biophys. 2007;458(1):48–56.
34. Abbatecola AM, Ferrucci L, Grella R, Bandinelli S, Bonafe M, Barbieri M, et al. Diverse effect of inflammatory markers on insulin resistance and insulin-resistance syndrome in the elderly. J Am Geriatr Soc. 2004;52(3):399–404.
35. Festa A, D'Agostino Jr R, Howard G, Mykkanen L, Tracy RP, Haffner SM. Chronic subclinical inflammation as part of the insulin resistance syndrome: the insulin resistance atherosclerosis study (IRAS). Circulation. 2000;102(1):42–7.
36. Yudkin JS, Stehouwer CD, Emeis JJ, Coppack SW. C-reactive protein in healthy subjects: associations with obesity, insulin resistance, and endothelial dysfunction: a potential role for cytokines originating from adipose tissue? Arterioscler Thromb Vasc Biol. 1999;19(4):972–8.
37. Liu S, Tinker L, Song Y, Rifai N, Bonds DE, Cook NR, et al. A prospective study of inflammatory cytokines and diabetes mellitus in a multiethnic cohort of postmenopausal women. Arch Intern Med. 2007;167(15):1676–85.
38. Wellen KE, Hotamisligil GS. Inflammation, stress, and diabetes. J Clin Invest. 2005;115(5):1111–9.
39. Pickup JC. Inflammation and activated innate immunity in the pathogenesis of type 2 diabetes. Diabetes Care. 2004;27(3):813–23.
40. Hu FB, Meigs JB, Li TY, Rifai N, Manson JE. Inflammatory markers and risk of developing type 2 diabetes in women. Diabetes. 2004;53(3):693–700.

41. Krakoff J, Funahashi T, Stehouwer CD, Schalkwijk CG, Tanaka S, Matsuzawa Y, et al. Inflammatory markers, adiponectin, and risk of type 2 diabetes in the Pima Indian. Diabetes Care. 2003;26(6):1745–51.
42. Duncan BB, Schmidt MI, Pankow JS, Ballantyne CM, Couper D, Vigo A, et al. Low-grade systemic inflammation and the development of type 2 diabetes: the atherosclerosis risk in communities study. Diabetes. 2003;52(7): 1799–805.
43. Pradhan AD, Manson JE, Rifai N, Buring JE, Ridker PM. C-reactive protein, interleukin 6, and risk of developing type 2 diabetes mellitus. JAMA. 2001;286(3):327–34.
44. Han TS, Sattar N, Williams K, Gonzalez-Villalpando C, Lean ME, Haffner SM. Prospective study of C-reactive protein in relation to the development of diabetes and metabolic syndrome in the Mexico City Diabetes Study. Diabetes Care. 2002;25(11):2016–21.
45. Spranger J, Kroke A, Mohlig M, Hoffmann K, Bergmann MM, Ristow M, et al. Inflammatory cytokines and the risk to develop type 2 diabetes: results of the prospective population-based European prospective investigation into cancer and nutrition (EPIC)-Potsdam Study. Diabetes. 2003;52(3):812–7.
46. Boos CJ, Lip GY. Is hypertension an inflammatory process? Curr Pharm Des. 2006;12(13):1623–35.
47. Muir KW, Tyrrell P, Sattar N, Warburton E. Inflammation and ischaemic stroke. Curr Opin Neurol. 2007;20(3):334–42.
48. Ridker PM, Hennekens CH, Buring JE, Rifai N. C-reactive protein and other markers of inflammation in the prediction of cardiovascular disease in women. N Engl J Med. 2000;342(12):836–43.
49. Koenig W, Sund M, Frohlich M, Fischer HG, Lowel H, Doring A, et al. C-Reactive protein, a sensitive marker of inflammation, predicts future risk of coronary heart disease in initially healthy middle-aged men: results from the MONICA (monitoring trends and determinants in cardiovascular disease) Augsburg Cohort Study, 1984 to 1992. Circulation. 1999;99(2):237–42.
50. Pearson TA, Mensah GA, Alexander RW, Anderson JL, Cannon 3rd RO, Criqui M, et al. Markers of inflammation and cardiovascular disease: application to clinical and public health practice: a statement for healthcare professionals from the Centers for Disease Control and Prevention and the American Heart Association. Circulation. 2003;107(3):499–511.
51. Kumar V. Robbins basic pathology, 7th ed updated. Philadelphia: Elsevier Science, 2005.
52. Hotamisligil GS, Shargill NS, Spiegelman BM. Adipose expression of tumor necrosis factor-alpha: direct role in obesity-linked insulin resistance. Science. 1993;259(5091):87–91.
53. Song Y, Manson JE, Tinker L, Rifai N, Cook NR, Hu FB, et al. Circulating levels of endothelial adhesion molecules and risk of diabetes in an ethnically diverse cohort of women. Diabetes. 2007;56(7):1898–904.
54. Ridker PM, Buring JE, Cook NR, Rifai N. C-reactive protein, the metabolic syndrome, and risk of incident cardiovascular events: an 8-year follow-up of 14 719 initially healthy American women. Circulation. 2003;107(3):391–7.
55. Shoelson SE, Lee J, Goldfine AB. Inflammation and insulin resistance. J Clin Invest. 2006;116(7):1793–801.
56. Dandona P, Aljada A, Bandyopadhyay A. Inflammation: the link between insulin resistance, obesity and diabetes. Trends Immunol. 2004;25(1):4–7.
57. Ridker PM. Cardiology patient page. C-reactive protein: a simple test to help predict risk of heart attack and stroke. Circulation. 2003;108(12):e81–5.
58. Endemann DH, Schiffrin EL. Endothelial dysfunction. J Am Soc Nephrol. 2004;15(8):1983–92.
59. Albelda SM, Smith CW, Ward PA. Adhesion molecules and inflammatory injury. FASEB J. 1994;8(8):504–12.
60. Rohde LE, Hennekens CH, Ridker PM. Cross-sectional study of soluble intercellular adhesion molecule-1 and cardiovascular risk factors in apparently healthy men. Arterioscler Thromb Vasc Biol. 1999;19(7):1595–9.
61. Ferri C, Desideri G, Valenti M, Bellini C, Pasin M, Santucci A, et al. Early upregulation of endothelial adhesion molecules in obese hypertensive men. Hypertension. 1999;34(4 Pt 1):568–73.
62. Abe Y, El-Masri B, Kimball KT, Pownall H, Reilly CF, Osmundsen K, et al. Soluble cell adhesion molecules in hypertriglyceridemia and potential significance on monocyte adhesion. Arterioscler Thromb Vasc Biol. 1998;18(5):723–31.
63. Hackman A, Abe Y, Insull Jr W, Pownall H, Smith L, Dunn K, et al. Levels of soluble cell adhesion molecules in patients with dyslipidemia. Circulation. 1996;93(7):1334–8.
64. Weyer C, Yudkin JS, Stehouwer CD, Schalkwijk CG, Pratley RE, Tataranni PA. Humoral markers of inflammation and endothelial dysfunction in relation to adiposity and in vivo insulin action in Pima Indians. Atherosclerosis. 2002;161(1):233–42.
65. Meigs JB, Hu FB, Rifai N, Manson JE. Biomarkers of endothelial dysfunction and risk of type 2 diabetes mellitus. JAMA. 2004;291(16):1978–86.
66. Schram MT, Stehouwer CD. Endothelial dysfunction, cellular adhesion molecules and the metabolic syndrome. Horm Metab Res. 2005;37 Suppl 1:49–55.
67. Price DT, Loscalzo J. Cellular adhesion molecules and atherogenesis. Am J Med. 1999;107(1):85–97.
68. Nawawi H, Osman NS, Annuar R, Khalid BA, Yusoff K. Soluble intercellular adhesion molecule-1 and interleukin-6 levels reflect endothelial dysfunction in patients with primary hypercholesterolaemia treated with atorvastatin. Atherosclerosis. 2003;169(2):283–91.

69. Witte DR, Broekmans WM, Kardinaal AF, Klopping-Ketelaars IA, van Poppel G, Bots ML, et al. Soluble intercellular adhesion molecule 1 and flow-mediated dilatation are related to the estimated risk of coronary heart disease independently from each other. Atherosclerosis. 2003;170(1):147–53.
70. Galland L. Magnesium and immune function: an overview. Magnesium. 1988;7(5–6):290–9.
71. McCoy H, Kenney MA. Magnesium and immune function: recent findings. Magnes Res. 1992;5(4):281–93.
72. Malpuech-Brugere C, Nowacki W, Daveau M, Gueux E, Linard C, Rock E, et al. Inflammatory response following acute magnesium deficiency in the rat. Biochim Biophys Acta. 2000;1501(2–3):91–8.
73. Weglicki WB, Phillips TM. Pathobiology of magnesium deficiency: a cytokine/neurogenic inflammation hypothesis. Am J Physiol. 1992;263(3 Pt 2):R734–7.
74. Bussiere FI, Tridon A, Zimowska W, Mazur A, Rayssiguier Y. Increase in complement component C3 is an early response to experimental magnesium deficiency in rats. Life Sci. 2003;73(4):499–507.
75. Levy AP, Hochberg I, Jablonski K, Resnick HE, Lee ET, Best L, et al. Haptoglobin phenotype is an independent risk factor for cardiovascular disease in individuals with diabetes: The Strong Heart Study. J Am Coll Cardiol. 2002;40(11):1984–90.
76. Blum S, Vardi M, Brown JB, Russell A, Milman U, Shapira C, et al. Vitamin E reduces cardiovascular disease in individuals with diabetes mellitus and the haptoglobin 2-2 genotype. Pharmacogenomics. 2010;11(5):675–84.
77. Alcock NW, Shils ME, Lieberman PH, Erlandson RA. Thymic changes in the magnesium-depleted rat. Cancer Res. 1973;33(9):2196–204.
78. Malpuech-Brugere C, Nowacki W, Gueux E, Kuryszko J, Rock E, Rayssiguier Y, et al. Accelerated thymus involution in magnesium-deficient rats is related to enhanced apoptosis and sensitivity to oxidative stress. Br J Nutr. 1999;81(5):405–11.
79. Malpuech-Brugere C, Kuryszko J, Nowacki W, Rock E, Rayssiguier Y, Mazur A. Early morphological and immunological alterations in the spleen during magnesium deficiency in the rat. Magnes Res. 1998;11(3):161–9.
80. Nishio A, Ishiguro S, Miyao N. Toxicological and pharmacological studies on magnesium deficiency in rats: histamine-metabolizing enzymes in some tissues of magnesium-deficient rats. Nihon Juigaku Zasshi. 1983;45(6):699–705.
81. Ishiguro S, Nishio A, Miyao N, Morikawa Y, Takeno K, Yanagiya I. Studies on histamine containing cells in the spleen of the magnesium-deficient rats. Nihon Yakurigaku Zasshi. 1987;90(3):141–6.
82. Bernardini D, Nasulewic A, Mazur A, Maier JA. Magnesium and microvascular endothelial cells: a role in inflammation and angiogenesis. Front Biosci. 2005;10:1177–82.
83. Maier JA, Malpuech-Brugere C, Zimowska W, Rayssiguier Y, Mazur A. Low magnesium promotes endothelial cell dysfunction: implications for atherosclerosis, inflammation and thrombosis. Biochim Biophys Acta. 2004;1689(1):13–21.
84. Fullerton DA, Hahn AR, Agrafojo J, Sheridan BC, McIntyre Jr RC. Magnesium is essential in mechanisms of pulmonary vasomotor control. J Surg Res. 1996;63(1):93–7.
85. Altura BM, Gebrewold A, Zhang A, Altura BT. Low extracellular magnesium ions induce lipid peroxidation and activation of nuclear factor-kappa B in canine cerebral vascular smooth muscle: possible relation to traumatic brain injury and strokes. Neurosci Lett. 2003;341(3):189–92.
86. Rayssiguier Y, Gueux E, Bussiere L, Durlach J, Mazur A. Dietary magnesium affects susceptibility of lipoproteins and tissues to peroxidation in rats. J Am Coll Nutr. 1993;12(2):133–7.
87. Bo S, Durazzo M, Guidi S, Carello M, Sacerdote C, Silli B, et al. Dietary magnesium and fiber intakes and inflammatory and metabolic indicators in middle-aged subjects from a population-based cohort. Am J Clin Nutr. 2006;84(5):1062–9.
88. Lopez-Garcia E, Schulze MB, Fung TT, Meigs JB, Rifai N, Manson JE, et al. Major dietary patterns are related to plasma concentrations of markers of inflammation and endothelial dysfunction. Am J Clin Nutr. 2004;80(4):1029–35.
89. Ma Y, Griffith JA, Chasan-Taber L, Olendzki BC, Jackson E, Stanek 3rd EJ, et al. Association between dietary fiber and serum C-reactive protein. Am J Clin Nutr. 2006;83(4):760–6.
90. Ma Y, Hebert JR, Li W, Bertone-Johnson ER, Olendzki B, Pagoto SL, et al. Association between dietary fiber and markers of systemic inflammation in the Women's Health Initiative Observational Study. Nutrition. 2008;24(10):941–9.
91. Chacko SA, Song Y, Nathan L, Tinker L, de Boer IH, Tylavsky F, Wallace R, Liu S. Relations of dietary magnesium intake to biomarkers of inflammation and endothelial dysfunction in an ethnically diverse cohort of postmenopausal women. Diabetes Care. 2010;33(2):304–310.
92. Song Y, Li TY, van Dam RM, Manson JE, Hu FB. Magnesium intake and plasma concentrations of markers of systemic inflammation and endothelial dysfunction in women. Am J Clin Nutr. 2007;85(4):1068–74.
93. King DE, Mainous 3rd AG, Geesey ME, Woolson RF. Dietary magnesium and C-reactive protein levels. J Am Coll Nutr. 2005;24(3):166–71.
94. Kim DJ, Xun P, Liu K, Loria C, Yokota K, Jacobs Jr DR, et al. Magnesium intake in relation to systemic inflammation, insulin resistance, and the incidence of diabetes. Diabetes Care. 2010;33(12):2604–10.

95. de Oliveira Otto MC, Alonso A, Lee DH, Delclos GL, Jenny NS, Jiang R, et al. Dietary micronutrient intakes are associated with markers of inflammation but not with markers of subclinical atherosclerosis. J Nutr. 2011;141(8):1508–15.
96. Rodriguez-Moran M, Guerrero-Romero F. Serum magnesium and C-reactive protein levels. Arch Dis Child. 2008;93(8):676–80.
97. Guerrero-Romero F, Rodriguez-Moran M. Relationship between serum magnesium levels and C-reactive protein concentration, in non-diabetic, non-hypertensive obese subjects. Int J Obes Relat Metab Disord. 2002;26(4): 469–74.
98. Rodriguez-Moran M, Guerrero-Romero F. Elevated concentrations of TNF-alpha are related to low serum magnesium levels in obese subjects. Magnes Res. 2004;17(3):189–96.
99. Almoznino-Sarafian D, Berman S, Mor A, Shteinshnaider M, Gorelik O, Tzur I, et al. Magnesium and C-reactive protein in heart failure: an anti-inflammatory effect of magnesium administration? Eur J Nutr. 2007;46(4): 230–7.
100. Nielsen FH, Johnson LK, Zeng H. Magnesium supplementation improves indicators of low magnesium status and inflammatory stress in adults older than 51 years with poor quality sleep. Magnes Res. 2010;23(4):158–68.
101. Barbagallo M, Dominguez LJ, Galioto A, Pineo A, Belvedere M. Oral magnesium supplementation improves vascular function in elderly diabetic patients. Magnes Res. 2010;23(3):131–7.
102. Shechter M, Sharir M, Labrador MJ, Forrester J, Silver B, Bairey Merz CN. Oral magnesium therapy improves endothelial function in patients with coronary artery disease. Circulation. 2000;102(19):2353–8.
103. Nielsen FH. Magnesium, inflammation, and obesity in chronic disease. Nutr Rev. 2010;68(6):333–40.
104. Agus MS, Agus ZS. Cardiovascular actions of magnesium. Crit Care Clin. 2001;17(1):175–86.
105. Nielsen FH, Milne DB, Gallagher S, Johnson L, Hoverson B. Moderate magnesium deprivation results in calcium retention and altered potassium and phosphorus excretion by postmenopausal women. Magnes Res. 2007;20(1): 19–31.
106. Malpuech-Brugere C, Rock E, Astier C, Nowacki W, Mazur A, Rayssiguier Y. Exacerbated immune stress response during experimental magnesium deficiency results from abnormal cell calcium homeostasis. Life Sci. 1998;63(20):1815–22.
107. Bussiere FI, Gueux E, Rock E, Mazur A, Rayssiguier Y. Protective effect of calcium deficiency on the inflammatory response in magnesium-deficient rats. Eur J Nutr. 2002;41(5):197–202.
108. Dunn MJ, Walser M. Magnesium depletion in normal man. Metabolism. 1966;15(10):884–95.
109. Shils ME. Experimental human magnesium depletion. Medicine (Baltimore). 1969;48(1):61–85.
110. Classen HG. Magnesium and potassium deprivation and supplementation in animals and man: aspects in view of intestinal absorption. Magnesium. 1984;3(4–6):257–64.
111. Kramer JH, Mak IT, Phillips TM, Weglicki WB. Dietary magnesium intake influences circulating pro-inflammatory neuropeptide levels and loss of myocardial tolerance to postischemic stress. Exp Biol Med (Maywood). 2003;228(6):665–73.
112. Dustin ML, Springer TA. Lymphocyte function-associated antigen-1 (LFA-1) interaction with intercellular adhesion molecule-1 (ICAM-1) is one of at least three mechanisms for lymphocyte adhesion to cultured endothelial cells. J Cell Biol. 1988;107(1):321–31.
113. Yamamoto Y, Gaynor RB. Role of the NF-kappaB pathway in the pathogenesis of human disease states. Curr Mol Med. 2001;1(3):287–96.
114. U.S. Department of Agriculture and U.S. Department of Health and Human Services. 2010 dietary guidelines for Americans. 7th ed. Washington, DC: U.S. Department of Agriculture and U.S. Department of Health and Human Services; 2010.

Chapter 3
Magnesium Role in Cytokine Regulation of Hypoxic Placentas Related to Certain Placental Pathology

Tamar Eshkoli, Valeria Feinshtein, Alaa Amash, Eyal Sheiner, Mahmoud Huleihel, and Gershon Holcberg

Key Points

- Normal implantation and placentation is essential for successful pregnancy.
- Oxygen tension plays an important role in guiding the differentiation process that leads to cytotrophoblast invasion to the uterus.
- Multiple signals, including cytokines, are needed to synchronize blastocyst maturation and uterine receptivity in order to complete normal implantation and placentation.
- Proinflammatory cytokines are thought to link placental ischemia with cardiovascular and renal dysfunction.
- Abnormal placentation has been suggested as a factor in adverse pregnancy outcomes.
- $MgSO_4$ has influence on placental function. It has tocolytic, neuroprotective, and anticonvulsant effects.

Key words $MgSO_4$ • Hypoxic placenta • IL-1 • IL-6 • TNF-α • ET-1 • Placental perfusion system

T. Eshkoli, M.D.
Obstetrics and Gynecology, Soroka University Medical Center, PO Box 151, Beer Sheva E84101, Israel
e-mail: esh.tamar@gmail.com

V. Feinshtein, M.Sc.
Department of Clinical Pharmacology, Ben-Gurion University of the Negev, Beer Sheva, Israel

A. Amash, Ph.D. • M. Huleihel, Ph.D.
Department of Microbiology & Immunology, Ben-Gurion University of the Negev, Beer Sheva, Israel
e-mail: Huleihel@bgumail.bgu.ac.il

E. Sheiner, M.D., Ph.D. • G. Holcberg, M.D. (✉)
Department of Obstetrics and Gynecology, Soroka University Medical Center,
Ben-Gurion University of the Negev, PO Box 151, Beer Sheva E84101, Israel
e-mail: holcberg@bgu.ac.il

R.R. Watson et al. (eds.), *Magnesium in Human Health and Disease*, Nutrition and Health,
DOI 10.1007/978-1-62703-044-1_3, © Springer Science+Business Media New York 2013

Introduction

This chapter deals with the issue of magnesium and placenta. Our group studied the area of proinflammatory cytokines in pathological conditions during pregnancy and the influence of $MgSO_4$.

After discussing the placental structure and what is known about hypoxic placenta, we mentioned some important facts about preeclampsia and described different cytokines and their role in the hypoxic placenta. The chapter continues with information about $MgSO_4$ and its functions. The last issue is our experiment model and description of our studies about proinflammatory cytokines and the influence of $MgSO_4$.

Placental Structure

The placenta is derived from the trophectoderm of the implanting blastocyst [1]. The human placenta is of the hemochorial type in which the fetal tissue is in direct contact with maternal blood. The membrane separating the maternal and fetal compartments consists of only three layers: the syncytiotrophoblast, a thin layer of connective tissue, and the vascular endothelium. The syncytiotrophoblast is formed by the fusion of mononuclear cytotrophoblasts (Langhans cells), which lie beneath the syncytium. The cytotrophoblasts are joined to each other and to the syncytiotrophoblast by desmosomes. They are situated on a basement membrane encapsulating a central core of mesenchymal cells, macrophages (Hofbauer cells), and capillary endothelium. The syncytiotrophoblast is the functional unit of the placenta. The plasma membrane of the syncytiotrophoblast is polarized, consisting of the brush-border membrane that is in direct contact with maternal blood and the basal membrane that faces the fetal circulation. The brush-border membrane possesses a microvillous structure that effectively amplifies the surface area, whereas the basal membrane lacks this structural organization. These two membranes are further differentiated from each other by their protein composition. Various enzymes, hormone receptors, and transporters are differentially distributed between the brush-border membrane and the basal membrane [1].

Hypoxic Placenta

Normal implantation and placentation is essential for successful pregnancy. Recent data demonstrates that multiple signals are needed to synchronize blastocyst maturation and uterine receptivity, including sex steroid and peptide hormones, growth factors, cytokines, and immunological factors, in order to complete that task [2–4].

Special interest arises from the effect of hypoxic environment on trophoblast invasion and placentation. Oxygen tension plays an important role in guiding the differentiation process that leads to cytotrophoblast invasion to the uterus. Angiogenesis is regulated by variations in O_2 tension and metabolic factors.

Cytokines has a major role during inflammation. They help to orchestrate the microbicidal activities of phagocytes, contribute to the recruitment of leukocytes, enhance hematopoiesis, and induce fever. For example, TNF-α and IL-1β are highly important for immune cell activation and, thus, the successful pathogen defense.

A long-standing assumption was that the primary function of the placenta is to supply the fetus with as much O_2 as possible. Recent data suggests that it is true only for the second half of pregnancy when fetal weight gain is predominant over organogenesis. Jauniaux and associates [5, 6], based on their in vivo and in vitro data, hypothesized that the placenta limits, rather than facilitates, O_2 supply to the fetus during the period of organogenesis. The earliest stages of development therefore take place in a low-O_2 environment. In mammals, the placenta is the essential interface between the maternal circulation carrying and the fetal circulation [6–8], making it the main controller of fetal O_2 tension. Kingdom and Kaufmann provided evidence that hypoxia has significant effects on placental development, causing hypercapillarization of the villous vasculature and similar to the changes seen in placentas at high altitude and maternal anemia [9]. Hypoxia is not the sole mechanism which effects the placental development. The changing environment from low- to high-O2 tension and the trophoblast ability to control oxidative stress can effect placental development and function. Mainly because of the high rate of cell division, oxidative stress can result in a harmful effect to the developing placenta and fetus, by interfering angiogenesis.

Abnormal placentation, especially restricted endovascular invasion, has been suggested as a factor in adverse pregnancy outcomes such as preeclampsia (PET), preterm premature rupture of membranes (PPROM), preterm labor (PTL), and intrauterine growth restriction (IUGR) [10–14]. Miscarriages, vasculopathies, and fetal structural defects [15] are associated with maternal diseases such as diabetes and diseases with known increased oxidative stress.

According to Tuuli et al. [16], Hypoxia activate pathways and mechanisms involved in trophoblast metabolism and cell death. Hypoxia is associated with a shift toward proliferation in first trimester specimens and reduced differentiation of cytotrophoblasts from both first and third trimester placentas. Phenotypic change of villi in response to hypoxia and hypoxia-reoxygenation in vitro mimics some of the change described in placentas from pregnancies complicated by preeclampsia and IUGR. Placental cells from pregnancies complicated by preeclampsia and IUGR demonstrate exaggerated responses to hypoxia in vitro [16].

Preeclampsia

Preeclampsia, a syndrome unique to the human, is one of the most common complications of pregnancy, affecting 5–7% of pregnant women and is associated with significant fetal and maternal morbidity and mortality worldwide [17]. Preeclampsia is a syndrome characterized by the onset of hypertension and proteinuria after 20 weeks of gestation. Additional signs and symptoms that can occur include visual disturbances, headache, epigastric pain, thrombocytopenia, and abnormal liver function [18]. These clinical manifestations result from mild to severe microangiopathy of target organs, including the brain, liver, kidney, and placenta [19].

The etiology of preeclampsia is thought to be a combination of immunological, environmental, and genetic factors resulted in impaired trophoblast invasion and defective placentation. As a result, a reduction in uteroplacental perfusion occurred, leading to placental ischemic/hypoxic conditions. The ischemic conditions in the placenta, during the late stages of gestation, initiates induced release of different angiogenic factors including proinflammatory cytokines into the maternal circulation leading to systemic endothelial dysfunction and finally to the clinical manifestations seen in preeclampsia [20, 21].

Cytokines and Placental Ischemia

Proinflammatory cytokines are thought to link placental ischemia with cardiovascular and renal dysfunction. The placenta is an integral component of this inflammatory response as it actively produces a variety of cytokines and immunomodulatory hormones [22, 23]. Blood pressure regulatory systems, such as the renin-angiotensin system (RAS) and the sympathetic nervous system, interact with proinflammatory cytokines, which affect angiogenic and endothelium-derived factors regulating endothelial function [22–24].

Tumor necrosis factor (TNF)-α is a well-known member of the TNF superfamily that is involved in numerous cellular processes. TNF-α known activities, mediated through two distinct receptors TNFR1 and TNFR2, include regulation of cytokines expression, immune receptors, proteases, growth factors, and cell cycle genes which in turn regulate inflammation, survival, apoptosis, cell migration, proliferation, and differentiation [25]. A number of groups have reported that TNF-α circulating levels are increased in women with preeclampsia [26, 27], suggesting its possible involvement in the pathogenesis of that disorder. However, there is a controversy about the expression of placental TNF-α levels in preeclampsia [28–30].

Interleukin (IL)-6 is one of the main proinflammatory cytokines that is produced by wide range of immune and nonimmune cells, including monocytes, macrophages, lymphocytes, endothelial cells, and different placental cells [31–33]. IL-6 expression is upregulated by other proinflammatory cytokines, such as IL-1 and TNF-α, and by lipopolysacharide (LPS). IL-6 is active mainly in an endocrinic manner, and its circulating levels are rapidly increased in response to inflammatory conditions. Plasma IL-6 levels reflects the severity as well as the prognosis of the disease [34, 35]. IL-6 is one of the proinflammatory cytokines suggested to be involved in the pathogenesis of preeclampsia [23]. The maternal circulating levels of IL-6, which are already more elevated in healthy pregnant women compared to nonpregnant controls, are further raised in patients with preeclampsia [27, 36, 37]. Recently, we have shown that IL-6 is secreted by the human placenta into the fetal and maternal circulations during normal pregnancy [31], suggesting a role for placental IL-6 in regulating the maternal immune response during pregnancy. Moreover, IL-6, as well as TNF-α and IL-1, was suggested as potential specific markers for preeclampsia, and they were reported as inducers of endothelial-free oxygen radicals, such as H_2O_2 and O_2^- [38]. However, the data regarding placental IL-6 expression in preeclampsia is still conflicted. While placental IL-6 levels were shown to be unchanged in preeclampsia [29], others detected decreased placental IL-6 levels in preeclampsia [39]. Increased IL-6 levels in amniotic fluids of second trimester pregnancies were detected in women who developed preeclampsia later [40]. However, others reported no differences in biologically active IL-6 levels in amniotic fluids from preeclamptic pregnancies at delivery [41].

Interleukin-1 (IL-1) is a proinflammatory cytokine produced by wide range of immune and nonimmune cells, including macrophages, monocytes, NK cells, epithelial cells, and fibroblasts. Its effects include regulation of adhesion molecules expression and cellular division, differentiation, and function of immune cells [35]. However, the IL-1 system is also suggested to play a key role in different human diseases such as preeclampsia. IL-1 has been suggested as a possible specific marker for preeclampsia, and it was described as an inducer for free radicals such as O_2^- and H_2O_2 in endothelial cells [42]. Also, IL-1Ra was reported as involved in preeclampsia since IL-1Ra gene polymorphism reported to be associated with the disease [43].

ET-1 is a product of endothelial and many other cell types that possesses a wide range of actions, including vasoconstriction, bronchoconstriction, and mitogenic activity on smooth muscle cells and fibroblasts [44]. ET-1, being a powerful vasoconstrictor, could cause significant vasoconstriction in the placental vasculature, and alterations in endothelin-1 levels in placental vasculature may therefore have a role in the pathogenesis of preeclampsia [45]. ET-1 levels were significantly higher in the placental tissues from women with preeclampsia [46]. ET-1 may contribute to the vasospasm associated with preeclampsia and lend further support to the involvement of endothelial cell dysfunction in the pathophysiology of this disorder [46]. However, the plasma levels of endothelin are highest during the later stage of the disease [47, 48], suggesting that endothelin may not be involved in the initiation of preeclampsia but rather in the progression of the disease into the malignant hypertensive phase [49, 50]. It was suggested that endothelin plays a major role in mediating the hypertension produced by chronic reduction of uterine perfusion pressure in pregnant rats [51].

ET-1 is known to interact with ET_A and ET_B receptors [52]. The interaction of endothelin with specific ET_A and ET_B receptors in smooth muscle initiates a cascade of biochemical events leading to smooth muscle contraction [52–56]. Endothelin may also interact with specific ET_B receptors in the endothelium leading to activation of ET_B-mediated vascular relaxation pathways [55, 56]; these studies suggest that during preeclampsia an increase in endothelin production and activation of ET_B-mediated vascular relaxation pathways may serve as a rescue mechanism against the excessive increases in vascular resistance and arterial pressure.

In agreement, ET-1 could induce cytokine release via nuclear factor kappa B (NF-kB) activation [57]. On the other hand, exposing cultured endothelial cells to TNF-α enhanced ET-1 secretion [58]. IL-1β increased the release of ET-1 by primary endothelial cells in a dose-dependent manner; as well, it reduced expression of the ET_B receptor on endothelial cells and increased expression of the ET_A receptor on vascular smooth muscle cells [59]. Moreover, LaMarca et al. and Gadonski et al. reported recently that serum levels of TNF-α and IL-6 are elevated after reductions in uterine perfusion pressure in rats, and chronic infusion of TNF-α and IL-6 into pregnant rats increases arterial pressure and decreases renal plasma flow and glomerular filtration rate [60, 61]. In addition, Roberts et al. reported recently that serum from pregnant rats exposed to reductions in uterine perfusion enhances endothelin production by endothelial cells via by AT1 receptor activation [51].

Magnesium Sulfate ($MgSO_4$)

Magnesium sulfate is the drug of choice for the treatment of severe preeclampsia [62], prevention of eclampsia [63], and prevention of recurrent eclamptic seizures [64], and it is also been used as a tocolytic agent for prevention of preterm birth [65]. Its efficacy and safety in these conditions is well described [66, 67]. The normal maternal serum magnesium level is 2.1 ± 0.3 mg/dl. Maternal administration of standard doses of magnesium sulfate (6 g load then 2–3 g/h) for preeclampsia or preterm tocolysis result in levels 4.8–8.4 mg/dl [68]. Magnesium has been shown to readily cross the placenta [69]. Fetal levels and maternal levels correlate [70] and equilibrium is thought to occur within two hours [71]. Hallak has shown that fetal serum levels increase within one hour while amniotic fluid in three hours after maternal magnesium intravenous administration [70].

Epidemiological data suggests that maternal administration of $MgSO_4$ may decrease the risk of cerebral palsy and intraventricular hemorrhage in surviving preterm neonates. Nelson and Grether reported a decreased risk of CP with magnesium therapy in neonates weighing <1,500 g. at birth with follow-up to 3 years [71]. Hauth evaluated CP rates for neonates with birth weight <1,000 g. and also noted an inverse relationship [72].

$MgSO_4$ as Anticonvulsant

Although the mechanism of action of $MgSO_4$ as an anticonvulsant agent in preeclampsia/eclampsia is still not clearly understood, some possible mechanisms, including vasodilatation of cerebral vasculature, inhibition of platelet aggregation, protection of endothelial cells from damage by free radicals, prevention of calcium ion entry into ischemic cells, decreasing the release of acetylcholine at motor end plates within the neuromuscular junction, and as a competitive antagonist to the glutamate N-methyl-D-aspartate receptor (which is epileptogenic), have been proposed [73].

$MgSO_4$ as Neuroprotective

The exact mechanism of the potential neuroprotective effects of $MgSO_4$ remains unclear. With regard to the use of magnesium sulfate for the prevention and treatment of eclampsia, there is evidence to a potential central anticonvulsant effect [74, 75]. However, Belfort found that magnesium vasodilates smaller diameter cerebral vessels [76]. This may be equally important due to the low tolerance of the brain of the preterm neonate to hypoperfusion.

The potential protective effects of magnesium during perinatal hypoxic-ischemia and other maternal insults continue to be studied. Magnesium has been shown to be protective against neuronal injury in animals. Magnesium may prevent injury and improve neonatal outcome through its action on placental vessels. Three possible mechanisms of action have been postulated: (1) Magnesium may dilate spiral arteries in the same fashion it vasodilates smaller diameter cerebral vessels [76], (2) magnesium may stabilize vascular tone and thereby decrease endothelial injury, and (3) magnesium may protect against proinflammatory-mediated damage by reducing cytokines and bacterial toxin synthesis [77, 78]. Magnesium deficiency has been associated with production of reactive oxygen species, cytokines, as well as vascular compromise in vivo [79]. Although magnesium deficiency-induced inflammatory change occurs during chronic magnesium deficiency in vivo, acute magnesium deficiency may also affect the vasculature and, consequently, predispose endothelial cells to perturbations associated with chronic magnesium deficiency [80]. Pathological conditions of placentas such as IUGR may enhance the secretion of proinflammatory cytokines, which may enhance the vasoconstriction of fetal-placental vascular bed. Our studies were concentrated in this area, and we have shown, for example, that $MgSO_4$ and Ang II have different effects on the capacity of fetal and maternal compartments of normal human placenta to secrete TNF-α and IL-6 [81].

Three meta-analyses have evaluated the neuroprotective effects of magnesium sulfate when given to women at risk of preterm birth; all included the same five trials and came to similar conclusions [82–84]. The following data from a Cochrane review are representative examples of these findings.

The Cochrane review of five trials included 6,145 infants; 1,493 infants were delivered preterm [82]. Antenatal magnesium sulfate therapy given to women at risk of preterm birth substantially reduced the risk of cerebral palsy in their child. There was also a significant reduction in the rate of substantial gross motor dysfunction. No statistically significant effect of antenatal magnesium sulfate therapy was detected on pediatric mortality or on other neurological impairments or disabilities in the first few years of life.

$MgSO_4$ as Tocolytic

Magnesium sulfate has been shown to have similar efficacy to ritodrine and other beta-agonists. However, the combination of magnesium sulfate and ritodrine can lead to dangerous maternal side effects, namely, chest pain and ECG changes suggestive of ischemia. Magnesium sulfate may be an effective tocolytic, but may require an intolerably high magnesium concentration to be effective [85].

It seems that there is not enough evidence to show any difference between magnesium maintenance therapy compared with either placebo or no treatment, or alternative therapies (ritodrine or terbutaline) in preventing preterm birth after an episode of threatened preterm labor [85].

Our Experiment Model

Our group was concentrated in studying the placenta at the end of pregnancy, by evaluating the interleukins according to pathological conditions such as preeclampsia. We also evaluate the influence of $MgSO_4$ on the placenta and the interleukins.

We use the placental perfusion system, which is described below, and evaluate interleukins such as IL-1, IL-6, and TNF-α.

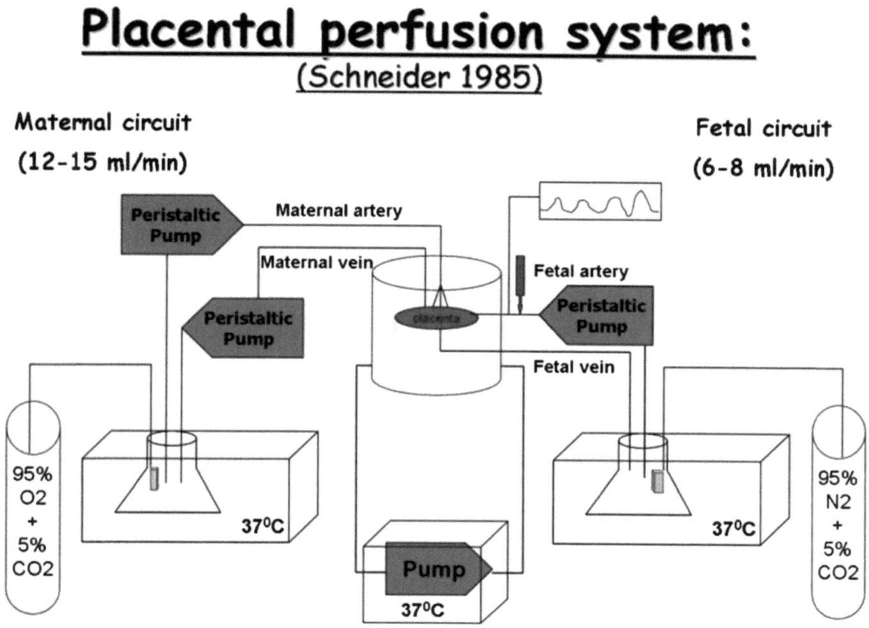

Fig. 3.1 Placental perfusion system

Placental Perfusion System

Placentas from preeclamptic and normotensive term pregnancies were collected immediately after vaginal delivery and placed in physiological saline at room temperature. The cord was clamped to maintain the dilatation of the fetal vascular system. The perfusion experiments were performed using the method originally described by Schneider and Huch [86] and modified by Holcberg et al. [81] (Fig. 3.1). After careful inspection of the chorionic and decidual surface, an intact cotyledon was selected.

A fetal artery and corresponding vein from a single cotyledon were cannulated within 15–20 min of delivery. Following successful establishment of the fetal circulation, the placenta was mounted in a perfusion chamber, and the maternal circulation was simulated by placing four catheters into the intervillous space of the lobe, corresponding to the isolated perfused cotyledon. Maternal perfusate that returned from the intervillous space was continuously drained by a maternal venous catheter, placed at the lowest level on the maternal decidual surface, to avoid significant pooling of perfusate.

The two reservoirs, containing the perfusion medium for the maternal and the fetal circuit, were placed into heated water baths at 37 °C and were equilibrated with a prehumidified gas mixture of 95% oxygen and 5% carbon dioxide on the maternal site and 95% N_2 and 5% CO_2 on the fetal site. Perfusion pressure is 20–30 mmHg, giving a flow rate of 6–8 ml/min and 10–12 ml/min in the fetal and maternal circulation, respectively. The venous return was recycled into the respective reservoir, giving a closed circuit perfusion.

Our Studies About Proinflammatory Cytokines and the Effect of $MgSO_4$

Selective Vasodilator Effect of Magnesium Sulfate in Human Placenta

Thromboxane was shown to induce vasoconstriction of human placenta in a dose-dependent manner, when the highest induction was examined at the highest concentrations of the TX used [87]. Addition of $MgSO_4$ to the TX-induced placenta did not significantly affect the TX-induced vasoconstriction levels.

Endothelin-1 was shown to induce vasoconstriction of human placenta in a dose-dependent manner, when the highest induction was examined at the highest concentrations of ET-1 used. Addition of $MgSO_4$ to the ET-1-induced placentas significantly reduces the ET-1-induced vasoconstrictor effect at all concentrations.

Ag II was shown to induce vasoconstriction capacity of human placenta only at the highest concentration used. Addition of $MgSO_4$ to the Ag II-induced placenta significantly reduced the AII-induced vasoconstriction effect at the highest induction concentration.

Thromboxane was shown to induce IL-1β secretion by human placental tissue at the concentration of 10 M. The secretion of IL-1β was shown to be time-course dependent when maximal IL-1 β levels were detected after 9 h of perfusion. Addition of $MgSO_4$ was shown to completely reduce the TX-induced IL-1β secretion.

Angiotensin II was shown to induce IL-1β secretion mainly at the concentration of 10^{-5} M. The secretion levels of IL-1β, from the fetal side, was shown to be time-course dependent when maximal IL-1 β levels were detected after 7 h of perfusion. Addition of $MgSO_4$ to the Ag II-induced placenta was shown to completely reduce their capacity to secrete IL-1β from the fetal side.

Our results demonstrated that $MgSO_4$ significantly decreases vasoconstriction of placental vasculature induced by ET-1 and Ag II, but was not capable to reduce TX-induced vasoconstriction capacity. The mechanism involved in this regulation could be mediated by reduction of IL-1b secretion by the placental tissues.

Different Effects of Magnesium Sulfate and Angiotensin II on the Capacity of the Fetal and Maternal Compartments of Normal Human Placenta to Secrete TNF-α and IL-6

TNF-α and IL-6 are pleiotrpic cytokines that mediate many diseases associated with inflammation, cachexia, shock, and tissue injury [81].

We showed that $MgSO_4$ is capable to increase TNF and IL-6 secretion from the maternal compartment of the human placenta and that AII significantly attenuates this effect.

However, in the fetal placental compartment, the interaction of $MgSO_4$ with AII leads to a significant increase in TNF and IL-6 levels, and $MgSO_4$ alone has no significant effect on TNF or IL-6 production by fetal placental compartment. This may reflect the different response or function of the fetal and maternal sites of the placenta to these agents.

After exposure of the placental tissue to AII in presence of $MgSO_4$, as it may occur in preeclamptic patients, TNF and IL-6 secretion is significantly attenuated in the maternal site. However, under the same condition, the fetal site responses differently, as TNF and IL-6 levels are significantly increased by AII in presence of $MgSO_4$ which could be harmful for the fetal brain.

Our results show that $MgSO_4$ is feasible to a significant increase in the TNF-α levels secreted by the human placenta and bolus injection of AII differently affect the cytokine production from maternal versus fetal sites. Of special interest is that the maternal site of the perfused placental tissue produces more TNF-α and IL-6 than the fetal site. However, the contribution of placental fetal site to TNF-α and IL-6 secretion may be crucial in this situation. The fact that the fetal compartment could secrete more prominent amounts of TNF-α, after exposure to $MgSO_4$ and AII, may suggest the possibility of harmful effect of TNF-α to fetal brain which may affect late stages of his development.

Downregulation of Placental Tumor Necrosis Factor-α Secretion: A Possible Therapeutic Effect of Magnesium Sulfate in Preeclampsia

TNF-α levels in the fetal and maternal circulation of preeclamptic placentas perfused with control medium (without $MgSO_4$) increased with time, as compared to TNF-α levels in the fetal circulation of normotensive placentas [88].

TNF-α levels in the fetal and maternal circulations of normotensive and preeclamptic placentas perfused with control medium were increased with time, reaching significantly higher levels of TNF-α in the maternal circulations as compared to the fetal circulations, at the end of perfusion.

This study suggests that preeclamptic placenta secretes higher levels of the inflammatory cytokine TNF-α into both the fetal and the maternal circulation, as compared to normotensive placentas.

Secreted TNF-α levels into the fetal or the maternal circulations of normotensive placenta was not affected by the addition of $MgSO_4$ to the maternal reservoir of the perfused normotensive placentas. However, addition of $MgSO_4$ to the maternal reservoir of preeclamptic placentas resulted in a significant decrease in secreted TNF-α levels into the fetal and the maternal circulations

TNF-α levels in the fetal and maternal compartments of preeclamptic placentas perfused with $MgSO_4$ were significantly lower, as compared to preeclamptic placentas perfused with control medium.

This effect of $MgSO_4$ suggests a possible therapeutic effect for this agent in reducing maternal endothelial dysfunction and in improving neonatal outcome in preeclampsia, by reducing TNF-α levels in maternal and fetal circulations. The unaffected TNF-α secretion by the normotensive placenta suggests that the effect of $MgSO_4$ is specific to the preeclamptic placenta and may indicate a different mechanism of regulation of TNF-α secretion in preeclampsia as compared to normotensive pregnancies.

Placental Secretion of Interleukin-1 and Interleukin-1 Receptor Antagonist in Preeclampsia; Effect of Magnesium Sulfate (in Press)

Preeclamptic placentas secreted increased levels of IL-1β and its natural inhibitor IL-1RA into the fetal and the maternal circulations, as compared to normotensive placentas. However, comparison of IL-1β:IL-1RA ratio in the fetal and the maternal circulations of preeclamptic placentas versus normotensive placentas did not reveal any significant differences between the two groups. Moreover, normotensive as well as preeclamptic placentas secreted significantly higher levels of IL-1β and IL-1RA into the maternal circulations as compared to the levels of IL-1β and IL-1RA in the corresponding fetal circulations. Similarly, IL-1β:IL-1RA ratios in the maternal circulations of normotensive and preeclamptic placentas were higher, as compared to the corresponding fetal circulations.

Addition of $MgSO_4$ into the maternal circulation of normotensive placentas did not affect IL-1β levels secreted into the fetal or the maternal circulations. However, administration of $MgSO_4$ in preeclamptic placentas did not affect IL-1β levels in the fetal circulation on one hand, but resulted in decreased IL-1β secretion into the maternal circulation, on the other hand. Furthermore, administration of $MgSO_4$ in the maternal circulation differently affected IL-1RA secretion by normotensive and preeclamptic placentas. In normotensive placentas, exposure to $MgSO_4$ resulted in increased IL-1RA levels in the fetal as well as the maternal circulations as compared to IL-1RA levels in the fetal and the maternal circulations of normotensive placentas perfused with medium alone. On the other hand, exposure of preeclamptic placentas to $MgSO_4$ resulted in decreased IL-1RA levels in the fetal and the maternal circulation as compared to IL-1RA levels in the fetal and the maternal circulations of preeclamptic placentas perfused with medium alone.

The IL-1β:IL-1RA ratio seems to be not affected by $MgSO_4$, suggesting that placental-derived IL-1β and IL-1RA may not have a critical role in the development of cytokine imbalance and maternal symptoms of preeclampsia.

Magnesium Sulfate Normalizes Placental Interleukin-6 Secretion in Preeclampsia

IL-6 levels in the fetal circulation of preeclamptic placentas increased with time as compared to IL-6 levels in the fetal circulation of normotensive placentas [89]. Similarly, IL-6 levels in the maternal circulation of preeclamptic placentas increased with time as compared to IL-6 levels in the maternal circulation of normotensive placentas. Furthermore, normotensive placentas, as well as preeclamptic placentas perfused with control medium, secreted significantly increased levels of IL-6 into the maternal circulations, as compared to IL-6 levels in the fetal circulations.

Addition of $MgSO_4$ into the maternal reservoir of the perfused normotensive placentas did not affect IL-6 secretion levels into the fetal or the maternal circulations. However, exposure of perfused preeclamptic placentas to $MgSO_4$ differently affected IL-6 secretion levels into the fetal and the maternal circulations. In the fetal circulation, a tendency toward decreased secretion levels of IL-6, in presence of $MgSO_4$, was detected through all the perfusion period, although this tendency did not reach statistical significant. On the other hand, IL-6 levels in the maternal circulation of preeclamptic placentas were significantly lower in presence of $MgSO_4$, as compared to IL-6 levels in the maternal circulation of the preeclamptic control group. Furthermore, IL-6 levels detected at the end of perfusion, in the fetal and maternal circulations of preeclamptic placentas exposed to $MgSO_4$, were statistically equal to IL-6 levels in the fetal and maternal circulations of normotensive placentas perfused with control medium.

Our results show an increased secretion of IL-6 by the preeclamptic placenta, into the maternal and the fetal circulations, compared with that in normotensive placentas. These results indicate that the placenta may contribute, at least partially, to the elevation in circulating IL-6 levels in preeclampsia. Furthermore, these data may confirm the theory about the role of the ischemic placenta in the maternal endothelial dysfunction and in the onset of clinical manifestations in preeclampsia.

In the current study, we demonstrated that exposure of preeclamptic placentas to $MgSO_4$ during perfusion normalizes placental IL-6 secretion levels into the maternal circulation of these placentas. These data indicate an anti-inflammatory effect of $MgSO_4$ on preeclamptic placental tissue, suggesting that $MgSO_4$ may improve maternal endothelial function by preventing the enhanced placental secretion of IL-6 into the maternal circulation.

Conclusion

Pro- as well as anti-inflammatory placental cytokines may play an important physiological role in maintaining the fetal maternal interface during pregnancy since they are produced by the fetal and maternal compartments of ischemic placentas.

$MgSO_4$ differently controls the release and the effect of cytokines in human placenta.

The knowledge gained from our studies increases our understanding of the pathophysiology of vascular activity of several autacoids in the presence of MgSO4 as a protective agent for placental vessels and aid the new strategies for diagnosis and treatment disorders related to placental vasoconstriction such as preeclampsia and IUGR.

References

1. Eshkoli T, Sheiner E, Ben-Zvi Z, Feinstein V, Holcberg G. Drug transport across the placenta. Curr Pharm Biotechnol. 2011;12:707–14.
2. Hoozemans DA, Schats R, Lambalk CB, et al. Human embryo implantation: current knowledge and clinicai implications in assisted reproductive technology. Reprod Biomed Online. 2004;9:692–715.

3. Jauniaux E. Design, beauty and differentiation: the human fetus during the lirst trimester of gestation. Reprod Biomed Online. 2000;1:107–8.
4. Norwitz ER, Schust DJ, Fisher SJ. Implantation and the survival of early pregnancy. N Eng J Med. 2001;345:1400–8.
5. Jauniaux E, Gulbis B, Burton GJ. The human first trimester gestational sac limits rather than facilitates oxygen transfer to the foetus--a review. Placenta. 2003;24(Suppl A):S86–93.
6. Jauniaux E, Hempstock J, Greenwold N, Burton GJ. Trophoblastic oxidative stress in relation to temporal and regional differences in maternal placental blood flow in normal and abnormal early pregnancies. Am J Pathol. 2003;162:115–25.
7. Burton GJ, Hempstock J, Jauniaux E. Oxygen, early embryonic metabolism and free radical-mediated embryopathies. Reprod Biomed Online. 2003;6:84–96.
8. Jauniaux E, Watson A, Burton G. Evaluation of respiratory gases and acid-base gradients in human fetal fluids and uteroplacental tissue between 7 and 16 weeks' gestation. Am J Obstet Gynecol. 2001;184:998–1003.
9. Kingdom JC, Kaufmann P. Oxygen and placental villous development: origins of fetal hypoxia. Placenta. 1997;18:613–21. discussion 623-6.
10. Brosens IA. Morphological changes in the utero-placental bed in pregnancy hypertension. Clin Obstet Gynaecol. 1977;4:573–93.
11. Kim YM, et al. Failure of physiologic transformation of the spiral arteries in patients with preterm labor and intact membranes. Am J Obstet Gynecol. 2003;189:1063–9.
12. Meekins JW, Pijnenborg R, Hanssens M, McFadyen IR, van Asshe A. A study of placental bed spiral arteries and trophoblast invasion in normal and severe pre-eclamptic pregnancies. Br J Obstet Gynaecol. 1994;101:669–74.
13. Ozturk O, Bhattacharya S, Saridogan E, Jauniaux E, Templeton A. Role of utero-ovarian vascular impedance: predictor of ongoing pregnancy in an IVF-embryo transfer programme. Reprod Biomed Online. 2004;9:299–305.
14. Zhou Y, Damsky CH, Fisher SJ. Preeclampsia is associated with failure of human cytotrophoblasts to mimic a vascular adhesion phenotype. One cause of defective endovascular invasion in this syndrome? J Clin Invest. 1997;99:2152–64.
15. Jawerbaum A, et al. Eicosanoid production by uterine strips and by embryos obtained from diabetic pregnant rats. Prostaglandins. 1993;45:487–95.
16. Tuuli MG, Longtine MS, Nelson DM. Review: oxygen and trophoblast biology--a source of controversy. Placenta. 2011;32 Suppl 2:S109–18. Epub 2011 Jan 7.
17. Myatt L, Miodovnik M. Prediction of preeclampsia. Semin Perinatol. 1999;23(1):45–57.
18. ACOG practice bulletin. Diagnosis and management of preeclampsia and eclampsia. Number 33. Obstet Gynecol. 2002;99:159–67.
19. Lain KY, Roberts JM. Contemporary concepts of the pathogenesis and management of preeclampsia. JAMA. 2002;287:3183–6.
20. Kharfi A, Giguère Y, Sapin V, Massé J, Dastugue B, Forest JC. Trophoblastic remodeling in normal and preeclamptic pregnancies: implication of cytokines. Clin Biochem. 2003;36:323–31.
21. Page NM. The endocrinology of pre-eclampsia. Clin Endocrinol (Oxf). 2002;57:413–23.
22. LaMarca BD, Ryan MJ, Gilbert JS, Murphy SR, Granger JP. Inflammatory cytokines in the pathophysiology of hypertension during preeclampsia. Curr Hypertens Rep. 2007;9:480–5.
23. Rusterholz C, Hahn S, Holzgreve W. Role of placentally produced inflammatory and regulatory cytokines in pregnancy and the etiology of preeclampsia. Semin Immunopathol. 2007;29:151–62.
24. Brewster JA, Orsi NM, Gopichandran N, McShane P, Ekbote UV, Walker JJ. Gestational effects on host inflammatory response in normal and pre-eclamptic pregnancies. Eur J Obstet Gynecol Reprod Biol. 2008;140:21–6.
25. Haider S, Knofler M. Human tumour necrosis factor: physiological and pathological roles in placenta and endometrium. Placenta. 2008. doi:10.1016/j.placenta.2008.10.012.
26. Conrad KP, Benyo DF. Placental cytokines and the pathogenesis of preeclampsia. Am J Reprod Immunol. 1997;37:240–9.
27. Conrad KP, Miles TM, Benyo DF. Circulating levels of immunoreactive cytokines in women with preeclampsia. Am J Reprod Immunol. 1998;40:102–11.
28. Wang Y, Walsh SW. TNF alpha concentrations and mRNA expression are increased in preeclamptic placentas. J Reprod Immunol. 1996;32:157–69.
29. Benyo DF, Smarason A, Redman CW, Sims C, Conrad KP. Expression of inflammatory cytokines in placentas from women with preeclampsia. J Clin Endocrinol Metab. 2001;86:2505–12.
30. Hayashi M, Ueda Y, Yamaguchi T, Sohma R, Shibazaki M, Ohkura T, et al. Tumor necrosis factor-alpha in the placenta is not elevated in pre-eclamptic patients despite its elevation in peripheral blood. Am J Reprod Immunol. 2005;53:113–9.
31. Akira S, Hirano T, Taga T, Kishimoto T. Biology of multifunctional cytokines: IL-6 and related molecules (IL-1 and TNF). FASEB J. 1990;4:2860–7.

32. Bowen JM, Chamley L, Mitchell MD, Keelan JA. Cytokines of the placenta and extra-placental membranes: biosynthesis, secretion and roles in establishment of pregnancy in women. Placenta. 2002;23:239–56.
33. Holcberg G, Amash A, Sapir O, Sheiner E, Levy S, Huleihel M. Perfusion with lipopolysaccharide differently affects the secretion of tumor necrosis factor-alpha and interleukin-6 by term and preterm human placenta. J Reprod Immunol. 2007;74:15–23.
34. Dinarello CA, Moldawer LL. Pro-inflammatory and anti-inflammatory cytokines in rheumatoid arthritis. 2nd ed. Thousand Oaks: AMGEN; 2000. p. 1–98. Chapter 1-4.
35. Gosain A, Gamelli RL. A primer in cytokines. J Burn Care Rehabil. 2005;26:7–12.
36. Greer IA, Lyall F, Perera T, Boswell F, Macara LM. Increased concentrations of cytokines interleukin-6 and interleukin-1 receptor antagonist in plasma of women with preeclampsia: a mechanism for endothelial dysfunction? Obstet Gynecol. 1994;84:937–40.
37. Vince GS, Starkey PM, Austgulen R, Kwiatkowski D, Redman CW. Interleukin-6, tumour necrosis factor and soluble tumour necrosis factor receptors in women with pre-eclampsia. Br J Obstet Gynaecol. 1995;102:20–5.
38. Tolando R, Jovanovic A, Brigelius-Flohe R, Ursini F, Maiorino M. Reactive oxygen species and proinflammatory cytokine signaling in endothelial cells: effect of selenium supplementation. Free Radic Biol Med. 2000;28:979–86.
39. Kauma SW, Wang Y, Walsh SW. Preeclampsia is associated with decreased placental interleukin-6 production. J Soc Gynecol Investig. 1995;2:614–7.
40. Nakabayashi M, Sakura M, Takeda Y, Sato K. Elevated IL-6 in midtrimester amniotic fluid is involved with the onset of preeclampsia. Am J Reprod Immunol. 1998;39:329–34.
41. Opsjon SL, Austgulen R, Waage A. Interleukin-1, interleukin-6 and tumor necrosis factor at delivery in preeclamptic disorders. Acta Obstet Gynecol Scand. 1995;74:19–26.
42. Tolando R, Jovanovic A, Brigelius-Flohe R, et al. Reactive oxygen species and proinflammatory cytokine signaling in endothelial cells: effect of selenium supplementation. Free Radic Biol Med. 2000;28:979–86.
43. Faisel F, Romppanen EL, Hiltunen M, et al. Polymorphism in the interleukin 1 receptor antagonist gene in women with preeclampsia. J Reprod Immunol. 2003;60:61–70.
44. Schiffrin EL, Touyz RM. Vascular biology of endothelin. J Cardiovasc Pharmacol. 1998;32:S2–13.
45. Krebs C, Macara LM, Leiser R, Bowman AW, Greer IA, Kingdom JC. Intrauterine growth restriction with absent end-diastolic flow velocity in the umbilical artery is associated with maldevelopment of the placental terminal villous tree. Am J Obstet Gynecol. 1996;175(6):1534–42.
46. Singh HJ, Rahman A, Larmie ET, Nila A. Endothelin-1 fetoplacental tissues from normotensive pregnant women and women with pre-eclampsia. Acta Obstet Gynecol Scand. 2001;80(2):99–103.
47. Taylor RN, Varma M, Teng NN, Roberts JM. Women with preeclampsia have higher plasma endothelin levels than women with normal pregnancies. J Clin Endocrinol Metab. 1990;71:1675–7.
48. Nova A, Sibai BM, Barton JR, Mercer BM, Mitchell MD. Maternal plasma level of endothelin is increased in preeclampsia. Am J Obstet Gynecol. 1991;165:724–7.
49. Clark BA, Halvorson L, Sachs B, Epstein FH. Plasma endothelin levels in preeclampsia: elevation and correlation with uric acid levels and renal impairment. Am J Obstet Gynecol. 1992;166:962–8.
50. Dekker GA, Sibai BM. Etiology and pathogenesis of preeclampsia: current concepts. Am J Obstet Gynecol. 1998;179:1359–75.
51. Roberts L, LaMarca BB, Fournier L, Bain J, Cockrell K, Granger JP. Enhanced endothelin synthesis by endothelial cells exposed to sera from pregnant rats with decreased uterine perfusion. Hypertension. 2006;47(3):615–8.
52. Seo B, Oemar BS, Siebenmann R, von Segesser L, Luscher TF. Both ETA and ETB receptors mediate contraction to endothelin-1 in human blood vessels. Circulation. 1994;89:1203–8.
53. Sumner MJ, Cannon TR, Mundin JW, White DG, Watts IS. Endothelin ETA and ETB receptors mediate vascular smooth muscle contraction. Br J Pharmacol. 1992;107:858–60.
54. LaDouceur DM, Flynn MA, Keiser JA, Reynolds E, Haleen SJ. ETA and ETB receptors coexist on rabbit pulmonary artery vascular smooth muscle mediating contraction. Biochem Biophys Res Commun. 1993;196:209–15.
55. Pollock DM, Keith TL, Highsmith RF. Endothelin receptors and calcium signaling. FASEB J. 1995;9:1196–204.
56. Schiffrin EL. Endothelin and endothelin antagonists in hypertension. J Hypertens. 1998;16:1891–5.
57. Virdis A, Schiffrin EL. Vascular inflammation: a role in vascular disease in hypertension? Curr Opin Nephrol Hypertens. 2003;12(2):181–7.
58. Corder R, Carrier M, Khan N, Klemm P, Vane JR. Cytokine regulation of endothelin-1 release from bovine aortic endothelial cells. J Cardiovasc Pharmacol. 1995;26 Suppl 3:S56–8.
59. Nowicki PT. IL-1beta alters hemodynamics in newborn intestine: role of endothelin. Am J Physiol Gastrointest Liver Physiol. 2006;291(3):G404–13.
60. LaMarca BB, Bennett WA, Alexander BT, Cockrell K, Granger JP. Hypertension produced by reductions in uterine perfusion in the pregnant rat: role of tumor necrosis factor-alpha. Hypertension. 2005;46(4):1022–5.
61. Gadonski G, LaMarca BB, Sullivan E, Bennett W, Chandler D, Granger JP. Hypertension produced by reductions in uterine perfusion in the pregnant rat: role of interleukin 6. Hypertension. 2006;48(4):711–6.

62. Witlin AG, Sibai BM. Magnesium sulfate therapy in preeclampsia and eclampsia. Obstet Gynecol. 1998;92: 883–9.
63. Lucas MJ, Leveno KJ, Cunningham FG. A comparison of magnesium sulfate with phenytoin for the prevention of eclampsia. N Engl J Med. 1995;333:201–5.
64. Duley L, Henderson-Smart D. Magnesium sulphate versus phenytoin for eclampsia. Cochrane Database Syst Rev. 2003(4).
65. Hankins GDV, Hammond TL, Yeomans ER. Amniotic cavity accumulation of magnesium with prolonged magnesium sulfate tocolysis. J Reprod Med. 1991;36:446–9.
66. Cotton DB, Gonik B, Dorman KF. Cardiovascular alteration in severe pregnancy-induced hypertension: acute effects of intravenous magnesium sulfate. Am J Obstet Gynecol. 1984;148:162–5.
67. Sibai BM. Magnesium sulfate is the ideal anticonvulsant in preeclampsia-eclampsia. Am J Obstet Gynecol. 1990;162:1141–5.
68. Sibai BM. Preeclampsia-eclampsia. Curr Prob Obstet Gynecol Fertil. 1990;13:1–45.
69. Hallak M, Berry SM, Madincea F, et al. Fetal serum and amniotic fluid magnesium concentrations with maternal treatment. Obstet Gynecol. 1993;81:185–8.
70. Pritchard JA. The use of the magnesium ion in the management of eclamptic toxemias. Surg Gynecol Obstet. 1955;100:131–40.
71. Nelson KB, Grether JK. Can magnesium sulfate reduce the risk of cerebral palsy in very low birth weight infants? Pediatrics. 1995;95:263–7.
72. Hauth JC, Goldenberg RL, Nelson KG, et al. Reduction of cerebral palsy with maternal magnesium sulfate treatment in newborns weighing 500-1000 G. Am J Obstet Gynecol. 1995;172:419. Abstract #581.
73. Roberts JM. Magnesium for preeclampsia and eclampsia. N Engl J Med. 1995;333:250–1.
74. Pritchard JA, Cunningham FG, Pritchard SA. The parkland memorial hospital protocol for treatment of eclampsia: evaluation of 245 cases. Am J Obstet Gynecol. 1984;148:951–63.
75. Cotton DB, Hallak M, Janusz C, et al. Central anticonvulsant effects of magnesium sulfate on N-methyl-D-aspartate-induced seizures. Am J Obstet Gynecol. 1993;168:974–8.
76. Belfort MA, Moise KJ. Effect of magnesium sulfate on maternal brain blood flow in preeclampsia: a randomized, placebo-controlled study. Am J Obstet Gynecol. 1992;167:661–6.
77. Kass EH, Schlievert PM, Parsonnet J, et al. Effect of magnesium on production of toxic-shock-syndrome toxin-1. J Infect Dis. 1988;158:44–51.
78. Weglicki WB, Phillips TM, Freedman AM, Cassidy AM, Dickens BF. Magnesium-deficiency elevates circulating levels of inflammatory cytokines and endothelin. Mol Cell Biochem. 1992;110:169–73.
79. Wiles ME, Wagner TL, Weglicki WB. Effect of acute magnesium deficiency (MgD) on aortic endothelial cell (EC) oxidant production. Life Sci. 1997;60:221–36.
80. Weglicki WB, Phillips TM. Pathobiology of magnesium deficiency: a cytokine/neurogenic inflammation hypothesis. Am J Physiol. 1992;263:734–7.
81. Holcberg G, Amash A, Sapir O, Hallak M, Ducler D, Katz M, et al. Different effects of magnesium sulfate and angiotensin II on the capacity of the fetal and maternal compartments of normal human placenta to secrete TNF-α and IL-6. J Reprod Immunol. 2006;69:115–25.
82. Doyle LW, Crowther CA, Middleton P, et al. Magnesium sulphate for women at risk of preterm birth for neuroprotection of the fetus. Cochrane Database Syst Rev. 2009;1:CD004661.
83. Costantine MM, Weiner SJ. Eunice Kennedy Shriver National Institute of Child Health and Human Development Maternal-Fetal Medicine Units Network. Effects of antenatal exposure to magnesium sulfate on neuroprotection and mortality in preterm infants: a meta-analysis. Obstet Gynecol. 2009;114:354.
84. Doyle LW, Crowther CA, Middleton P, Marret S. Antenatal magnesium sulfate and neurologic outcome in preterm infants: a systematic review. Obstet Gynecol. 2009;113:1327.
85. Blumenfeld YJ, Lyell DJ. Curr Opin Obstet Gynecol. 2009;21:136–41.
86. Schneider H, Huch A. Dual in vitro perfusion of an isolated lobe of human placenta: method and instrumentation. Contrib Gynecol Obstet. 1985;13:40–7.
87. Holcberg G, Sapir O, Hallak M, Alaa A, Shorok HY, David Y, et al. Selective vasodilator effect of magnesium sulfate in human placenta. Am J Reprod Immunol. 2004;51(3):192–7.
88. Amash A, Weintraub AY, Sheiner E, Zeadna A, Huleihel M, Holcberg G. Possible therapeutic effect of magnesium sulfate in pre-eclampsia by the down-regulation of placental tumor necrosis factor-alpha secretion. Eur Cytokine Netw. 2010;21(1):58–64.
89. Amash A, Holcberg G, Sheiner E, Huleihel M. Magnesium sulfate normalizes placental interleukin-6 secretion in preeclampsia. J Interferon Cytokine Res. 2010;30(9):683–90.

Section B
Magnesium Status in Disease

Chapter 4
Magnesium Links to Asthma Control

Alexandra Kazaks

Key Points

- Asthma is a serious health problem characterized by symptoms that can vary from mild chest tightness, shortness of breath, and coughing or wheezing to respiratory failure and death.
- Currently, there is no cure for asthma. In addition, current therapies for asthma have limitations, and chronic symptoms remain a problem even with pharmacologic treatment.
- Degree of asthma severity is assessed with tests of pulmonary function and bronchial hyperreactivity, indices of inflammation, and questionnaires to gauge quality of life and asthma control.
- Evidence suggests that magnesium can directly influence lung function as it is implicated in regulating smooth muscle contractility, neuromuscular excitability, immune function, and inflammation and oxidative stress.
- Research indicates that adults with mild-to-moderate asthma may benefit from taking magnesium supplements as most Americans consume less than recommended amount.
- A trial of 6-month supplementation with oral magnesium citrate showed significant improvement in both objective (methacholine challenge) and subjective measures (questionnaires) of and asthma control in the treatment group. No such changes were seen in the placebo group.
- Magnesium supplementation could be a useful complement to asthma treatment as it may reduce airway hyperresponsiveness and result in a subjective perception of improved quality of life and asthma control.

Keywords Asthma • Magnesium supplement • Dietary magnesium • Magnesium status • Methacholine • Quality of life

A. Kazaks, Ph.D. (✉)
Department of Nutrition and Exercise Science, Bastyr University,
14500 Juanita Drive N.E., Kenmore, WA 98028-4966, USA
e-mail: akazaks@bastyr.edu

R.R. Watson et al. (eds.), *Magnesium in Human Health and Disease*, Nutrition and Health,
DOI 10.1007/978-1-62703-044-1_4, © Springer Science+Business Media New York 2013

Asthma Epidemiology, Pathology, Assessment, and Therapy

Significance and Pathology of Asthma

Asthma is a significant public health burden that affects 25.6 million people in the United States [1] and more than 200 million people worldwide [2]. The disease is a major cause of illness and disability. Annual costs due to asthma in the United States have been estimated at more than $20 billion in direct costs for medications and physician and hospital visits and in indirect costs such as lost productivity. In 2008, asthma accounted for 14.2 million lost workdays [3]. Asthma is a disease characterized by episodes of reversible narrowing of the airways in response to a wide range of endogenous and environmental triggers. It occurs in individuals who are predisposed to develop the disease as a result of genetic and environmental factors. Symptoms, caused by inflammation and smooth muscle contraction in the bronchioles, can vary from mild chest tightness, shortness of breath, and coughing or wheezing to respiratory failure and death [4]. Currently, there is no cure for asthma. The disease is managed by taking appropriate medication and minimizing contact with environmental triggers such as air pollution, tobacco smoke, pets, dust mites, cockroach allergens, and mold. Other common causes of asthma exacerbations include respiratory infections, stress, and even exercise [5].

Asthma Assessment

Because no single measurement of asthma control is accurate by itself, multiple measures including tests of pulmonary function and bronchial hyperreactivity, indices of inflammation, and questionnaires to assess quality of life and asthma control should be used to diagnose and evaluate asthma.

Pulmonary Function Testing

Asthma is generally diagnosed based on patterns of symptoms and pulmonary function tests that quantify the volume and speed of inhaled and exhaled breath. Asthma classification is determined by the persistence of symptoms (persistent or intermittent) and their severity (mild to severe). Table 4.1 displays the guidelines for classification of asthma severity from the National Asthma Education and Prevention Program (NAEPP), a group administered and coordinated by the National Heart, Lung, and Blood Institute of the National Institutes of Health.

The pulmonary function tests in the NAEPP classification use spirometry to measure air flow and volume. Customary outcome measures are:

1. Forced vital capacity (FVC)—the maximum volume of air that can be inhaled or exhaled
2. Peak expiratory flow rate (PEFR)—the maximum flow rate that can be generated during a forced exhalation
3. Forced expiratory volume (FEV1)—the maximum volume of air expired in 1 s

If the airways are obstructed, these measurements fall below normal levels.

Because bronchiole narrowing and restricted air movement is the most recognizable feature of asthma, the condition has historically been described as a disease of smooth muscle dysfunction. That concept has evolved, and asthma is now characterized as a disease caused by chronic inflammation of the airways. Asthma breathing difficulties usually happen in discreet episodes called attacks or exacerbations; however, the inflammation underlying asthma is an early and persistent component of the

Table 4.1 Classification of asthma severity by the National Asthma Education and Prevention Program

Classification of asthma severity (youths ≥12 years of age and adults)				
		Persistent		
	Intermittent	Mild	Moderate	Severe
Symptoms	≤2 days/week	>2 days/week but not daily	Daily	Throughout the day
Nighttime awakenings	≤2×/month	3–4×/month	>1×/week but not nightly	Often 7×/week
Medication use for symptom control	≤2 days/week	>2 days/week but not >1×/day	Daily	Several times per day
Interference with normal activity	None	Minor limitation	Some limitation	Extremely limited
Lung function	Normal FEV1 between exacerbations			
Normal FEV1/FVC				
8–19 years 85% 20–39 years 80%	FEV1 >80% Predicted	FEV1 ≥80% Predicted	FEV1 >60% but < 80% predicted	FEV1 <60% Predicted
40–59 years 75% 60–80 years 70%	FEV1/FVC normal	FEV1/FVC normal	FEV1/FVC reduced 5%	FEV1/FVC reduced >5%

Table adapted from Expert Panel Report 3: Guidelines for the Diagnosis and Management of Asthma National Asthma Education and Prevention Program [4]

disease [4]. The NAEPP emphasizes that all individuals with asthma express the inflammatory component of the disease, with specific immune cells as mediators of symptoms. The resulting bronchial hyperresponsiveness, defined as an exaggerated bronchoconstrictor response to a wide variety of stimuli, is the hallmark of the disease. Airway hyperresponsiveness can be measured by inhalation challenge testing with measured doses of drugs such as methacholine or histamine that increase airway irritability and resistance when inflammation is present. The resulting response is expressed as $PC_{20}FEV_1$ which is the concentration of drug that causes a 20% fall in FEV_1 when compared to baseline. These bronchial provocation tests are useful to diagnose and assess severity of asthma [6, 7].

Exhaled Nitric Oxide

While pulmonary function tests can be used to assess airflow and bronchial hyperresponsiveness, they do not directly gauge the degree of underlying airway inflammation. Measurement of exhaled nitric oxide (eNO) has been suggested as a noninvasive and rapid tool to evaluate airway inflammation. The idea that eNO measurements could be useful in asthma treatment came about following observations in the early 1990s that patients with asthma had significantly higher levels of nitric oxide (NO) in exhaled breath. The gas is produced in the respiratory tract by a wide variety of cell types in both the upper and lower respiratory tract [8]. Once produced, NO may diffuse into the airway where it can be measured in exhaled breath [9]. Significant correlations between eNO and lung function parameters [10] and airway hyperresponsiveness [11] have been described in asthma. Other studies, however, found no relationship between eNO and lung function and inflammation [12, 13]. The explanation and physiological relevance of elevated levels of NO in asthma have not been elucidated. It is not known whether NO is protective, harmful, or merely a surrogate marker of some other unknown mechanism in the pathogenesis of asthma.

Quality of Life and Asthma Control

Persistent symptoms such as breathlessness, impaired sleep, or reduced ability to be physically active impact quality of life. The Asthma Quality of Life Questionnaire (AQLQ) and the Asthma Control Questionnaire (ACQ) were developed to measure patient-centered outcomes in an objective and standardized way. The AQLQ is a 32-item questionnaire that measures functional impairments most important for adults with asthma. The items are in four domains that include symptoms, emotions, exposure to environmental stimuli, and activity limitations. Independent studies have demonstrated the validity of the AQLQ [14, 15], and it has been verified that AQLQ strongly correlates with FEV_1 and bronchial hyperreactivity [16]. The Asthma Control Questionnaire is based on criteria (symptoms, airway caliber, beta-agonist use, and FEV_1 % predicted) put forth by international guidelines committees, along with validated markers for determining how well asthma is controlled in asthma patients [17, 18].

Asthma Treatment

Since there is no cure for asthma, current medical treatment focuses on symptom control. Quick relief, or rescue, medications are intended to open airways to immediately improve breathing during an acute exacerbation. Long-term control drugs are taken even when no symptoms are present, to minimize on-going inflammation in the airways. The current therapies for asthma have limitations, and chronic symptoms remain a problem even with pharmacologic treatment. Moreover, common asthma medications such inhaled beta-agonists may be associated with important side effects such as cardiac arrhythmia and other cardiovascular disturbances as well as headache, shakiness, and trouble sleeping. These adverse effects may cause asthma patients to limit the dosage and frequency of medication use. Many individuals with asthma turn to complementary and alternative medical therapies in addition to traditional medical treatments. Dietary and nutritional approaches are the most prevalent and are perceived as the most useful alternative treatment options for asthma [19].

Proposed Mechanisms for Benefits of Mg in Asthma Treatment

Epidemiologic studies have examined the potential effects of various minerals, vitamins, and fatty acids on pulmonary health, and some research groups have attempted to show that the inflammation associated with asthma can be moderated by dietary intervention [20, 21]. At this time, the relative importance of individual nutrients in asthma control is unclear, but Mg is one of a number of dietary constituents that may be significant. Mg can participate in a multitude of physiologic roles as Mg-linked ATP processes activate enzymes which are involved in functions such as intracellular mineral transport, nerve impulse generation, cell membrane electrical potential, muscle contraction, and blood vessel tone. When considering the links between Mg and asthma, evidence suggests that Mg can directly influence lung function as it is implicated in smooth muscle contractility [22], neuromuscular excitability [23], immune function [24], and inflammation and oxidative stress [25]. The interactions of Mg with calcium and its influence on cell membrane properties are likely to be responsible for its bronchodilation and anti-inflammatory effects in asthma.

Mg and Bronchial Relaxation

Increased bronchial hyperreactivity with consequent bronchial smooth muscle contractility is a characteristic of asthma. Mg causes bronchial smooth muscle relaxation in vitro by its action as a calcium antagonist. Mg decreases intracellular calcium by blocking its entry and release from the endoplasmic reticulum and by activating sodium-calcium pumps. Inhibiting interaction of calcium with myosin can result in muscle cell relaxation [22, 26]. Mg has been shown to be a bronchodilator in vivo [27, 28], and clinical data have established that intravenously administered Mg sulfate has a myorelaxant effect during bronchospasm [29–31]. Mg is associated with inhibition of acetylcholine release at motor nerve terminals and reduced muscle fiber excitability [32]. This mechanism may also contribute, at least in part, to the beneficial effect of Mg in asthma.

Mg and Inflammation

Experimental animal models have provided evidence for the anti-inflammatory action of Mg. During Mg deficiency, inflammatory vasoactive agents including interleukin-1, tumor necrosis factor alpha, and interferon gamma are produced in rodents. These immune system agents can stimulate oxygen radical production from endothelial and smooth muscle cells and leukocytes [24]. Mak et al. showed that Mg deficiency in rats results in a rise in the level of circulating substance P, a potent vasodilator whose actions are mediated in part by stimulation of NO production [33, 34]. Another group showed a significant increase in plasma interleukin-6 and of acute phase proteins in Mg-deficient rats when compared to rats fed a control diet [25]. Bois demonstrated that Mg-deficient rats had increased histamine release from mast cells. Mg stabilizes T cells and inhibits mast cell degranulation, thereby reducing inflammatory mediators [35].

In human studies, Cairns demonstrated that clinically relevant concentrations of Mg attenuated the oxidative burst in neutrophils from adult asthma patients [36]. Mak et al. found specific antioxidant enzyme activities in erythrocytes from individuals with asthma as erythrocyte superoxide dismutase and catalase activities were significantly increased compared with controls [34]. They concluded that asthma is accompanied by alterations in systemic antioxidant status due to oxidative stress caused by increased amounts of reactive oxygen species. Increased C-reactive protein (CRP), a systemic marker of inflammation, has been associated with asthma. Data from the Third National Health and Nutrition Examination Survey (NHANES) indicated that individuals with asthma have increased serum CRP concentrations when compared with participants without asthma [37]. Studies suggest that systemic inflammation may be modulated by Mg intake. After taking other risk factors into account, adults with dietary intakes below the Recommended Daily Allowance (RDA) for Mg were as much as 75% more likely to have elevated CRP levels than those who met the recommendation [38]. Data from nearly 12,000 women participating in the Women's Health Study showed an inverse association with Mg intake and systemic inflammation in middle-age and older women [39].

Studies of Mg in Asthma Treatment

Intravenous and Nebulized Mg

Use of Mg for asthma treatment was first published in 1940 as anecdotal reports describing beneficial effects of intravenously (IV) administered Mg during acute asthma exacerbations [40]. More recent studies with IV or inhaled nebulized Mg in acute asthma treatment point to Mg as a possible adjunct for asthma control, although there are contrasting studies that show no effect. Included here are examples of those studies.

IV Mg Positive

In 1990, Noppen et al. reported that patients with severe asthma who had normal serum and intracellular Mg experienced significant bronchodilation when they were infused with Mg sulfate ($MgSO_4$)

[28]. Patients also experienced additive benefit of $MgSO_4$ used with beta-agonist treatment [31]. Rolla and Bucca demonstrated that patients with clinical asthma experienced increased FEV_1 with IV $MgSO_4$ [41]. In a meta-analysis of IV Mg use in children with moderate-to-severe asthmatic attacks, IV Mg did show a benefit in preventing hospitalization in conjunction with standard bronchodilators and steroids [42].

An overall message taking into account results from various studies is that, in severe acute asthma attacks, IV Mg treatment may improve bronchial relaxation when used in addition to conventional treatment [43, 44].

IV Mg Negative

In a large study, Green and Rothrock randomized patients with asthma who were unresponsive to beta-agonist treatment to receive a $MgSO_4$ infusion in addition to standard treatment with steroids and beta-agonists. They found no significant differences in hospitalization rates or in PEFR between the Mg and control groups [45]. Bernstein found that IV Mg was not effective as a bronchodilator medication in patients with chronic, stable asthma nor in normal adults [46]. A review of studies of $MgSO_4$ administered for exacerbations of acute asthma seen in emergency departments found that evidence does not support routine use of IV Mg in all patients with asthma symptoms. It does appear to be safe and beneficial for cases of severe acute asthma by improving peak expiratory flow rate and forced expiratory volume in 1 s [43].

Nebulized Mg

Two double-blind, randomized, controlled trials done by Rolla and Bucca showed that bronchoconstriction and bronchial hyperresponsiveness were inhibited by inhaled $MgSO_4$ [47, 48]. In contrast, Chande and Skoner demonstrated that Mg administered in nebulized form had no bronchodilating effect [49]. In a 2005 Cochrane review of inhaled $MgSO_4$ in acute asthma, aerosolized Mg in combination with a beta-agonist improved pulmonary function and was associated with a trend toward fewer hospital admissions [50].

These studies assessing the capacity of Mg therapy to reduce asthma symptoms have given conflicting results. IV $MgSO_4$ appears to improve pulmonary function in acute asthma, but is not useful in patients with more moderate episodes. It is unclear whether Mg administration merely corrects an underlying deficiency state or whether it has a specific pharmacologic effect.

Mg Status

Some currently used indices of Mg status—total and ionized serum Mg, intracellular or erythrocyte Mg, and IV Mg load testing—have been correlated with asthma occurrence and symptoms. Evidence supporting a role for Mg in the pathogenesis of asthma has been shown by studies demonstrating reduced levels of serum and erythrocyte Mg in subjects with asthma when compared to controls [51]. Patients with chronic asthma and with low serum Mg tended to have more hospitalizations than those with normal Mg levels. Hypomagnesemia was also associated with more severe asthma symptoms [52]. A study by Dominguez in 1998 reported a correlation between low intracellular Mg levels and airway reactivity in subjects with asthma [53]. Lower levels of Mg in skeletal muscle [54] and in polymorphonuclear cells [55] from subjects with asthma have also been reported. Increased retention after a Mg loading test was correlated with increased bronchial reactivity, suggesting either a Mg deficiency or an increased requirement in those with asthma [56].

Other researchers have found no difference in serum or intracellular Mg levels between asthma and control subjects [55, 57]. Falkner et al. measured serum Mg levels in patients during acute asthma exacerbations and found no difference when compared to a control population. The authors concluded that serum Mg measurements are not clinically useful for predicting asthma severity. Altura did not find differences in total serum Mg but did find lower ionized Mg in patients with asthma [58]. Because assessment of Mg status in humans is challenging as common biomarkers are limited with regard to their sensitivity, it may be worthwhile to evaluate a combination of measures.

Epidemiologic studies suggest that environmental factors, including diet, may be implicated in asthma occurrence and severity. Certain population groups, including individuals with asthma, may not be obtaining adequate dietary Mg, and as a result, they have compromised Mg status. There is cross-sectional epidemiologic evidence for protective effects of dietary Mg against asthma. Britton et al. reported that dietary Mg intake is independently related to lung function, airway hyperreactivity, and self-reported wheezing in the general population [59]. In a case-control study done in Scotland, low intakes of Mg were significantly associated with an increased risk of bronchial hyperreactivity [60]. As a result of current food preferences and processing methods, inadequate dietary Mg intake is common. According to the NHANES for 1999–2000, less than half of American adults consumed recommended levels of Mg [61].

Dietary Mg Supplementation

Evidence suggests that high dietary Mg intake may be associated with better lung function and reduced bronchial reactivity, while studies examining the effects of oral supplements to increase dietary Mg in adults with asthma have shown little or no improvement in pulmonary function or measures of inflammation. A 3-week intervention using daily 400-mg oral Mg supplementation demonstrated improvement only in subjective symptoms and bronchodilator use that were of borderline clinical significance as indicators of asthma control [62]. However, neither this study nor a 12-week trial of 450-mg oral Mg per day showed improvement in lung function or bronchial reactivity. The authors concluded that oral Mg added no clinical benefit to standard outpatient therapy for chronic stable asthma in adults [63]. It is possible that these study periods may not have been of sufficient length to see changes in Mg stores required to affect asthma control.

Mg Status in People with Asthma Versus Healthy Controls

Following here are accounts of a Mg status study and a Mg supplement study that were undertaken in order to more fully understand the role of dietary Mg in people with mild-to-moderate asthma. In the Mg status study, the aim was to investigate five measures of Mg status in people with asthma compared to healthy controls. Details about the study participants and methods are available in the published article [64]. Briefly, Mg status was measured with total and ionized serum Mg, erythrocyte Mg, Mg loading test, and dietary intake of Mg from food and water. Asthma status was assessed with pulmonary function tests including methacholine challenge testing (MCCT).

Average ionized Mg and erythrocyte Mg levels and IV Mg load retention in all subjects were within the expected normal reference ranges. However, 9% of asthma subjects and 12% of controls had serum Mg levels of less than 0.70 mmol/L; therefore, they can be described as being Mg-deficient [65]. When the serum Mg data were divided into tertiles (low < 0.64 mmol/L; high > 0.80 mmol/L), there were no significant differences between the low and high tertiles in any lung function parameter when groups were analyzed separately or together. Because dietary requirements for Mg differ according to age and sex, the percentage Recommended Daily Allowance (RDA) for Mg is a more meaningful indicator of adequate intake than milligrams Mg per day. Although there was a wide variation in Mg intake, values indicated that, on average, both the asthma and control subjects consumed less than the RDA.

Table 4.2 Magnesium status indices in healthy controls and participants with asthma

Mg status index	Control	Asthma
iMg (mmol/L)	0.58±0.01	0.57±0.01
S-Mg (mmol/L)	0.77±0.01	0.76±0.01
R-Mg (mmol/L)	2.8±0.1	2.8±0.1
IV Mg load retention (%)	−19±5	−24±4
Average diet Mg (mg)	326±15	306±17
% Mg RDA	94±5	89±5

Mean±SE
iMg ionized magnesium, *S-Mg* total serum magnesium, *R-Mg* erythrocyte magnesium, *% Mg RDA* percentage of recommended daily allowance for age and sex
Adapted from Kazaks et al. [64]

As shown in Table 4.2, there were no significant differences in any biochemical or dietary measure of Mg status between asthma and control subjects. These results support the findings of other studies that showed no association of serum or erythrocyte Mg with asthma [55, 57, 66, 67]; however, they were in contrast with studies that reported lower Mg status in people with asthma [51, 56, 68, 69]. Additionally, asthma control measured by pulmonary function tests was not correlated with any measure of Mg status. While the results were similar to research that showed no Mg-asthma relationship [51, 69], they differed from studies that reported a negative correlation between airway reactivity and erythrocyte Mg [53] and between serum Mg and severity of asthma [52].

The reasons for the inconsistency in results from various studies may be due to differences in age, sex, and severity of asthma among subjects. Participants in the Mg status study were matched for sex and age, and an MCCT was used to help confirm presence or absence of asthma. Asthma subjects had mild-to-moderate asthma that was controlled only with inhaled beta-agonists or inhaled corticosteroids, and they may have been medically managed to the point that variation in Mg status did not have an obvious effect on lung function.

Mg Supplementation Versus Placebo in People with Asthma

The aim of the Mg supplement study was to examine the effect of oral Mg supplementation on asthma control and Mg status. The hypothesis was that 6.5 months of treatment with oral Mg would show improved markers of asthma control and increased measures of Mg status in men and women with mild-to-moderate asthma. Asthma control was measured with a combination of subjective and objective assessments including quality of life and asthma control questionnaires, pulmonary function tests, bronchial reactivity to methacholine, and bronchial and systemic indices of inflammation. In-depth information about the participant selection and methods are described elsewhere [70]. In summary, the study was a placebo-controlled, double-blind, parallel group trial in which subjects were randomly assigned to consume 340 mg (14.0 mmol) of Mg as Mg citrate per day or a placebo. The primary outcome was change in bronchial responsiveness from baseline measured by an MCCT. Secondary outcomes were changes in pulmonary function tests, subjective measures of asthma control, indices of bronchial and systemic inflammation, and Mg status.

Bronchial Responsiveness and Pulmonary Function

After 6.5 months of Mg supplementation, the within individual decrease in bronchial responsiveness to methacholine observed in the Mg-supplemented group was statistically and clinically significant. As shown in Fig. 4.1, there was also a significant between-group difference in methacholine PC_{20} over the study period. These data indicate that participants who received the Mg supplement had a

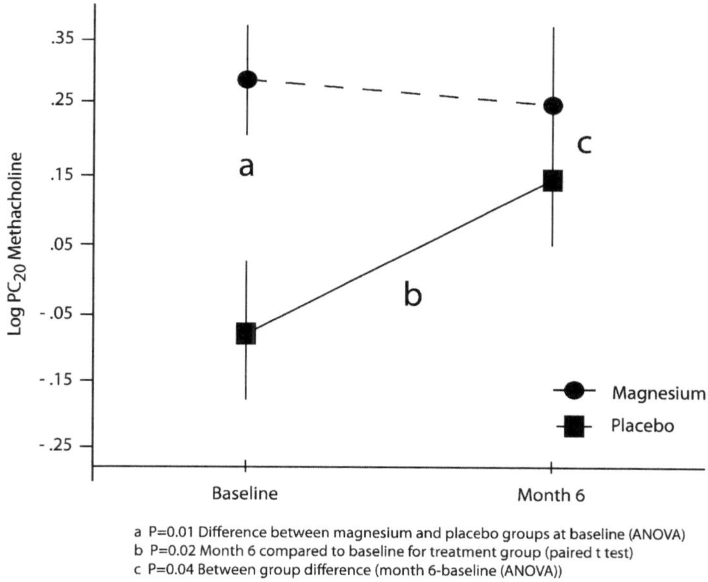

Fig. 4.1 Methacholine challenge tests differences between placebo group and magnesium-treated participants with asthma measured at baseline and month 6

reduction in airway reactivity while there was no significant change in bronchial responsiveness to methacholine among the placebo subjects.

The Mg supplement study results are consistent with another trial that reported a decrease in bronchial reactivity in a 2-month randomized placebo-controlled trial using 400 mg/day oral Mg supplementation in 37 children with moderate, persistent asthma [71]. The results are at variance with previous 3-week [62] and 16-week [63] studies of Mg supplementation in adults with asthma that showed no improvement in MCCT results.

Pulmonary Function

While there were no significant changes in FEV1 or FVC, long-term Mg supplementation increased predicted PEFR by 5.8% over time with no difference in the placebo group, providing further evidence of improved airway resistance in the Mg-treated group. The PEFR is an important and frequently used clinical measure of airway obstruction as it gauges ability to efficiently expel air from the lungs. There were no other significant pulmonary function changes in either group, which is consistent with other reports of Mg supplementation in adults and children with asthma [62, 63, 71]. Given that mean pulmonary test values for both the Mg and placebo groups at enrollment indicated that study participants already had reasonably well-managed asthma, this level of asthma control may have limited the potential of Mg supplementation to significantly improve pulmonary function.

Assessment of Quality of Life and Asthma Control

Physiologic appraisals alone cannot provide a complete view of the impact of asthma on activity limits, sleep impairment, and emotional functioning [72]. Patients with asthma vary in their personal perception of severity of symptoms compared with quantitative measures. For those reasons,

United States and international guidelines on asthma treatment stress that assessment of control should include subjective information about patient level of function and asthma symptoms in addition to objective measures of lung function such as spirometry. During the Mg supplement study, the ACQ and the AQLQ instruments were employed as two measures of the perceived effect of asthma on quality of life and individual control of the disease. After 6.5 months of Mg supplementation, there were significant differences between Mg and placebo subjects in control of morning symptoms, nocturnal waking, activity limitations, shortness of breath, wheezing, and beta-agonist use as measured by the ACQ.

The improvement in these symptoms was consistent with a study by Hill et al. [62] in which symptom scores were significantly lower during the Mg treatment period in subjects with stable asthma. In responses on the AQLQ, all participants at enrollment indicated that asthma caused some limitation to their quality of life "only a little of the time." During the trial, both groups reported increased exposure to asthma exacerbation triggers (described as dust, smoke, weather, and air pollution). Despite this increase in environmental stimulation of inflammation and airway responsiveness, subjects who were supplemented with Mg had significantly improved quality of life measured by the AQLQ. As has been reported in previous research [16], this study data showed that increases in AQLQ scores were correlated with improvements in bronchial hyperreactivity.

Measures of Inflammation

Since elevated concentrations of airway nitric oxide have been associated with asthma exacerbations and disease severity [73], eNO was used as a noninvasive marker to estimate the effect of Mg supplementation on airway inflammation. At baseline, 80% of subjects receiving placebo and 84% of subjects receiving Mg had eNO concentrations above the 20-ppb reference value associated with presence of asthma [74]. While the results showed an inverse correlation between eNO and bronchial responsiveness for all subjects at enrollment, there was no relationship at month 6.5 in either study group. In addition, there were no significant differences or changes within or between groups in eNO concentrations. Attributing an eNO response to a dietary intervention such as Mg is complicated. Because NO is a free radical that reacts with oxidants and antioxidants, its presence in exhaled breath reflects a balance between rates of NO production and consumption [75], and factors other than asthma inflammation such as sex or increased body fat may play a role [76].

As there were no prospective asthma studies that correlated Mg supplementation with CRP levels, high-sensitivity C-reactive protein measures were used as an additional method to assess systemic inflammation. Mean baseline CRP concentrations for both the placebo and Mg groups implied some increased inflammation status but, at month 6.5, there were no significant differences within or between groups in CRP concentrations. CRP is probably too broad a test of inflammation to detect modest effects of Mg on airway inflammation.

Measures of Mg Status

Mean serum Mg values were within normal limits for both groups in the Mg supplement study. Thirteen percent of participants (3 in the Mg group and 4 in the placebo group) had serum Mg below <0.7 mmol/L at baseline. Although there was no improvement in measures of Mg status in participants after 6.5 months of Mg supplementation, finding no change in individuals who generally have adequate Mg status is consistent with past research. Feillet-Coudray reported no significant improvement in serum or erythrocyte Mg after 8 weeks of 366 mg per day Mg supplementation in healthy women. However, there was a small increase in plasma ionized Mg [77]. Similarly, there was no significant alteration in serum Mg in a study of Mg supplementation in children after consumption of

300 mg of Mg per day for 2 months [71], nor in a trial of 300 men and women who received 450 mg/day Mg chelate for 16 weeks [63]. Different results were obtained in a crossover study of 17 people with asthma who were restricted to a low Mg diet (100–200 mg/day). Serum Mg was higher at the end of the 400 mg/day Mg treatment than after the placebo period [62].

Mg Intake from Food

The average percentage RDA obtained from food was 84% at baseline for all participants in the Mg supplement study, and only 12 of 52 participants consumed 100% of the RDA for Mg. The treatment dose of 340-mg Mg combined with dietary intake predictably put all subjects in the supplemented group above 100% of the Mg RDA. (The tolerable upper intake level (UL) for Mg is set at 350 mg per day; however, this refers only to the amount taken in dietary supplements and not that obtained from food.) In contrast, in the placebo group, 4 of 25 subjects met 100% of the RDA at baseline and 6 of 25 met it at month 6.5. In neither group did the daily Mg intake from food change significantly over the period of the study. It is challenging to explain the absence of a relationship between Mg supplementation and indices of Mg status in view of the improvement in airway resistance and subjective measures of asthma control and asthma quality of life. Participants showed no significant change in Mg status over the course of the study; nevertheless, those in the treatment group still had an apparent benefit from Mg supplementation. The asthma disease process may create an increased Mg need above normal stores, or Mg may be shifted to pools that were not assessed in this trial. Bronchial and quality of life improvements shown in this study may be due to some therapeutic effect of consistent and adequate Mg exposure.

Conclusion

The pathophysiology of asthma is not fully understood, considering that it is a complex disease with genetic, immune system, environmental, behavioral, and nutritional components. Epidemiologic studies have implicated Mg in asthma management as the mineral is associated with bronchodilation, immune function, and anti-inflammatory properties. While studies of Mg supplementation have shown mixed results in ability to modify Mg status and control of asthma symptoms, Mg status is not clinically easy to detect, and increased need for Mg may exist in people with asthma. Mg may be a useful complement to medical treatment of asthma as it may reduce airway hyperresponsiveness, increase PEFR, and result in a subjective perception of improved quality of life and asthma control.

References

1. NCHS. Asthma prevalence, health care use, and mortality: United States, 2005–2009. 2011. http://www.cdc.gov/nchs/data/nhsr/nhsr032.pdf. Accessed 30 July 2011.
2. WHO. Asthma fact sheet. 2011;Number 307. http://www.who.int/mediacentre/factsheets/fs307/en/index.html. Accessed 20 Aug 2011.
3. ALA. Asthma in adults fact sheet. 2010. http://www.lungusa.org/lung-disease/asthma/resources/facts-and-figures/asthma-in-adults.html. Accessed 20 Aug 2011.
4. NAEPP. National Asthma Education and Prevention Program. Expert panel report 3: guidelines for the diagnosis and management of asthma. 2007(NIH Pub No 07–4051). http://www.nhlbi.nih.gov/guidelines/asthma/. Accessed 10 Aug 2011.
5. EPA. Asthma prevalence 2007. 2011. http://cfpub.epa.gov/eroe/index.cfm?fuseaction=detail.viewInd&lv=list.listbyalpha&r=235294&subtop=381. Accessed 2 June 2011.

6. Crapo RO, Casaburi R, Coates AL, et al. Guidelines for methacholine and exercise challenge testing-1999. This official statement of the American Thoracic Society was adopted by the ATS Board of Directors, July 1999. Am J Respir Crit Care Med. 2000;161(1):309–29.
7. Hunter CJ, Brightling CE, Woltmann G, Wardlaw AJ, Pavord ID. A comparison of the validity of different diagnostic tests in adults with asthma. Chest. 2002;121(4):1051–7.
8. Leone AM, Gustafsson LE, Francis PL, Persson MG, Wiklund NP, Moncada S. Nitric oxide is present in exhaled breath in humans: direct GC-MS confirmation. Biochem Biophys Res Commun. 1994;201(2):883–7.
9. Dweik RA, Comhair SA, Gaston B, et al. NO chemical events in the human airway during the immediate and late antigen-induced asthmatic response. Proc Natl Acad Sci U S A. 2001;98(5):2622–7.
10. Kharitonov SA, Barnes PJ. Clinical aspects of exhaled nitric oxide. Eur Respir J. 2000;16(4):781–92.
11. Dupont LJ, Rochette F, Demedts MG, Verleden GM. Exhaled nitric oxide correlates with airway hyperresponsiveness in steroid-naive patients with mild asthma. Am J Respir Crit Care Med. 1998;157(3 Pt 1):894–8.
12. Berlyne GS, Parameswaran K, Kamada D, Efthimiadis A, Hargreave FE. A comparison of exhaled nitric oxide and induced sputum as markers of airway inflammation. J Allergy Clin Immunol. 2000;106(4):638–44.
13. Franklin PJ, Turner SW, Le Souef PN, Stick SM. Exhaled nitric oxide and asthma: complex interactions between atopy, airway responsiveness, and symptoms in a community population of children. Thorax. 2003;58(12):1048–52.
14. Juniper EF, Guyatt GH, Cox FM, Ferrie PJ, King DR. Development and validation of the mini asthma quality of life questionnaire. Eur Respir J. 1999;14(1):32–8.
15. Juniper EF, Wisniewski ME, Cox FM, Emmett AH, Nielsen KE, O'Byrne PM. Relationship between quality of life and clinical status in asthma: a factor analysis. Eur Respir J. 2004;23(2):287–91.
16. Emel'ianov AV, Zinakova MK, Krasnoshchekova OI, Sinitsina TM, Rudinskii KA, Fedulov AV. Quality of life and characteristics of the external respiration function in patients with bronchial asthma. Ter Arkh. 2001;73(12): 63–5.
17. Juniper EF, O'Byrne PM, Ferrie PJ, King DR, Roberts JN. Measuring asthma control. Clinic questionnaire or daily diary? Am J Respir Crit Care Med. 2000;162(4 Pt 1):1330–4.
18. Juniper EF, O'Byrne PM, Roberts JN. Measuring asthma control in group studies: do we need airway calibre and rescue beta2-agonist use? Respir Med. 2001;95(5):319–23.
19. Blanc PD, Trupin L, Earnest G, Katz PP, Yelin EH, Eisner MD. Alternative therapies among adults with a reported diagnosis of asthma or rhinosinusitis: data from a population-based survey. Chest. 2001;120(5):1461–7.
20. Smit HA, Grievink L, Tabak C. Dietary influences on chronic obstructive lung disease and asthma: a review of the epidemiological evidence. Proc Nutr Soc. 1999;58(2):309–19.
21. Fogarty A, Britton J. Nutritional issues and asthma. Curr Opin Pulm Med. 2000;6(1):86–9.
22. Spivey WH, Skobeloff EM, Levin RM. Effect of magnesium chloride on rabbit bronchial smooth muscle. Ann Emerg Med. 1990;19(10):1107–12.
23. Rude RK. Magnesium deficiency: a cause of heterogeneous disease in humans. J Bone Miner Res. 1998;13(4):749–58.
24. Weglicki WB, Phillips TM, Freedman AM, Cassidy MM, Dickens BF. Magnesium-deficiency elevates circulating levels of inflammatory cytokines and endothelin. Mol Cell Biochem. 1992;110(2):169–73.
25. Malpuech-Brugere C, Nowacki W, Daveau M, et al. Inflammatory response following acute magnesium deficiency in the rat. Biochim Biophys Acta. 2000;1501(2–3):91–8.
26. Iseri LT, French JH. Magnesium: nature's physiologic calcium blocker. Am Heart J. 1984;108(1):188–93.
27. Skobeloff EM, Spivey WH, McNamara RM, Greenspon L. Intravenous magnesium sulfate for the treatment of acute asthma in the emergency department. JAMA. 1989;262(9):1210–3.
28. Noppen M, Vanmaele L, Impens N, Schandevyl W. Bronchodilating effect of intravenous magnesium sulfate in acute severe bronchial asthma. Chest. 1990;97(2):373–6.
29. Okayama H, Aikawa T, Okayama M, Sasaki H, Mue S, Takishima T. Bronchodilating effect of intravenous magnesium sulfate in bronchial asthma. JAMA. 1987;257(8):1076–8.
30. Rolla G, Bucca C. Magnesium, beta-agonists, and asthma. Lancet. 1988;1(8592):989.
31. Bloch H, Silverman R, Mancherje N, Grant S, Jagminas L, Scharf SM. Intravenous magnesium sulfate as an adjunct in the treatment of acute asthma. Chest. 1995;107(6):1576–81.
32. Del Castillo J, Engbaek L. The nature of the neuromuscular block produced by magnesium. J Physiol. 1954;124(2): 370–84.
33. Mak IT, Komarov AM, Wagner TL, Stafford RE, Dickens BF, Weglicki WB. Enhanced NO production during Mg deficiency and its role in mediating red blood cell glutathione loss. Am J Physiol. 1996;271(1 Pt 1):C385–90.
34. Mak JC, Leung HC, Ho SP, et al. Systemic oxidative and antioxidative status in Chinese patients with asthma. J Allergy Clin Immunol. 2004;114(2):260–4.
35. Bois P. Effect of magnesium deficiency on mast cells and urinary histamine in rats. Br J Exp Pathol. 1963;44:151–5.
36. Cairns CB, Kraft M. Magnesium attenuates the neutrophil respiratory burst in adult asthmatic patients. Acad Emerg Med. 1996;3(12):1093–7.

37. Ford ES. Asthma, body mass index, and C-reactive protein among US adults. J Asthma. 2003;40(7):733–9.
38. King DE, Mainous 3rd AG, Geesey ME, Woolson RF. Dietary magnesium and C-reactive protein levels. J Am Coll Nutr. 2005;24(3):166–71.
39. Song Y, Ridker PM, Manson JE, Cook NR, Buring JE, Liu S. Magnesium intake, C-reactive protein, and the prevalence of metabolic syndrome in middle-aged and older U.S. women. Diabetes Care. 2005;28(6):1438–44.
40. Haury V. Blood serum magnesium in bronchial asthma and its treatment by the administration of magnesium sulphate. J Lab Clin Med. 1940;26:340–1.
41. Rolla G, Bucca C, Brussino L, Colagrande P. Effect of intravenous magnesium infusion on salbutamol-induced bronchodilatation in patients with asthma. Magnes Res. 1994;7(2):129–33.
42. Cheuk DK, Chau TC, Lee SL. A meta-analysis on intravenous magnesium sulphate for treating acute asthma. Arch Dis Child. 2005;90(1):74–7.
43. Rowe BH, Bretzlaff JA, Bourdon C, Bota GW, Camargo Jr CA. Intravenous magnesium sulfate treatment for acute asthma in the emergency department: a systematic review of the literature. Ann Emerg Med. 2000;36(3):181–90.
44. Alter HJ, Koepsell TD, Hilty WM. Intravenous magnesium as an adjuvant in acute bronchospasm: a meta-analysis. Ann Emerg Med. 2000;36(3):191–7.
45. Green SM, Rothrock SG. Intravenous magnesium for acute asthma: failure to decrease emergency treatment duration or need for hospitalization. Ann Emerg Med. 1992;21(3):260–5.
46. Bernstein WK, Khastgir T, Khastgir A, et al. Lack of effectiveness of magnesium in chronic stable asthma. A prospective, randomized, double-blind, placebo-controlled, crossover trial in normal subjects and in patients with chronic stable asthma. Arch Intern Med. 1995;155(3):271–6.
47. Rolla G, Bucca C, Arossa W, Bugiani M. Magnesium attenuates methacholine-induced bronchoconstriction in asthmatics. Magnesium. 1987;6(4):201–4.
48. Rolla G, Bucca C, Bugiani M, Arossa W, Spinaci S. Reduction of histamine-induced bronchoconstriction by magnesium in asthmatic subjects. Allergy. 1987;42(3):186–8.
49. Chande VT, Skoner DP. A trial of nebulized magnesium sulfate to reverse bronchospasm in asthmatic patients. Ann Emerg Med. 1992;21(9):1111–5.
50. Blitz M, Blitz S, Hughes R, et al. Aerosolized magnesium sulfate for acute asthma: a systematic review. Chest. 2005;128(1):337–44.
51. Zervas E, Papatheodorou G, Psathakis K, Panagou P, Georgatou N, Loukides S. Reduced intracellular Mg concentrations in patients with acute asthma. Chest. 2003;123(1):113–8.
52. Alamoudi OS. Hypomagnesaemia in chronic, stable asthmatics: prevalence, correlation with severity and hospitalization. Eur Respir J. 2000;16(3):427–31.
53. Dominguez LJ, Barbagallo M, Di Lorenzo G, et al. Bronchial reactivity and intracellular magnesium: a possible mechanism for the bronchodilating effects of magnesium in asthma. Clin Sci (Lond). 1998;95(2):137–42.
54. Gustafson T, Boman K, Rosenhall L, Sandstrom T, Wester PO. Skeletal muscle magnesium and potassium in asthmatics treated with oral beta 2-agonists. Eur Respir J. 1996;9(2):237–40.
55. Fantidis P, Ruiz Cacho J, Marin M, Madero Jarabo R, Solera J, Herrero E. Intracellular (polymorphonuclear) magnesium content in patients with bronchial asthma between attacks. J R Soc Med. 1995;88(8):441–5.
56. Emelyanov A, Fedoseev G, Barnes PJ. Reduced intracellular magnesium concentrations in asthmatic patients. Eur Respir J. 1999;13(1):38–40.
57. de Valk HW, Kok PT, Struyvenberg A, et al. Extracellular and intracellular magnesium concentrations in asthmatic patients. Eur Respir J. 1993;6(8):1122–5.
58. Altura BM, Altura BT. Role of magnesium in patho-physiological processes and the clinical utility of magnesium ion selective electrodes. Scand J Clin Lab Invest Suppl. 1996;224:211–34.
59. Britton J, Pavord I, Richards K, et al. Dietary magnesium, lung function, wheezing, and airway hyperreactivity in a random adult population sample. Lancet. 1994;344(8919):357–62.
60. Soutar A, Seaton A, Brown K. Bronchial reactivity and dietary antioxidants. Thorax. 1997;52(2):166–70.
61. Ford ES, Mokdad AH. Dietary magnesium intake in a national sample of US adults. J Nutr. 2003;133(9):2879–82.
62. Hill J, Micklewright A, Lewis S, Britton J. Investigation of the effect of short-term change in dietary magnesium intake in asthma. Eur Respir J. 1997;10(10):2225–9.
63. Fogarty A, Lewis SA, Scrivener SL, et al. Oral magnesium and vitamin C supplements in asthma: a parallel group randomized placebo-controlled trial. Clin Exp Allergy. 2003;33(10):1355–9.
64. Kazaks AG, Uriu-Adams JY, Albertson TE, Stern JS. Multiple measures of magnesium status are comparable in mild asthma and control subjects. J Asthma. 2006;43(10):783–8.
65. Elin RJ. Assessment of magnesium status. Clin Chem. 1987;33(11):1965–70.
66. Falkner D, Glauser J, Allen M. Serum magnesium levels in asthmatic patients during acute exacerbations of asthma. Am J Emerg Med. 1992;10(1):1–3.
67. Picado C, Deulofeu R, Lleonart R, et al. Dietary micronutrients/antioxidants and their relationship with bronchial asthma severity. Allergy. 2001;56(1):43–9.

68. Hashimoto Y, Nishimura Y, Maeda H, Yokoyama M. Assessment of magnesium status in patients with bronchial asthma. J Asthma. 2000;37(6):489–96.
69. Oladipo OO, Chukwu CC, Ajala MO, Adewole TA, Afonja OA. Plasma magnesium in adult asthmatics at the Lagos University Teaching Hospital, Nigeria. East Afr Med J. 2003;80(9):488–91.
70. Kazaks AG, Uriu-Adams JY, Albertson TE, Shenoy SF, Stern JS. Effect of oral magnesium supplementation on measures of airway resistance and subjective assessment of asthma control and quality of life in men and women with mild to moderate asthma: a randomized placebo controlled trial. J Asthma. 2010;47(1):83–92.
71. Gontijo-Amaral C, Ribeiro MA, Gontijo LS, Condino-Neto A, Ribeiro JD. Oral magnesium supplementation in asthmatic children: a double-blind randomized placebo-controlled trial. Eur J Clin Nutr. 2007;61(1):54–60.
72. Masoli M, Fabian D, Holt S, Beasley R. The global burden of asthma: executive summary of the GINA Dissemination Committee report. Allergy. 2004;59(5):469–78.
73. Zeidler MR, Kleerup EC, Tashkin DP. Exhaled nitric oxide in the assessment of asthma. Curr Opin Pulm Med. 2004;10(1):31–6.
74. Olivieri M, Talamini G, Corradi M, et al. Reference values for exhaled nitric oxide (reveno) study. Respir Res. 2006;7:94.
75. Dweik RA, Sorkness RL, Wenzel S, et al. Use of exhaled nitric oxide measurement to identify a reactive, at-risk phenotype among patients with asthma. Am J Respir Crit Care Med. 2010;181(10):1033–41.
76. Kazaks A, Uriu-Adams JY, Stern JS, Albertson TE. No significant relationship between exhaled nitric oxide and body mass index in people with asthma. J Allergy Clin Immunol. 2005;116(4):929–30. author reply 930.
77. Feillet-Coudray C, Coudray C, Tressol JC, et al. Exchangeable magnesium pool masses in healthy women: effects of magnesium supplementation. Am J Clin Nutr. 2002;75(1):72–8.

Chapter 5
Magnesium and Kidney Disease

Ioannis P. Tzanakis and Dimitrios G. Oreopoulos

Key Points

- Magnesium plays an important role in the regulation of the neuromuscular and the cardiovascular system.
- The kidneys are of the main determinants of magnesium metabolism.
- Available data suggest that low magnesium is associated with atherosclerosis and cardiovascular diseases in general population and in patients with renal disease.
- Large clinical studies are needed to prove if high magnesium has a cardio-protective action.

Key words Magnesium • Kidney • Kidney disease • Atherosclerosis

Introduction

Magnesium plays a critical role in cellular physiology in humans, regulating many fundamental functions. The kidneys are the main determinants of magnesium metabolism modulating along with the alimentary ingestion and the intestinal absorption, the total body magnesium burden as well as its serum levels in humans, both in healthy people and in patients with renal disease. In the general population, its deficiency is associated with many disorders, the most significant being cardiovascular diseases. Patients with end-stage renal disease (ESRD) are usually hypermagnesaemic but without clinical manifestations. The role of magnesium in many aspects of ESRD may be also very important, although this has not been explored in depth. The most significant of these roles is the use of magnesium salts as phosphate binders and its ability to inhibit the development or progression of vascular

I.P. Tzanakis, M.D., Ph.D. (✉)
Department of Nephrology, General Hospital of Chania, Greece,
18, Michali Mefa st, PC 73100, Chania, Crete, Greece

Nephrological Department, Gen. Hospital of Chania, PC 73300, Mournies, Greece
e-mail: ioatzan@gmail.com

D.G. Oreopoulos, M.D., Ph.D., FRCPC, FACP
Department of Nephrology, University Health Network and University of Toronto,
399 Bathurst st., Toronto, ON M5T 2S8, Canada
e-mail: dgo@teleglobal.ca

R.R. Watson et al. (eds.), *Magnesium in Human Health and Disease*, Nutrition and Health,
DOI 10.1007/978-1-62703-044-1_5, © Springer Science+Business Media New York 2013

calcification and atherosclerosis. Data provided by recent studies on these issues have promoted promising renewed interest in the role of magnesium in ESRD and its possible favourable therapeutic application in these patients. Further large studies are needed to establish its efficacy and safety.

General

Magnesium (Mg) is the fourth abundant extracellular cation and the second abundant intracellular cation in humans [1]. The total body magnesium of an adult is approximately 21–28 g of Mg [1]. Of the body's Mg, the vast majority of this ion is in the intracellular compartment (99%), and the remaining 1% is in the extracellular fluid [2]. The distribution of the intracellular magnesium in the several tissues is as follows: bones, 60–65%; skeletal muscle, 25–30%; and other nonmuscle soft tissues, 10–15% [2]. In the bones, two distinct pools, cortical and trabecular, have been described. It is thought that Mg forms a surface constituent of the hydroxyapatite mineral component. Initially, much of this magnesium is readily exchangeable with serum and therefore represents a moderately accessible magnesium store, which can be drawn on in times of hypomagnesaemia. The majority of the intracellular Mg is bound to several chelators, such as citrate, proteins, adenosine diphosphate and triphosphate, RNA and DNA. Only 5–10% is free, which is essential for the regulation of the intracellular Mg content [3]. In the clinical practice, intracellular magnesium may be calculated in the skeletal muscles, in bones, in platelets and in peripheral lymphocytes [4].

Of the serum magnesium, 60–65% is free-ionized and filterable and the rest 35–40% is bound with proteins, phosphates, citrates, etc., so is not filterable [5]. The ionized fraction of the serum magnesium is the biologically active form of this element.

Serum magnesium concentrations can be reported as mmol/l, meq/l or mg/dl. The normal serum values of magnesium range 0.75–1.00 mmol/l = 1.50–2.00 mEq/l = 1.80–2.40 mg/dl. Of note that like all predominately intracellular ions, the serum magnesium concentration does not necessarily reflect the total body storages.

Biological action of magnesium has three important roles in human biological systems [6, 7]:

1. It is the basic "biological competitor" of calcium, antagonizing it in binding in cellular membranes and proteins.
2. It is a fundamental cofactor in more than 300 essential enzymatic reactions related to the transportation, storage and use of energy.
3. It regulates the passage of electrolytes and other substances through cellular membranes.

As a consequence of these actions at the cellular level, magnesium has multiple physiological functions. It interferes with the neuromuscular conductivity, muscular contraction and relaxation and arterial pressure and peripheral blood flow [6, 7].

Magnesium is a substance handled mainly by the kidneys so it should be of interest to nephrologists. The kidneys are the main determinants of magnesium metabolism regulating along with the alimentary ingestion and the intestinal absorption, the total body magnesium burden as well as its serum levels in humans, both in healthy people and in patients with renal disease.

Approximately 60–65% of Mg in plasma is filtered by the glomeruli, of which 15–20% is reabsorbed in the proximal tubule and 65–75% is reabsorbed in the thick ascending limb by a passive paracellular mechanism. Another 5–10% is absorbed by an active transcellular mechanism in the distal convoluted tubule. Only 3–5% is excreted in the urine. This active reabsorption of Mg is affected by the intracellular and extracellular Mg concentrations, the acid–base balance and drugs including estradiols, cyclosporines, thiazide diuretics, etc. [8]. The ultimate proportion of the filtered magnesium which is capable of being excreted by the kidneys (fractional excretion) may decrease to 0.5%

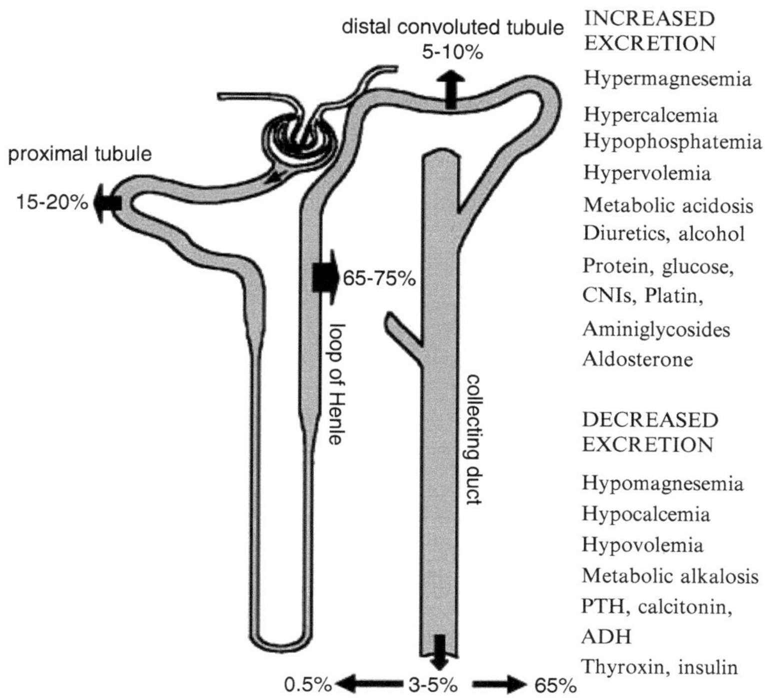

Fig. 5.1 Schematic representation of renal magnesium transfer

Table 5.1 Magnesium and "nephrology interest"

MG in ESRD-hypermagnesaemia
ECG abnormalities and clinical signs if sMG >2.00 mmol/l or 4.80 mg/dl
MG salts as phosphate binders
MG and atherosclerosis
MG and arterial calcifications
MG and arterial pressure, diabetes, lipid disorders and metabolic syndrome
MG and PTH excretion
MG and intradialytic haemodynamic stability
Mg and calcineurin inhibitors

in case of magnesium depletion or increase up to 65% in chronic kidney disease (CKD) [9]. Representation of magnesium renal transport is seen in Fig. 5.1.

Magnesium plays a critical role in cellular physiology in humans, regulating many fundamental functions. In the general population, its deficiency is associated with many clinical disorders, one of the most significant being cardiovascular diseases. The role of magnesium in many aspects of ESRD may be also very important, although this has not been explored in depth. Few of these issues have been investigated adequately. The most significant are magnesium as a phosphate binder and the relationship of magnesium with cardiac annular, vascular and coronary artery calcification, with intradialytic haemodynamic stability, with parathyroid hormone (PTH) and with calcineurin inhibitors in transplanted patients (see "Table 5.1"). Moreover, magnesium deficiency has been found to be associated with diabetes mellitus, hypertension, lipid disorders and metabolic syndrome; the entire above are established risk factors for cardiovascular as well as renal disease.

Magnesium in Chronic Kidney Disease

Magnesium Balance in Chronic Kidney Disease

In CKD, serum magnesium is maintained in normal ranges until glomerular filtration rate (GFR) falls below 30 ml/min due to the increased renal Mg fractional excretion [9, 10]. As CKD progresses and GFR is decreased under 30 ml/min, urinary Mg excretion is usually inadequate to balance the intestinal Mg absorption, so hypermagnesaemia may develop [2, 8]. In ESRD, both serum and total magnesium are increased [11]. Renal failure is the most common cause of hypermagnesaemia, which is usually mild and asymptomatic even in ESRD patients [12]. Occasionally, hypermagnesaemia may also develop in patients with normal renal function in whom large amount of magnesium is administered parenterally for therapeutic reasons.

In ESRD patients, either on haemodialysis or on continuous ambulatory peritoneal dialysis (CAPD), serum magnesium levels are usually higher than 1.00 mmol/l (2.00 mEq/l or 2.40 mg/dl). However, there is a greater rise in total than that in ionic Mg in the ESRD patients; this may occur because of a higher level of Mg complexing with proteins, phosphate and other anions in this setting [5, 11, 13]. On the opposite, in another study, Dewitte et al. demonstrated that the ionized fraction of total serum magnesium was not different in the haemodialysis patients compared to healthy controls [14]. The mean serum total magnesium value of 646 prevalent haemodialysis patients of our renal unit of the last 15 years is 2.71 ± 0.24 mg (anecdotal data); the distribution of these mean serum total Mg values is seen in “Fig. 5.2”.

Studies evaluating tissue magnesium content in uremic patients have shown conflicting results. It has been found that serum magnesium levels are not correlated with any tissue concentration except those of the interstitial space and bones [15].

The concentration of Mg in different tissues, such as skeletal muscle, has been reported to be either low, normal or elevated [4]. Nilsson et al. found normal Mg levels in lymphocytes of haemodialysis patients, in spite of high serum Mg concentrations [16]. In a study from our unit, we calculated serum and peripheral lymphocytes magnesium in 93 haemodialysis patients and in 29 healthy subjects [17]. Haemodialysis patients had significantly higher serum magnesium levels than normal controls, whereas intracellular magnesium values were similar in both groups. Neither in haemodialysis patients nor in normal subjects there exist any correlation between intracellular and extracellular magnesium levels (see “Fig. 5.3”). On the opposite to these findings, increased bone Mg content is a universal

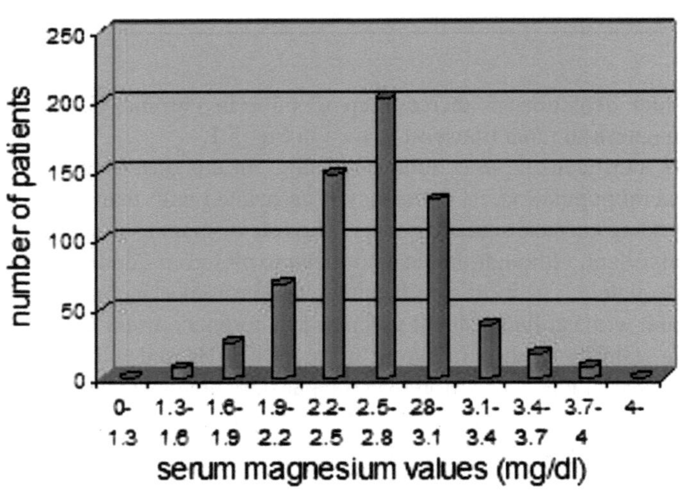

Fig. 5.2 Distribution of serum magnesium values in haemodialysis patients

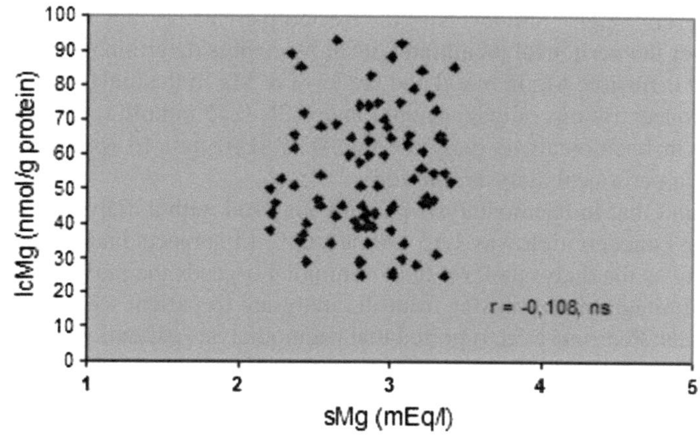

Fig. 5.3 Correlation between intracellular (*IcMg*) and serum magnesium (*sMg*) in haemodialysis

finding with this excess of Mg distributed in both the rapidly exchangeable and the non-exchangeable pools [4, 5].

Despite the high serum magnesium in the ESRD patients, electrocardiographic (ECG) abnormalities and clinical signs of hypermagnesaemia are presented only in patients with extreme high values of serum Mg, more than 2.00 mmol/l (or 4.80 mg/dl) [5]. The ECG changes are similar to those of hyperkalaemia. The clinical signs of hypermagnesaemia arise from the neuromuscular and the cardiovascular system. At levels between 2.00 and 3.00 mmol/l hyporeflexia, drowsiness and lethargy may occur [18]. Hypermagnesaemia with serum Mg level between 3 and 5 mmol / l can result in somnolence, loss of deep tendon reflexes, hypocalcaemia and cardiovascular effects including hypotension, bradycardia and ECG changes such as prolongation of the P-R interval, increase in QRS duration and Q-T interval [5]. At even higher levels of serum Mg, the clinical picture gradually becomes more serious: generalized muscular paresis, tetraplegia, paralytic ileus, apnea and finally death.

In ESRD, serum Mg depends on the Mg dialysate concentration, drug administration (phosphate binders, laxatives and vitamin D), the adequacy of dialysis and the alimentary ingestion of magnesium.

Under normal conditions, dietary Mg intake is a major determinant of serum and total body Mg levels. Gastrointestinal absorption is not linear and can adapt to intake; therefore, under low Mg intake, a high proportion of Mg can be absorbed. In a study, Fine et al. found that as net magnesium ingestion increased from 37 to 973 mg, net absorption increased from 24 mg to only 109 mg, representing a decrease in fractional absorption from 65% to 11% [19]. In patients with renal failure, because they do not have alternative routes for Mg excretion, the role of digestive Mg absorption is controversial [20, 21]. Generally, it is felt that gastrointestinal magnesium absorption is normal in ESRD patients [22], although in a small old study of five dialysis patients, small intestinal magnesium absorption was decreased compared with normal controls [23]; the authors speculated that 1,25 dihydroxyvitamin D deficiency, which is usually present in dialysis patients, may have accounted for the finding [23]. However, no more data are available to confirm this speculation nor are there data evaluating magnesium absorption in vitamin D-replete dialysis patients. Other balance studies in uraemic patients on different protein-restricted diets have revealed that intestinal absorption of Mg is reduced during periods of low-dietary protein intake [24]. Last, there are no reports about Mg excretion through the gut in ESRD.

The composition of the dialysis fluids, both peritoneal dialysis solutions and haemodialysis dialysates, plays an important role in regulating the acid–base balance and the serum concentration of electrolytes in ESRD patients who are on maintenance dialytic therapies for the substitution of renal function. By altering the concentration of the dialysis fluids for several ions, we can add or remove amounts of these substances to or from the patients. Ionized Mg is a basic component of these solutions, and its concentration parallels with the serum magnesium levels of the treated patients, as magnesium readily crosses the dialysis membrane or the peritoneal membrane; its movement is determined by the gradient between the concentration of diffusible Mg in blood and the level of Mg in the dialysis solutions.

Nowadays, the concentrations of magnesium range 0.35–0.75 mmol/l (predominant concentration 0.45–0.50 mmol/l) in haemodialysis dialysate and 0.25–0.80 mmol/l (predominant concentration 0.30–0.40 mmol/l) in peritoneal dialysis solutions.

Navaro et al. found that in haemodialysis patients dialyzed with a dialysate Mg of 0.75 mmol /l, the mean serum Mg concentration was 1.15 mmol/l; 68% of subjects had hypermagnesaemia [25]. This is not surprising as the dialysate level of 0.75 mmol/l exceeds the normal diffusible Mg level in blood, resulting in a magnesium transfer from the dialysate to patient's plasma. Two more studies showed similar results: Pedrozzi et al. reported that haemodialysis patients dialyzed against a magnesium of the same concentration 0.75 mmol/l had increased plasma levels to a total of 1.15 mmol/l and ionized magnesium 0.71 mmol/l compared with control subjects (total, 0.81; ionized, 0.54 mmol/l, respectively) [13]. Truttmann et al., using an identical dialysate magnesium composition, found that magnesium levels decreased from 2.86 predialysis to 2.64 mg/dl postdialysis, while ionized magnesium decreased from 1.70 to 1.56 mg/dl. Both of these values remained above those seen in normal controls (total 1.97 mg/dl and ionized 1.37 mg/dl) [11]. In both of these studies [11, 13], the ionized fraction was slightly, although significantly, lower in dialysis patients (61% vs. 67% and 68%, respectively), suggesting increased protein binding.

Other authors have reported that a reduction in the Mg dialysate results to a significant decrease in the serum Mg concentration as early as the first month [26–28], while in contrast, in another study, by changing the dialysate Mg from 0.75 to 1.5 mmol/l, the mean serum Mg concentration increased from 1.25 to 1.7 mmol/l [29].

Kelber et al. [30], in a more systematic study, studied three groups of haemodialysis patients using high-efficiency dialysis and dialysate magnesium concentrations of 0.90, 0.25 and 0 mmol/l. They found that serum Mg levels fell from 3.3 to 1.6 mg/dl on the 0 mmol/l dialysate and from 3.4 to 2.1 mg/dl on the 0.25 mmol/l dialysate; serum Mg did not change significantly on the 0.90 mmol/l dialysate [30]. Net magnesium removal was 486, 306 and 56 mg on the 0, 0.25 and 0.90 mmol/l baths, respectively. Additionally, these authors noted that the zero magnesium dialysate was poorly tolerated due to severe leg cramps in 8 of 15 patients.

Magnesium concentration in dialysis fluid also influences serum Mg concentration in patients undergoing peritoneal dialysis (PD). Katopodis et al. [31] retrospectively studied 34 CAPD patients who were divided into two groups: those dialyzed with a dialysis solution with Mg concentration 0.75 mmol/L (19 patients) and those dialyzed with a solution with Mg concentration 0.50 mmol/L (15 patients). They found that serum Mg level was significantly higher in patients dialyzed with 0.75 mmol/L versus those dialyzed with 0.50 mmol/L ($p<0.01$). Ejaz et al. [32] and Hutchison et al. [33] reported similar findings. Serum Mg in CAPD patients depends on the concentration gradient between serum and dialysis fluid, and the use of Mg-free dialysate leads to hypomagnesaemia [20].

All these data taken into account clearly demonstrate that the Mg concentration of the dialysis fluids plays a major role in maintaining magnesium homeostasis in ESRD patients treated with haemodialysis or PD [5].

The presence of residual renal function does not influence the serum Mg level. In a study of 100 PD patients with weekly Kt/V of more than 2.1, Page et al. found no significant correlation between residual renal function and serum Mg [34].

Magnesium as Phosphate Binder

Hyperphosphataemia is a stable finding in patients with ESRD and has been recognized to be associated with a spectrum of disease states in patients with CKD [35, 36]. Until the end of 1990s, it was believed that the hyperphosphataemia in ESRD was implicated only in the development of secondary hyperparathyroidism and the subsequent skeleton disorders, the so-called renal osteodystrophy. However, epidemiological studies of the last decade have clearly demonstrated that elevated serum phosphate is also linked with both cardiovascular and all cause morbidity and mortality in dialysis patients [37, 38]. High levels of serum phosphate are associated with the development of vascular calcification in conduit arteries with attendant functional consequences [39]. These changes, at least partially, stem from direct effects of elevated serum phosphate on the biology of vascular smooth muscle cells [40]. These large vessel changes can also be seen in conjunction with small vessel calcifications, resulting to the devastating complication of calcific uremic arteriolopathy [41]. Nowadays, the complex of the disorders arising from the hyperphosphataemia and its related metabolic abnormalities in the patients with terminal renal disease is termed as chronic kidney disease-related bone and mineral disorders (CKD-BMD).

The control of serum phosphate has therefore become a key therapeutic target in the management of patients receiving maintenance dialysis and may become an increasingly important issue in patients with less severe impairment of their renal function. In addition to optimal dialysis treatment and dietary restrictions, oral phosphate binders are the treatment of choice in renal patients with hyperphosphataemia [42].

Four decades after the introduction of chronic haemodialysis in the early 1960s, we have not yet found the ideal phosphate binder(s) in terms of combined efficacy, safety and low cost. Aluminium and calcium salts, and non-aluminium and non-calcium agents, for example, sevelamer HCl and lanthanum carbonate, all have advantages and disadvantages. The affinity of magnesium ions (Mg^{2+}) for phosphorus is weaker than of the aluminium (Al^{3+}), hydrogen (H^+) and calcium (Ca^{2+}) ions, but the poorer intestinal absorption of magnesium leaves more elemental magnesium for phosphate binding [19, 43]. Magnesium-based phosphate binders have been tried since the aluminium era. In the "early period" from the 1980s until the middle of the 1990s, magnesium carbonate ($MgCO_3$) and magnesium hydroxide ($Mg(OH)_2$) were administered either alone or in combination with calcium salts in studies involving small series of both haemodialysis and peritoneal dialysis patients, with good results [44–50]. However, these compounds were not widely used in ESRD patients because nephrologists had an inordinate fear of hypermagnesaemia and a belief that magnesium administration is frequently accompanied by gastrointestinal disorders.

In recent years, there has been a reawakening of interest in magnesium containing phosphate binders. We will report more extensively on three recent studies (see "Table 5.2"). In a randomized controlled trial, we studied the efficacy and the safety of magnesium carbonate as a phosphate binder in 46 haemodialysis patients [51]. The patients administered either $MgCO_3$ or calcium carbonate ($CaCO_3$). The concentration of magnesium in the dialysate bath was 0.48 mmol/l in the CaCO3 group and lower, 0.30 mmol/l in the $MgCO_3$ group The serum phosphate level was reduced equally in both groups, but patients receiving MgCO3 were more likely to have serum calcium levels within the K/DQOI guidelines, 74% versus 65%. Mean serum calcium values were higher in the $CaCO_3$ group; mean serum magnesium was slightly but not significantly higher in the $MgCO_3$ group, whereas iPTH levels did not differ in the two groups. Only 2 of 25 patients discontinued MgCO3 because of complications: one (4%) because of persistent diarrhoea and one (4%) because of recurrent, but asymptomatic, hypermagnesaemia (serum magnesium defined as >4 mg/dl).

Spiegel et al., in a prospective, randomized (2:1), open-label trial, administered tablets containing a mixture of $MgCO_3$ and $CaCO_3$ to 20 haemodialysis patients and calcium acetate alone to ten controls for 12 weeks [52]. Both regimens provided equal control of serum phosphorous. Serum calcium was significantly higher in patients taking calcium acetate. Three of the 20 patients on $MgCO_3/CaCO_3$ withdrew the medication because of worsening of diarrhoea, whereas no patient experienced symptoms related to hypermagnesaemia. There were no differences between groups with regard to serum iPTH or bicarbonate concentrations. The authors concluded that magnesium carbonate calcium/ carbonate mixture was generally well tolerated and was effective in controlling serum phosphorus while reducing elemental calcium ingestion.

Table 5.2 A synopsis of 3 recent RCTs evaluating Mg-based compounds as phosphate binders in ESRD patients

Authors	Clinical setting	Conclusions
Tzanakis et al. [51] 2008	Efficacy and safety of MgCO3 in comparison to CaCO3 in 46 HD patients	Both regimens provided equal control of serum phosphorus Less hypercalcaemic episodes, minor GIs with MgCO3 One patient discontinued MgCO3 due to recurrent hypermagnesaemia
Spiegel et al. [52] 2007	Efficacy and safety of a mixture of $MgCO_3$ and $CaCO_3$ in comparison to calcium acetate in 30 HD patients	Both regimens provided equal control of serum phosphorus Serum calcium was significantly higher in patients taking calcium acetate Three of the 20 patients on $MgCO_3/CaCO_3$ withdrew because of worsening of diarrhoea No patient experienced symptoms related to hypermagnesaemia
CALMAG study [53] 2010	Efficacy and safety of a mixture of $MgCO_3$ and calcium acetate in comparison with sevelamer HCl in 255 patients	Both regimens provided equal control of serum phosphorus Lower serum calcium, more GIs in sevelamer group No patient experienced symptoms related to hypermagnesaemia

Last, in a large polycentric randomized clinical trial including 255 patients from five European countries (The CALMAG study) [53], the investigators compared a regimen containing $MgCO_3$ and calcium acetate (CaMg) with sevelamer HCl. Serum phosphorus levels had decreased significantly with both drugs. Ionized serum calcium did not differ between groups; total serum calcium increased in the CaMg group (treatment difference 0.0477 mmol/L; $P=0.0032$) but was not associated with a higher risk of hypercalcaemia. An asymptomatic increase in serum magnesium occurred in CaMg-treated patients (treatment difference 0.2597 mmol/L, $P<0.0001$). Serum iPTH decreased in a larger degree in the CaMg group ($P<0.05$). Adverse events, definitively related to the study drug, were significantly more frequent in the sevelamer HCl group and were gastrointestinal disorders.

In the answer to the question *do magnesium-based phosphate binders have a role in ESRD patients?* our answer is yes, because they have satisfactory efficacy, they manifest minor complications and their use minimizes calcium overload. A significant proportion of dialysis patients need more than one category of binding agents to achieve satisfactory phosphate binding. In such cases, magnesium salts, especially MgCO3, ideally combined with a calcium-containing salt, could complement the phosphate-binding regimen. An additional advantage is that MgCO3 is much less expensive than the newest sevelamer HCl and lanthanum carbonate. Furthermore, there are, currently, adequate data supporting the observation that magnesium has a beneficial action on the cardiovascular system of these patients. However, when administering these agents to patients, we must look for possible hypermagnesaemia by regular serum magnesium measurements and probably by using a low dialysate magnesium concentration.

Serum Magnesium, Parathyroid Hormone and Bone Disease in ESRD

Bone is the main store of body magnesium, and probably its content is the more reliable indicator of the total body magnesium burden [22]. Bone magnesium exits in two pools, a surface limited exchangeable pool and a larger pool associated with the apatite crystals [54]. The exchangeable bone, which represents the 20–30% of the total bone magnesium, acts as a reservoir both for a rapid release of magnesium in plasma in case of acute deficiency and for sinking the excess magnesium in case of overload [54]. The rest, stable not exchangeable, is complexed along with calcium, in a Ca/Mg proportion approximately 50:1, with the apatite crystals.

Furthermore, magnesium is involved in the bone disorders that are developed in patients suffering from ESRD by two means: first, by affecting directly the bone histomorphology and second, indirectly, by influencing the synthesis and release of the parathyroid hormone (PTH). Early studies in the 1970s demonstrated high bone Mg content in dialysis patients which was correlated with mineralization defects [55, 56]. These findings led to the suggestion that high bone Mg content is implicated in the pathogenesis of osteomalacic renal osteodystrophy and guided the nephrologists to decrease the Mg concentration of the dialysis solutions [57]. Currently, the hypothesis of a direct effect of magnesium on the bone histology in the renal patients is weak. The renal low turnover bone disease, which includes the adynamic bone disease and the osteomalacia, is a sinuous disorder caused by a plenty of factors such as phosphate binders and vitamin D administration, overtreatment of the uremic secondary hyperparathyroidism, hypercalcaemia, deposition and incorporation of iron and other metals in the bones, disturbances of the thyroid function and the uraemia per se.

The relationship between Mg and PTH is complex. Parathyroid hormone increases serum Mg (like calcium) by increasing its gastrointestinal absorption, bone resorption and renal reabsorption [2]. On the other hand, Mg is essential for synthesis, release and adequate tissue sensitivity to PTH. Hypermagnesaemia, similar to hypercalcaemia, inhibits PTH secretion. Experimental studies in both animals and humans have shown that perfusion of isolated parathyroid glands with varying concentrations of Mg showed that acute elevations of Mg levels inhibited PTH secretion [58].

Serum magnesium has the same effect as serum-ionized calcium in suppressing PTH secretion in humans with normal renal function, although the action of magnesium is a factor of 3–5 weaker than that of calcium [4]. However, until some years ago, the exact relationship between serum magnesium and PTH levels in ESRD patients was not clear because serum levels of PTH in ESRD patients are affected by many confounding factors. Initial studies showed either a positive or a negative or a neutral effect of serum magnesium on PTH levels. In later years, a highly significant inverse correlation between serum Mg and PTH in dialysis patients has been clearly demonstrated. In "Table 5.3", all published studies on the relationship between serum magnesium and parathyroid hormone in dialysis patients are summarized [21, 25, 31, 44, 59–73]. Ten of thirteen studies among patients on haemodialysis and five of six studies among patients on peritoneal dialysis showed a significant inverse relationship between serum Mg and serum intact parathyroid hormone. On the basis of these findings, one could speculate that an elevated serum Mg may suppress PTH synthesis and that a low serum Mg may stimulate PTH synthesis and/or secretion in ESRD. The effect of serum magnesium on the secretion of PTH is one of the determinants that must be taken into account when deciding dialysate magnesium composition in order to minimize the risk of either hyperparathyroidism or adynamic bone disease. Currently, the concentration of magnesium in haemodialysis solutions ranges from 0.35 to 0.75 mmol/l and that in CAPD solutions from 0.25 to 0.80 mmol/l.

Magnesium and Intradialytic Hemodynamic Stability

Symptomatic intradialytic hypotension is a common complication of haemodialysis. The incidence of hypotension in the dialysis population is quite frequent, occurring in 20–50% of haemodialysis treatments. Intradialytic hypotension has a negative impact on patient's health-related quality of life, affects the adequacy of the delivered dose of dialysis, may cause vascular access thrombosis and in general constitutes an important cause of morbidity and mortality in chronic haemodialysis patients. The frequency and the severity vary with the age and sex of the patient, with the greatest number of episodes being seen in older patients and in women. Several dialytic- and patient-related factors

Table 5.3 Studies reporting the relationship between serum magnesium and parathyroid hormone in patients on dialysis

Author (year)	Patients (*n*)	*r* value	*P* value
Hemodialysis studies			
Pletka et al. (1974) [59]	26		<0.001
Parsons et al. (1980) [60]	18		<0.05
Gonella et al. (1981) [61]	22		>0.05
McGonigle et al. (1984) [62]	20		<0.001
O'Donovan et al. (1986) [44]	28		>0.05
Kenny et al. (1987) [63]	16	– 0.833	<0.001
Oe et al. (1987) [64]	18		<0.05
Navarro et al. (1988) [65]	41	–0.60	<0.001
Bellucci et al. (1998) [66]	14		0.01
Navarro et al. (1999) [67]	110	–0.58	<0.001
Gohda et al. (2002) [68]	86	–0.28	<0.001
Guh et al. (2002) [69]	126		<0.05
Tzanakis et al. (2002) [70]	75		0.48
CAPD studies			
Saha et al. (1997) [71]	26	–0.42	<0.05
Navarro et al. (1998) [21]	20	–0.63	<0.001
Navarro et al. (1999) [25]	51	–0.70	<0.001
Cho et al. (2002) [72]	56	–0.295	<0.05
Katopodis et al. (2003) [31]	23		>0.05
Wei et al. (2006) [73]	46	–0.357	<0.05

influence blood pressure drop during the haemodialysis session: the ultrafiltration and the subsequent water and sodium removal, the rapid plasma alteration in the concentration of several ions, the uremic and/or diabetic autonomic neuropathy, the pre-existence cardiac dysfunction, several medications and the haemodialysis procedure itself are some of the most known factors that provoke an intradialytic haemodynamic instability and hypotension.

Magnesium plays a role in maintaining intradialytic haemodynamic stability in haemodialysis patients. Several investigators have demonstrated that the magnesium bath concentration is inversely associated with the manifestation of hypotensive episodes during the dialysis sessions. Kyriazis et al. [74] investigated the potential role of Mg as a modulator of the cardiovascular response during haemodialysis. They found that intradialytic changes in serum Mg had a significant and independent effect on systemic haemodynamics. Increasing dialysate Mg level to 0.75 mmol/l could prevent intradialytic hypotension regardless of the dialysate Ca concentration.

These associations have been confirmed by more recent studies where the investigators reassured that intradialytic lowering in serum magnesium were significantly correlated with hypotension during the dialysis session [75, 76].

Magnesium exerts a direct modulatory action on cardiac excitability and vascular smooth muscle contraction and relaxation. It has been systematically shown that a rapid magnesium salt infusion lowers blood pressure by reduction of total peripheral resistance, despite a moderate increase in cardiac output. The latter is accomplished via increases in heart rate and coronary flow or a direct myocardial effect. It seems that the substantial intradialytic fall of arterial pressure seen with a dialysis solution containing low Mg and Ca concentrations is because of unusual impairment of myocardial contractility not compensated by increased total peripheral resistance [74]. Given that most guidelines suggest a low dialysate Ca bath of 1.25 mmol/l, it is of great importance to keep dialysate Mg at relatively high levels. So, the recently published European Best Practice Guidelines on haemodialysis suggest that "In patients with frequent episodes of intradialytic hypotension, low (0.25 mmol/l) magnesium in dialysate should be avoided, especially in combination with low-calcium dialysate (Level II)", Guideline 3.2.4a [77].

Hypomagnesaemia in Transplanted Patients

The use of the calcineurin inhibitors (CNIs), which are widely administered to renal graft recipients as immunosuppressants after renal transplantation, is frequently associated with hypomagnesaemia [78]. It concerns both cyclosporine and tacrolimus but is more profound in patients on tacrolimus [78, 79]. This side effect is mainly due to a decrease of Mg reabsorption in the distal tubule through the reduction of TRPM6 expression and due to the shifting of magnesium ions from plasma to other tissues [80, 81]. The increased renal magnesium loss and the subsequent decrease of serum magnesium occur early, from the first week after the transplantation [81], are observed more frequently in diabetic patients [82]. It has also been found that hypomagnesaemia is an independent predictor of new-onset diabetes after transplantation (NODAT) in renal transplant recipients [83, 84], although other studies have not confirmed it [78]. Hypomagnesaemia enhances the nephrotoxic action of the CNIs, which in turn increases the renal magnesium loss, leading to a vicious cycle. The CNI-induced hypomagnesaemia may cause a functional GFR reduction due to an excretion of vasoconstrictive substances (renin, endothelin, etc.). Furthermore, it also contributes to the development of the chronic fibrotic lesions that the CNIs trigger in the interstitial via a direct activation of the known fibrinogenic molecules (cortical mRNA of transforming growth factor beta, plasminogen activator inhibitor, tissue inhibitor of matrix metalloproteinase-1, etc.) [85].

Experimental studies in animals have shown that magnesium supplementation to the depleted transplants reduces the release of the above-mentioned vasoconstrictive and fibrinogenic factors resulting to prevention, even to attenuation, of the functional and the fibrotic tubulointerstitial disorders of the renal graft [85–87].

In a retrospective study including a large population of 320 renal graft recipients, Holzmacher et al. [88] demonstrated that low serum Mg levels were associated with a greater rate of decline in kidney allograft function and with an increased rate of graft loss in renal transplant recipients with chronic cyclosporine nephropathy.

Moreover, hypomagnesaemia has been associated with abnormal lipid metabolism in nontransplant patients. In a prospective short-term pilot trial, Gupta et al. [89] investigated the effect of oral Mg supplementation on lipid profile in stable hypomagnesaemic renal transplant patients. They found that correction of serum Mg in nondiabetic renal-transplant patients is associated with reduced serum total cholesterol, LDL cholesterol and the total cholesterol/HDL cholesterol ratio.

Taking all the above in to consideration, it is suggested to calculate urine and serum magnesium to all renal transplants during their regular follow-up.

Magnesium and Cardiovascular Diseases

General Population

Magnesium is a natural calcium antagonist and modulates vasomotor tone, blood pressure and peripheral blood flow. As a consequence of these actions, magnesium has an important role in the normal function of the cardiovascular system, and its deficiency is associated with many cardiovascular disorders [90, 91]. It has been found that hypomagnesaemia is related with a wide broad of cardiovascular abnormalities such as arrhythmias, acute myocardial ischaemia, congestive heart failure, cardiopulmonary bypass, peripheral arterial ischaemic disease, calcifications of arteries and cardiac valves, ischaemic brain disease, hypertension and coronary artery disease.

The first indications regarding the possible role of magnesium in the pathogenesis of cardiovascular diseases came up into the light in the 1970s and 1980s when it was found that populations of white South-Africans of Anglo-Saxon origin that manifested an increased frequency of coronary artery disease

Table 5.4 Epidemiologic studies evaluating the association of Mg with CVD

Authors, year	Study type	n	Results
Gartside P and Glueck C (NHANES I) [104], 1995	PES	8,251	Low sMg was correlated with increase frequency of CAD, hospitalization and cardiovascular mortality
Ford E. (NHANES II) [105], 1999	PES	25,292	Serum magnesium levels were inversely correlated with both CAD mortality and all cause mortality
Liao et al. (ARIC) [106], 1998	PES	13,922	Subjects who manifested CAD had lower sMg Low sMg was an independent prognostic risk factor for CAD in females
Ohirra T et al. (ARIC) [107], 2009	PES	14,221	Low sMg levels could be associated with increased risk of ischaemic stroke, in part, via effects on hypertension and diabetes
Peacock et al. (ARIC) [108], 2010	PES	14,232	Low levels of serum Mg may be an important predictor of sudden cardiac death

PES prospective epidemiologic study, *CVD* cardiovascular diseases, *CAD* coronary artery disease

compared with other Anglo-Saxons in other countries consumed water with a lower magnesium concentration [92]. At the same time, experimental studies showed that diets containing high amounts of magnesium in foods and drinking water provided protection from cardiovascular diseases in the examined animals [93–96]. Commenting on the above results, Laurant et al. [96] noted: *these findings suggest that Mg deficiency modifies the mechanical properties of the common carotid artery in young rats. Since Mg deficiency is considered a risk factor, these mechanical alterations could contribute to the development of atherosclerosis, hypertension and cardiovascular diseases.*

In humans, low concentration of magnesium has been found in myocardial cells of humans who died from infarction, as well as in serum and in cerebrospinal fluid in patients with stroke [97, 98].

In clinical studies, Amighi et al. [99] prospectively studied 323 patients with symptomatic peripheral artery disease and intermittent claudication for a median of 20 months and found that patients with Mg serum values <0.76 mmol/l exhibited a 3.29-fold increased adjusted risk (95% CI, 1.34 to 7.90; P=0.009) for neurological events, but patients with Mg serum values of 0.76 mmol/l to 0.84 mmol/l had no increased risk, while Schechter et al. [100], in a very sophisticated study, demonstrated that oral magnesium therapy improved endothelial function in patients with coronary artery disease. Ma et al., in the context of the ARIC epidemiologic study, found that an inverse association between sMg and carotid intima-media thickness existed in the 13,922 subjects who were included in the study. [101]. Adamopoulos et al. [102] reported that a serum magnesium level of 2 mEq/l or less was associated with increased cardiovascular mortality in 1,569 chronic systolic and diastolic heart failure patients with normal sinus rhythm who participated in the Digitalis Investigation Group trial.

Magnesium supplementation is the treatment of choice for the life-threatening ventricular arrhythmias such as torsades de pointes, although a causal role of hypomagnesaemia has not been proven [103].

Large epidemiological studies from the USA including thousands of patients showed clearly that there is an inverse association between low serum magnesium and the frequency of coronary artery disease, stroke and sudden death in the general population [104–108]. A summary of these studies is shown in "Table 5.4".

Last, there is a lot of evidence that hypomagnesaemia coexists or is associated with the development of hypertension, diabetes mellitus, lipid disorders and metabolic syndrome, all of which are established cardiovascular risk factors [109, 110].

Taking into account the existing data, Maier [91] postulated that magnesium *may be a missing link between diverse cardiovascular risk factors and atherosclerosis.*

However, despite the huge body of data, there are not yet randomized clinical trials to convincingly address the question if magnesium has a protective cardiovascular effect.

End-Stage Kidney Disease Patients

Arterial Calcifications, Atherosclerosis

Cardiovascular disease (CVD) is the most common cause of morbidity and mortality among patients with end-stage renal disease (ESRD). Cardiovascular mortality in dialysis patients is 10–20 times greater compared to the general population [111].

The aetiology of the extremely high cardiovascular mortality of haemodialysis patients remains an enigma. Traditional and emerging cardiovascular risk factors cannot explain adequately the excess CVD in ESRD patients. One of the most important structural components of arterial atherosclerosis in these patients is medial vascular calcification, the crystalline structure of which is hydroxyapatite. Vascular calcification in ESRD is a dynamic process of multifactor origin. The role of abnormal metabolism of calcium and phosphate in the development of arterial calcification is well established. There is, furthermore, much evidence that hypomagnesaemia also plays a significant role in the pathogenesis of cardiovascular diseases and that a high serum magnesium level may retard the development and/or acceleration of arterial calcification and, subsequently, atherosclerosis in ESRD patients. Many experimental studies in animal models have clearly shown a relationship of low Mg with vascular calcification and atherosclerotic lesions [96, 112–114].

In clinical studies, Meema et al. [115] first evaluated the role of low magnesium on the progression of the arterial calcifications in ESRD patients and found that serum Mg was significantly lower in PD patients with progressive arterial calcifications. Earlier, Izawa et al. [116] had described a haemodialysis patient in whom soft-tissue calcification resolved after treatment with a dialysate with a high concentration of magnesium. Subsequently, other clinical studies demonstrated a clear association between low magnesium and arterial calcifications or other atherosclerotic lesions in dialysed renal patients [115–124]. In a number of clinical studies, a correlation of low serum magnesium with either vascular calcifications or carotid intima-media thickness (an index of atherosclerosis) has been demonstrated in ESRD patients. A summary of all these published studies is seen in "Table 5.5".

However, most of these trials were observational in nature, so they cannot establish if hypermagnesaemia has a protective role in the progression of vascular calcification in ESRD. Only well-designed prospective controlled trials can address this issue.

How Does Magnesium Affect Cardiovascular Calcification and Atherosclerosis?

Magnesium is regarded as a natural biological calcium antagonist inhibiting its entrance into endothelial and smooth muscle cells [7, 90, 125]; by these means it regulates blood pressure and endothelial function, an effect which is essential in the atherosclerotic process. Furthermore, low magnesium promotes endothelial inflammation via oxidation of HDL cholesterol, formation of oxygen radicals and proinflammatory agents, changes in membrane fatty acid saturation and platelets' aggregation [90, 91, 126, 127]. It has also been found that low magnesium coexists with or even predisposes to diabetes, dyslipidaemia and metabolic syndrome, which are known atherosclerotic risk factors [109, 110]. Last but not least, there is much evidence that magnesium is an inhibitor of the calcification process and consequently low magnesium promotes vascular calcification. The underlying pathogenetic mechanism(s) of the latter action are not well understood, but there are three possible explanations:

1. First, as has been shown in experimental models, ambient Mg^{2+} can prevent early calcium phosphate hydroxyapatite crystal growth by affecting crystal solubility in biological fluids [128].
2. Second, magnesium is a cofactor of alkaline phosphatase which is present in tissue of vascular calcification. Thus, magnesium concentration may affect alkaline phosphatase activity leading to modulation of the calcification mechanism in ESRD patients [129].

Table 5.5 Studies on the relationship between magnesium and vascular lesions in patients on dialysis

Authors, year	Study type	Clinical setting	n	Conclusions
Izawa et al. [116], 1974	Case report	HD patient	1	Amelioration of soft tissue calcification after dialysis with a dialysate with high Mg concentration
Meema et al. [115], 1987	RS	Arterial calcifications in patients on PD	44	Serum Mg was significantly lower in patients with progressive arterial calcifications
Nakagawa [117], 1997	RS	Aortic calcification in patients on HD	28	Treatment with Mg inhibited elevation of the calcium content of the aorta
Tzanakis et al. [118], 1997	RS	MAC detected by echocardiography in HD patients	56	Serum Mg levels were significantly lower in patients with MAC
Tamashiro et al. [119], 2001	RS	CAC detected by EBT in HD patients	24	No difference in serum Mg among slow and rapid progressors
Tzanakis et al.[120], 2004	RS	C-IMT detected by B-mode ultrasound in HD patients	93	Inverse association of both serum and lymphocytes Mg with C-IMT
Ishimura et al. [121], 2007	RS	Arterial calcifications in HD patients	390	Low serum Mg was significantly associated with the presence of vascular calcification of the hand arteries, independent of serum calcium and phosphate levels
Turgut et al. [122], 2008	PS	Mg supplementation to evaluate the C-IMT in HD patients	47	cIMT was significantly improved in patients treated with Mg
Spiegel et al. [123], 2009	PS	Mg supplementation as binders to evaluate CAC by EBT in HD patients	7	Minimal progression of CAC at 18 months
Okasha et al. [124], 2010	RS	Arterial calcifications in HD patients	65	Vascular calcification is associated with a lower serum Mg level

n patients' number, *RS* retrospective study, *PS* prospective study, *HD* haemodialysis, *PD* peritoneal dialysis, *MAC* mitral annular calcification, *CAC* coronary arteries calcification, *EBT* electron beat tomography, *C-IMT* carotid intima-media thickness complex

3. Third, chronic hypermagnesaemia may suppress PTH excretion in ESRD patients, which is implicated in the development of soft-tissue calcification, including the vascular calcification [59].

A schematic presentation of the implicated mechanisms of the low magnesium-induced atherosclerosis is seen in Fig. 5.4.

Magnesium and Survival on Haemodialysis

Some data show a relationship between magnesium and haemodialysis patients' survival. In a prospective study [130], we followed up for 60 months the 93 haemodialysis patients in the earlier mentioned study (Ref.123) to assess the long-term prognostic effect of intracellular and serum magnesium on their outcome. We found that lymphocyte magnesium was an independent prognostic factor for improved survival (OR 0.0169, 95% CI 0.0018–0.0321, P=0.029) and that the relationship for serum magnesium was similar but weaker (OR 0.084, 95% CI 0.007–1.026, P=0.069) [130].

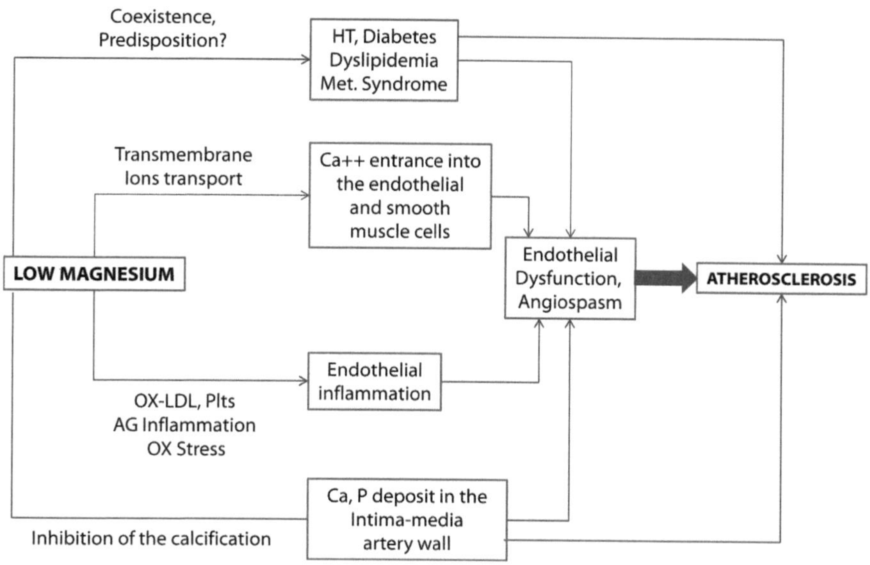

Fig. 5.4 Schematic presentation of the implicated mechanisms of the low magnesium-induced atherosclerosis

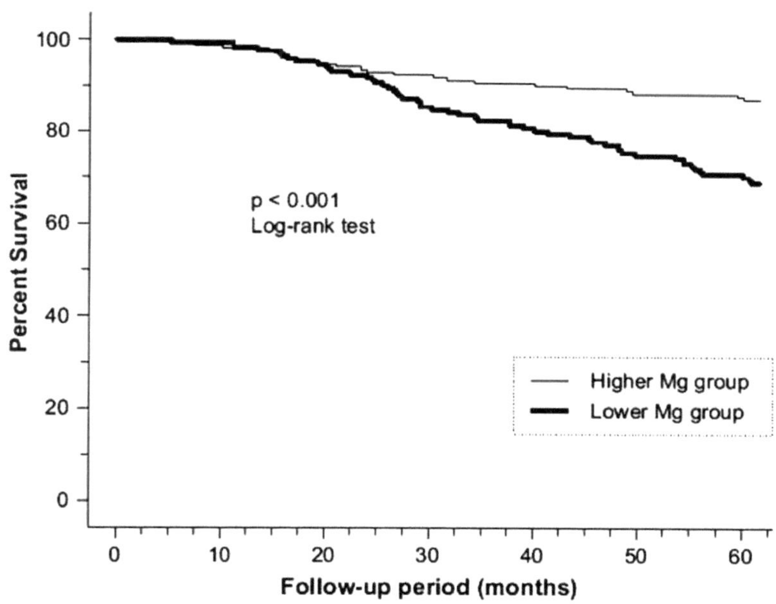

Fig. 5.5 Patient's survival rate according to serum magnesium levels. Higher Mg group: sMg > 2.77, lower Mg group: sMg < 2.77 mg/dl (From ref. 134 with permission)

Similarly, Ishimura et al. [131] followed up a large cohort of 515 haemodialysis patients for a median time of 51 ± 17 months and demonstrated that lower serum magnesium level was a significant independent predictor of mortality in haemodialysis patients, particularly non-cardiovascular mortality (OR 0.485, 955 CI 0.241–0.975, P = 0.0424), see Fig. 5.5.

Last, in an observational retrospective analysis of a large population of 27,544 prevalent haemodialysis patients, Lacson et al. [132] found that high normal and elevated serum Mg levels were associated with lower risk of mortality among the subjects. However, cause and effect were not established in this study.

Currently, there is evidence suggesting a possible protective role of high magnesium on the cardiovascular disease of ESRD patients. This fact raises three significant questions, two of theoretical and one of clinical interest:

1. First, is hypomagnesaemia one more (modifiable) cardiovascular risk factor?
2. Second, is hypomagnesaemia the *missing link* between the known risk factors and the cardiovascular disease?
3. Third, how can we translate a favourable action of magnesium into a therapeutic application in ESRD patients who are usually already hypermagnesaemic?

In our opinion, existing data allow us to suggest a positive answer to the first two questions. The reply to the last question is not easy and awaits prospective controlled studies to be proven. However, we have some notions on this issue.

The results of the studies we presented above suggest that the possible protective effect of magnesium exists in a linear form, covering all the spectrum of serum Mg levels; one should postulate that *the higher magnesium the better for the patients*.

Serum Mg levels between 1.00 and 1.80 mmol/l which are considered as “hypermagnesaemia” are usual in dialysis patients without causing any problem.

As known, the ionized fraction of total serum magnesium is its biologically active form; in ESRD patients, this fraction is lower than in healthy subjects [5, 11, 14] that means that hypermagnesaemia probably is overestimated in these patients.

References

1. Kelepouris E, Agus ZS. Hypomagnesemia: renal magnesium handling. Semin Nephrol. 1998;18:58–73.
2. Kanbay M, Goldsmith D, Uyar ME, Turgut F, Covic A. Magnesium in chronic kidney disease: challenges and opportunities. Blood Purif. 2010;29:280–92.
3. Gunther T. Mechanisms of regulation of Mg^{2+} efflux and Mg^{2+} influx. Miner Electrolyte Metab. 1993;19:259–65.
4. Mountokalakis TD. Magnesium metabolism in chronic renal failure. Magnes Res. 1990;3:121–7.
5. Navarro-Gonzαlez JF, Mora-Fernαndez C, Garcva-Pιrez J. Clinical implications of disordered magnesium homeostasis in chronic renal failure and dialysis. Semin Dial. 2009;22:37–44.
6. Tzanakis I, Oreopoulos D. Beneficial effects of magnesium in chronic renal failure: a foe no longer. Int Urol Nephrol. 2009;41:363–71.
7. Altura BM, Altura BT. Role of magnesium in the pathogenesis of hypertension updated: relationship to its actions on cardiac, vascular smooth muscle and endothelial cells. In: Laragh JH, Brenner BM, editors. Hypertension: pathophysiology, diagnosis and management. 2nd ed. New York: Raven Press; 1995. p. 1213–42.
8. Alexander RT, Hoenderop JG, Bindels RJ. Molecular determinants of magnesium homeostasis: insights from human disease. J Am Soc Nephrol. 2008;19(8):1451–8.
9. Steele T, Weng S, Evenson M, et al. The contributions of the chronically diseased kidney to magnesium homeostasis in man. J Lab Clin Med. 1968;71:455.
10. Okuno S, Inaba N. Magnesium in hemodialysis patients. In: Nishizawa Y, Morii H, Durlach J, editors. New perspectives in magnesium research, Nutrition and health. 2nd ed. London: Springer; 2007. p. 316–29.
11. Truttmann AC, Faraone R, Von Vigier RO, Nuoffer JM, Pfister R, Bianchetti MG. Maintenance hemodialysis and circulating ionized magnesium. Nephron. 2002;92:616–21.
12. Spiegel DM. The role of magnesium binders in chronic kidney disease. Semin Dial. 2007;20:333–6.
13. Pedrozzi NE, Truttmann AC, Faraone R, et al. Circulating ionized and total magnesium in end-stage kidney disease. Nephron. 1998;79:288–92.
14. Dewitte K, Dhondt A, Lameire N, Stöckl D, Thienpont LM. The ionized fraction of serum total magnesium in hemodialysis patients: is it really lower than in healthy subjects? Clin Nephrol. 2002;58(3):205–10.

15. Hutchison AJ. Serum magnesium and end-stage renal disease. Perit Dial Int. 1997;17:327–9.
16. Nilsson P, Johansson SG, Danielson BG. Magnesium studies in hemodialysis patients before and after treatment with low dialysate magnesium. Nephron. 1984;37:25–9.
17. Tzanakis I, Virvidakis K, Tsomi, Papadaki A, Kallivretakis N, Mountokalakis T. Relationship between intracellular and extracellular magnesium in haemodialysis patients and normal subjects. Proceedings of the 5th Bantao Congress, Thessaloniki; 2001. p. 149–150.
18. Covic AC. Magnesium in chronic kidney disease – more than just phosphate binding. Eur Nephrol. 2010;4:3–11.
19. Fine KD, Santa Ana CA, Porter JL, Fordtran JS. Intestinal absorption of magnesium from food and supplements. J Clin Invest. 1991;88:396–402.
20. Wei M, Esbaei K, Bargman J, Oreopoulos DG. Relationship between serum magnesium, parathyroid hormone, and vascular calcification in patients on dialysis: a literature review. Perit Dial Int. 2006;26:366–73.
21. Navarro-Gonzalez JF. Magnesium in dialysis patients: serum levels and clinical implications. Clin Nephrol. 1998;49(6):373–8.
22. Spiegel DM. Magnesium in chronic kidney disease: unanswered questions. Blood Purif. 2011;31:172–6.
23. Brannan PG, Vergne-Marini P, Pak CYC, Hull AR, Fordtran JS. Magnesium absorption in the human small intestine. Results in normal subjects, patients with chronic renal disease, and patients with absorptive hypercalciuria. J Clin Invest. 1976;57:1412–8.
24. Kopple JD, Coburn JW. Metabolic studies of low protein diets in uremia. II. Calcium, phosphorus and magnesium. Medicine. 1973;52(6):597–607.
25. Navarro J, Mora C, Macia M, Garcia J. Serum magnesium concentration is an independent predictor of parathyroid hormone levels in peritoneal dialysis patients. Perit Dial Int. 1999;19:455–61.
26. Gonella M, Moriconi L, Betti G, et al. Serum levels of PTH, Mg, Ca, inorganic phosphorus and alkaline phosphatase in uremic patients on different Mg dialysis. Proc Eur Dial Transplant Assoc. 1980;17:362–6.
27. Nilsson P, Johansson SG, Danielson BG. Magnesium studies in hemodialysis patients before and after treatment with low dialysate magnesium. Nephron. 1989;37:25–9.
28. Kancir CB, Wanscher M. Effect of magnesium gradient concentration between plasma and dialysate on magnesium variations induced by hemodialysis. Magnesium. 1989;89:132–6.
29. Nair KS, Holdaway IM, Evans MC, Cameron AD. Influence of Mg on the secretion and action of parathyroid hormone. J Endocrinol Invest. 1979;2:267–70.
30. Kelber J, Slatopolsky E, Delmez JA. Acute effects of different concentrations of dialysate magnesium during high-efficiency dialysis. Am J Kidney Dis. 1994;24:453–60.
31. Katopodis KP, Koliousi EL, Andrikos EK, Pappas MV, Elisaf MS, Siamopoulos KC. Magnesium homeostasis in patients undergoing continuous ambulatory peritoneal dialysis: role of the dialysate magnesium concentration. Artif Organs. 2003;27(9):853–7.
32. Ejaz AA, McShane AP, Gandhi VC, Leehey DJ, Ing TS. Hypomagnesemia in continuous ambulatory peritoneal dialysis patients dialyzed with a low-magnesium peritoneal dialysis solution. Perit Dial Int. 1995;15(1):61–4.
33. Hutchison AJ, Were AJ, Boulton HF, Mawer EB, Laing I, Gokal R. Hypercalcaemia, hypermagnesaemia, hyperphosphataemia and hyperaluminaemia in CAPD: improvement in serum biochemistry by reduction in dialysate calcium and magnesium concentrations. Nephron. 1996;72(1):52–8.
34. Page DE, Knoll GA, Cheung V. The relationship between residual renal function, protein catabolic rate, and phosphate and magnesium levels in peritoneal dialysis patients. Adv Perit Dial. 2002;18:189–91.
35. Covic A, Kothawala P, Bernal M, et al. Systematic review of the evidence underlying the association between mineral metabolism disturbances and risk of all-cause mortality, cardiovascular mortality and cardiovascular events in chronic kidney disease. Nephrol Dial Transplant. 2009;24:1506–23.
36. Lezaic V, Tirmenstajn-Jankovic B, Bukvic D, et al. Efficacy of hyperphosphatemia control in the progression of chronic renal failure and the prevalence of cardiovascular calcification. Clin Nephrol. 2009;71:21–9.
37. Block GA, Klassen PS, Lazarus JM, Ofsthun N, Lowrie EG, Chertow GM. Mineral metabolism, mortality, and morbidity in maintenance hemodialysis. J Am Soc Nephrol. 2004;15:2208–18.
38. Tentori F, Blayney MJ, Albert JM, et al. Mortality risk for dialysis patients with different levels of serum calcium, phosphorus, and PTH: the dialysis outcomes and practice patterns study (DOPPS). Am J Kidney Dis. 2008;52:519–30.
39. London GM, Marchais SJ, Guérin AP, Métivier F. Arteriosclerosis, vascular calcifications and cardiovascular disease in uremia. Curr Opin Nephrol Hypertens. 2005;14(6):525–31.
40. Lau WL, Pai A, Moe SM, Giachelli CM. Direct effects of phosphate on vascular cell function. Adv Chronic Kidney Dis. 2011;18(2):105–12.
41. Cozzolino M, Brenna I, Ciceri P, Volpi E, Cusi D, Brancaccio D. Vascular calcification in chronic kidney disease: a changing scenario. J Nephrol. 2011;Suppl 18:S3–10.
42. Martin KJ, González EA. Prevention and control of phosphate retention/hyperphosphatemia in CKD-MBD: what is normal, when to start, and how to treat? Clin J Am Soc Nephrol. 2011;6(2):440–6.

43. Sheikh MS, Maguire JA, Emmett M, et al. Reduction of dietary phosphorus absorption by phosphorus binders. A theoretical, in vitro, and in vivo study. J Clin Invest. 1989;83(1):66–73.
44. O'Donovan R, Baldwin D, Hammer M, Moniz C, Parsons V. Substitution of aluminium salts by magnesium salts in control of dialysis hyperphosphataemia. Lancet. 1986;1(84–86):880–2.
45. Parsons V, Baldwin D, Moniz C, Marsden J, Ball E, Rifkin I. Successful control of hyperparathyroidism in patients on continuous ambulatory peritoneal dialysis using magnesium carbonate and calcium carbonate as phosphate binders. Nephron. 1993;63(4):379–83.
46. Guillot AP, Hood VL, Runge CF, Gennari FJ. The use of magnesium-containing phosphate binders in patients with end-stage renal disease on maintenance hemodialysis. Nephron. 1982;30(2):114–7.
47. Roujouleh H, Lavaud S, Toupance O, Melin JP, Chanard J. Magnesium hydroxide treatment of hyperphosphatemia in chronic hemodialysis patients with an aluminium overload. Nephrologie. 1987;8(2):45–50.
48. Shah GM, Winer RL, Cutler RE, et al. Effects of a magnesium-free dialysate on magnesium metabolism during continuous ambulatory peritoneal dialysis. Am J Kidney Dis. 1987;10(4):268–75.
49. Moriniere P, Vinatier I, Westeel PF, et al. Magnesium hydroxide as a complementary aluminium-free phosphate binder to moderate doses of oral calcium in uremic patients on chronic haemodialysis: lack of deleterious effect on bone mineralization. Nephrol Dial Transplant. 1998;3(5):651–6.
50. Delmez JA, Kelber J, Norword KY, Giles KS, Slatopolsky E. Magnesium carbonate as a phosphorus binder: a prospective, controlled, crossover study. Kidney Int. 1996;49(1):163–7.
51. Tzanakis I, Papadaki A, Wei M, et al. Magnesium carbonate for phosphate control in patients on hemodialysis. A randomized controlled trial. Int Urol Nephrol. 2008;40:193–201.
52. Spiegel DM, Farmer B, Smits G, Chonchol M. Magnesium carbonate is an effective phosphate binder for chronic hemodialysis patients: a pilot study. J Ren Nutr. 2007;17(6):416–22.
53. de Francisco AL, Leidig M, Covic AC, et al. Evaluation of calcium acetate/magnesium carbonate as a phosphate binder compared with sevelamer hydrochloride in haemodialysis patients: a controlled randomized study (CALMAG study) assessing efficacy and tolerability. Nephrol Dial Transplant. 2010;25(11):3707–17.
54. Alfrey AC, Miller NL, Trow R. Effect of age and magnesium depletion on bone magnesium pools in rats. J Clin Invest. 1974;54:1074–81.
55. Alfrey AC, Miller NL. Bone magnesium pools in uremia. J Clin Invest. 1973;52:3019–23.
56. Contiguglia SR, Alfrey AC, Miller N, Butkus D. Total-body magnesium excess in chronic renal failure. Lancet. 1972;ii:1300–2.
57. Gonella M, Ballanti P, Della Roca C, et al. Improved bone morphology by normalizing serum magnesium in chronically hemodialyzed patients. Miner Electrolyte Metab. 1988;14:240–5.
58. Cholst IN, Steinberg SF, Tropper PJ, Fox HE, Segre GV, Bilezikian JP. The influence of hypermagnesemia on serum calcium and parathyroid hormone levels in human subjects. N Engl J Med. 1984;310:1221–5.
59. Pletka P, Bernstein DS, Hampers CL, Merrill JP, Sherwood LM. Relationship between magnesium and secondary hyperparathyroidism during long-term hemodialysis. Metabolism. 1974;23(7):619–30.
60. Parsons V, Papapoulos SE, Weston MJ, Tomlinson S, O'Riordan JL. The long-term effect of lowering dialysate magnesium on circulating parathyroid hormone in patients on regular haemodialysis therapy. Acta Endocrinol. 1980;93(4):455–60.
61. Gonella M, Bonaguidi F, Buzzigoli G, Bartolini V, Mariani G. On the effect of magnesium on the PTH secretion in uremic patients on maintenance hemodialysis. Nephron. 1981;27(1):40–2.
62. McGonigle RJ, Weston MJ, Keenan J, Jackson DB, Parsons V. Effect of hypermagnesemia on circulating plasma parathyroid hormone in patients on regular hemodialysis therapy. Magnesium. 1984;3(1):1–7.
63. Kenny MA, Casillas E, Ahmad S. Magnesium, calcium and PTH relationships in dialysis patients after magnesium repletion. Nephron. 1987;46(2):199–205.
64. Oe PL, Lips P, van der Meulen J, de Vries PM, van Bronswijk H, Donker AJ. Long-term use of magnesium hydroxide as a phosphate binder in patients on hemodialysis. Clin Nephrol. 1987;28(4):180–5.
65. Navarro JF, Macia ML, Gallego E, et al. Serum magnesium concentration and PTH levels. Is long-term chronic hypermagnesemia a risk factor for adynamic bone disease? Scand J Urol Nephrol. 1997;31(3):275–80.
66. Bellucci G, Alessandri M, Buracchi P, et al. Serum magnesium concentration and serum parathormone in hemodialysis patients. Ital J Miner Elect M. 1998;12(3–4):77–9.
67. Navarro JF, Mora C, Jimenez A, Torres A, Macia M, Garcia J. Relationship between serum magnesium and parathyroid hormone levels in hemodialysis patients. Am J Kidney Dis. 1999;34(1):43–8.
68. Gohda T, Shou I, Fukui M, et al. Parathyroid hormone gene polymorphism and secondary hyperparathyroidism in hemodialysis patients. Am J Kidney Dis. 2002;39(6):1255–60.
69. Guh JY, Chen HC, Chuang HY, Huang SC, Chien LC, Lai YH. Risk factors and risk for mortality of mild hypoparathyroidism in hemodialysis patients. Am J Kidney Dis. 2002;39(6):1245–54.
70. Tzanakis I, Papadaki A, Spandidakis V, et al. Does serum magnesium influence PTH release in long term basis in hemodialysis patients? Paper presented at the 14th Pan-Hellenic Congress of Nephrology, Chalchidiki, 2006 May 31–June 3.

71. Saha HH, Harmoinen AP, Pasternack AI. Measurement of serum ionized magnesium in CAPD patients. Perit Dial Int. 1997;17(4):347–52.
72. Cho MS, Lee KS, Lee YK, et al. Relationship between the serum parathyroid hormone and magnesium levels in continuous ambulatory peritoneal dialysis (CAPD) patients using low-magnesium peritoneal dialysate. Korean J Intern Med. 2002;17(2):114–21.
73. Wei M, Esbaei K, Bargman JM, Oreopoulos DG. Inverse correlation between serum magnesium and parathyroid hormone in peritoneal dialysis patients: a contributing factor to a dynamic bone disease? Int Urol Nephrol. 2006;38:317–22.
74. Kyriazis I, Kalogeropoulou K, Bilirakis L, et al. Dialysate magnesium level and blood pressure. Kidney Int. 2004;66(3):1221–31.
75. Elsharkawy MM, Youssef AM, Zayoon MY. Intradialytic changes of serum magnesium and their relation to hypotensive episodes in hemodialysis patients on different dialysates. Hemodial Int. 2006;10 Suppl 2: S16–23.
76. Pakfetrat M, Roozbeh Shahroodi J, Malekmakan L, et al. Is there an association between intradialytic hypotension and serum magnesium changes? Hemodial Int. 2010;14(4):492–7.
77. Kooman J, Basci A, Pizzarelli F, et al. EBPG guideline on hemodynamic instability. Nephrol Dial Transplant. 2007;22 Suppl 2:22–44.
78. Osorio JM, Bravo J, Pérez A, Ferreyra C, Osuna A. Magnesemia in renal transplant recipients: relation with immunosuppression and posttransplant diabetes. Transplant Proc. 2010;42(8):2910–3.
79. Filler G. Calcineurin inhibitors in pediatric renal transplant recipients. Paediatr Drugs. 2007;9(3):165–74.
80. Navaneethan SD, Sankarasubbaiyan S, Gross MD, Jeevanantham V, Monk RD. Tacrolimus- associated hypomagnesemia in renal transplant recipients. Transplant Proc. 2006;38:1320–2.
81. Aisa Y, Mori T, Nakazato T, et al. Effects of immunosuppressive agents on magnesium metabolism early after allogeneic hematopoietic stem cell transplantation. Transplantation. 2005;80(8):1046–50.
82. Vannini SD, Mazzola BL, Rodini L, et al. Permanently reduced plasma ionized magnesium among renal transplant recipients on cyclosporine. Transplant Int. 1999;12(4):244–9.
83. Balla A, Chobanian M. New-onset diabetes after transplantation: a review of recent literature. Curr Opin Organ Transplant. 2009;14(4):375–9.
84. Van Laecke S, Van Biesen W, Verbeke F, De Bacquer D, Peeters P, Vanholder R. Post transplantation hypomagnesemia and its relation with immunosuppression as predictors of new-onset diabetes after transplantation. Am J Transplant. 2009;9(9):2140–9.
85. Miura K, Nakatani T, Asai T, et al. Role of hypomagnesemia in chronic cyclosporine nephropathy. Transplantation. 2002;73(3):340–7.
86. Asai T, Nakatani T, Yamanaka S, et al. Magnesium supplementation prevents experimental chronic cyclosporine a nephrotoxicity via renin-angiotensin system independent mechanism. Transplantation. 2002;74(6):784–91.
87. Yuan J, Zhou J, Chen BC, et al. Magnesium supplementation prevents chronic cyclosporine nephrotoxicity via adjusting nitric oxide synthase activity. Transplant Proc. 2005;37:1892–5.
88. Holzmacher R, Kendziorski C, Hofman R, Jaffery J, Becker B, Djamali A. Low serum magnesium is associated with decreased graft survival in patients with chronic cyclosporin nephrotoxicity. Nephrol Dial Transplant. 2005;20(7):1456–62.
89. Gupta BK, Glicklich D, Tellis VA. Magnesium repletion therapy improved lipid metabolism in hypomagnesemic renal transplant recipients: a pilot study. Transplantation. 1999;67:1485–7.
90. Shechter M. Magnesium and cardiovascular system. Magnes Res. 2010;23(2):60–72.
91. Maier J. Low magnesium and atherosclerosis: an evidence-based link. Mol Aspects Med. 2003;24:137–46.
92. Leary WP. Content of magnesium in drinking water and deaths from ischemic heart disease in white South Africans. Magnesium. 1986;5(3–4):150–3.
93. Ahsan SK. Magnesium in health and disease. J Pak Med Assoc. 1998;48(8):246–50.
94. Bloom S. Coronary arterial lesions in Mg-deficient hamsters. Magnesium. 1985;4(2–3):82–95.
95. Rayssiguier Y, Gueux E. Magnesium and lipids in cardiovascular disease. J Am Coll Nutr. 1986;5(6):507–11.
96. Laurant P, Hayoz D, Brunner H, Berthelot A. Dietary magnesium intake can affect mechanical properties of rat carotid artery. Br J Nutr. 2000;84(5):757–64.
97. Altura BM, Altura BT. Magnesium and the cardiovascular system: experimental and clinical aspects updated. In: Sigel H, Sigel A, editors. Compendium on magnesium and its role in biology, nutrition, and physiology. New York: Marcel Dekker; 1990. p. 359–416.
98. Borowik H, Pryszmont M. Concentration of magnesium in serum and cerebrospinal fluid in patients with stroke. Neurol Neurochir Pol. 1998;32(6):1377–83. abstract in English.
99. Amighi J, Sabeti S, Schlager O, et al. Low serum magnesium predicts neurological events in patients with advanced atherosclerosis. Stroke. 2004;35(1):22–7.
100. Schechter M, Sharir M, Labrador MJ, et al. Oral magnesium therapy improves endothelial function in patients with coronary artery disease. Circulation. 2000;102:2353–8.

101. Ma J, Folsom A, Melnick S, et al. Associations of serum and dietary magnesium with cardiovascular disease, hypertension, diabetes, insulin, and carotid arterial wall thickness: the ARIC study. Atherosclerosis risk in communities study. J Clin Epidemiol. 1995;48(7):927–40.
102. Adamopoulos C, Pitt B, Sui X, et al. Low serum magnesium and cardiovascular mortality in chronic heart failure: a propensity-matched study. Int J Cardiol. 2009;136:270–7.
103. Cupta A, Lawrence A, Krishan K, Kasinsky C, Trohman K. Current concepts in the mechanisms and management of drug-induced QT prolongation and torsade de pointes. Am Heart J. 2007;153:891–9.
104. Kartside PS, Glueck CJ. The important role of modifiable dietary and behavioural characteristics in the causation and prevention of coronary heart disease hospitalization and mortality: the prospective NHANES I follow-up study. J Am Coll Nutr. 1995;14:71–9.
105. Ford ES. Serum magnesium and ischaemic heart disease: findings from a national sample of US adults. Int J Epidemiol. 1999;28(4):645–51.
106. Liao F, Folsom AR, Brancati FL. Is low magnesium concentration a risk factor for coronary heart disease? the atherosclerosis risk in communities study. Am Heart J. 1998;136:480–90.
107. Ohira T, Peacock J, Iso H, Chambless L, Rosamond W, Folsom A. Serum and dietary magnesium and risk of ischemic stroke the atherosclerosis risk in communities study. Am J Epidemiol. 2009;169:1437–44.
108. Peacock J, Ohira T, Post W, Sotoodehnia N, Rosamond W, Folsom A. Serum magnesium and risk of sudden cardiac death in the atherosclerosis risk in communities (ARIC) study. Am Heart J. 2010;16(3):467–70.
109. Ueshima K. Magnesium and ischemic heart disease: a review of epidemiological, experimental, and clinical evidences. Magnes Res. 2005;18(4):275–84.
110. Mooren FC, Krüger K, Völker K, Golf SW, Wadepuhl M, Kraus A. Oral magnesium supplementation reduces insulin resistance in non-diabetic subjects - a double-blind, placebo-controlled, randomized trial. Diabetes Obes Metab. 2011;13(3):281–4.
111. Collins AJ, Li S, Ma JZ, Herzog C. Cardiovascular disease in end-stage renal disease patients. Am J Kidney Dis. 2001;38(4 suppl 1):S26–9.
112. Cohen H, Sherer Y, Shaish A, et al. Atherogenesis inhibition induced by magnesium-chloride fortification of drinking water. Biol Trace Elem Res. 2002;90(1–3):251–9.
113. Schwille PO, Schmiedl A, Schwille R, et al. Media calcification, low erythrocyte magnesium, altered plasma magnesium, and calcium homeostasis following grafting of the thoracic aorta to the infrarenal aorta in the rat differential preventive effects of long-term oral magnesium supplementation alone and in combination with alkali. Biomed Pharmacother. 2003;57(2):88–97.
114. King J, Miller R, Blue J, O'Brien W, Erdman J. Inadequate dietary magnesium intake increases atherosclerotic plaque development in rabbits. Nutr Res. 2009;29:343–9.
115. Meema HE, Oreopoulos DG, Rapoport A. Serum magnesium level and arterial calcification in end-stage renal disease. Kidney Int. 1987;32(3):388–94.
116. Izawa H, Imura M, Kuroda M, Takeda R. Effect of magnesium on secondary hyperparathyroidism in chronic hemodialysis: a case with soft tissue calcification improved by high Mg dialysate. Calcif Tissue Res. 1974;15(2):162.
117. Nakagawa K. A study of aortic calcification uremia (in Japanese). Nippon Jinzo Gakkai Shi. 1997;39:135–43.
118. Tzanakis I, Pras A, Kounali D, et al. Mitral annular calcifications in haemodialysis patients: a possible protective role of magnesium. Nephrol Dial Transplant. 1997;12(9):2036–7.
119. Tamashiro M, Iseki K, Sunagawa O, et al. Significant association between the progression of coronary artery calcification and dyslipidemia in patients on chronic hemodialysis. Am J Kidney Dis. 2001;38:64–9.
120. Tzanakis I, Virvidakis K, Tsomi A, et al. Intra- and extracellular magnesium levels and atheromatosis in haemodialysis patients. Magnes Res. 2004;17(2):102–8.
121. Ishimura E, Okuno S, Kitatani K, et al. Significant association between the presence of peripheral vascular calcification and lower serum magnesium in hemodialysis patients. Clin Nephrol. 2007;68:222–7.
122. Turgut F, Kanbay M, Metin MR, Uz E, Akcay A, Covic A. Magnesium supplementation helps to improve carotid intima media thickness in patients on hemodialysis. Int Urol Nephrol. 2008;40:1075–82.
123. Spiegel D, Farmer B. Long term effects of magnesium carbonate on coronary artery calcification and bone mineral density in hemodialysis patients: a pilot study. Hemodial Int. 2009;13:453–9.
124. Okasha K, Bendary A, Mourad A. Evaluation of peripheral vascular calcification and serum magnesium level in a group of Egyptian hemodialysis patients. Arab J Nephrol Transplant. 2010;3(1):11–6.
125. Gums JG. Magnesium in cardiovascular and other disorders. Am J Health Syst Pharm. 2004;61(15):1569–76.
126. Alturra BM, Altura BT. Magnesium and cardiovascular biology: an important link between cardiovascular risk factors and atherogenesis. Cell Mol Biol Res. 1995;41:347–59.
127. Sherer Y, Bitzur R, Cohen H, et al. Mechanisms of action of the anti-atherogenic effect of magnesium: lessons from a mouse model. Magnes Res. 2001;14:173–9.
128. Bennett RM, Lehr JR, McCarty DJ. Factors affecting the solubility of calcium pyrophosphate dehydrate crystals. J Clin Invest. 1975;56:1571–9.

129. Giachelli CM, Jono S, Shioi A, Nishizawa Y, Mori K, Morii H. Vascular calcification and inorganic phosphate. Am J Kidney Dis. 2001;38:34–7.
130. Tzanakis I, Tsomi A, Mantakas E, Papadaki AN, Kalliretakis N. The higher intracellular magnesium, the better long-term survival on hemodialysis. Paper presented at the 39th ASN congress, San Diego, 2006 Nov 16–21.
131. Ishimura E, Okuno S, Yamakawa T, Inaba M, Nishizawa Y. Serum magnesium concentration is a significant predictor of mortality in maintenance hemodialysis patients. Magnes Res. 2007;20(4):237–44.
132. Lacson E, Wang W, Lazarus M, Hakim R. Magnesium and Mortality Risk in Hemodialysis Patients. Paper presented at the 42th ASN congress San Diego, 2009 Oct 27–Nov 1.

Chapter 6
Magnesium Intake, Genetic Variants, and Diabetes Risk

Yiqing Song, Cuilin Zhang, Lu Wang, Qi Dai, and Simin Liu

Key Points

- Accumulating evidence suggests that adequate magnesium intake is important in maintaining glucose and insulin homeostasis and thereby in protecting against the development of type 2 DM.
- Prospective data suggest a consistent and inverse association between magnesium intake and risk of type 2 DM in a dose–response manner.
- Genetic defects of some candidate genes may contribute to disturbance of magnesium metabolism, which is one potential mechanism underlying the pathogenesis of type 2 DM.
- The precise mechanisms for the regulatory role of *TRPM6* and *TRPM7* in magnesium homeostasis remain largely undefined; *TRPM6* and *TRPM7* play central roles in magnesium absorption, transport, and renal handling.
- Previous population genetic studies did not find robust and significant association between genetic variants of *TRPM6* and *TRPM7* and diabetes risk but provided suggestive evidence that two common nonsynonymous TRPM6 coding region variants, Ile1393Val and Lys1584Glu polymorphisms, might confer susceptibility to type 2 DM in women with low magnesium intake.
- Further investigation in large and well-defined population studies is warranted to help to decipher the mechanisms underlying the magnesium-type 2 DM association, optimize magnesium requirement for health effects, identify diabetes high-risk population, and tailor prediction and intervention strategies in diabetes research.

Key words Type 2 diabetes mellitus • Cardiometabolic disease • Metabolic-related disorders • Magnesium intake

Y. Song, M.D., Sc.D. (✉) • L. Wang, M.D., Ph.D.
Division of Preventive Medicine, Brigham and Women's Hospital, Harvard Medical School, 900 Commonwealth Avenue East, Boston, MA 02148, USA
e-mail: ysong3@rics.bwh.harvard.edu

C. Zhang, M.D., Ph.D.
Epidemiology Branch, Division of Epidemiology, Statistics, and Prevention Research, Eunice Kennedy Shriver National Institute of Child Health and Human Development, National Institutes of Health, Bethesda, MD, USA

Q. Dai, M.D., Ph.D.
Vanderbilt Epidemiology Center, Institute for Medicine and Public Health, Nashville, TN, USA

S. Liu, M.D., Sc.D., MPH
UCLA Departments of Epidemiology, Medicine, and Obstetrics & Gynecology, Center for Metabolic Disease Prevention, 650 Charles E. Young Drive South, CHS 71-254, Box 951772, Los Angeles, CA 90095, USA
e-mail: siminliu@ucla.edu

R.R. Watson et al. (eds.), *Magnesium in Human Health and Disease*, Nutrition and Health,
DOI 10.1007/978-1-62703-044-1_6, © Springer Science+Business Media New York 2013

Abbreviations

Type 2 DM	Type 2 diabetes mellitus
TRPM6	Transient receptor potential membrane melastatin 6
TRPM7	Transient receptor potential membrane melastatin 7
PLCN-1	Paracellin-1
SLC41A1	Solute carrier family 41 member 1
MRS2	Mitochondrial RNA splicing 2
FXYD2	Sodium-potassium-ATPase, gamma-1 polypeptide
SLC12A3	Solute carrier family 12 (sodium/chloride transporter)

Introduction

Type 2 diabetes mellitus (DM) is a major global public health burden, affecting more than 170 million individuals worldwide [1]. Overall, it is estimated that the prevalence of diabetes will increase by 42% among adults living in developed countries and by 170% among adults in developing countries by 2025 [2, 3]. More alarmingly, by the time type 2 DM is diagnosed, most individuals have developed complications such as peripheral artery disease, renal failure, and neuropathy, and the vast majority of diabetic patients die of these complications [4]. Given the rising global burden of type 2 DM and its devastating complications, there is a great urgency to develop effective strategies to curb the epidemic by identifying individuals at high risk and optimizing prevention and early treatment. The predisposition to type 2 DM varies widely in the population and is largely determined by complex gene–environment interactions.

Accumulating evidence suggests that certain foods and dietary factors may be associated with type 2 DM. In particular, high consumption of whole grains, nuts, fruits, and vegetables has been associated with a lower risk of type 2 DM in various populations. Some micronutrients rich in these foods may contribute to the protective effects. Magnesium is one of essential minerals abundant in whole grains, green leafy vegetables, legumes, and nuts. National survey data indicate that dietary magnesium intake is inadequate in the US general population, particularly among adolescent girls, adult women, and the elderly [5, 6]. Adequate magnesium intake is important in maintaining magnesium status in the human body [7]. Homeostasis of magnesium is tightly regulated and depends on the balance between intestinal absorption and renal excretion. Accumulating evidence suggests that adequate magnesium intake is important in maintaining glucose and insulin homeostasis and in protecting against the development of type 2 DM (Fig. 6.1). However, details on the association between magnesium intake and diabetes risk factors remain largely unclear. Cross-sectional epidemiological studies have shown that magnesium intake correlates significantly with features of the metabolic syndrome (or insulin resistance syndrome), including adiposity, hyperinsulinemia, insulin resistance, hypertriglyceridemia, and low HDL cholesterol and hypertension [8, 9]. The metabolic syndrome, defined as a cluster of metabolic abnormalities including obesity, hyperglycemia, hypertension, and dyslipidemia, is now reaching epidemic proportions worldwide and may reflect a common underlying pathophysiology related to insulin resistance. Moreover, in prospective observational studies, dietary magnesium intake has been inversely associated with the incidence of the metabolic syndrome [10] and related chronic diseases including type 2 DM [11–13], cardiovascular disease [14–16], hypertension [17, 18], and colorectal cancer [19, 20].

Magnesium homeostasis in the human body is tightly regulated and may involve the as-yet-unidentified mechanism underlying the balance between intestinal absorption and renal excretion. Many genes are involved in magnesium uptake, distribution, and metabolism in the human body. Of them, ion channel transient receptor potential membrane melastatin 6 and 7 (*TRPM6* and *TRPM7*) play a central role in magnesium homeostasis. TRPM6 is a magnesium-permeable channel protein

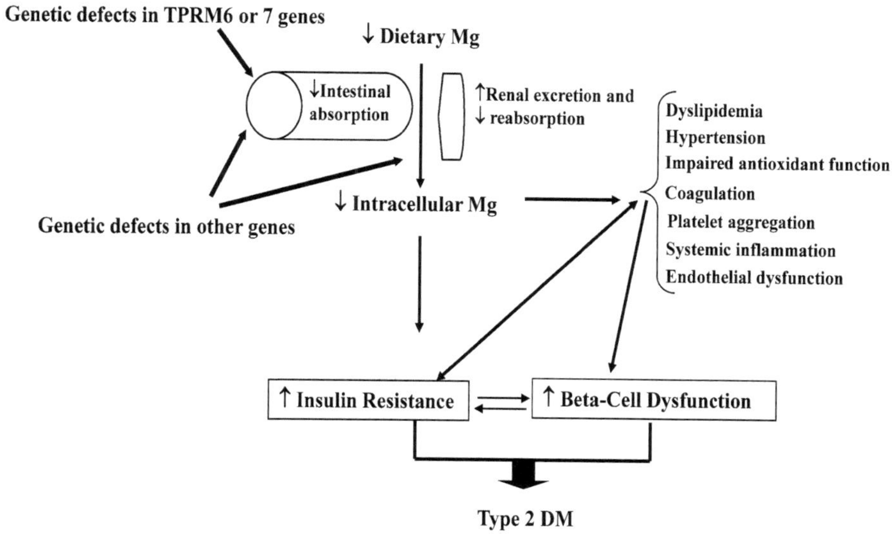

Fig. 6.1 Scheme showing the hypothetic mechanisms for a link between magnesium intake and type 2 DM

primarily expressed in intestinal epithelia and kidney tubules that may play an important role in intestinal and renal magnesium handling [21, 22]. Several loss-of-function mutations in *TRPM6* have been identified among patients with autosomal-recessive familial hypomagnesemia with secondary hypocalcemia [21–23]. Furthermore, a magnesium-deficient diet was shown to upregulate *TRPM6* mRNA expression in mice [24]. Low serum magnesium levels caused by *TRPM6* mutations among HSH patients can be ameliorated by oral supplementation of high doses of magnesium, indicating a potential gene-diet interaction on magnesium homeostasis [25]. *TRPM7* is ubiquitously expressed in various tissues or cell lines [26, 27] and may be a part of a magnesium-sensing and/or uptake mechanism underlying cellular magnesium homeostasis [26, 27]. Recent evidence from human population data suggested that common variants of *TRPM6* and *TRPM7* may confer a discernable susceptibility to type 2 DM, and such a relation was modified by dietary magnesium intake with the association being significant only among individuals with inadequate magnesium intake. Further exploration of potential gene–gene and gene–environmental interactions would contribute to a deeper understanding of the relation between magnesium homeostasis and the etiology of type 2 DM and more importantly the identification of individualized intervention strategy targeted at dietary magnesium intake for the prevention of type 2 DM. This review aims to summarize both epidemiological and experimental observations in relation to magnesium intake and genetic variants to type 2 DM.

Magnesium Intake and Metabolic Phenotypes Related to Type 2 DM

A large body of in vitro and in vivo studies has accumulated for decades, implicating a pivotal role of magnesium intake in cardiometabolic systems. The beneficial effect of magnesium intake on cardiometabolic disease (including hypertension, type 2 DM, and cardiovascular disease) may be explained by several mechanisms, including improvement of glucose and insulin homeostasis, oxidative stress, lipid metabolism, vascular or myocardial contractility, endothelium-dependent vasodilation, antiarrhythmic effects, anticoagulant or antiplatelet effects [28, 29], and anti-inflammatory effects [9].

Of particular note, magnesium may play a critical role in glucose homeostasis, insulin action in peripheral tissues, and pancreatic insulin secretion [7, 28]. Although the exact mechanisms are not well understood, several mechanisms have been proposed. First, magnesium functions as a cofactor for several enzymes critical for glucose metabolism, utilizing high-energy phosphate bonds [7]. Diminished levels of magnesium were observed to decrease tyrosine-kinase activity at insulin receptors [30] and to increase intracellular calcium levels [28], leading to an impairment in insulin signaling. Thus, intracellular magnesium levels have been hypothesized to be important for maintaining insulin sensitivity in skeletal muscle or adipose tissue [28, 31]. Additionally, intracellular magnesium levels may also influence glucose-stimulated insulin secretion in pancreatic β-cells through altered cellular ion metabolism [28], oxidative stress [32], and proinflammatory response [33, 34].

Epidemiologic data provide further support for a pivotal role of magnesium in insulin sensitivity. Some cross-sectional studies have shown an inverse association between plasma or erythrocyte magnesium levels and fasting insulin levels in both diabetic patients and apparently healthy individuals [35, 36]. Several other studies have also found an association between dietary magnesium intake and insulin homeostasis [8, 13, 35, 37]. Overall, cross-sectional evidence has shown that magnesium intake correlates significantly with features of the metabolic syndrome (or insulin resistance syndrome), including adiposity, hyperinsulinemia, insulin resistance, hypertriglyceridemia, and low HDL cholesterol and hypertension [9]. The metabolic syndrome, defined as a cluster of metabolic abnormalities including obesity, hyperglycemia, hypertension, and dyslipidemia, is now reaching epidemic proportions worldwide and may reflect a common underlying pathophysiology related to type 2 DM. Of note, most of epidemiologic studies are ecologic or cross-sectional by design and are potentially confounded by other aspects of diet, lifestyle, or socioeconomic factors. Their results need to be interpreted cautiously, although they may help generate hypotheses.

In human intervention studies, a randomized double-blind and placebo-controlled trial is considered the best approach to examine a cause–effect relation. However, available controlled trials are usually small with short-term follow-up and performed in the secondary prevention setting of chronic disease because of cost and logistical considerations. Numerous small clinical trials have assessed the therapeutic effect of magnesium supplements in individual metabolic syndrome components but yielded inconsistent results. Many factors may have contributed to the inconsistency in findings across these trials, including small sample size, incomplete randomization, lack of blinding in design, variable duration of follow-up, high rates of noncompliance, and differences in magnesium treatment protocols, magnesium formulation and dose, and study populations. Some [38, 39] but not all [40] trials suggest that magnesium supplementation can improve insulin resistance and lower fasting insulin and glucose concentrations among nondiabetic and diabetic subjects. Several short-term metabolic studies and small randomized trials have specifically examined the efficacy of magnesium supplementation in improving insulin sensitivity among nondiabetic individuals, but the results have varied. A meta-analysis of nine randomized double-blind controlled trials reported that a 4- to 16-week magnesium supplementation was effective in reducing fasting plasma glucose levels and raising HDL-C among diabetics; however, no effects were observed on total cholesterol, low-density lipoprotein cholesterol, or triglycerides [41]. Long-term benefits and safety of magnesium treatment in managing metabolic abnormalities remain unclear with the lack of reliable data from large-scale, well-designed randomized controlled trials with long-term follow-up.

Magnesium Intake and the Development of Type 2 DM

Accumulating evidence from both animal models and epidemiological studies supports a potentially important role of magnesium in the development of type 2 DM. Magnesium supplementation has been observed to prevent fructose-induced insulin resistance [42] and to delay the onset of spontaneous type 2 DM in rat models [43].

In earlier clinical studies, hypomagnesemia was frequent among patients with diabetes, especially those with poor metabolic control [28]. Several cross-sectional studies have documented an inverse

Table 6.1 Prospective data on the association between dietary magnesium intake and incident type 2 DM

Study	Population	Follow-up, years	Multivariate RR and p for trend[a]
Nurses' Health Study	85,060	18	0.66 (0.60–0.73), p<0.001
Health Professionals Follow-up Study	42,872	12	0.67 (0.56–0.80), p<0.001
Atherosclerosis Risk in Communities Study	12,128	6	0.95 (0.52–1.74) for Blacks, p=0.47 0.80 (0.56–1.14) for Whites, p=0.49
Iowa Women's Health Study	35,988	6	0.67 (0.55–0.82) P=0.0003
Women's Health Study	39,345	6	BMI>=25: 0.77 (0.61–0.98) p=0.02 BMI<25: 1.77 (0.95–3.32) p=0.29
Melbourne Collaborative Cohort Study	31,641	4	0.55 (0.32–0.97) p=?
EPIC-Potsdam Study	25,067	7	0.99 (0.78–1.26) p=0.87
Black Women's Health Study	41,186	8	0.81 (0.68–0.97) p=0.02
Shanghai Women's Health Study	64,191	6.9	0.80 (0.68–0.93) p<0.0001
The Multiethnic Cohort (MEC) (Hawaii component)	75,512	14	Men: 0.77 (0.70–0.85) p<0.0001 Women: 0.84 (0.76–0.93) p=0.0003
Coronary Artery Risk Development in Young Adults (CARDIA) Study	4,497	20	0.53 (0.32–0.86) p<0.01
Japan Public Health Center-Based Prospective Study	33,919	5	Men: 0.86 (0.63–1.16) p=0.18 Women: 0.92 (0.66–1.28) p=0.96
Japan Collaborative Cohort Study for Evaluation of Cancer Risk	17,592	5	Men: 0.64 (0.38–1.09) p=0.10 Women: 0.68 (0.38–1.22) p=0.20
Meta-analysis (Larsson SC. et al. 2007)	286,668	4–17 (median=6)	0.86 (0.77–0.95)
Meta-analysis (Schulze MB. et al. 2007)	271,869	4–18 (median=6)	0.77 (0.72–0.84)
Meta-analysis (Dong JY et al. 2011)	536,318	4–20 (median=6)	0.78 (0.73–0.84)

[a]Multivariate-adjusted RR represented those comparing the highest category with the lowest category of magnesium intake; p for trend indicated the linear trend across quartiles or quintiles of magnesium intake

association between plasma or erythrocyte magnesium levels and fasting insulin levels in both diabetic patients and apparently healthy individuals [8, 13, 35, 36]. Other cross-sectional studies have also shown an inverse association between serum or plasma concentrations of magnesium and prevalence of type 2 DM [31, 35, 44]. However, it should be noted that evidence from cross-sectional studies cannot imply any causal relation between hypomagnesemia and type 2 DM.

The evidence from prospective studies on magnesium intake and the incidence of type 2 DM appeared to be consistent and paralleled the findings from metabolic studies supporting the role of adequate magnesium intake in maintaining insulin sensitivity in both healthy and diabetic participants [31, 45]. Table 6.1 summarizes prospective evidence for the association between dietary magnesium intake and the risk of type 2 DM. When the data from these independent prospective cohorts were pooled together using meta-analysis analytic strategies by three independent groups, the pooled estimates of RRs comparing the highest category of dietary magnesium intake with the lowest category of intake were 0.86 (95% CI: 0.77–0.95) [46], 0.77 (95% CI: 0.72–0.84) [47], and 0.78 (95% CI: 0.73–0.84) [48], respectively. Taken together, available prospective data clearly demonstrate a significant and inverse association between magnesium intake and risk of type 2 DM in a dose–response manner.

Genetic Mechanism Underlying Magnesium Homeostasis

Magnesium homeostasis in the human body is tightly regulated and may involve the as-yet-unidentified mechanism underlying the balance between intestinal absorption and renal excretion. Intestinal magnesium absorption occurs via two different pathways: a saturable active transcellular transport and a nonsaturable paracellular passive transport [22]. In the kidney, about 80% of total serum magnesium is filtered in the glomeruli, with 70% reabsorbed in the loop of Henle and 5–10% in the distal convoluted tubule (DCT) [49]. Genetic defects in both renal magnesium reabsorption and intestinal absorption result in the hypomagnesemia were observed in some familial diseases [21, 22, 50]. In addition, recent data from in vitro and in vivo studies have identified several genes encoding magnesium transport proteins in both prokaryotic and eukaryotic cells (summarized in Table 6.2), highlighting a genetic component in the regulation mechanisms underlying cellular magnesium homeostasis.

Highly Selective Magnesium-Permeable Channel Proteins

Transient receptor potential membrane melastatin 6 (*TRPM6*): *TRPM6*, located on chromosome 9q21.13, comprises 39 exons spanning 163 kb of genomic sequence that encodes 2,022 amino acids (Fig. 6.2a). *TRPM6* mRNA is expressed primarily along the intestine (duodenum, jejunum, ileum, and colon) and the DCT of the kidney [51]. *TRPM6* is a magnesium-permeable channel protein localized to the apical domain of the DCT and the brush-border membrane of the absorptive cells in the duodenum, the main sites of transepithelial magnesium transport [51]. Thus, *TRPM6* may play an important role in intestinal and renal magnesium handling. *TRPM6* and *TRPM7* are unique channel kinases with both ion channel and protein kinase activity [52]. *TRPM6* forms functional homomeric channels by itself and functional heteromeric *TRPM6/7* complexes with *TRPM7*. *TRPM6*, *TRPM7*, and *TRPM6/7* are three distinct types of channels and display different divalent cation permeability profiles [53]. *TRPM6* and *TRPM7* have similar electrophysiological properties, but their functions are not redundant [54]. Mutations in *TRPM6*, either incomplete or complete loss-of-function, cause autosomal-recessive hypomagnesemia with secondary hypocalcemia (HSH). Individuals with HSH display seizures and muscle spasms or tetany in early infancy, which may lead to permanent neurological damage or death if untreated. Laboratory evaluations reveal extremely low serum magnesium and calcium levels. For HSH patients, supplementation with high oral doses of magnesium can ameliorate the symptoms associated with *TRPM6* mutations and help elevate serum magnesium levels [21, 22, 55, 56]. The dysfunction of the *TRPM6*-dependent magnesium influx pathway in patients with HSH provides a genetic model for severe hypomagnesemia. It is therefore plausible that genetic variation in *TRPM6* may be important in regulating intestinal and renal magnesium handling in the general population.

TRPM7 (Fig. 6.2b), located on chromosome 15q21, spans 127 kb of DNA sequence and encodes a protein of 1,863 amino acids [57]. *TRPM7* is ubiquitously expressed in various tissues or cell lines. *TRPM7* is a highly regulated transmembrane magnesium transporter functionally important in regulating cellular magnesium influx and maintaining intracellular magnesium levels [57]. Electrophysiological analyses have shown that *TRPM7* is a divalent cation-permeable channel that preferentially allows magnesium flux under physiologic conditions [54, 58]. Furthermore, *TRPM7* is inhibited by cytosolic magnesium levels as well as by intracellular levels of magnesium and magnesium-ATP complexes [53, 59]. Targeted gene deletion of *TRPM7* in DT40 B-lymphocyte cell lines led to growth arrest and lethality [60]. However, both the viability and the proliferation of *TRPM7*-deficient B cells were observed to be rescued by supplementing the growth medium with magnesium, indicating an essential role of *TRPM7*-dependent magnesium influx for cell viability [60]. In vitro studies have shown that *TRPM6* and *TRPM7* are not functionally redundant; the trafficking of *TRPM6* to the cell surface requires *TRPM7* co-expression, and *TRPM6* can modulate *TRPM7* function, but not vice versa [54]. In light of its ubiquitous expression pattern and its constitutive activity, *TRPM7* may act as a magnesium-sensing and/or uptake mechanism required for cellular magnesium homeostasis [57].

The precise mechanisms for the regulatory role of *TRPM6* and *TRPM7* in magnesium homeostasis remain largely undefined. It has been hypothesized that intestinal magnesium absorption occurs via two different pathways: an active transcellular transport and a passive paracellular transport [22].

Table 6.2 A list of potential candidate genes for magnesium homeostasis in human

		Biological mechanism[a]		
Gene	Initials	Transporter type	Location	Function
Transient receptor potential membrane melastatin 6	TRPM6	Channel	Cell membrane	Intestinal absorption and renal reabsorption
Transient receptor potential membrane melastatin 7	TRPM7	Channel	Cell membrane	
Paracellin-1 or claudin-16	PCLN-1	Channel	Cell membrane	Paracellular calcium and magnesium fluxes in the nephron
Claudin-19	Claudin-19	Channel	Cell membrane	Renal calcium and magnesium reabsorption
Magnesium transporter 1	MagT1	Channel	Cell membrane	Cellular magnesium transport
Solute carrier family 41 member 1	SLC41A1	Carrier	Cell membrane	Relatively nonselective cellular magnesium and other divalent cation transport
Solute carrier family 41 member 2	SLC41A2	Carrier	Cell membrane	Cellular magnesium and other divalent cation transport
Solute carrier family 12 member 3	SLC12A3	Carrier	Cell membrane	Renal sodium, chloride, and magnesium reabsorption
Ancient conserved domain-containing protein 1	ACDP1	Carrier	Cell membrane	Cellular divalent cation transport
Ancient conserved domain-containing protein 2	ACDP2	Carrier	Cell membrane	Cellular divalent cation transport
Non-imprinted in Prader-Willi/Angelman syndrome 1	NIPA1 (SPG6)	Carrier	Cell membrane	Cellular magnesium transport
Non-imprinted in Prader-Willi/Angelman syndrome 1	NIPA2	Carrier	Cell membrane	Cellular magnesium transport
Huntingtin-interacting protein 14	HIP14	Carrier	Cell membrane	Cellular magnesium transport
Huntingtin-interacting protein 14-like protein	H1P14L	Carrier	Cell membrane	Cellular magnesium transport
Mitochondrial RNA splicing 2	Mrs2	Channel	Mitochondria	Mitochondrial magnesium homeostasis
Membrane magnesium transporter 1	MMgT1	Channel	Golgi system	Regulation of magnesium-dependent enzymes
Membrane magnesium transporter 1	MMgT2	Channel	Golgi system	Cellular magnesium uptake
Sodium-potassium-ATPase, gamma-1 polypeptide	FXYD2 or ATP1G1	Channel	Cellular member in the nephron	Renal sodium and magnesium handling
Sodium-dependent exchanger (sodium/magnesium exchanger)	ND	Exchanger	Cell membrane	Cellular magnesium extrusion

[a]Information is based on either publicly accessible function databases or published papers

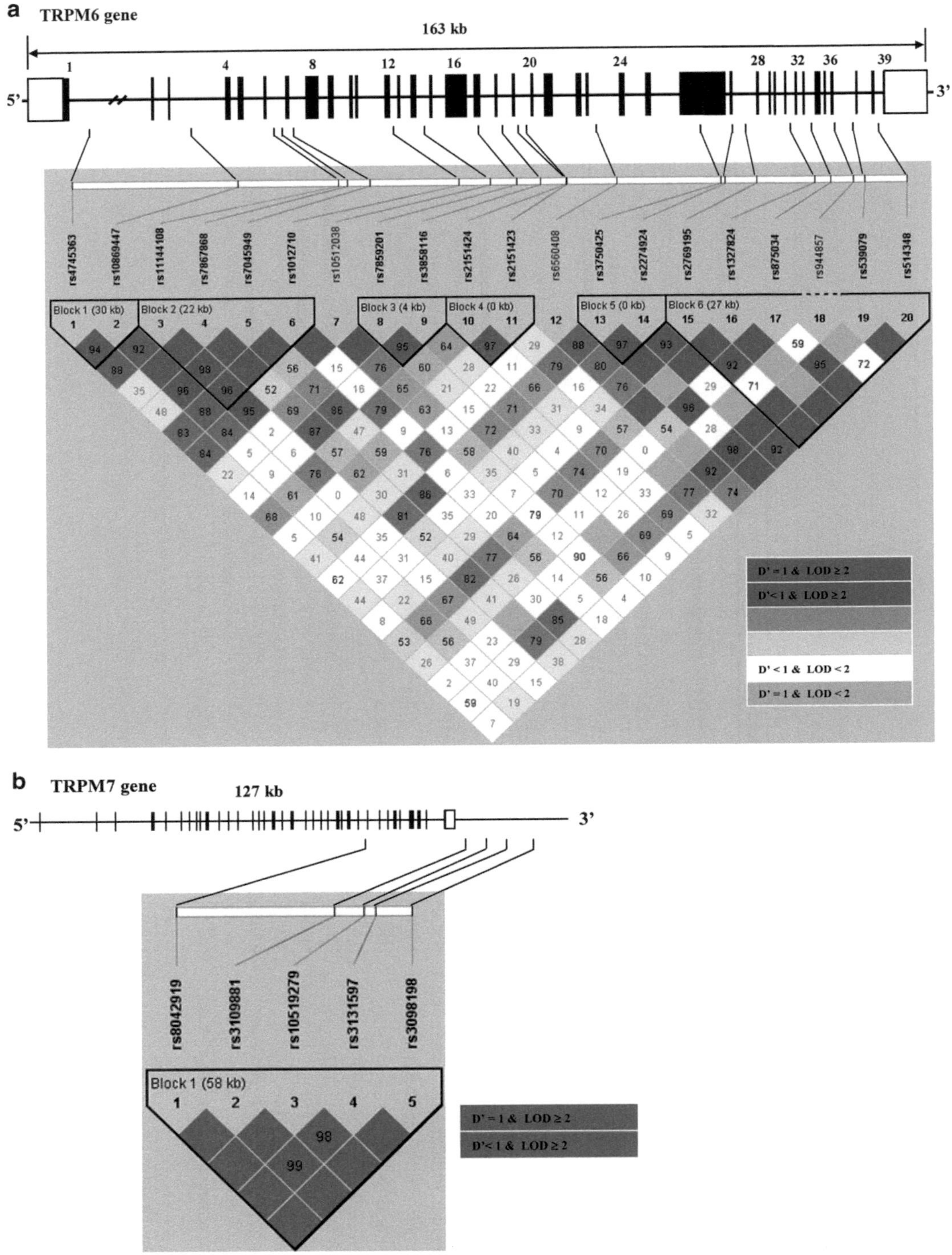

Fig. 6.2 Location and linkage disequilibrium (LD) map of 20 SNPs in *TRPM6* (**a**) and 5 SNPs in *TRPM7* (**b**). Published figure from the article by Song et al., BMC Med Gen. 2009;10:4 (Reproduced with permission from BioMed Central)

High magnesium concentrations in intestinal tracts may overcome the genetic defects of magnesium absorption and independently increase magnesium absorption via the passive paracellular pathway. In contrast, when dietary magnesium intake is inadequate, the function of *TRPM6* in active intestinal magnesium absorption and renal reabsorption becomes very important.

Paracellin-1 (claudin-16, *PCLN1*): *PCLN1* is a new member of the claudin tight junction protein family and mediates paracellular magnesium and calcium reabsorption through tight junctions in renal tubular epithelia. *PCLN1* is located on chromosome 3q27, spans 23 kb of DNA sequence (5 exons), and encodes a protein of 305 amino acids [61]. Exon–intron analysis of both human and mouse genes revealed a 100% homology of coding exon lengths and splice-site loci, indicating evolutionary sequence conservation [61]. Paracellin-1 is a renal tight junction protein and is exclusively identified in the kidney, especially in the thick ascending limb of the loop of Henle [61, 62]. In the kidney, the vast majority of filtered magnesium (65–75%) is reabsorbed in the thick ascending limb via the passive paracellular route, mediated by paracellin-1 [62]. Loss-of-function mutations in *PCLN1* were associated with familial hypomagnesemia with hypercalciuria and nephrocalcinosis (FHHNC, OMIM 248250), a hereditary renal disease with urinary magnesium and calcium loss, and progression to end-stage renal failure during childhood or adolescence [50, 63]. Although the precise role of *PCLN1* in magnesium homeostasis is still not clear, it is likely that *PCLN1* plays a role in regulating magnesium absorption in intestines and kidney.

Other Specific Magnesium Transport Proteins

Solute carrier family 41 member 1 (*SLC41A1*): *SLC41A1* is a member of the human solute carrier family, as designated by the human genome nomenclature committee. *SLC41A1* comprises 10 putative transmembrane domains, 2 of which are highly homologous to the integral membrane part of the prokaryote MgtE protein family, a magnesium transporter found in certain bacteria (an overall similarity of 15%) [64]. Human *SLC41A1* is located on chromosome 1q32.1 and comprises 11 exons spanning 25 kb of genomic sequence that encodes 513 amino acids. The mouse *SLC41A1* cDNA is similar to the human sequence (98%) [64]. In mice, *SLC41A1* mRNA is expressed in the brain, kidney, liver, colon, and small intestine [65]. *SLC41A1* transcription is upregulated by low magnesium: *SLC41A1* mRNA significantly increased in the kidney (4.0-fold), colon (2.1-fold), and heart (2.0-fold) in mice fed low-magnesium diets as compared with those maintained on normal diets for 5 days [65]. Although *SLC41A1* is a relatively nonselective divalent cation transporter, it prefers magnesium cation to other divalent cations as a substrate [65]. *SLC41A1* does not regulate calcium transport and is not influenced by calcium concentrations [65]. Despite its unknown functions, *SLC41A1* transporter is a likely candidate transporter contributing to cellular magnesium homeostasis.

Mitochondrial RNA splicing 2 (*MRS2*): *MRS2* gene is the first mitochondrial magnesium transporter gene identified in bacteria and plants [66]. A homologue of the yeast *MRS2* protein, the mammalian *MRS2* gene has recently been cloned [67]. Because of its structural similarity to the bacterial magnesium transport protein CorA, mammalian *MRS2* has been hypothesized to exhibit a similar effect on magnesium influx into the organelle. As an integral protein of the inner mitochondrial membrane, *MRS2* protein may form a homooligomeric magnesium channel to mediate magnesium influx in mammalian mitochondria driven by the inner membrane potential. The human *MRS2* gene is located on chromosome 6 (6p22.1–22.3) and is composed of 11 exons [67]. Despite little evidence for the function of human *MRS2* protein, it is presumed to be essential for the magnesium influx system in mitochondria (like its yeast homologue) and may thus be a candidate for genetic disturbances in cellular magnesium homeostasis. Recent research on the interrelationships among mitochondrial dysfunction, hypomagnesemia, and type 2 DM has also shed light on the underlying molecular basis for mitochondrial-dependent magnesium transport [68].

Magnesium Homeostasis-Related Channel Proteins

Sodium-Potassium-ATPase, Gamma-1 Polypeptide (*FXYD2* or *ATP1G1*)

The *FXYD2* gene is located on chromosome 11q23 and encodes a gamma subunit of the basolateral sodium-potassium-ATPase (Na-K-ATPase) in the DCT segment of the nephron (FXYD2a and FXYD2b). *FXYD2* is predominantly expressed in the kidney [69]. FXYD2a and FXYD2b colocalize with Na-K-ATPase in the basolateral membrane of renal epithelial cells and regulate the sodium affinity of Na-K-ATPase [70]. Although several members of the FXYD protein family are involved in the regulation of renal sodium and potassium handling, only a mutation in *FXYD2* has been linked to cases of human hypomagnesemia. A heterozygous mutation, 121 G-A (a substitution of arginine for glycine, G41R), in the *FXYD2* gene, accounted for dominant isolated renal magnesium loss in the Dutch patients [71].

Solute Carrier Family 12 (Sodium/Chloride transporter; *SLC12A3*)

SLC12A3 is a member of the human solute carrier family encoding the renal thiazide-sensitive sodium/chloride (Na-Cl) cotransporter at the apical membrane of the distal convoluted tubule (DCT). Human *SLC12A3* is located on chromosome 16q13 and comprises 26 exons spanning 49 kb of genomic sequence that encodes 1,021 amino acids [72]. *SLC12A3* mRNA is expressed primarily in the apical membrane of distal convoluted tubule (DCT) of kidney and is believed to be the principal mediator of sodium and chloride reabsorption in this nephron segment. More than 100 different mutations have been reported, and most of them seemed to be loss-of-function. Missense mutations of the *SLC12A3* gene were found in patients with Gitelman syndrome, in which hypomagnesemia is seen in combination with hypocalciuria, hypokalemia, and metabolic alkalosis [73–75].

Despite recent findings of magnesium-related genes, the precise mechanism underlying magnesium homeostasis in the human body still remains uncertain. So far, limited information is available about the genuine function of other genes that may be involved in cellular magnesium intake, transport, and extrusion.

Interactions Between Common Genetic Variants of *TRPM6* and *TRPM7* and Magnesium Intake in Association with the Risk of Type 2 Diabetes

As described above, ion channel transient receptor potential membrane melastatin 6 and 7 (*TRPM6* and *TRPM7*) play a central role in magnesium homeostasis, which is critical for maintaining normal glucose and insulin metabolism. However, it is unclear whether common genetic variation in *TRPM6* and *TRPM7* contributes to the risk of type 2 diabetes. Recently, we conducted a nested case–control study of 359 incident diabetes cases and 359 matched controls in the Women's Health Study to address this question [76]. On the basis of the publicly accessible dataset, we focused on common SNPs including intronic SNPs, synonymous SNPs, and nonsynonymous SNPs to characterize genetic variation spanning the *TRPM6* gene, including at least 30 kb upstream and downstream of the coding regions. We analyzed 20 haplotype-tagging single nucleotide polymorphisms (SNPs) in *TRPM6* and 5 common SNPs in *TRPM7* for their association with diabetes risk (Fig. 6.2). Overall, there was no robust and significant association between any single SNP and diabetes risk. Neither was there any evidence of association between common *TRPM6* and *TRPM7* haplotypes and diabetes risk. Our haplotype analyses suggested a significant higher risk of type 2 diabetes among carriers of both the rare alleles from two nonsynonymous SNPs in *TRPM6* (Val1393Ile in exon 29 [rs3750425] and

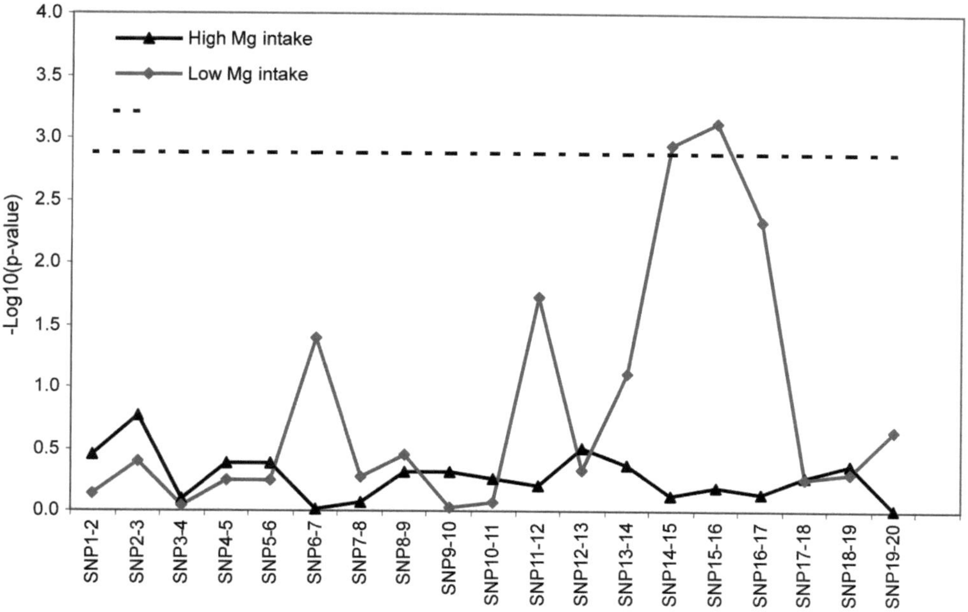

Fig. 6.3 Sliding window analysis for the association of haplotypes of TRPM6 gene and type 2 DM. Haplotype was reconstructed by 2-SNP sets. The omnibus test was used for the *TRPM6* haplotype-type 2 diabetes associations for each sliding window set. Empirical p-value was based on permutation test. The dashed line represents a $-\log_{10}P$ value of 2.88 (corresponding to P=0.0013), which was used as the global significance threshold by Bonferroni correction for 19 window frames. Published figure from the article by Song et al., BMC Med Gen. 2009;10:4 (Reproduced with permission from BioMed Central)

Lys1584Glu in exon 30 [rs2274924]) when their magnesium intake was lower than 250 mg/day [76] (Fig. 6.3). Compared with noncarriers, women who were homozygous carriers of the haplotype 1393Ile-1584Glu had an increased risk of type 2 diabetes (OR, 4.92, 95% CI, 1.05–23.0) only when they had low magnesium intake (<250 mg/day) (Fig. 6.4). Our results provide suggestive evidence that two common nonsynonymous *TRPM6* coding region variants, Ile1393Val and Lys1584Glu polymorphisms, might confer susceptibility to type 2 diabetes in women with low magnesium intake [76]. In our study population, approximately 33% are heterozygous for the rare C allele of rs2274924 and 18% heterozygous for the allele T of rs3750425. Both SNPs are also very common in populations with various genetic backgrounds, with the frequencies of 0.07–0.21 for the rs3750425 T allele (1393 Ile) and 0.15–0.36 for the rs2274924 C allele (1584Glu) in four ethnic groups from Hapmap database. Two changed amino acids are located between the coiled region and kinase near the C-terminal [23] and are thus unlikely responsible for direct regulation of the *TRPM6* channel transmembrane structures and kinase activity. However, we cannot rule out the possibility that they may lead to changes in protein conformation and thus reduce *TRPM6* channel activity. If our finding is replicated in future studies, it will suggest that common genetic variants in the *TRPM6* locus known to harbor severe mutations causing monogenic magnesium deficiency confers a modest susceptibility to the risk of type 2 diabetes in a small subgroup of the general population.

Given a limited number of SNPs (n=5) across *TRPM7* (128 kb) [76], it is likely that *TRPM7*, as a housekeeping gene regulating cellular magnesium metabolism, may truly have limited genetic variability. Biologically, *TRPM7* is ubiquitously expressed, and its constitutive activation is required for cellular survival. Animal studies showed that dietary magnesium restriction did not alter *TRPM7* mRNA expression in mouse kidney and colon [24].

As described above, several other candidate genes involved in magnesium bioavailability in the human body may also affect the risk of type 2 DM, but the relative importance of each gene involved

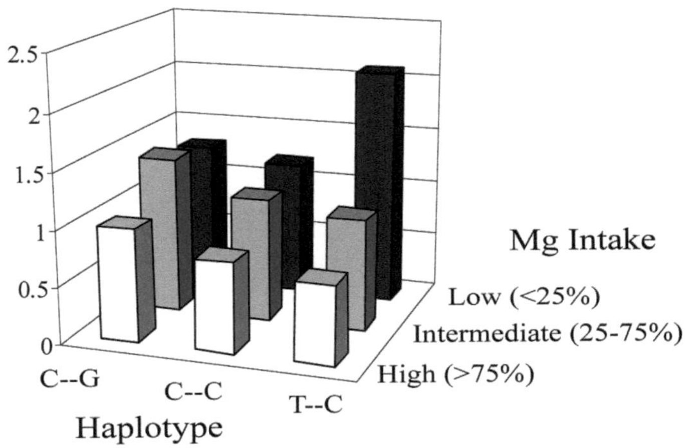

Fig. 6.4 Joint association between two missense SNP-based haplotype of TRPM6 gene and the risk of type 2 DM stratified by levels of magnesium intake. This figure was created based on the values from the article by Song et al., BMC Med Gen. 2009;10:4. The haplotype based on two nonsynonymous SNPs: Ile1393Val (C/T) in exon 29 and Lys1584 Glu (G/C) in exon 30

in magnesium metabolism pathways has not yet been clarified. Future studies in this research area will have to include both genetic variants and magnesium status (dietary or biochemical marker of magnesium) to unravel the relative importance of common genetic variation in each candidate gene for diabetes risk.

Implications and Future Research

Type 2 DM is a complex disorder caused by the interplay between genetic susceptibility and environment factors (such as diets and medications). Genome-wide association studies (GWAS) approach for type 2 DM has achieved great success in identifying novel susceptibility variants [77], but it remains uncertain whether many more genetic variants with smaller effects that are not being identified in the GWAS could contribute to the "missing" heritability due to their interactions with environment factors. Magnesium is one of the most promising nutritional factor for prevention of type 2 DM. Despite a century of research on potential health impact of magnesium, there has been a longstanding debate over the inconsistent results of dietary or supplemental magnesium and type 2 DM risk from observational studies. Moreover, due to the lack of robust evidence for gene-magnesium interactions, it remains uncertain whether magnesium intake may benefit or risk individuals only under specific genetic backgrounds. Growing evidence suggests that variants in many genes that are involved in magnesium uptake, distribution, metabolism, and other molecular pathways may be important determinants for magnesium metabolism in the human. Future epidemiological studies integrating both genetic information and nutrition data have the potential to facilitate better understanding biological mechanisms underlying health impact of magnesium intake. Of note, measurement errors in the dietary assessment, including potential dietary change over the course of follow-up and residual confounding by poorly measured or unmeasured variables and highly correlated nutrients, may have substantially limited the ability of large cohort studies to elucidate the causal effect of any single nutrient on disease risk. Serum magnesium concentrations are still the most commonly used metric to define magnesium deficiency in humans and have been shown to have a heritable component. Recent GWAS identified common genetic variants in at least six genomic regions (including *TRPM6*) that are reproducibly associated with serum magnesium levels and clinically defined hypomagnesemia [78].

Although these SNPs explained about 1.6% of variation in serum magnesium levels [78], further investigation of these newly implicated genomic loci may provide invaluable insights into the regulatory mechanisms of magnesium homeostasis in human. From a mechanistic perspective, there is compelling need for the development of a reliable method to measure total body magnesium store and levels of intracellular magnesium or biologically active ionized or free magnesium. More accurate, reliable, and affordable methods to assess individual magnesium status in large population studies would provide more informative answers regarding the role of magnesium intake in the development of metabolic-related disorders.

Summary

Emerging evidence has suggested that several genetic factors play central roles in magnesium metabolism in the human. It has become increasingly important to consider molecular and genetic variations in the homeostatic regulation of magnesium metabolism and their roles in the etiology of type 2 DM. Recent studies have implicated the interaction of genetic variants in *TRPM7* or *TRPM6* with magnesium intake in association with type 2 DM. The findings, though preliminary, may advance our understanding of the pathophysiological role of magnesium homeostasis in the pathogenesis of type 2 DM. Further investigation is warranted to decipher the mechanisms underlying the magnesium-type 2 DM association, optimize the amount of magnesium requirement for health effects, identify individuals who are at high risk and may benefit most from increasing magnesium intake, and tailor intervention strategies for the prevention of type 2 DM.

References

1. Stumvoll M, Goldstein BJ, van Haeften TW. Type 2 diabetes: principles of pathogenesis and therapy. Lancet. 2005;365:1333–46.
2. King H, Aubert RE, Herman WH. Global burden of diabetes, 1995–2025: prevalence, numerical estimates, and projections. Diabetes Care. 1998;21:1414–31.
3. Wild S, Roglic G, Green A, Sicree R, King H. Global prevalence of diabetes: estimates for the year 2000 and projections for 2030. Diabetes Care. 2004;27:1047–53.
4. Nathan DM. Long-term complications of diabetes mellitus. N Engl J Med. 1993;328:1676–85.
5. Ford ES, Mokdad AH. Dietary magnesium intake in a national sample of us adults. J Nutr. 2003;133:2879–82.
6. Vaquero MP. Magnesium and trace elements in the elderly: intake, status and recommendations. J Nutr Health Aging. 2002;6:147–53.
7. Saris NE, Mervaala E, Karppanen H, Khawaja JA, Lewenstam A. Magnesium. An update on physiological, clinical and analytical aspects. Clin Chim Acta. 2000;294:1–26.
8. Fung TT, Manson JE, Solomon CG, Liu S, Willett WC, Hu FB. The association between magnesium intake and fasting insulin concentration in healthy middle-aged women. J Am Coll Nutr. 2003;22:533–8.
9. Song Y, Ridker PM, Manson JE, Cook NR, Buring JE, Liu S. Magnesium intake, c-reactive protein, and the prevalence of metabolic syndrome in middle-aged and older U.S. Women. Diabetes Care. 2005;28:1438–44.
10. He K, Liu K, Daviglus ML, Morris SJ, Loria CM, Van Horn L, et al. Magnesium intake and incidence of metabolic syndrome among young adults. Circulation. 2006;113:1675–82.
11. Colditz GA, Manson JE, Stampfer MJ, Rosner B, Willett WC, Speizer FE. Diet and risk of clinical diabetes in women. Am J Clin Nutr. 1992;55:1018–23.
12. Salmeron J, Ascherio A, Rimm EB, Colditz GA, Spiegelman D, Jenkins DJ, et al. Dietary fiber, glycemic load, and risk of NIDDM in men. Diabetes Care. 1997;20:545–50.
13. Song Y, Manson JE, Buring JE, Liu S. Dietary magnesium intake in relation to plasma insulin levels and risk of type 2 diabetes in women. Diabetes Care. 2004;27:59–65.
14. Abbott RD, Ando F, Masaki KH, Tung KH, Rodriguez BL, Petrovitch H, et al. Dietary magnesium intake and the future risk of coronary heart disease (the Honolulu heart program). Am J Cardiol. 2003;92:665–9.

15. Al-Delaimy WK, Rimm EB, Willett WC, Stampfer MJ, Hu FB. Magnesium intake and risk of coronary heart disease among men. J Am Coll Nutr. 2004;23:63–70.
16. Song Y, Manson JE, Cook NR, Albert CM, Buring JE, Liu S. Dietary magnesium intake and risk of cardiovascular disease among women. Am J Cardiol. 2005;96:1135–41.
17. Ascherio A, Rimm EB, Giovannucci EL, Colditz GA, Rosner B, Willett WC, et al. A prospective study of nutritional factors and hypertension among us men. Circulation. 1992;86:1475–84.
18. Song Y, Sesso HD, Manson JE, Cook NR, Buring JE, Liu S. Dietary magnesium intake and risk of incident hypertension among middle-aged and older us women in a 10-year follow-up study. Am J Cardiol. 2006;98:1616–21.
19. Folsom AR, Hong CP. Magnesium intake and reduced risk of colon cancer in a prospective study of women. Am J Epidemiol. 2006;163:232–5.
20. Larsson SC, Bergkvist L, Wolk A. Magnesium intake in relation to risk of colorectal cancer in women. JAMA. 2005;293:86–9.
21. Walder RY, Landau D, Meyer P, Shalev H, Tsolia M, Borochowitz Z, et al. Mutation of trpm6 causes familial hypomagnesemia with secondary hypocalcemia. Nat Genet. 2002;31:171–4.
22. Schlingmann KP, Weber S, Peters M, Niemann Nejsum L, Vitzthum H, Klingel K, et al. Hypomagnesemia with secondary hypocalcemia is caused by mutations in trpm6, a new member of the trpm gene family. Nat Genet. 2002;31:166–70.
23. Schlingmann KP, Konrad M, Seyberth HW. Genetics of hereditary disorders of magnesium homeostasis. Pediatr Nephrol. 2004;19:13–25.
24. Groenestege WM, Hoenderop JG, van den Heuvel L, Knoers N, Bindels RJ. The epithelial mg2+ channel transient receptor potential melastatin 6 is regulated by dietary mg2+ content and estrogens. J Am Soc Nephrol. 2006;17:1035–43.
25. Shalev H, Phillip M, Galil A, Carmi R, Landau D. Clinical presentation and outcome in primary familial hypomagnesaemia. Arch Dis Child. 1998;78:127–30.
26. Takezawa R, Schmitz C, Demeuse P, Scharenberg AM, Penner R, Fleig A. Receptor-mediated regulation of the trpm7 channel through its endogenous protein kinase domain. Proc Natl Acad Sci USA. 2004;101:6009–14.
27. Montell C. The trp superfamily of cation channels. Sci STKE. 2005;2005:re3.
28. Barbagallo M, Dominguez LJ, Galioto A, Ferlisi A, Cani C, Malfa L, et al. Role of magnesium in insulin action, diabetes and cardio-metabolic syndrome x. Mol Aspects Med. 2003;24:39–52.
29. Chakraborti S, Chakraborti T, Mandal M, Mandal A, Das S, Ghosh S. Protective role of magnesium in cardiovascular diseases: a review. Mol Cell Biochem. 2002;238:163–79.
30. Suarez A, Pulido N, Casla A, Casanova B, Arrieta FJ, Rovira A. Impaired tyrosine-kinase activity of muscle insulin receptors from hypomagnesaemic rats. Diabetologia. 1995;38:1262–70.
31. Paolisso G, Barbagallo M. Hypertension, diabetes mellitus, and insulin resistance: the role of intracellular magnesium. Am J Hypertens. 1997;10:346–55.
32. Giugliano D, Paolisso G, Ceriello A. Oxidative stress and diabetic vascular complications. Diabetes Care. 1996;19:257–67.
33. Weglicki WB, Phillips TM, Freedman AM, Cassidy MM, Dickens BF. Magnesium-deficiency elevates circulating levels of inflammatory cytokines and endothelin. Mol Cell Biochem. 1992;110:169–73.
34. Kurantsin-Mills J, Cassidy MM, Stafford RE, Weglicki WB. Marked alterations in circulating inflammatory cells during cardiomyopathy development in a magnesium-deficient rat model. Br J Nutr. 1997;78:845–55.
35. Ma J, Folsom AR, Melnick SL, Eckfeldt JH, Sharrett AR, Nabulsi AA, et al. Associations of serum and dietary magnesium with cardiovascular disease, hypertension, diabetes, insulin, and carotid arterial wall thickness: the aric study. Atherosclerosis risk in communities study. J Clin Epidemiol. 1995;48:927–40.
36. Rosolova H, Mayer O, Reaven GM. Insulin-mediated glucose disposal is decreased in normal subjects with relatively low plasma magnesium concentrations. Metabolism. 2000;49:418–20.
37. Manolio TA, Savage PJ, Burke GL, Hilner JE, Liu K, Orchard TJ, et al. Correlates of fasting insulin levels in young adults: the cardia study. J Clin Epidemiol. 1991;44:571–8.
38. Guerrero-Romero F, Tamez-Perez HE, Gonzalez-Gonzalez G, Salinas-Martinez AM, Montes-Villarreal J, Trevino-Ortiz JH, et al. Oral magnesium supplementation improves insulin sensitivity in non-diabetic subjects with insulin resistance. A double-blind placebo-controlled randomized trial. Diabetes Metab. 2004;30:253–8.
39. Paolisso G, Sgambato S, Gambardella A, Pizza G, Tesauro P, Varricchio M, et al. Daily magnesium supplements improve glucose handling in elderly subjects. Am J Clin Nutr. 1992;55:1161–7.
40. Purvis JR, Cummings DM, Landsman P, Carroll R, Barakat H, Bray J, et al. Effect of oral magnesium supplementation on selected cardiovascular risk factors in non-insulin-dependent diabetics. Arch Fam Med. 1994;3:503–8.
41. Song Y, He K, Levitan EB, Manson JE, Liu S. Effects of oral magnesium supplementation on glycaemic control in type 2 diabetes: a meta-analysis of randomized double-blind controlled trials. Diabet Med. 2006;23:1050–6.
42. Balon TW, Jasman AP, Scott S, Meehan WP, Rude RK, Nadler JL. Dietary magnesium prevents fructose-induced insulin insensitivity in rats. Hypertension. 1994;23:1036–9.

43. Balon TW, Gu JL, Tokuyama Y, Jasman AP, Nadler JL. Magnesium supplementation reduces development of diabetes in a rat model of spontaneous NIDDM. Am J Physiol. 1995;269:E745–52.
44. Kao WH, Folsom AR, Nieto FJ, Mo JP, Watson RL, Brancati FL. Serum and dietary magnesium and the risk for type 2 diabetes mellitus: the atherosclerosis risk in communities study. Arch Intern Med. 1999;159:2151–9.
45. de Lourdes LM, Cruz T, Carreiro Pousada J, Rodrigues LE, Barbosa K, Canguca V. The effect of magnesium supplementation in increasing doses on the control of type 2 diabetes. Diabetes Care. 1998;21:682–6.
46. Larsson SC, Wolk A. Magnesium intake and risk of type 2 diabetes: a meta-analysis. J Intern Med. 2007;262:208–14.
47. Schulze MB, Schulz M, Heidemann C, Schienkiewitz A, Hoffmann K, Boeing H. Fiber and magnesium intake and incidence of type 2 diabetes: a prospective study and meta-analysis. Arch Intern Med. 2007;167:956–65.
48. Dong JY, Xun P, He K, Qin LQ. Magnesium intake and risk of type 2 diabetes: meta-analysis of prospective cohort studies. Diabetes Care. 2011;34:2116–22.
49. Dai LJ, Ritchie G, Kerstan D, Kang HS, Cole DE, Quamme GA. Magnesium transport in the renal distal convoluted tubule. Physiol Rev. 2001;81:51–84.
50. Simon DB, Lu Y, Choate KA, Velazquez H, Al-Sabban E, Praga M, et al. Paracellin-1, a renal tight junction protein required for paracellular mg2+ resorption. Science. 1999;285:103–6.
51. Voets T, Nilius B, Hoefs S, van der Kemp AW, Droogmans G, Bindels RJ, et al. Trpm6 forms the mg2+ influx channel involved in intestinal and renal mg2+ absorption. J Biol Chem. 2004;279:19–25.
52. Nilius B, Voets T, Peters J. Trp channels in disease. Sci STKE. 2005;2005:re8.
53. Li M, Jiang J, Yue L. Functional characterization of homo- and heteromeric channel kinases trpm6 and trpm7. J Gen Physiol. 2006;127:525–37.
54. Schmitz C, Dorovkov MV, Zhao X, Davenport BJ, Ryazanov AG, Perraud AL. The channel kinases trpm6 and trpm7 are functionally nonredundant. J Biol Chem. 2005;280:37763–71.
55. Schlingmann KP, Sassen MC, Weber S, Pechmann U, Kusch K, Pelken L, et al. Novel trpm6 mutations in 21 families with primary hypomagnesemia and secondary hypocalcemia. J Am Soc Nephrol. 2005;16:3061–9.
56. Chubanov V, Waldegger S, Mederos y Schnitzler M, Vitzthum H, Sassen MC, Seyberth HW, et al. Disruption of trpm6/trpm7 complex formation by a mutation in the trpm6 gene causes hypomagnesemia with secondary hypocalcemia. Proc Natl Acad Sci USA. 2004;101:2894–9.
57. Wolf FI. Trpm7: channeling the future of cellular magnesium homeostasis? Sci STKE. 2004;2004:pe23.
58. Jiang J, Li M, Yue L. Potentiation of trpm7 inward currents by protons. J Gen Physiol. 2005;126:137–50.
59. Demeuse P, Penner R, Fleig A. Trpm7 channel is regulated by magnesium nucleotides via its kinase domain. J Gen Physiol. 2006;127:421–34.
60. Schmitz C, Perraud AL, Johnson CO, Inabe K, Smith MK, Penner R, et al. Regulation of vertebrate cellular mg2+ homeostasis by trpm7. Cell. 2003;114:191–200.
61. Weber S, Schlingmann KP, Peters M, Nejsum LN, Nielsen S, Engel H, et al. Primary gene structure and expression studies of rodent paracellin-1. J Am Soc Nephrol. 2001;12:2664–72.
62. Hou J, Paul DL, Goodenough DA. Paracellin-1 and the modulation of ion selectivity of tight junctions. J Cell Sci. 2005;118:5109–18.
63. Kausalya PJ, Amasheh S, Gunzel D, Wurps H, Muller D, Fromm M, et al. Disease-associated mutations affect intracellular traffic and paracellular mg2+ transport function of claudin-16. J Clin Invest. 2006;116:878–91.
64. Wabakken T, Rian E, Kveine M, Aasheim HC. The human solute carrier slc41a1 belongs to a novel eukaryotic subfamily with homology to prokaryotic mgte mg2+ transporters. Biochem Biophys Res Commun. 2003;306:718–24.
65. Goytain A, Quamme GA. Functional characterization of human slc41a1, a mg2+ transporter with similarity to prokaryotic mgte mg2+ transporters. Physiol Genomics. 2005;21:337–42.
66. Kolisek M, Zsurka G, Samaj J, Weghuber J, Schweyen RJ, Schweigel M. Mrs2p is an essential component of the major electrophoretic mg2+ influx system in mitochondria. EMBO J. 2003;22:1235–44.
67. Zsurka G, Gregan J, Schweyen RJ. The human mitochondrial mrs2 protein functionally substitutes for its yeast homologue, a candidate magnesium transporter. Genomics. 2001;72:158–68.
68. Wilson FH, Hariri A, Farhi A, Zhao H, Petersen KF, Toka HR, et al. A cluster of metabolic defects caused by mutation in a mitochondrial trna. Science. 2004;306:1190–4.
69. Sweadner KJ, Wetzel RK, Arystarkhova E. Genomic organization of the human fxyd2 gene encoding the gamma subunit of the na, k-atpase. Biochem Biophys Res Commun. 2000;279:196–201.
70. Geering K. Fxyd proteins: new regulators of na-k-atpase. Am J Physiol Renal Physiol. 2006;290:F241–50.
71. Meij IC, Koenderink JB, van Bokhoven H, Assink KF, Groenestege WT, de Pont JJ, et al. Dominant isolated renal magnesium loss is caused by misrouting of the na(+), k(+)-atpase gamma-subunit. Nat Genet. 2000;26:265–6.
72. Gamba G, Saltzberg SN, Lombardi M, Miyanoshita A, Lytton J, Hediger MA, et al. Primary structure and functional expression of a cdna encoding the thiazide-sensitive, electroneutral sodium-chloride cotransporter. Proc Natl Acad Sci USA. 1993;90:2749–53.

73. Mastroianni N, Bettinelli A, Bianchetti M, Colussi G, De Fusco M, Sereni F, et al. Novel molecular variants of the na-cl cotransporter gene are responsible for Gitelman syndrome. Am J Hum Genet. 1996;59:1019–26.
74. Simon DB, Nelson-Williams C, Bia MJ, Ellison D, Karet FE, Molina AM, et al. Gitelman's Variant of Bartter's syndrome, inherited hypokalaemic alkalosis, is caused by mutations in the thiazide-sensitive na-cl cotransporter. Nat Genet. 1996;12:24–30.
75. Takeuchi K, Kure S, Kato T, Taniyama Y, Takahashi N, Ikeda Y, et al. Association of a mutation in thiazide-sensitive na-cl cotransporter with familial Gitelman's syndrome. J Clin Endocrinol Metab. 1996;81:4496–9.
76. Song Y, Hsu YH, Niu T, Manson JE, Buring JE, Liu S. Common genetic variants of the ion channel transient receptor potential membrane melastatin 6 and 7 (trpm6 and trpm7), magnesium intake, and risk of type 2 diabetes in women. BMC Med Genet. 2009;10:4.
77. De Silva NM, Frayling TM. Novel biological insights emerging from genetic studies of type 2 diabetes and related metabolic traits. Curr Opin Lipidol. 2010;21:44–50.
78. Meyer TE, Verwoert GC, Hwang SJ, Glazer NL, Smith AV, van Rooij FJ, et al. Genome-wide association studies of serum magnesium, potassium, and sodium concentrations identify six loci influencing serum magnesium levels. PLoS Genet. 2010;6:e1001045.

Chapter 7
Magnesium Deficiency in Type 2 Diabetes

Dharam Paul Chaudhary

Key Points

- Magnesium is the second most abundant intracellular cation which is widely distributed in plant and animal foods.
- Magnesium deficiency is actually fairly common, and low magnesium intake may play a role in the development of type 2 diabetes.
- Magnesium supplementation has been shown to improve insulin sensitivity and diabetes mellitus.
- Magnesium deficiency is associated with the development of atherosclerotic cardiovascular diseases.

Keywords Magnesium deficiency • Insulin resistance • Diabetes

Introduction

Magnesium is a ubiquitous element in nature and forms an estimated 2.1% of the earth's crust. Like calcium, it is an alkali earth metal. Magnesium plays an important role in a wide range of fundamental biological reactions. Magnesium is the second most abundant intracellular cation and the fourth most abundant cation in the body. The normal adult contains about 22–24 g of magnesium [1]. About 60% of the magnesium is present in the bones, of which 30% is exchangeable and functions as reservoir to stabilize serum concentration. About 20% is in skeletal muscles and 19% in other soft tissues. Skeletal muscle and liver contain between 7 and 9 mmol/kg wet tissue; between 20% and 30% of this is relatively exchangeable. Magnesium is principally an intracellular cation, with less than 1% in the extracellular fluids. Most intracellular magnesium exists in bound form and only 0.25–1 nmol is present as free Mg^{2+}. Free magnesium constitutes only 1–5% of total cellular magnesium. In normal adults, approximately 20% of serum magnesium is protein-bound, 65% is ionized, and the rest is complexed with various anions such as phosphate and citrate [1]. Of the protein-bound fraction, 60–70% is associated with albumin, and the rest is bound to globulins [2]. Free ionized magnesium is the physiologically active form of the element. The intracellular levels of the free magnesium serve to regulate intermediate metabolism through activation of rate-limiting enzymes such as hexokinase,

D.P. Chaudhary, Ph.D. (✉)
Department of Biochemistry, Directorate of Maize Research, Pusa Campus, New Delhi, Delhi 110012, India
e-mail: chaudharydp@gmail.com

R.R. Watson et al. (eds.), *Magnesium in Human Health and Disease*, Nutrition and Health,
DOI 10.1007/978-1-62703-044-1_7, © Springer Science+Business Media New York 2013

pyruvate dehydrogenase, enolase, or creatine phosphokinase [3]. In enzymatic reactions magnesium interacts either by binding to the substrate or directly to the enzyme. All of the enzymatic reactions that hydrolyze and transfer enzymatic groups, involving those associated with the reactions involving adenosine triphosphate (ATP), show an absolute requirement for magnesium [4]. There is great volume of research that suggests an association between reduced magnesium intake and insulin resistance, central to type 2 diabetes. It is the purpose of this chapter to associate reduced magnesium intake to type 2 diabetes.

Magnesium is widely distributed in plant and animal foods. Almost all of our magnesium supplies come from the vegetable kingdom, though seafood has fairly high amounts. Since it is a component of chlorophyll, therefore the dark green vegetables are good source of magnesium. Most nuts, seeds, and legumes have high amounts of magnesium; soy products, especially soy flour and tofu, and nuts such as almonds, pecans, cashews, and Brazil nuts are good examples. The whole green, particularly wheat (especially the bran and germ), millet, and brown rice, and fruits such as avocado and dried apricot are other sources. Hard water can also be a valuable source of magnesium. Dolomite and bone meal are good source of magnesium. Bioavailability of magnesium is on average around 30% but depends strongly on the amount of magnesium in the meal, the presence of inhibitory or enhancing food components, and magnesium status. Refining, processing, and preparing food may cause a substantial loss of magnesium.

The recommended daily dietary allowance (RDA) for magnesium is 6 mg/kg/d. This means 400–420 mg/d for adult men and 310–320 mg/d for adult women (with greater requirement for pregnant or lactating women) [5]. By these standards, which have been promulgated by the National Academy of Science and the Institute of Medicine after great deliberation, research, and literature review, an estimated 50–85% of the population of United States of America is receiving an inadequate magnesium intake. The refining and processing of food causes a substantial loss of magnesium. For example, the refining and processing of wheat to flour, rice to polished rice, and corn to starch depletes magnesium by 82%, 83%, and 97%, respectively [6]. Drinking soft water also decreases magnesium intake. Thus, modern food technologies partially explain why a significant segment of population has intake of magnesium below recommended dietary amounts and may be predisposed to chronic, latent magnesium deficiency. On the other hand, diuretic drugs cause magnesium loss, as do alcohol, caffeine, and sugar. Decreased blood and tissue levels of magnesium have been shown to be related to high blood pressure, kidney stones, heart diseases, and particularly heart attacks due to coronary artery spasm (magnesium helps relax and dilates coronary arteries). Magnesium's role in alleviating premenstrual syndrome (PMS) has made big news as well [7].

Meals high in protein or fat, a diet high in phosphorus or calcium, or alcohol use may decrease magnesium absorption [8]. It has been observed that magnesium absorption in humans occurs uniformly throughout the small intestine [9]. An inverse relationship was reported between intake of magnesium and its fractional absorption, which ranges from 65% absorption at low intake to 11% absorption at high intake [10]. Magnesium balance is mainly regulated in the body by kidneys. Normal intake of magnesium is about 300 mg/d. About 1/3 of the intake is absorbed by the gastrointestinal tract. Over a 24-h period, in humans, about 100–150 ml magnesium is excreted per day, an amount equivalent to that absorbed by the gastrointestinal tract each day [11]. Most magnesium reabsorption takes place in the loop of Henle [12]. Several drugs, particularly diuretics, thiazides, cisplatin, gentamicin, and cyclosporine, cause magnesium loss in urine by inhibiting magnesium reabsorption in the kidney [13, 14]. Lipid-lowering drugs have also been reported to cause magnesium loss [15]. Persons with alcoholism represent the second largest group of people with hypomagnesemia. This is due in fact to the inherent effect of alcohol on magnesium homeostasis and in part to the consequences of the poor diet typical of alcohol abusers. Actually, alcohol increases urinary magnesium excretion by as much as 260% above baseline values [16]. Until recently, the function of magnesium in biological processes was largely ignored, and it was treated as a forgotten ion. However, in recent years a great volume of research has been focused in the physiological and therapeutic properties of this ion.

Magnesium Deficiency

Magnesium deficiency is actually fairly common; however, it is usually not looked for and therefore not found or corrected. The first observation of magnesium deficiency in animal studies was made in 1932 by Kruse et al. [17]. In man, the first description of clinical depletion was published in 1934 [18]. Deficiency is more likely in those who eat processed food diet; in people who cook or boil all foods, especially vegetables; in those who drink soft water; in alcoholics; and in people who eat food grown in magnesium-deficient soils, where synthetic fertilizers containing no magnesium are often used [16, 19]. Deficiency is also more common when magnesium absorption is decreased, such as after burns, serious injuries, or surgery and in patients with diabetes, liver disease, or malabsorption problems, and when magnesium elimination is increased, as in people who use alcohol, caffeine, or excess sugar or who take diuretics or birth control pills [20].

Magnesium deficiency is usually secondary to another disease or a therapeutic agent; hence, features of a primary disease may mask or complicate magnesium deficiency [21]. Magnesium deficiency was first induced in humans by dietary restrictions during 1960s [22]. Marked symptoms of magnesium deficiency appeared, and the early symptoms of magnesium deficiency can include fatigue, anorexia, irritability, insomnia, and muscle tremors or twitching [23]. Psychological changes, such as apathy, apprehension, decreased learning ability, confusion, and poor memory, may occur. Tachycardia (rapid heartbeat) and other cardiovascular changes are likely with moderate deficiency, while severe magnesium deficiency may lead to numbness, tingling, and tetany (sustained contraction) of the muscle, as well as delirium and hallucination [24]. The details of various signs and symptoms of magnesium deficiency are listed in Table 7.1 [21, 25].

Magnesium deficit can be categorized into two types: magnesium deficiency and magnesium depletion. Magnesium depletion may be due to dysregulation of factors controlling magnesium status: intestinal hypoabsorption of magnesium, reduced uptake and mobilization of bone magnesium, sometimes urinary leakage, hyperadrenoglucocorticism by decreased adaptability to stress, insulin resistance, and adrenergic hyporeceptivity [26]. Magnesium deficiency in aging largely results from various pathologies and treatments to elderly persons, i.e., diabetes mellitus and use of hypermagnesuric diuretics [27]. Magnesium deficiency has been demonstrated in 7–11% of hospitalized patients [28]. The prevalence of hypomagnesemia in critically ill patients is even higher [29], ranging from 20% to 65%. Hypomagnesemia in intensive care patients is associated with increased mortality [30]. Magnesium deficiency in the body is reflected by fall in serum magnesium levels. Serum magnesium levels are considered to be the first indicator of magnesium deficiency. Although magnesium

Table 7.1 Symptoms and signs of Mg deficiency (Adapted from data in [21, 25])

Neuromuscular	*Cardiovascular*
Positive Chvostek's and Trousseau's sign	Arrhythmias
Spontaneous carpopedal spasm (tetany)	ECG changes
Seizures	Myocardial infarction
Vertigo, tremor, ataxia, nystagmus	Hypertension
Muscular weakness, tremor, fasciculation, wasting	Atherosclerotic vascular disease
Psychiatric: depression, psychosis	
Gastrointestinal	*Bone metabolism*
Anorexia	Hypocalcemia
Nausea and vomiting	Impaired PTH secretion
	Renal and skeletal resistance to PTH
	Resistance to vitamin D
Potassium homeostasis	
Hypokalemia	
Renal K wasting, decreased intracellular K	

Table 7.2 Laboratory tests for the assessment of Mg status (Adapted from data in [36])

Tissue Mg	Physiologic assessment	Free magnesium
Serum/plasma	Mg load/retention test	Extracellular free Mg
Erythrocytes	Renal excretion of Mg	(Serum)
Mononuclear blood cells	Balance studies	Intracellular free Mg
Muscle	Isotope studies	

is principally an intracellular cation, with less than 0.3% of total body content present in serum, serum and plasma magnesium measurement is the most available and most common employed test of magnesium status because it is easy and inexpensive to perform. Shils [22] reported that during experimentally induced magnesium deficiency, the first change appears to be a fall in serum magnesium concentration. Some previous studies have reported similar findings and suggested that serum magnesium falls rapidly during magnesium depletion in humans [31] and animals [32]. Alfray and coworkers [33] have correlated serum magnesium with bone magnesium status during both hypo- and hypermagnesemia and suggested that for clinical purposes serum magnesium is a suitable indicator of total body magnesium. Pham et al. [34] have recently suggested that in patients with suspected magnesium deficiency, a low serum magnesium concentration is sufficient to confirm the diagnosis. Similarly, erythrocyte magnesium content is reported to be another reliable indicator of total body stores of magnesium. Elin et al. [35] have found marked reduction in plasma and erythrocyte magnesium content in magnesium-deficient animals. Elin [36] divides magnesium assessment tests into three categories: tissue magnesium, physiologic assessment of magnesium, and free magnesium (Table 7.2). The equilibrium between some tissue pools occurs slowly, and the biological half-life for large part of magnesium in the body is more than 1,000 h [37]. Therefore, determining magnesium in one tissue may not provide information about magnesium status in other tissues [36].

Magnesium Deficiency and Type 2 Diabetes

Magnesium plays an important role in carbohydrate metabolism. Low magnesium intake may play a role in the development of type 2 diabetes. It may influence the release and activity of the hormones that help control blood glucose levels [38]. Low blood levels of magnesium are frequently observed in individuals with type 2 diabetes [39]. As early as 1946, hypomagnesemia was noted in patients with diabetic ketoacidosis [40]. In 1968, a survey of 5,100 consecutive patients at a diagnostic clinic in the USA revealed diabetes to be the most common condition associated with hypomagnesemia [41]. Twenty percent of the patients with hypomagnesemia (Mg<0.74 mmol) were diabetic. Hypomagnesemia may worsen insulin resistance, a condition that often precedes diabetes, or may be a consequence of insulin resistance. Individuals with insulin resistance cannot use insulin efficiently and require greater amounts of insulin to maintain blood glucose within normal levels. Kidneys possibly lose their ability to retain magnesium during periods of severe hyperglycemia. The increased loss of magnesium in urine may then result in lower blood levels of magnesium in older adults; correcting magnesium depletion may improve insulin response and action [39]. In a study conducted by the Nurses' Health Study (NHS) and Health Professionals' Follow-up Study (HFS), over 127,932 research subjects (85,060 women and 42,872 men) with no history of diabetes, cardiovascular diseases, or cancer at baseline were examined for risk factors for developing type 2 diabetes. Women were followed for 18 years, and men were followed for 12 years. Over time, the risk of developing type 2 diabetes was greater in men and women with lesser magnesium intake. This study supports the dietary recommendation to increase consumption of foods with higher magnesium content, such as whole grains, nuts, and green leafy vegetables [42]. Researchers from the Iowa Women's Health Study examined the association between women's risk of developing type 2 diabetes and intake of magnesium. Based on the baseline dietary intake assessment only, researchers' findings suggested that greater intake of whole grains, dietary fiber, and magnesium decreased the risk of developing diabetes in women [43].

On the other hand, Atherosclerosis Risk in Communities Study (ARIC) did not find any association between dietary magnesium intake and the risk of type 2 diabetes. During 6 years of follow-up, ARIC researchers studied the risk for type 2 diabetes in over 12,000 middle-aged adults without diabetes at baseline examination. It was reported that there was no statistical association between dietary magnesium intake and incidence of type 2 diabetes in black or white research subjects [44]. Researchers attributed this contradiction to the difference in population studied compared with the other studies. Recently, in a study conducted on indigenous Australians, the researchers found a significant correlation between the magnesium content of the municipal water supplies and the age-standardized deaths due to diabetes [45]. Most studies, therefore, confirm that a high-magnesium diet has a beneficial effect on reducing the risk of developing diabetes. Earlier studies suggest several possible mechanisms whereby low serum magnesium levels may lead to development of type 2 diabetes. First of all, it is an essential cofactor in reactions involving phosphorylation; thus, magnesium deficiency could impair the insulin signal transduction pathway [46, 47]. Second, low serum or erythrocyte magnesium levels may affect the interaction between insulin and insulin receptor by decreasing hormone receptor affinity or by increasing membrane microviscosity [48]. Finally, magnesium can also be a limiting factor in carbohydrate metabolism, since many of the enzymes in this process require magnesium as a cofactor during reactions that utilize phosphorus bond [49, 50]. Diabetes mellitus has been suggested to be the most common metabolic disorder associated with magnesium deficiency, having 25–39% prevalence [11]. In a study carried out in our own laboratory, we observed significant hyperglycemia, hyperinsulinemia, and hypertriglyceridemia in rats consuming low magnesium high-sucrose diet for a period of 3 months [51]. We have further reported that a high-sucrose low-magnesium diet, fed for a 3-month period, reduces in vitro glucose uptake in target tissues of rats [52]. Hwang et al. [53] and others [54] produced type 2 diabetes and its associated diseases in normal Sprague–Dawley rats. The animals were fed a supposedly adequate diet for 2 weeks in which 66% of the caloric content consists of D-glucose. Balon et al. [55] observed that this high-fructose diet contained only 0.06% magnesium rather than 0.24% in the control diet. On the other hand, hypomagnesemia has been reported to occur in 13.5–47.7% of nonhospitalized patients with type 2 diabetes compared with 2.5–15% among their counterparts without diabetes [34]. Hypomagnesemia was observed in 23.2% and intracellular magnesium depletion in 36.1% of patients with metabolic syndrome [56]. A strong inverse relationship between magnesium levels in serum and presence of metabolic syndrome was observed by Evangelopoulos and coworkers [57]. In a recently conducted study, it was reported that diabetic patients have lower levels of magnesium and are considered at increased risk of complications [58].

Magnesium supplementation has been shown to improve insulin sensitivity and diabetes mellitus [59]. Diabetic patients tend to have low magnesium levels. Magnesium leads to improved insulin production in elderly people with NIDDM [60]. According to some studies, elders without diabetes can also produce more insulin because of magnesium supplementation [39]. Insulin requirements are lower in people with IDDM, who are supplemented with magnesium. In a study in type 1 diabetic patients, oral replacement with magnesium hydroxide at a dosage of 250 mg twice daily resulted in increased levels of magnesium in the skeletal muscles [61]. This was associated with decreased insulin requirements but no reduction in glycosylated hemoglobin levels in diabetic patients. In a double-blind, randomized crossover study, Paolisso and colleagues [39] investigated the effect of magnesium supplementation in elderly subjects with insulin resistance on the handling of glucose following an intravenous glucose and euglycemic hyperinsulinemic clamp procedure. Magnesium pidolate at 4.5 g/day (15.8 mmol/day) for 4 weeks significantly improved insulin action and oxidative glucose metabolism, increased erythrocyte magnesium concentration, and decreased erythrocyte membrane microviscosity. Olatunji et al. [62] have reported that dietary magnesium supplementation significantly improved the impaired glucose tolerance and dyslipidemia in a female rat model. Similarly, McKeown and colleagues [63] demonstrated that magnesium intake is inversely associated with prevalence of the metabolic syndrome in older adults. They further recommended that older adults should be encouraged to eat foods rich in magnesium such as green vegetables, legumes, and whole grains. It was reported that hypomagnesemia is independently associated with the development of impaired glucose tolerance (IGT), impaired fasting glucose (IFG)+IGT, and type 2 diabetes but not with the development of IFG [64]. Schulze et al. [65] examined associations between magnesium intake and risk of type 2 diabetes and summarized existing prospective studies by meta-analysis. The outcome strongly supports that higher magnesium intake may decrease diabetes risk. In a recent study, it was reported that magnesium intake may protect against the development of type 2 diabetes in Chinese women [66].

At the same time, there exists substantial evidence linking magnesium deficiency and atherosclerotic cardiovascular diseases [67]. Experimental studies suggest that magnesium deficiency may play a role in the pathogenesis of atherosclerosis [68]. Magnesium deficiency may contribute to the progression of atherosclerosis by its effect on lipid metabolism, platelet aggregation, and blood pressure. Experimental magnesium deficiency is characterized by increased triglycerides, cholesterol, VLDL, LDL, apolipoprotein B and triglyceride-rich lipoproteins A1, and plasma lecithin-cholesterol acyltransferase activity [69]. Magnesium can prevent atherosclerotic disease by counteracting the adverse effect of excessive intracellular calcium, retaining intracellular potassium, and contributing both to stabilize the plasma membrane and to maintain the integrity of subcellular structures [70, 71]. Magnesium supplementation of hyperlipidemic subjects has been shown to cause a reduction in total and LDL cholesterol and apolipoprotein B and an increase in HDL cholesterol and triglycerides [67].

To summarize, it could be concluded that magnesium plays a very important role in the development of type 2 diabetes. Magnesium deficiency is a common finding in patients with type 2 diabetes. Disorders of magnesium metabolism are common and are usually unrecognized. We could conclude that strong evidence exists for associating decreased intracellular concentration of magnesium with the etiology of type 2 diabetes mellitus. Hypomagnesemia, though less frequent, can also lead to cardiovascular manifestations. Early recognition of the disordered magnesium metabolism is necessary to avoid these complications.

References

1. Saris NE, Mervaala E, Karppanen H, Khawaja JA, Lewenstam A. Magnesium. An update on physiological, clinical and analytical aspects. Clin Chim Acta. 2000;294:1–26.
2. Kroll MH, Elin RJ. Relationship between magnesium and protein concentrations in serum. Clin Chem. 1985;31:244–6.
3. Altura BM, Altura BT. Role of magnesium in patho-physiological processes and the clinical utility of magnesium ion selective electrodes. Scand J Clin Lab Invest Suppl. 1996;224:211–34.
4. Dresoti E. Magneium status and health. Nutr Rev. 1995;53:S23.
5. Academy N. Dietary reference intakes for calcium, phosphorus, magnesium, vitamin D, and fluoride. Washington: National Academy Press; 1997.
6. Marrier JR. Magnesium content of the food supply in the modern day world. Magnesium. 1986;5:1–8.
7. Quaranta S, Buscaglia MA, Meroni MG, Colombo E, Cella S. Pilot study of the efficacy and safety of a modified-release magnesium 250 mg tablet (sincromag) for the treatment of premenstrual syndrome. Clin Drug Investig. 2007;27:51–8.
8. Brink EJ, Beynen AC. Nutrition and magnesium absorption: a review. Prog Food Nutr Sci. 1992;16:125–62.
9. Quamme GA. Renal magnesium handling: new insights in understanding old problems. Kidney Int. 1997;52:1180–95.
10. Hardwick LL, Jones MR, Brautbar N, Lee DB. Magnesium absorption: mechanisms and the influence of vitamin D, calcium and phosphate. J Nutr. 1991;121:13–23.
11. Rude RK. Magnesium deficiency and diabetes mellitus. Causes and effects. Postgrad Med. 1992;92:217–9.
12. Quamme GA. Control of magnesium transport in the thick ascending limb. Am J Physiol. 1989;256:F197–210.

13. Lajer H, Daugaard G. Cisplatin and hypomagnesemia. Cancer Treat Rev. 1999;25:47–58.
14. Rob PM, Lebeau A, Nobiling R, Schmid H, Bley N, Dick K, et al. Magnesium metabolism: basic aspects and implications of cyclosporine toxicity in rats. Nephron. 1996;72:59–66.
15. Haenni A, Ohrvall M, Lithell H. Serum magnesium status during lipid- lowering drug treatment in non-insulin-dependent diabetic patients. Metabolism. 2001;50:1147–51.
16. Rivlin RS. Magnesium deficiency and alcohol intake: mechanisms, clinical significance and possible relation to cancer development (a review). J Am Coll Nutr. 1994;13:416–23.
17. Kruse HD, Orent ER, McCollum EV. Studies on magnesium deficiency in animals – I. Symptomatology resulting from magnesium deprivation. J Biol Chem. 1932;96:519–39.
18. Hirschfelder AD, Haury VG. Clinical manifestations of high and low plasma magnesium: dangers of Epsom salt purgation in nephritis. JAMA. 1934;102:1138–41.
19. Wilkinson SR, Stuedemann JA, Grunes DL, Devine OJ. Relation of soil and plant magnesium to nutrition of animals and man. Magnesium. 1987;6:74–90.
20. Oralewska B, Zawadzki J, Jankowska I, Popinska K, Socha J. Disorders of magnesium homeostasis in the course of liver disease in children. Magnes Res. 1996;9:125–8.
21. Rude RK. Magnesium disorders. In: Kokko JP, Tannen RL, editors. Fluids and electrolytes. Philadelphia: Saunders; 1996. p. 421–45.
22. Shils ME. Experimental human magnesium depletion. Medicine. 1969;48:61–85.
23. Durlach J, Bac P, Durlach V, Bara M, Guiet-Bara A. Neurotic, neuromuscular and autonomic nervous form of magnesium imbalance. Magnes Res. 1997;10:169–95.
24. Hisakawa N, Yasuoka N, Itoh H, Takao T, Jinnouchi C, Nishiya K, et al. A case of Gitelman's syndrome with chondrocalcinosis. Endocr J. 1998;45:261–7.
25. Shils ME. Magnesium. In: Shils ME, Olson JE, Shike M, Ross AC, editors. Modern nutrition in health & disease. Baltimore: Williams & Wilkins; 1998. p. 169–92.
26. Durlach J, Bac P, Durlach V, Rayssiguier Y, Bara M, Guiet-Bara A. Magnesium status and ageing: an update. Magnes Res. 1998;11:25–42.
27. Durlach J, Durlach V, Bac P, Rayssiguier Y, Bara M, Guiet-Bara A. Magnesium and ageing. II. Clinical data: aetiological mechanisms and pathophysiological consequences of magnesium deficit in the elderly. Magnes Res. 1993;6:379–94.
28. Hayes JP, Ryan MF, Brazil N, Riordan TO, Walsh JB, Coakley D. Serum hypomagnesaemia in an elderly day-hospital population. Ir Med J. 1989;82:117–9.
29. Noronha JL, Matuschak GM. Magnesium in critical illness: metabolism, assessment, and treatment. Intensive Care Med. 2002;28:667–79.
30. Rubeiz GJ, Thill-Baharozian M, Hardie D, Carlson RW. Association of hypomagnesaemia and mortality in acutely ill medical patients. Crit Care Med. 1993;21:203–9.
31. Dunn MJ, Walser M. Magnesium depletion in normal man. Metabolism. 1966;15:884–95.
32. Martindale L, Heaton FW. Magnesium deficiency in the adult rat. Biochem J. 1964;92:119–26.
33. Alfrey AC, Miller NL, Butkus D. Evaluation of body magnesium stores. J Lab Clin Med. 1974;84:153–62.
34. Pham PC, Pham PM, Pham SV, Miller JM, Pham PT. Hypomagnesemia in patients with type 2 diabetes. Clin J Am Soc Nephrol. 2007;2:366–73.
35. Elin RJ, Armstrong WD, Singer L. Body fluid electrolyte composition of chronically magnesium-deficient and control rats. Am J Physiol. 1971;220:543–8.
36. Elin RJ. Laboratory tests for the assessment of magnesium status in humans. Magnes Trace Elem. 1991;10:172–81.
37. Avioli LV, Berman M. Mg28 Kinetics in man. J Appl Physiol. 1966;21:1688–94.
38. Kobrin SM, Goldfarb S. Magnesium deficiency. Semin Nephrol. 1990;10:525–35.
39. Paolisso G, Sgambato S, Gambardella A, Pizza G, Tesauro P, Varricchio M, et al. Daily magnesium supplements improve glucose handling in elderly subjects. Am J Clin Nutr. 1992;55:1161–7.
40. Martin HE, Smith K, Wilson ML. Fluid and electrolyte therapy of severe diabetic acidosis and ketosis – study of 29 episodes (26 patients). Am J Med. 1958;24:376–89.
41. Jackson CE, Meier DW. Routine serum magnesium analysis. Correlation with clinical state in 5,100 patients. Ann Intern Med. 1968;69:743–8.
42. Lopez-Ridaura R, Willett WC, Rimm EB, Liu S, Stampfer MJ, Manson JE, et al. Magnesium intake and risk of type 2 diabetes in men and women. Diabetes Care. 2004;27:270–1.
43. Meyer KA, Kushi LH, Jacobs Jr DR, Slavin J, Sellers TA, Folsom AR. Carbohydrates, dietary fiber, and incident type 2 diabetes in older women. Am J Clin Nutr. 2000;71:921–30.
44. Kao WH, Folsom AR, Nieto FJ, Mo JP, Watson RL, Brancati FL. Serum and dietary magnesium and the risk for type 2 diabetes mellitus: the atherosclerosis risk in communities study. Arch Intern Med. 1999;159:2151–9.
45. Longstreet DA, Heath DL, Panaretto KS, Vink R. Correlations suggest low magnesium may lead to higher rates of type 2 diabetes in indigenous Australians. Rural Remote Health. 2007;7:843.

46. Elin RJ. Magnesium metabolism in health and disease. Dis Mon. 1988;34:161–218.
47. Styler L. Biochemistry. 3rd ed. New York: Freeman & Co; 1988.
48. Tongyai S, Rayssiguier Y, Motta C, Gueux E, Maurois P, Heaton FW. Mechanism of increased erythrocyte membrane fluidity during magnesium deficiency in weanling rats. Am J Physiol. 1989;257:C270–6.
49. Caro JF, Triester S, Patel VK, Tapscott EB, Frazier NL, Dohm GL. Liver glucokinase: decreased activity in patients with type II diabetes. Horm Metab Res. 1995;27:19–22.
50. Tosiello L. Hypomagnesemia and diabetes mellitus. A review of clinical implications. Arch Intern Med. 1996;156:1143–8.
51. Chaudhary DP, Boparai RK, Sharma R, Bansal DD. Studies on the development of an insulin resistance rat model by chronic feeding of high sucrose low magnesium diet. Magnes Res. 2004;17:293–300.
52. Chaudhary DP, Boparai RK, Bansal DD. Effect of low magnesium diet on *in-vitro* glucose uptake in sucrose fed rats. Magnes Res. 2007;20:187–95.
53. Hwang IS, Ho H, Hoffman BB, Reaven GM. Fructose induced insulin resistance and hypertension in rats. Hypertension. 1987;10:512–6.
54. Zavaroni I, Sander S, Scott S, Reaven GM. Effect of fructose feeding on insuin secretion and insulin action in the rats. Metabolism. 1980;29:970–3.
55. Balon TW, Jasman A, Scott S, Meehan WP, Rude RK, Nadler JL. Dietary magnesium prevents fructose induced insulin insenstivity in rats. Hypertension. 1994;23:1036–9.
56. Lima ML, Cruz T, Rodrigues LE, Bomfim O, Melo J, Correia R, et al. Serum and intracellular magnesium deficiency in patients with metabolic syndrome–evidences for its relation to insulin resistance. Diabetes Res Clin Pract. 2009;83:257–62.
57. Evangelopoulos AA, Vallianou NG, Panagiotakos DB, Georgiou A, Zacharias GA, Alevra AN, et al. An inverse relationship between cumulating components of the metabolic syndrome and serum magnesium levels. Nutr Res. 2008;28:659–63.
58. Seyoum B, Siraj ES, Saenz C, Abdulkadir J. Hypomagnesemia in Ethiopians with diabetes mellitus. Ethn Dis. 2008;18:147–51.
59. Kandeel FR, Balon E, Scott S, Nadler JL. Magnesium deficiency and glucose metabolism in rat adipocytes. Metabolism. 1996;45:838–43.
60. Paolisso G, Scheen A, Cozzolino D, Maro GD, Varricchio M, D'Onofrio F, et al. Changes in glucose turnover parameters and improvement of glucose oxidation after 4-week magnesium administration in elderly noninsulin-dependent (type II) diabetic patients. J Clin Endocrinol Metab. 1994;78:1510–4.
61. Sjögren A, Florén CH, Nilsson A. Oral administration of magnesium hydroxide to subjects with insulin-dependent diabetes mellitus: effects on magnesium and potassium levels and on insulin requirements. Magnesium. 1988;7:117–22.
62. Olatunji LA, Oyeyipo IP, Micheal OS, Soladoye AO. Effect of dietary magnesium on glucose tolerance and plasma lipid during oral contraceptive administration in female rats. Afr J Med Sci. 2008;37:135–9.
63. McKeown NM, Jacques PF, Zhang XL, Juan W, Sahyoun NR. Dietary magnesium intake is related to metabolic syndrome in older Americans. Eur J Nutr. 2008;47:210–6.
64. Guerrero-Romero F, Rascón-Pacheco RA, Rodríguez-Morán M, de la Peña JE, Wacher N. Hypomagnesaemia and risk for metabolic glucose disorders: a 10-year follow-up study. Eur J Clin Invest. 2008;38:389–96.
65. Schulze MB, Schulz M, Heidemann C, Schienkiewitz A, Hoffmann K, Boeing H. Fiber and magnesium intake and incidence of type 2 diabetes. Arch Intern Med. 2007;167:956–65.
66. Villegas R, Yang G, Gao Y, Cai H, Li H, Zheng W, et al. Dietary patterns are associated with lower incidence of type 2 diabetes in middle- aged women: the shanghai women's health study. Int J Epidemiol. 2010;39:889–99.
67. Fox C, Ramsoomair D, Carter C. Magnesium: its proven and potential clinical significance. South Med J. 2001;94:1195–201.
68. Nadler JL, Rude RK. Disorders of magnesium metabolism. Endocrinol Metab Clin North Am. 1995;24:623–41.
69. Nozue T, Kobayashi A, Uemasu F, Takagi Y, Sako A, Endoh H. Magnesium status, serum HDL cholesterol, and apolipoprotein a-1 levels. J Pediatr Gastroenterol Nutr. 1995;20:316–8.
70. Seelig MS, Heggtveit A. Magnesium interrelationships in ischemic heart disease: a review. Am J Clin Nutr. 1974;27:59–79.
71. Mather HM, Levin GE, Nisbet JA, Hadley LA, Oakley NM, Pilkington TR. Diurnal profiles of plasma magnesium and blood glucose in diabetes. Diabetologia. 1982;22:180–3.

Section C
Magnesium Supplementation and Disease

Chapter 8
Magnesium and Metabolic Disorders

Abigail E. Duffine and Stella Lucia Volpe

Key Points

- Magnesium, the second most abundant intracellular cation, is a vital macromineral in maintaining normal metabolism and health.
- The functions of magnesium are plentiful and widespread throughout the body. It is clear that the relationship between magnesium status and insulin resistance, type 2 diabetes, and metabolic syndrome is an important one.
- The negative and inverse correlation suggests that low magnesium levels are related to the development of these disorders.
- On the other hand, however, high magnesium levels may provide protective characteristics to these dangerous conditions. It is important to obtain an adequate amount of dietary magnesium to lessen the risk of insulin resistance, type 2 diabetes, and metabolic syndrome.

Keywords Metabolic syndrome • Insulin resistance • Type 2 diabetes mellitus

Introduction

Magnesium is the second most common intracellular cation, with only 1% of total magnesium found in the blood [1]. It has a wide range of functions throughout the body, acting as a cofactor for over 300 enzymatic reactions, affecting glucose transportation and influencing insulin sensitivity. Abundant research focuses on identifying the true functions and interactions of magnesium within the body. Evidence has shown that low magnesium levels are negatively correlated with the development of insulin resistance and increasing the risk of metabolic syndrome and type 2 diabetes mellitus. Hypomagnesemia is present in 10% of patients admitted into hospitals and as many as 65% of patients admitted to intensive care units [1]. Studying the prevention of these diseases is at the forefront of nutrition-focused research. The strong evidence connecting magnesium with metabolic syndrome and type 2 diabetes mellitus suggests magnesium supplementation may be an appropriate intervention technique to control the progression of these diseases. In the following chapter, the prevalence of these diseases and how magnesium is related will be discussed.

A.E. Duffine, M.S. • S.L. Volpe, Ph.D., R.D., LDN, FACSM (✉)
Department of Nutrition Science, Drexel University, 245 N. 15th Street,
Mail Stop 1030, Bellet 521, Philadelphia, PA 18938, USA
e-mail: aeduffine@gmail.com; Stella.L.Volpe@drexel.edu

R.R. Watson et al. (eds.), *Magnesium in Human Health and Disease*, Nutrition and Health,
DOI 10.1007/978-1-62703-044-1_8, © Springer Science+Business Media New York 2013

Metabolic Syndrome

Metabolic syndrome is a collection of risk factors that, in synthesis, greatly increase the risk of the development of certain diseases including type 2 diabetes mellitus, cardiovascular disease, and cerebrovascular accident (stroke). It is unclear whether metabolic syndrome is caused by a single underlying factor or a combination thereof. Several definitions of metabolic syndrome exist, but the common diagnostic criteria includes having three of the following five symptoms: elevated waist circumference (≥40 in. in men, ≥35 in. in women), elevated triglyceride concentrations (≥150 mg/dL or medication treatment for elevated triglycerides), reduced high-density lipoprotein cholesterol concentrations (HDL-C) (<40 mg/dL in men, <50 mg/dL in women, or medication treatment for reduced HDL-C), elevated blood pressure (≥130 mmHg systolic or ≥85 mmHg diastolic or medication treatment for hypertension), and elevated fasting glucose concentrations (≥100 mg/dL or drug treatment for elevated glucose) [2]. The increased prevalence of metabolic syndrome worldwide is related both to the increased prevalence of obesity and sedentary lifestyles [3]. Existence of the metabolic syndrome increases the risk of cardiovascular disease to twice that of individuals without the metabolic syndrome [3] and risk of developing type 2 diabetes mellitus at a fivefold increase [3]. Although controversy exists over the true existence of this syndrome or simply a conglomerate of symptoms, a linking factor is the possible development of insulin resistance.

The Centers for Disease Control and Prevention [4] indicates that approximately 34% of the population, 20 years of age and over, meets the criteria for metabolic syndrome. Among the five diagnostic criteria for metabolic syndrome, abdominal obesity, hypertension, and hyperglycemia are the most prevalent. These three diagnostic characteristics are greatly affected by magnesium status.

Hypertension

High blood pressure (hypertension) affects nearly one billion people throughout the world, two-thirds of whom live in developing countries [5]. Hypertension is a risk factor for cardiovascular disease, metabolic syndrome, and stroke above 140 mmHg systolic and a diastolic blood pressure at or above 90 mmHg. Almost 95% of all hypertension cases are considered primary or essential, defined as an elevated blood pressure with no identifiable cause. Secondary hypertension is defined as elevated blood pressure with an identifiable cause, such as renal disease or endocrine disturbances [6]. Several additive mechanisms may affect the development of essential hypertension. These include increased vascular resistance due to decreased diameters of arteries, arterioles, or both; malfunctioning sodium excretion; a disordered renin-angiotensin system; hyperactivity of the nervous system; or inappropriate levels of micronutrients [7]. Intracellular concentrations of magnesium, along with sodium, potassium, and calcium, play integral roles in the regulation of blood pressure.

Normal Insulin Response and Insulin Resistance

In healthy individuals, beta cells (β-cells) of the pancreatic islets of Langerhans release insulin in response to elevated blood glucose levels. Insulin exhibits its actions by binding to insulin receptors located on the cell membrane of skeletal and adipose tissue [8]. These receptors are composed of an α-(alpha) and β-(beta) subunit. Insulin binding induces a conformational change of the α-subunit, allowing the binding of adenosine triphosphate (ATP) and the phosphorylation of the β-subunit, which then allows the activity of tyrosine kinase [8]. This phosphorylation enables substrates to signal within the cell and further mediate the many actions of insulin [9]. One function of insulin is the activation of GLUT-4 receptors found on fat and muscle tissue. The translocation of GLUT-4 receptors allows glucose to enter the cell for metabolism. Glycolysis and glycogenolysis are stimulated, while the production of new glucose through gluconeogenesis is inhibited [9].

Insulin resistance, alternatively, is a condition in which the body produces insulin but cannot properly use it [10]. Triggered by an increase in visceral adiposity, insulin receptors begin to lose sensitivity and decrease glucose uptake into the cells [11]. The ensuing hyperglycemia induces a compensatory hyperinsulinemia needed to maintain normal blood glucose levels. The maintained state of hyperglycemia causes a constant production of insulin, but without the appropriate receptor response, a state of euglycemia is unachievable. This constant stimulation of the pancreatic β-cells may lead to eventual exhaustion of the cells, marked by a ceasing of insulin production. One measurement of insulin resistance is the homeostasis model assessment-insulin resistance index (HOMA-IR index), a mathematical manipulation of glucose and insulin levels. This model predicts a blood insulin and glucose level for a given sensitivity and insulin secretion. A normal insulin sensitivity is no greater than 2.71 [1].

If left untreated, insulin resistance can develop into type 2 diabetes mellitus. The disease affects 5.9% of the world's adult population, with almost 80% of the total in developing countries [12]. It is a progressive disease that can lead to several macro- and microvascular complications. Although intimately related to hormone interactions, especially insulin, type 2 diabetes mellitus is greatly affected by micronutrient concentrations. The following sections will illuminate the relationship between magnesium, insulin resistance, the metabolic syndrome, and type 2 diabetes mellitus.

Magnesium in Metabolic Syndrome and Type 2 Diabetes Mellitus

Magnesium is essential for the activation of over 300 enzymes, glucose transportation between membranes, glucose oxidation, all reactions involving phosphorylation, energy exchange, and for the proper activity of insulin [1]. It is the second most abundant intracellular cation in the body [13]. Approximately 99% of body magnesium is distributed in the intracellular compartment, and only 1% in the extracellular fluid [14]. Because magnesium is an intracellular cation, serum levels of the micronutrient may not accurately reflect magnesium status of the body [13]. Nonetheless, a decrease in serum magnesium levels does indicate a deficiency. These measurements are specific, but not sensitive, to magnesium deficiency [13]. Serum magnesium levels in healthy people are usually constant with a reference interval of 0.75–0.96 mmol/L, with a mean of 0.85 mmol/L [14]. A more accurate assessment of body magnesium is through the administration of an intravenous magnesium load, monitoring levels before and after the load [13]. Although this is the most accurate estimation of magnesium levels, most researchers assess serum magnesium concentrations typically because it is less expensive and less cumbersome than the magnesium loading test [13]. Magnesium is excreted through the kidneys via urine. During renal reabsorption, about 65% of the reabsorbed magnesium occurs in the loop of Henle, and another 20% in the proximal tubule [7]. Only about 5% of filtered magnesium is excreted in the urine [7]. Fluctuations in dietary intake and the use of medication can alter excreted magnesium concentration. Diuretic medications, as well as protein, alcohol, and caffeine consumption, can increase urinary magnesium excretion, whereas parathyroid hormone (PTH) inhibits magnesium excretion by increasing reabsorption in the kidney tubules [7].

Mechanisms

Insulin Regulates Magnesium Metabolism

Insulin is an anabolic hormone that increases the expression or activity of enzymes that catalyze the synthesis of glycogen, lipids, and proteins [7]. It is necessary for the transport of magnesium from the extracellular to intracellular space. Consistent intracellular magnesium concentrations are important for the maintenance of regular cellular activity. Nuclear magnetic resonance (NMR) studies in healthy individuals have shown that this action is specific, dose-related, and independent of cellular glucose uptake [14]. Following the ingestion of glucose, healthy individuals exhibited a shift of magnesium from the extracellular to the intracellular space, producing a significant decline in plasma concentrations [1]. Agents, such as insulin, that reduce the second messenger activity of cyclic adenosine monophosphate (cAMP) induce a marked accumulation of magnesium into cells [15]. This action may be related to the activation of the tyrosine kinase insulin receptor, evidenced by the abolishment of magnesium transport into the cell after the administration of a monoclonal antibody directed towards this tyrosine kinase receptor [5]. Intracellular magnesium accumulation might depend upon the activation of the tyrosine kinase insulin receptors [1]. The unique relationship of magnesium and the tyrosine kinase receptors will be discussed later in this chapter.

Magnesium Regulates Insulin Activity and Sensitivity

Intracellular magnesium levels may be predicative for insulin resistance because they may make cells less responsive to insulin stimulation [14]. It is evident that the lower the basal intracellular magnesium levels, the less responsible the cell is to insulin stimulation [14]. Insulin receptors are members of the tyrosine kinase family of receptors [16]. This receptor type is a heterotetrameric glycoprotein composed of four subunits, two α-subunits (alpha) and two β-subunits (beta), linked together by a disulfide bond [16]. The α-subunits are extracellular while the β-subunits are composed of an extracellular, transmembrane, and intracellular portion. The intracellular portion contains the tyrosine kinase activity [16]. In a healthy individual, phosphorylation of tyrosine kinase induces a cascade of events leading to the translocation of the GLUT-4 receptor to the cell membrane [16]. This allows glucose entrance into the cell for metabolism. Intracellular magnesium concentration is critical in the phosphorylation of the tyrosine kinase of the insulin receptor [14]. Without an adequate intracellular magnesium supply, appropriate phosphorylation of intracellular components is unattainable. Magnesium deficiency, therefore, may result in decreased postreceptorial activity of insulin receptors leading to insulin resistance and decreased glucose utilization [14].

Magnesium and Glucose Metabolism

Glycolysis is the metabolic pathway of converting glucose into two pyruvate molecules. In aerobic conditions with ample oxygen supply, pyruvate is then metabolized fully into carbon dioxide and water, accompanied by the production of energy as ATP. Under anaerobic conditions, pyruvate is converted to lactate. Lactate can then be used in several directions, depending on the body's needs. Glycolysis, however, is the common pathway needed to form the pyruvate molecules. Glycolysis is a process of ten sequential reactions, three of which are irreversible and require phosphorylation and energy transfer. Phosphorylation is the metabolic process of adding a phosphate group to an organic molecule. These steps are the phosphorylation of glucose to glucose-6-phophate, the phosphorylation of fructose-6-phosphospate to fructose-1,6-bisphosphate, and the donation of a phosphate group from phosphoenolpyruvate to adenosine diphosphate (ADP). These steps are rate-limiting and required for the proper metabolism of glucose [7].

Magnesium is a necessary cofactor for all phosphorylation processes and in all reactions that utilize and transfer energy as ATP [14]. Glycolysis, therefore, strongly depends on the utilization of intracellular magnesium. Magnesium plays a key role in regulating many, if not all, glycolytic enzymes. This dependence is seen with the removal of magnesium, followed by an inhibition of glycolysis and an accumulation of serum glucose [15]. Magnesium-ATP complexes activate all rate-limiting enzymes of glycolysis [14]. Thus, deficiency in intracellular magnesium limits the body's ability to properly conduct glycolysis. Unmetabolized serum glucose, therefore, begins to accumulate. The resulting hyperglycemia requires incrementally more insulin in order to evoke the same metabolic response. Such unresponsiveness to insulin secretion decreases insulin sensitivity [5], seen frequently in metabolic syndrome and type 2 diabetes mellitus.

Magnesium, Calcium, and Hypertension

Magnesium is an essential component of regulating vascular tone, contractility, growth, and reactivity. It helps regulate contractile proteins; modulates the transport of calcium, sodium, and potassium; and acts as an essential cofactor in the activation of ATPase [17]. An inverse relationship is seen between body magnesium levels and blood pressure [17]. Low serum magnesium levels are associated with increased smooth muscle tension, vasospasms, and higher blood pressure [18]. The first association of magnesium and blood pressure was evidenced in the early 1900s with the infusion of magnesium salts seen to lower peripheral vascular resistance, even when an increase of myocardial contractility was seen [17]. This may be due to the intimate relationship of intracellular magnesium and calcium.

Magnesium shares a common transmembrane channel as calcium for ion transport from the extracellular to intracellular spaces [15], antagonizing the activity of calcium on smooth muscle cells. With low magnesium, calcium ions are free to move through transmembrane channels and enter the cell. Shimosawa et al. [19] exhibited in rat models that magnesium and calcium ions share the N-type calcium channel. Experimental data show that reduced magnesium levels decrease the blockage on the shared critical channels. This, in turn, allows the influx of calcium and the release of substance P, an early-stage inflammatory protein. The increased levels of substance P induce nitric oxide and free radical production, as well as cytokine secretion. These steps initiate a cascade of inflammatory-oxidative events [20]. Calcium also plays an essential role in skeletal and smooth muscle contraction. Therefore, when magnesium depletion or insufficiency is evident, the increase of intracellular calcium may lead to muscle cramps, hypertension, coronary, and cerebral vasospasms [13]. It is also important to note that increases in intracellular free calcium may negatively impact the insulin sensitivity of adipocytes and skeletal muscle and may play an important role in the development of insulin resistance [13].

Magnesium, Inflammation, and Hypertension

C-reactive protein (CRP) is an acute-phase inflammatory protein produced by the liver that stimulates phagocytosis by white blood cells and activates complement proteins in the inflammation process [7]. Low-grade inflammation has a role in the impairment of endothelial function and atherosclerosis. It is associated with an increased of hypertension, cardiovascular disease, and diabetes mellitus [21]. In adults, CRP is the most sensitive acute-phase protein to systemic inflammation [21]. An inverse relationship exists between magnesium and CRP concentrations [13,21–24]. Ford et al. [23] examined the relationship between dietary magnesium and the prevalence of metabolic syndrome among US adults. They discovered a significant and inverse association between dietary magnesium intake and concentrations of C-reactive protein concentrations [23]. Kim et al. [24] also indicated the same significant and inverse relationship between dietary magnesium intake and C-reactive protein concentrations. The connection between low magnesium concentration and C-reactive protein may suggest a correlation between oxidative stress, low-grade inflammation, and magnesium deficiency. Researchers have suggested that, "the interaction of inflammation and oxidative stress is related and increases the risk for metabolic syndrome" [25].

Evidence of Relationship Between Magnesium Deficiency, Type 2 Diabetes Mellitus, and the Metabolic Syndrome

The elevated blood pressure and increased insulin resistance associated with magnesium deficiency are two primary risk factors for the development of type 2 diabetes mellitus and the metabolic syndrome. Ample research has highlighted an inverse correlation between magnesium status and occurrence of metabolic syndrome and type 2 diabetes mellitus. The true cause of magnesium deficiency in obesity, however, is uncertain; increased urinary excretion and deficient dietary intake are the principal explanations for magnesium deficiency in this population.

Changes in popular dietary habits have resulted in the reduction of magnesium intake to below the Dietary Reference Intake (DRI) recommendations (See Table 8.1) [14]. It is important to note, however, that DRI recommendations are indicated for healthy individuals only. Magnesium sources include unrefined whole grains, most green leafy vegetables, nuts, seeds, meats, poultries, fishes, dry beans, peas, lentils, and products derived from soy (see Table 8.2) [26, 34]. Research shows that there is a connection between decreased ingestion of magnesium rich foods and the development of insulin resistance, type 2 diabetes mellitus, and metabolic syndrome.

As discussed previously, hyperglycemia with associated insulin resistance is seen with magnesium deficiency. The hyperinsulinemia associated with elevated glucose levels also contributes to magnesium depletion [14]. In the absence of good metabolic control, increased urinary magnesium excretions are common [14]. Elevated urinary losses of magnesium are not an uncommon occurrence with diabetic patients. Kidney tubular reabsorption of magnesium, which decreases urinary losses, is lessened when severe hyperglycemia is present [27]. It is difficult to know for sure the causality of magnesium deficiency and hyperglycemia. Though magnesium deficiency is shown to induce hyperglycemia, hyperglycemia is also indicated to increase urinary excretion and further the state of deficiency. This cyclical expression makes the true cause of the deficiency elusive.

Researchers have evaluated the prevalence of magnesium deficiency in serum and mononuclear cells in patients with metabolic syndrome, but without diabetes mellitus, as indicated by fasting blood glucose levels [1]. Lima et al. [1] attempted to correlate their findings with insulin resistance, other

Table 8.1 Dietary Reference Intakes for Magnesium [33]

Age (years)	Male (mg/day)	Female (mg/day)	Pregnancy (mg/day)	Lactation (mg/day)
1–3	80	80	N/A	N/A
4–8	130	130	N/A	N/A
9–13	240	240	N/A	N/A
14–18	410	360	400	360
19–30	400	310	350	310
31+	420	320	360	320

Table 8.2 Magnesium content in foods (From Office of Dietary Supplements) [26]

Food	Magnesium (mg)
Wheat bran, 1/4 cup	89
Almonds, dry roasted, 1 ounce	80
Spinach, frozen, cooked, 1/2 cup	78
Soybeans, cooked, 1/2 cup	74
Nuts, mixed, dry roasted, 1 ounce	64
Oatmeal, instant, fortified, prepared with water, 1 cup	61
Peanuts, dry roasted, 1 ounce	50
Halibut, cooked, 6 ounces	48
Potato, baked with skin, 1 medium	48
Pinto beans, cooked, ½ cup	43
Rice, brown, long grained, cooked, ½ cup	42
Lentils, cooked, ½ cup	36
Yogurt, fruit, low fat, 8 fluid ounces	32
Bread, whole wheat	23

components of metabolic syndrome, and risk of other cardiovascular factors. Insulin resistance was evaluated using the HOMA-IR equation [1]. Over half of the metabolic-syndrome-affected research population presented with obesity (72.2%), high blood pressure (77.8%), and low high-density lipoprotein cholesterol (62.5%). Fifty percent of participants had insulin resistance, indicated by a HOMA-IR $\geq$ 2.7 [1]. Serum and intracellular magnesium levels were progressively lower according to weight variations based on body mass index (BMI): normal weight (19–24.9 kg/m^2), overweight (25–29.9 kg/m^2), and obese (>30 kg/m^2). Additionally, individuals with insulin resistance had lower magnesium levels than individuals without insulin resistance [1]. The results found in this research indicate that patients with metabolic syndrome have lower serum and intramononuclear magnesium concentrations as compared to subjects without the condition. The research also indicates that insulin-resistant patients have lower magnesium levels as compared to patients with higher insulin sensitivity [1].

Similar research shows a recurring association between magnesium levels and the instance of metabolic syndrome. Guerrero-Romero and Rodriguez-Moran [28] also evaluated the relationship between serum magnesium levels and metabolic syndrome in Mexican subjects. They found that subjects with metabolic syndrome had higher fasting serum insulin levels and HOMA-IR index than control subjects and exhibited lower serum magnesium levels [28]. They also found a strong and compelling relationship between dyslipidemia and magnesium deficiency [28].

Patients without metabolic syndrome, type 2 diabetes mellitus, history of cardiovascular disease, or cancer also exhibit this negative association between magnesium levels and risk of diabetes. Lopez-Riduara et al. [29] followed 85,060 women and 42,872 men to evaluate magnesium intake using validated food frequency questionnaires every 2–4 years. After 18 years of follow-up in women and 12 years in men, 20% of the women and 32% of the men had documented type 2 diabetes mellitus. After adjusting for age, BMI, physical activity, family history of diabetes, smoking, alcohol consumption, and history of hypertension and dyslipidemia at baseline, the relative risk for type 2 diabetes for women and men was 0.66 and 0.67, respectively, compared to the highest and lowest quintile of total magnesium intake [29]. This research indicates an inverse relationship between magnesium intake and the risk of type 2 diabetes mellitus [29]. It is interesting to note that the inverse relationship remained within subgroups even after adjusting for the main predictors of type 2 diabetes mellitus [29].

Further research confirms the relationship between dietary intake of magnesium and the prevalence of metabolic syndrome. Using a nationally representative sample of US men and women, Ford et al. [23] evaluated dietary intake through a single 24-h dietary recall in relation to existence of metabolic syndrome. The research showed that dietary intake of magnesium was significantly and inversely associated with concentrations of insulin, elevated concentrations of C-reactive protein, and percent kilocalories (kcals) from carbohydrates [23]. They found a similar inverse relationship between all five components of the metabolic syndrome in comparison to greater dietary magnesium intake; of these components, the inverse trend was significant for abdominal obesity (waist circumference >102 cm in men or >88 cm in women), low high-density lipoprotein cholesterol, and hyperglycemia [23]. These studies alone, and in combination, reveal a true association between magnesium intake and laboratory assessments compared to risk of metabolic syndrome and type 2 diabetes mellitus. It is evident that insufficient magnesium increases the risk of developing these conditions.

Effect of Magnesium Supplementation

Evidence of the correlation between magnesium deficiency and the incidence of insulin resistance, metabolic syndrome, and type 2 diabetes mellitus is strong. There is ample scientific research indicating the significant and inverse correlation between the vitamin and these conditions. It is prudent, then, to consider the effects of magnesium supplementation to decrease the discrepancies seen within the effected population.

Early signs of magnesium deficiency include loss of appetite, nausea, vomiting, fatigue, and weakness. People may experience numbness, tingling, muscle contractions and cramps, seizures, abnormal heart rhythms, and coronary spasms [30]. These symptoms may be ameliorated with the appropriate supplementation of magnesium. Oral magnesium supplementation in its elemental form is safe in adults when used in dosages below the upper intake level of 2.5 g per day (37), though there is no upper limit for dietary magnesium intake. Some common forms of magnesium supplements include magnesium oxide (61% elemental magnesium), magnesium hydroxide (42% elemental magnesium), and magnesium citrate (16% elemental magnesium) [30]. These formulas are generally nontoxic, though the most common side effects are headache, nausea, hypotension, and nonspecific slight abdominal and bone pain. These side effects generally do not require specific treatment or discontinuation of the magnesium salt supplementation (39).

Researchers in an Iowa Women's Health Study found that postmenopausal women showed a significant reduction in the relative risk of diabetes with an increased intake of whole grains and other food sources of magnesium [31]. This may indicate a protective factor of magnesium consumption. This has also been illuminated in a study investigating the oral supplementation of magnesium to nondiabetic subjects with insulin resistance [22]. Researchers sought to increase insulin sensitivity of nondiabetic patients with the daily supplementation of 2.5 g magnesium chloride ($MgCl_2$). All subjects were apparently healthy but presented with both a decreased serum magnesium level and insulin resistance. Results of this double-blind placebo-controlled trial indicated that magnesium supplementation did, in fact, increase serum magnesium concentrations and decreased fasting serum glucose, insulin levels, and insulin resistance [30]. The data presented suggest that the association between plasma magnesium concentrations and insulin resistance is, in fact, present in healthy individuals without type 2 diabetes mellitus [30]. The results found are promising, as they suggest the protective nature of magnesium supplementation.

Individuals already diagnosed with metabolic syndrome or type 2 diabetes mellitus may also benefit from oral supplementation. Similar research was conducted to elucidate the same positive response in type 2 diabetic subjects. A daily 2.5 g magnesium chloride ($MgCl_2$) supplement was administered to diabetic patients in a double-blind placebo-controlled trial to determine whether it improved insulin sensitivity and metabolic control [32]. Patients receiving active magnesium supplements exhibited improved serum magnesium concentrations and HbA_{1c} levels and a lower HOMA-IR index and fasting glucose concentrations compared to controls. These results confirm that magnesium supplementation improves metabolic control of diabetic subjects [32].

How to Increase Magnesium Intake

Adequate dietary magnesium intake is important for healthy living, with or without disease. It is found in a variety of foods and can be incorporated into a healthy and diverse diet. Nuts and seeds, especially pumpkin seeds, almonds, Brazil nuts, and sesame seeds, top the list of dietary magnesium content. (see Table 8.2) [26]. Eating a varied diet, full of legumes, vegetables, whole grains, dairy, and lean protein, helps ensure that the recommended amount of nutrients is satisfied. In the fast-paced society of today, processed and packaged foods fill the grocery shelves, but unfortunately, refining foods strips the magnesium from food products. Eating whole, unprocessed foods as part of a healthy diet can increase magnesium consumption and lessen the risk of developing type 2 diabetes and metabolic syndrome.

Conclusion

Magnesium, the second most abundant intracellular cation, is a vital macromineral in maintaining normal metabolism and health. The functions of magnesium are plentiful and widespread throughout the body. It is clear that the relationship between magnesium status and insulin resistance, type 2 diabetes, and metabolic syndrome is an important one. The negative and inverse correlation suggests that low magnesium levels are related to the development of these disorders. On the other hand, however, high magnesium levels may provide protective characteristics to these dangerous conditions. It is important to obtain an adequate amount of dietary magnesium to lessen the risk of insulin resistance, type 2 diabetes, and metabolic syndrome.

Keywords Defined

Metabolic Syndrome – Metabolic syndrome is a collection of diagnostic characteristics that increase risk of type 2 diabetes mellitus, hypertension, and cardiovascular disease. Diagnostic characteristics include having three of the five following traits: central adiposity, decreased high-density lipoproteins, elevated triglycerides, elevated fasting glucose levels, and elevated blood pressure.

Insulin resistance – Insulin resistance is a condition in which the body decreases sensitivity to insulin secretion. The pancreas produces ample insulin in response to elevated blood glucose concentrations, but the body does not react effectively to the action of insulin.

Type 2 diabetes mellitus –Type 2 diabetes mellitus is a condition where body cells no longer respond to insulin stimulation. This can lead to hyperglycemia and cell starvation.

References

1. Lima M, Cruz T, Rodriques L, et al. Serum and intracellular magnesium deficiency in patients with metabolic syndrome- evidences for its relation to insulin resistance. Diabetes Res Clin Pract. 2009;14:257–62.
2. Gundy S, Cleeman J, Daniels S, Donato K, Eckel H, et al. Diagnosis and management of the metabolic syndrome: an American heart association/national heart, lung, and blood institute scientific statement: executive summary. Circulation. 2005;112:285–91.
3. Alberti K, Eckel R, Grundy S, et al. Harmonizing the metabolic syndrome. Circulation. 2009;120:1640–5.

4. Centers for Disease Control and Prevention. Metabolic syndrome. In: National health statistics reports. 2009. http://www.cdc.gov/nchs/data/nhsr/nhsr013.pdf. Accessed 16 Feb 2012.
5. Hypertension-Fact sheet. World health organization. 2001. http://www.searo.who.int/linkfiles/non_communicable_diseases_hypertension-fs.pdf. Accessed 16 Feb 2012.
6. Dickinson HO, Nicloson D, Campbell F, Cook JV, Beyer FR, Ford GA, Mason J. Magnesium supplementation for the management of primary hypertension in adults (Review). Cochrane Database Syst Rev. 2006;3:1–55.
7. Gropper SS, Smith JL, Groff JL. Magnesium. In: Advanced nutrition and human metabolism. 5th ed. Belmont, CA: Thomas and Wadsworth Cengage Learning. 2009;447–51.
8. Wilcox G. Insulin and insulin resistance. Clin Biochem. 2005;26:19–39.
9. Kido Y, Nakae J, Accili D. The insulin receptor and its cellular targets. J Clin Endocrinol Metab. 2001;86(3): 972–29.
10. Insulin resistance and prediabetes. http://diabetes.niddk.nih.gov/dm/pubs/insulinresistance/. Accessed 6 Dec 2011.
11. Jeffrey A. Insulin resistance. Nurs Stand. 2003;17(32):47–53.
12. Agrawal P, Arora S, Singh B, Manamalli A, Dolia P. Association of macrovascular complications of type 2 diabetes mellitus with serum magnesium levels. Diabetes Metab Syndr Clin Res Rev. 2011;5:41–4.
13. Volpe SL. Magnesium, the metabolic syndrome, insulin resistance, and type 2 diabetes mellitus. Crit Rev Food Sci Nutr. 2008;48:1–8.
14. Barbagallo M, Dominguez L. Magnesium metabolism in type 2 diabetes mellitus, metabolic syndrome and insulin resistance. Arch Biochem Biophys. 2007;458:40–7.
15. Romani A. Regulation of magnesium homeostasis and transport in mammalian cells. Arch Biochem Biophys. 2007;50:90–102.
16. Sesti G. Pathophysiology of insulin resistance. Best Pract Res Clin Endocrinol Metab. 2006;50:665–79.
17. Sontia B, Touyz R. Role of magnesium in hypertension. Arch Biochem Biophys. 2007;458:33–9.
18. Das UN. Nutritional factors in the pathobiology of human essential hypertension. Nutrition. 2001;17:337–46.
19. Shimosawa T, Takano K, Ando K, Fujita T. Magnesium inhibits norepinephrine release by blocking N-type calcium channels at peripheral sympathetic nerve endings. Hypertension. 2004;44(6):897–902.
20. Guerrero-Romero F, Rodriguez-Moran M. Hypomagnesemia, oxidative stress, inflammation, and metabolic syndrome. Diabetes Metab Res Rev. 2006;22:471–6.
21. Rodriguez-Moran M, Guerrero-Romero F. Serum magnesium and C-reactive protein levels. Arch Dis Child. 2008;93:676–80.
22. Evangelopoulos A, Vallianou N, Panagiotakos D, Georgiou A, Zacharias G, Alvera A, et al. An inverse relationship between cumulating components of the metabolic syndrome and serum magnesium levels. Diabetology. 2008;28(10):659–63.
23. Ford E, Li C, McGuire L, Mokdad A, Liu S. Intake of dietary magnesium and the prevalence of the metabolic syndrome among U.S adults. Obesity. 2007;15:1139–46.
24. Kim D, Xn P, Liu K, Loria C, Yokota K, Jacobs D, et al. Magnesium intake in relation to systemic inflammation, insulin resistance, and the incidence of diabetes. Diabetes Care. 2010;33:2604–10.
25. Guerrero-Romero F, Rodriguez-Moran M. Relationship between serum magnesium levels and C-reactive protein concentration, in non-diabetic, non-hypertensive obese subjects. Diabetes. 2006;26:469–74.
26. Magnesium. In: Dietary supplement fact sheet. 2009. http://ods.od.nih.gov/factsheets/magnesium/. Accessed 16 Feb 2012.
27. Barbagallo M, Dominguez L, Galioto A, Ferlisi A, Cani C, Malfa L, et al. Role of magnesium in insulin action, diabetes and cardio-metabolic syndrome X. Mol Aspects Med. 2003;24:39–52.
28. Guerrero-Romero F, Rodriguez-Moran M. Low serum magnesium levels and metabolic syndrome. Acta Diabetol. 2002;29:209–13.
29. Lopez-Ridaura R, Willett W, Rimm E, Liu S, Stampper M, Manson J, et al. Magnesium intake and risk of type 2 diabetes in men and women. Diabetes Care. 2004;27:134–40.
30. Guerrera M, Volpe SL, Mao J. Therapeutic uses of magnesium. Am Fam Physician. 2009;80(2):157–62.
31. Meter KA, Kushi LH, Jacobs Jr DR, Slvin J, Sellers TA, Folsom AR. Carbohydrates, dietary fiber and incident of type 2 diabetes in older women. Am J Clin Nutr. 2000;71:921–30.
32. Guerrero-Romero F, Rodriguez-Moran M. Complementary therapies for diabetes. The case for chromium, magnesium and antioxidants. Arch Med Res. 2005;30:250–7.
33. Dietary reference intakes (DRIs): estimated average requirements. 2011. http://www.iom.edu/Activities/Nutrition/SummaryDRIs/~/media/Files/Activity%20Files/Nutrition/DRIs/5_Summary%20Table%20Tables%201-4.pdf. Accessed 16 Feb 2012.
34. Guerrero-Romero F, Tamez-Perez H, Gonzalez-Gonzalez G, Salinas-Martinez A, Villarreal J, Trevino-Ortiz J, et al. Oral magnesium supplementation improves insulin sensitivity in non-diabetic subjects with insulin resistance. A double-blind placebo-controlled randomized trial. Diabetes Metab. 2004;30:253–8.

Chapter 9
Magnesium and Diabetes Prevention

Akiko Nanri and Tetsuya Mizoue

Key Points

- While a meta-analytic study reported the association of high dietary magnesium intake with lower risks of type 2 diabetes, the findings of the three prospective studies dealing with Asian populations within the meta-analysis were inconsistent.
- Further, the principal sources of dietary magnesium intake differ between Japanese and Western populations, and therefore, further investigation in Asian populations (including Japanese) is needed to confirm whether magnesium intake contributes to the prevention of type 2 diabetes and whether it differs according to the source.
- In addition, studies measuring serum magnesium are also required to confirm the findings based on dietary assessment.

Keywords Magnesium supplements • Type 2 diabetes risk • Magnesium intake • Insulin actions • Japanese study

Introduction

In Japan, the respective prevalence of probable and possible diabetes increased respectively from 6.9 to 8.9 and from 6.8 to 13.2 million cases between 1997 and 2007 [1]. Probable diabetes cases are characterized by either high levels of glycated hemoglobin (≥ 6.1%) or by receiving medication for type 2 diabetes. Possible diabetes cases are characterized by glycated hemoglobin levels ranging from 5.6% to 6.1% [1]. Within populations of Japanese men and women aged over 50 years, the prevalence of probable and possible diabetes has been reported to be as high as 30–40% [1]. Although the prevalence of obesity (an important risk factor of type 2 diabetes) is lower in Japanese than in Western populations [2], the prevalence of type 2 diabetes is not dramatically lower [3]. While these epidemic proportions of type 2 diabetes among Japanese have mainly been attributed to genetic differences between Asian and Caucasian populations [4], some limited evidence also suggests the influence of environmental factors, such as diet. For example, magnesium is involved in glucose homeostasis and

A. Nanri, Ph.D. (✉) • T. Mizoue, M.D., Ph.D.
Department of Epidemiology and Prevention, Clinical Research Center,
National Center for Global Health and Medicine, 1-21-1 Toyama Shinjuku-ku, Tokyo 162-8655, Japan
e-mail: nanri@ri.ncgm.go.jp

R.R. Watson et al. (eds.), *Magnesium in Human Health and Disease*, Nutrition and Health,
DOI 10.1007/978-1-62703-044-1_9,

insulin action [5], and its protective effect against type 2 diabetes has been demonstrated by several prospective studies, including one involving a Japanese population.

Magnesium Intake in Japan

According to the National Health and Nutrition Survey conducted in Japan in 2008 [6], the daily magnesium intake for adult Japanese men and women is 268 mg and 239 mg, respectively. Although intake among men was slightly lower than the 270–310 mg/day recommended in 2010 by the dietary reference intakes for Japanese [7], intake among women fell within the recommended range (220–240 mg/day). The major sources of magnesium in the Japanese diet were determined to be cereals (17.6%), vegetables (15.2%), pulses (11.9%), spices and seasonings (including soy sauce and miso paste) (11.1%), and fishes and shellfishes (10.7%) [6]. The relative number of magnesium supplement users in Japan was extremely small, with a population-based study conducted among 1,152 men and 1,107 women aged 40–82 years by the National Institute for Longevity Sciences-Longitudinal Study of Aging showing that only 0.2% of men and 0.3% of women reported daily magnesium supplement use in the past 12 months [8]. In addition, seldom users (those who reported magnesium supplement use at least once a year but less than once a week for the past 12 months) accounted for 0.1% of men and 0.2% of women surveyed [8].

Magnesium Intake and Type 2 Diabetes Risk in the Japan Public Health Center-Based Prospective (JPHC) Study

Studies of the association between magnesium intake and type 2 diabetes have been conducted mainly among Western populations, with relatively few concentrating on Asian populations. In addition, the observed inverse association between magnesium intake and type 2 diabetes appears to be weaker among Asian than Western populations, suggesting an ethnic difference in the effect of magnesium on type 2 diabetes risk. In our previous study, we prospectively investigated the association of magnesium intake with the risk of developing type 2 diabetes using data from the large-scale population-based cohort study in Japan [9].

JPHC Study

The JPHC study was launched in 1990 for cohort I and in 1993 for cohort II [10]. The participants of cohort I included residents aged 40–59 years, in five Japanese public health center areas (Iwate, Akita, Nagano, Okinawa, and Tokyo); the participants of cohort II included residents aged 40–69 years in six Japanese public health center areas (Ibaraki, Niigata, Kochi, Nagasaki, Okinawa, and Osaka). A questionnaire survey was conducted at baseline (in 1990 for cohort I and in 1993 for cohort II), 5-year follow-up (in 1995 for cohort I and in 1998 for cohort II), and 10-year follow-up (in 2000 for cohort I and in 2003 for cohort II). Information on medical histories and health-related lifestyle, smoking, drinking, and dietary habits, was obtained at each survey.

Study Participants

From the study population at baseline (n = 140,420), we excluded those who resided in two public health center areas because of the differences in recruitment criteria. Of the remaining 116,672 eligible participants, 95,373 (81.7%) responded to the questionnaire survey at baseline. Of these, 80,128 (84.0%) also responded to the 5-year follow-up survey, which is the baseline of the present analysis. Of these, 71,075 (88.7%) responded to the 10-year follow-up survey. We excluded participants who reported a history of type 2 diabetes (n = 5,183) or severe disease (n = 6,284), including of cancer, cerebrovascular disease, myocardial infarction, chronic liver disease, and renal disease, at baseline or 5-year follow-up surveys. A further 590 participants who reported extreme total energy intake were excluded, leaving a total of 59,791 participants (25,872 men and 33,919 women) ultimately enrolled in this analysis.

Dietary Assessment

At baseline, 5-year follow-up, and 10-year follow-up surveys, participants completed a self-administered questionnaire on lifestyle and health. In this analysis, we used data from the 5-year follow-up survey as baseline data because the questionnaire used for the 5-year follow-up survey more comprehensively inquired about food intakes than that used for the baseline survey. At the 5-year follow-up survey, a food frequency questionnaire (FFQ) was used to assess the average intake of 147 food and beverage items over the previous year [11]. For most food items, participants were asked consumption frequency by choosing one of nine options (almost none, 1–3 times/month, 1–2 times/week, 3–4 times/week, 5–6 times/week, once/day, 2–3 times/day, 4–6 times/day, or ≥7 times/day). A standard portion size was specified for each food, and respondents were asked to choose their usual portion size from three options (less than one-half, standard, or more than 1.5 times). Daily food intake was calculated by multiplying daily consumption frequency by the typical portion size. Daily intake of nutrients was calculated by using the 5th revised and enlarged edition of the Standard Tables of Food Composition in Japan [12]. The validity and reproducibility of dietary magnesium intake were assessed in a subsample of the participants in the JPHC study cohort I and cohort II. For validity of the FFQ, Spearman's correlation coefficients between energy-adjusted intake estimated from the FFQ and those estimated from dietary records for magnesium were 0.45–0.46 in men and 0.42–0.45 in women [13]. With regard to the reproducibility of the FFQ, Spearman's correlation coefficients for energy-adjusted intake estimated from the two FFQs administered 1 year apart were 0.62–0.70 in men and 0.61 in women [13].

Ascertainment of Type 2 Diabetes

Type 2 diabetes newly diagnosed during the 5-year period after the 5-year follow-up survey was determined by a self-administered questionnaire at the 10-year follow-up survey. At the 10-year follow-up survey, study participants were asked if they had ever been diagnosed with diabetes, and if so, when the initial diagnosis had been made. Because the 5-year follow-up survey was used as the starting point of observation for the incidence of type 2 diabetes, only those participants who were diagnosed after 1995 for cohort I and 1998 for cohort II were regarded as incident cases during follow-up. In a previous study that we conducted, 94% of self-reported diabetes cases were confirmed as such by medical records [14]. We also conducted a cross-sectional survey in 1990 to examine the sensitivity of diagnosed diabetes according to the criteria at that time for a JPHC subpopulation (health checkup participants) whose plasma glucose data were available [14]. Of 6,118 participants with plasma glucose data, 248 participants had diagnosed diabetes. Among 5,870 participants who did not have diagnosed diabetes, 49 participants (0.83%) had diabetes according to the commonly used diagnostic standards in Japan in 1990 (fasting plasma glucose ≥7.8 mmol/L; casual plasma glucose ≥11 mmol/L [15]) based on a single measurement. Taking into account the above-mentioned positive predictive

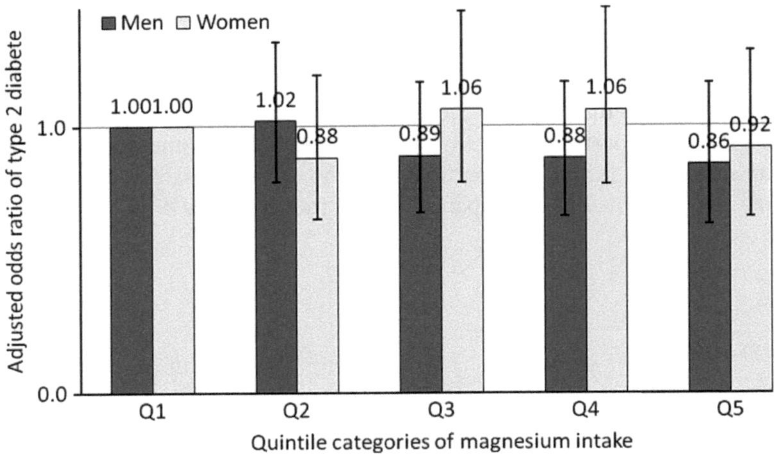

Fig. 9.1 *Multivariate-adjusted odds ratios of type 2 diabetes according to quintile categories of magnesium intake in the JPHC Study* [9] (Adjusted for age, study area, body mass index, smoking status, alcohol consumption, family history of diabetes mellitus, physical activity, history of hypertension, coffee consumption, calcium intake, and energy intake. Magnesium intake was adjusted for energy intake using residual method. *P* for trend was 0.18 in men and 0.96 in women based on multiple logistic regression analysis, assigning ordinal numbers 0–4 to the quintile categories of magnesium intake)

value, the sensitivity and specificity of diagnosed diabetes were 85.5% and 99.7%, respectively, in men and 79.3% and 99.7%, respectively, in women.

Results

During the 5-year period, 1,114 participants were newly diagnosed with diabetes (634 men and 480 women). Among men, the risk of type 2 diabetes tended to decrease with magnesium intake; the multivariate-adjusted odds ratios (95% confidence intervals) for the lowest through highest quintiles of energy-adjusted magnesium intake were 1.00 (reference), 1.02 (0.79–1.32), 0.89 (0.68–1.17), 0.88 (0.66–1.17), and 0.86 (0.63–1.16) (Fig. 9.1). However, the trend association was not statistically significant (*P* for trend=0.18). Among women, no association was also observed. In an analysis stratified by body mass index, there was also no association among both men and women with body mass index of ≥25 kg/m^2; the multivariate-adjusted odds ratios (95% confidence intervals) for the lowest through highest quintiles of magnesium intake were 1.00 (reference), 1.08 (0.75–1.56), 0.79 (0.53–1.19), 0.81 (0.54–1.23), and 0.89 (0.57–1.38) in men (*P* for trend=0.28) and 1.00 (reference), 0.87 (0.56–1.34), 1.30 (0.86–1.97), 1.12 (0.72–1.73), and 1.04 (0.65–1.66) in women (*P* for trend=0.57).

In this large-scale prospective study, although a small protective effect cannot be excluded, this study does not provide evidence to support a significant role of magnesium in the development of type 2 diabetes in Japanese adults.

Magnesium Intake and Type 2 Diabetes Risk in Another Japanese Study

In Japan, an association was observed between magnesium intake (assessed using a validated FFQ including 33 food items) and incidence of self-reported type 2 diabetes cases (237 men and 222 women) during 5-year follow-up among 6,480 men and 11,112 women aged 40–65 years who were initially free of type 2 diabetes or other chronic disease as determined in another population-based study, the Japan Collaborative Cohort Study for Evaluation of Cancer Risk (JACC study) [16]. Although no significant inverse association between magnesium intake and type 2 diabetes risk was observed in either men or women, the overall study population (combined men and women) showed a significant association between magnesium intake and decreased risk of type 2 diabetes, with odds ratios (95% confidence intervals) of type 2 diabetes for the lowest through highest quartiles of magnesium intake being calculated as 1.00 (reference), 0.83 (0.64–1.09), 0.79 (0.59–1.07), and 0.64 (0.44–0.94) after adjustment for age, sex, body mass index, family history of diabetes, smoking habit, alcohol consumption, physical activity (walking and sport), consumption of green tea and/or coffee, and energy intake (*P* for trend = 0.04).

Although the JACC study demonstrated a clear inverse association between magnesium intake and type 2 diabetes risk [16], the JPHC study revealed no such association [9]. Although this difference might be due to random variation, it may also be due to the number of food and beverage items included in FFQ (33 in the JACC study and 147 in the JPHC study) or the 5–10 years' lapse between the two studies (dietary intake was assessed from 1988 to 1990 in the JACC and 1995 to 1998 in the JPHC study). Indeed, the major source of magnesium intake may have changed over time, and a previous study [17] has suggested that the effect of magnesium intake on type 2 diabetes risk might differ according to the ingestion source (discussed later in this publication).

Magnesium Intake and Type 2 Diabetes Risk According to Ethnic Group

The prevalence of type 2 diabetes differs by country and ethnicity [3], possibly due in part to differences among ethnicities in the prevalence of obesity, insulin sensitivity levels, beta cell function, and lifestyle. In addition, differences by ethnicity may also influence the effect of magnesium intake on type 2 diabetes risk. In the multiethnic cohort study conducted in Hawaii, Hopping et al. [18] examined the association between magnesium intake and type 2 diabetes risk among Caucasians, Japanese Americans, and subjects with native Hawaiian ancestry. Among 36,256 men and 39,256 women aged 45–75 years who were free of type 2 diabetes at baseline, 8,587 incident cases of type 2 diabetes were identified during 14 years of follow-up. Magnesium intake was assessed using a calibrated quantitative FFQ. Daily magnesium intakes in both men and women were somewhat higher among Caucasians compared to Japanese Americans and native Hawaiians. Respective energy-adjusted medians as calculated by density methods (per 4,184 kJ) in men and women were 158 and 174 mg for Caucasians, 149 and 165 mg for Japanese Americans, and 140 and 153 mg for native Hawaiians.

After adjustment for ethnicity, body mass index, physical activity, education, and energy intake, magnesium intake was associated with decreased risk of type 2 diabetes in both men and women. Given the strong correlation noted between magnesium and dietary fiber intake, the observed beneficial effect on type 2 diabetes risk may be due to either or both of these compounds. However, the association between magnesium intake and type 2 diabetes risk remained significant even after additional adjustment for dietary fiber intake. Moreover, the observed inverse association between dietary fiber intake and type 2 diabetes risk was not significant after additional adjustment for magnesium intake.

The above-mentioned association between magnesium intake and type 2 diabetes risk differed by ethnic group in the multiethnic cohort study. Among women, a significant inverse association between magnesium intake and type 2 diabetes risk was observed among Caucasians but not Japanese Americans or native Hawaiians (Fig. 9.2). Similarly, among men, the association of magnesium intake with type 2 diabetes risk was stronger in Caucasians than in Japanese Americans or native Hawaiians. Multivariate-adjusted hazard ratios (95% confidence intervals) of type 2 diabetes for the highest

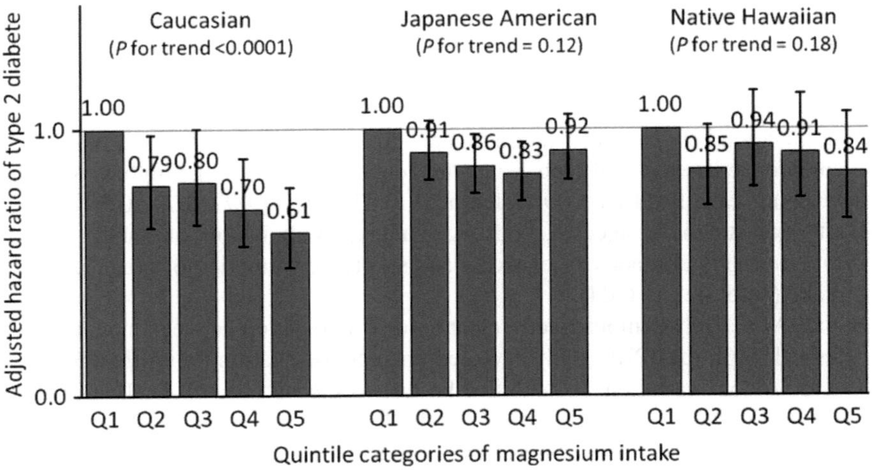

Fig. 9.2 *Multivariate-adjusted hazard ratios of type 2 diabetes according to quintile categories of magnesium intake by ethnic group in women in the multiethnic cohort in Hawaii* [18] (Adjusted for body mass index, physical activity, education level, and energy intake. Magnesium intake was adjusted for energy intake using density method)

versus lowest quintiles of magnesium intake were 0.65 (0.53–0.79; *P* for trend <0.0001) in Caucasians, 0.85 (0.75–0.96; *P* for trend = 0.004) in Japanese Americans, and 0.79 (0.61–1.03; *P* for trend = 0.08) in native Hawaiians.

Magnesium Intake from Different Sources and Type 2 Diabetes Risk

Magnesium may be found in a number of foods such as cereal, vegetables, beans and soy beans, nuts, and seafood, and its effect on type 2 diabetes may differ depending on the source. In the Shanghai Women's Health Study, Villegas et al. [17] examined the association between type 2 diabetes risk and magnesium intake from distinct animal, vegetable, or dairy sources. Among 64,191 women aged 40–70 years who were free of type 2 diabetes or other chronic disease at baseline, 2,270 incident cases of type 2 diabetes were identified during the 6.9 years of follow-up. Magnesium intake was assessed using a validated FFQ consisting of 77 food items and food groups. In this population, the major sources of magnesium intake were rice (33.3%), tofu (8.25%), seafood (6.9%), legumes (6.35%), and Chinese greens (4.35%). Magnesium intake from all sources was significantly associated with decreased risk of type 2 diabetes, with respective multivariate-adjusted relative risks (95% confidence intervals) for the lowest through highest quintiles of magnesium intake of 1.00 (reference), 0.80 (0.69–0.93), 0.85 (0.73–0.99), 0.73 (0.62–0.85), and 0.80 (0.68–0.93) (*P* for trend <0.0001) after adjustment for age, body mass index, waist-to-hip ratio, smoking status, alcohol consumption, physical activity, income level, education level, occupation, hypertension, and energy intake. Upon arrangement by source, the authors showed that magnesium from animal and dairy sources was significantly associated with decreased risk of type 2 diabetes, while no such association was found with magnesium sourced from vegetables (Fig. 9.3). The authors suggested that this may be due to the fact that vitamin D in animal and dairy food sources favorably affects the absorption of magnesium, thereby strengthening the protective association of dietary magnesium from animal or dairy sources with type 2 diabetes.

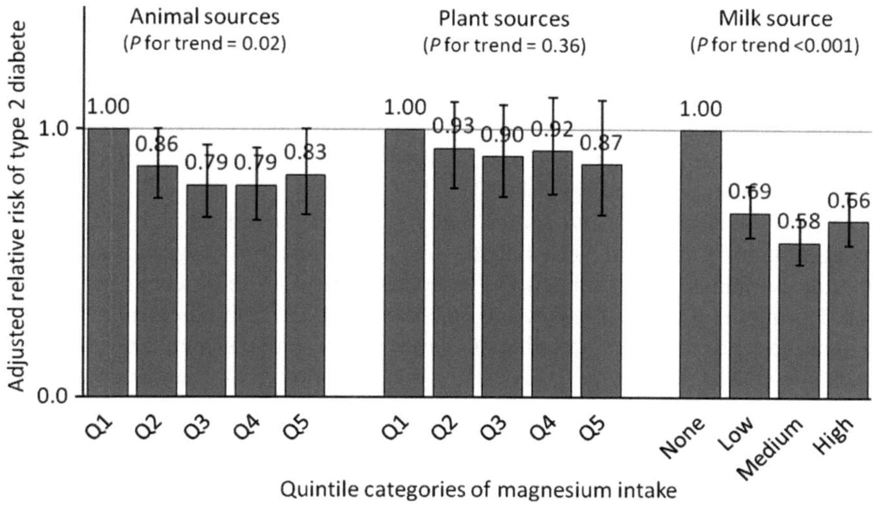

Fig. 9.3 *Multivariate-adjusted relative risks of type 2 diabetes according to quintile categories of magnesium intake from animal, plant, and dairy sources in the Shanghai Women's Health Study* [17] (Adjusted for age, body mass index, waist-to-hip ratio, smoking status, alcohol consumption, physical activity, income level, education level, occupation, hypertension, and energy intake. Magnesium intake was adjusted for energy intake using residual method)

Magnesium Intake and Type 2 Diabetes Risk: Result of Meta-Analysis

Three out of 13 prospective cohort studies examining the association between magnesium intake and type 2 diabetes risk that were reviewed in a recent meta-analysis [19] were conducted in Asia (including the two Japanese studies mentioned above), while the rest were conducted in the United States, Europe, and Australia. Among a total of 536,318 participants, 24,516 cases of type 2 diabetes were identified during the 4–20 years of follow-up. Nine of these studies observed a significant inverse association between magnesium intake and type 2 diabetes risk, whereas the remaining four found no such association. The summary relative risk of type 2 diabetes for the highest category of magnesium intake was 0.78 (95% confidence interval: 0.73–0.84) compared to the lowest category. A significant inverse association was observed among participants with a body mass index ≥25 kg/m^2, with a relative risk of type 2 diabetes for the highest versus lowest category of magnesium intake of 0.73 (95% confidence interval: 0.66–0.81). In contrast, among participants with body mass index of <25 kg/m^2, the corresponding value was 1.09 (95% confidence interval: 0.76–1.56; *P* for interaction between magnesium intake and body mass index = 0.13). These findings suggest that magnesium may have had a greater impact on insulin sensitivity in overweight individuals than in nonoverweight ones. Further, among the five studies that examined the association between magnesium intake and type 2 diabetes after adjustment for cereal fiber intake, the summary relative risk of type 2 diabetes for the highest versus lowest category of magnesium intake remained significant (relative risk 0.74, 95% confidence intervals: 0.68–0.80). This finding suggests that magnesium intake is associated with decreased risk of type 2 diabetes, independent of fiber intake.

Serum Magnesium Concentrations and Type 2 Diabetes Risk

All of the studies mentioned above used FFQ to estimate dietary magnesium intake, and while most used validated methods, the estimated magnesium intake might not reflect systemic body magnesium levels. However, two other prospective studies did investigate the association between serum magnesium concentration and type 2 diabetes risk [20, 21].

The Atherosclerosis Risk in Communities Study [21] categorized 2,622 black and 9,506 white participants aged 45–64 years who were diabetes-free at baseline into six groups according to serum magnesium concentrations and examined the association between these concentrations and type 2 diabetes risk. During the 6 years of follow-up, 367 black and 739 white incident cases of type 2 diabetes were identified. For the white participants, serum magnesium concentrations were significantly associated with decreased risk of type 2 diabetes, with odds ratios (95% confidence intervals) for type 2 diabetes of the lowest (0.25–0.70 mmol/L) through highest (0.95–11.30 mmol/L) categories of serum magnesium concentrations of 1.76 (1.18–2.61), 1.24 (0.86–1.78), 1.11 (0.80–1.56), 1.20 (0.86–1.68), 1.13 (0.79–1.61), and 1.00 (reference) after adjustment for age, sex, education, family history of diabetes, body mass index, waist-to-hip ratio, sports index, alcohol consumption, diuretic use, and serum calcium and potassium levels (P for trend=0.01). No such association was observed in black populations. In contrast, dietary magnesium intake was not associated with type 2 diabetes risk. The authors ascribed the discrepancies between serum magnesium concentrations and dietary magnesium intake to an imprecise method of dietary magnesium intake measurement. In addition, serum magnesium levels may reflect compartmentalization or renal handling of whole-body magnesium level rather than dietary intake of magnesium.

The other study [20] using serum magnesium measurement examined the association between hypomagnesemia (<0.74 mmol/L) and new-onset type 2 diabetes among 817 men and women in Mexico aged 20–65 years who were free of diabetes at baseline. Seventy-eight participants were identified as having new-onset diabetes during the follow-up period (10 years), and on comparison with participants without hypomagnesemia, those with hypomagnesemia had a 2.54-times higher risk of developing type 2 diabetes, independent of age, sex, family diabetes history, waist circumference, and homeostasis model assessment of insulin resistance (HOMA-IR) index.

Effect of Magnesium Supplementation on Type 2 Diabetes Risk

Few studies have examined the effect of oral magnesium supplementation on type 2 diabetes risk or insulin resistance among participants without type 2 diabetes. Out of four intervention studies investigating oral magnesium supplementation [22–25], three have noted beneficial effects on insulin resistance [22, 23, 25].

In a Mexican study [22], 60 apparently healthy participants with low serum magnesium concentrations and insulin resistance showed significant improvement in insulin resistance by HOMA-IR index due to magnesium supplementation (2.5 g $MgCl_2$/day) for 12 weeks (HOMA-IR index at baseline and 12 weeks: 4.6 and 2.6 in the intervention group and 5.2 and 5.3 in the placebo group, respectively). A Greek study showed that, among 48 participants with mild uncomplicated hypertension, oral glucose tolerance test–derived insulin sensitivity indices of Stumvoll, Matsuda, and Cederholm were increased in groups prescribed a magnesium supplement (600 mg/day) in addition to lifestyle recommendations for 12 weeks compared to the placebo group, which only received the lifestyle recommendations [23]. A recent German study investigating the effect of oral magnesium supplementation on insulin sensitivity in 52 normomagnesemic, overweight (body mass index ≥25 kg/m^2), insulin-resistant, nondiabetic subjects reported a significant improvement on insulin sensitivity indices in the intervention group that was treated with magnesium-aspartate-hydrochloride (365 mg/day for 6 months) compared to the placebo group [25]. In contrast, a Korean study focusing on 155 normomagnesemic, overweight (body mass index ≥23 kg/m^2), nondiabetic participants noted no beneficial effect of magnesium supplementation (300 mg/day for 12 weeks) on HOMA-IR index [24].

A meta-analysis of nine randomized double-blind controlled trials that examined the effect of oral magnesium supplementation on glycemic control in type 2 diabetes reported that participants in an intervention group that received oral magnesium supplementation (360 mg/day [median] during 4–16 weeks [median: 12 weeks]) had significantly decreased fasting glucose levels compared to the placebo group (weighted mean difference: 0.56, 95% confidence intervals: 1.10 to −0.01), whereas the changes in glycated hemoglobin levels between the two groups were not significant (weighted mean difference: 0.31, 95% confidence intervals: 0.81 to 0.19) [26].

Conclusions

While a meta-analytic study reported the association of high dietary magnesium intake with lower risks of type 2 diabetes, the findings of the three prospective studies dealing with Asian population within the meta-analysis were inconsistent. Further, the principal sources of dietary magnesium intake differ between Japanese and Western populations, and therefore, further investigation in Asian populations (including Japanese) is needed to confirm whether or not magnesium intake contributes to the prevention of type 2 diabetes and if it differs according to the source. In addition, studies measuring serum magnesium are also required to confirm the findings based on dietary assessment.

References

1. Kenko Eiyo Joho Kenkyukai. The National Health and Nutrition Survey in Japan, 2007. Tokyo: Daiichi-shuppan; 2010.
2. Huxley R, Omari A, Caterson ID. Obesity and diabetes. In: Ekoe JM, Rewers M, Williams R, Zimmet P, editors. The epidemiology of diabetes mellitus. 2nd ed. Chichester: Wiley-Blackwell; 2008.
3. International Diabetes Federation. IDF diabetes atlas. 4th ed. Brussels: International Diabetes Federation; 2009.
4. Yazaki Y, Kadowaki T. Combating diabetes and obesity in Japan. Nat Med. 2006;12:73–4.
5. Paolisso G, Scheen A, D'Onofrio F, Lefebvre P. Magnesium and glucose homeostasis. Diabetologia. 1990;33:511–4.
6. Kenko Eiyo Joho Kenkyukai. The National Health and Nutrition Survey in Japan, 2008. Tokyo: Daiichi-shuppan; 2011.
7. Ministry of Health Labour and Welfare, Japan. Dietary reference intakes for Japanese, 2010. Tokyo: Daiichi-shuppan; 2009.
8. Imai T, Nakamura M, Ando F, Shimokata H. Dietary supplement use by community-living population in Japan: data from the National Institute for Longevity Sciences Longitudinal Study of Aging (NILS-LSA). J Epidemiol. 2006;16:249–60.
9. Nanri A, Mizoue T, Noda M, et al. Magnesium intake and type II diabetes in Japanese men and women: the Japan Public Health Center-based Prospective Study. Eur J Clin Nutr. 2010;64:1244–7.
10. Tsugane S, Sobue T. Baseline survey of JPHC study–design and participation rate. Japan Public Health Center-based Prospective Study on Cancer and Cardiovascular Diseases. J Epidemiol. 2001;11:S24–9.
11. Sasaki S, Kobayashi M, Ishihara J, Tsugane S. Self-administered food frequency questionnaire used in the 5-year follow-up survey of the JPHC Study: questionnaire structure, computation algorithms, and area-based mean intake. J Epidemiol. 2003;13:S13–22.
12. Council for Science and Technology; Ministry of Education, Culture, Sports, Science and Technology, Japan. Standard tables of food composition in Japan. 5th revised and enlarged ed. Tokyo: National Printing Bureau; 2005.
13. Ishihara J, Inoue M, Kobayashi M, et al. Impact of the revision of a nutrient database on the validity of a self-administered food frequency questionnaire (FFQ). J Epidemiol. 2006;16:107–16.

14. Kato M, Noda M, Inoue M, Kadowaki T, Tsugane S. Psychological factors, coffee and risk of diabetes mellitus among middle-aged Japanese: a population-based prospective study in the JPHC study cohort. Endocr J. 2009;56:459–68.
15. Kosaka K. Diagnostic criteria for diabetes mellitus in Japan–from a report of the Japan Diabetes Society (JDS) Committee on the Diagnosis of Diabetes Mellitus, 1982. Diabetes Res Clin Pract. 1994;24(Suppl):S59–62.
16. Kirii K, Iso H, Date C, Fukui M, Tamakoshi A. Magnesium intake and risk of self-reported type 2 diabetes among Japanese. J Am Coll Nutr. 2010;29:99–106.
17. Villegas R, Gao YT, Dai Q, et al. Dietary calcium and magnesium intakes and the risk of type 2 diabetes: the Shanghai Women's Health Study. Am J Clin Nutr. 2009;89:1059–67.
18. Hopping BN, Erber E, Grandinetti A, Verheus M, Kolonel LN, Maskarinec G. Dietary fiber, magnesium, and glycemic load alter risk of type 2 diabetes in a multiethnic cohort in Hawaii. J Nutr. 2009;140:68–74.
19. Dong JY, Xun P, He K, Qin LQ. Magnesium intake and risk of type 2 diabetes: meta-analysis of prospective cohort studies. Diabetes Care. 2011;34:2116–22.
20. Guerrero-Romero F, Rascon-Pacheco RA, Rodriguez-Moran M, de la Pena JE, Wacher N. Hypomagnesaemia and risk for metabolic glucose disorders: a 10-year follow-up study. Eur J Clin Invest. 2008;38:389–96.
21. Kao WH, Folsom AR, Nieto FJ, Mo JP, Watson RL, Brancati FL. Serum and dietary magnesium and the risk for type 2 diabetes mellitus: the Atherosclerosis Risk in Communities Study. Arch Intern Med. 1999;159:2151–9.
22. Guerrero-Romero F, Tamez-Perez HE, Gonzalez-Gonzalez G, et al. Oral magnesium supplementation improves insulin sensitivity in non-diabetic subjects with insulin resistance. A double-blind placebo-controlled randomized trial. Diabetes Metab. 2004;30:253–8.
23. Hadjistavri LS, Sarafidis PA, Georgianos PI, et al. Beneficial effects of oral magnesium supplementation on insulin sensitivity and serum lipid profile. Med Sci Monit. 2010;16:CR307–12.
24. Lee S, Park HK, Son SP, Lee CW, Kim IJ, Kim HJ. Effects of oral magnesium supplementation on insulin sensitivity and blood pressure in normo-magnesemic nondiabetic overweight Korean adults. Nutr Metab Cardiovasc Dis. 2009;19:781–8.
25. Mooren FC, Kruger K, Volker K, Golf SW, Wadepuhl M, Kraus A. Oral magnesium supplementation reduces insulin resistance in non-diabetic subjects – a double-blind, placebo-controlled, randomized trial. Diabetes Obes Metab. 2011;13:281–4.
26. Song Y, He K, Levitan EB, Manson JE, Liu S. Effects of oral magnesium supplementation on glycaemic control in type 2 diabetes: a meta-analysis of randomized double-blind controlled trials. Diabet Med. 2006;23:1050–6.

Chapter 10
Magnesium Supplementation and Bone

Hasan Aydın

Key Points

- Magnesium deficiency affects almost all phases of bone metabolism.
- There is a strong relationship between magnesium deficiency and bone loss, osteopenia, and osteoporosis.
- Supplementation not only restores body magnesium stores but also has beneficial effects on bone metabolism as well as bone mineral density.

Keywords Magnesium deficiency • Bone metabolism • Osteoporosis • Bone turnover • Magnesium supplementation

Introduction

Magnesium is the second most abundant intracellular cation in the body. Of the total body magnesium, about 50–60% is found in the bones where it accounts for about 1% of bone ash. Within the bone, magnesium forms a surface constituent of the hydroxyapatite mineral component. It functions as a readily accessible magnesium reservoir. One third of it is easily exchangeable with serum for maintaining a normal extracellular magnesium concentration. The remainder is complexed with crystalline structure of bone mineral within the hydroxyapatite lattice [1, 2].

Magnesium is an essential element for bone and plays a major role in bone and mineral homeostasis by functioning on all phases of skeletal metabolism. It directly affects bone cell function, influences hydroxyapatite crystal formation and growth, takes part in normal activity of calcitropic hormones, and plays an important role in calcium metabolism [3]. It regulates PTH secretion and action and vitamin D activation through interacting with calcium-sensing receptors on parathyroid glands and renal tubular cells [4–6]. Experimental studies have shown that acute reduction in magnesium levels can stimulate PTH secretion while acute elevation suppresses it in a manner less potent than calcium [4, 5, 7, 8]. Short-term dietary magnesium deficiency, shown by Rude et al., on the other hand, resulted in relative hypoparathyroidism and concomitant hypocalcaemia in 80% of patients [5]. These results

H. Aydın, M.D. (✉)
Department of Internal Medicine, Section of Endocrinology and Metabolism, Yeditepe University Hospital, Devlet Yolu Ankara Caddesi No: 102-104 Kozyatagi, Istanbul 34752, Turkey
e-mail: haydin@yeditepe.edu.tr

R.R. Watson et al. (eds.), *Magnesium in Human Health and Disease*, Nutrition and Health,
DOI 10.1007/978-1-62703-044-1_10, © Springer Science+Business Media New York 2013

mandate magnesium levels to be constant for proper calcium metabolism. Unless magnesium status is normal, adequate calcium intake by itself may not be enough for bone health.

Unless there is a significant loss from the body like malabsorption from prolonged diarrhea and urinary loss, magnesium deficiency is uncommon in adults and rarely produces clinical symptoms. On the other hand, in Western countries, magnesium intake of individuals stays below the recommended daily allowances [9, 10]. Suboptimal intake of magnesium results in a state of mild magnesium deficiency and threatens the bone health. During severe magnesium deficiency, many phases of bone metabolism including an arrest in bone growth, reduced bone volume, increased fragility of bones, and decrease in osteoblastic activity and bone formation had been reported [3].

Relationship Between Magnesium Deficiency and Osteoporosis

Bone is one of the tissues mostly affected by magnesium deficiency as it is the major storage site for magnesium in the body. Many studies reported a close association of magnesium deficiency to osteoporosis. (See Table 10.1)

Magnesium Content of the Body in Osteoporosis

Bone magnesium concentration fell with increasing age probably due to enlargement of bone crystals with advancing age that contracts surface area and decreases surface limited fraction of magnesium [11, 12]. Decreased trabecular bone magnesium content in association with larger and more perfect crystals in trabecular bone had been reported in a group of postmenopausal women [13]. During magnesium deficiency, both exchangeable and non-exchangeable magnesium pools decrease in the body but largely in the fractional change in the exchangeable pool [12].

Fractional change in bone magnesium pools results in a deficiency in circulating magnesium levels. It has been reported in a small group of postmenopausal osteoporotic women that serum magnesium concentrations but not red blood cell or urine magnesium were significantly reduced compared to healthy controls [14]. Another study using the method of absorption spectrophotometry revealed a reduced level of magnesium in red blood cells in 63.6% of senile, 66.7% postmenopausal, and 22.2% drug-induced osteoporosis [15]. Postmenopausal women with osteoporosis have been reported to be deficient in serum magnesium levels [16]. Chronic alcoholism, gluten-sensitive enteropathy, and post-cardiac transplant patients who are at increased risk of osteoporosis have also been shown to have a significant decline in serum magnesium concentrations and increased excretion of magnesium in urine [17–19]. Hypomagnesemic state affects various phases of bone metabolism. Serum PTH concentrations, rate of bone loss, and bone turnover were found to be lower in patients with low serum magnesium levels [20].

Table 10.1 The effects of chronic magnesium deficiency on bone

Decrease in magnesium content of bone
Decrease in osteoblastic activity and bone formation
Increase in osteoclastic activity and bone resorption
Decrease in iPTH secretion
Arrest in bone growth and bone volume
Decrease in bone mineralization
Loss of trabecular bone volume
Decrease in bone mineral density and content
Development of osteopenia and osteoporosis
Increase in fragility of bones

Not only circulating magnesium concentrations but also bone magnesium content was found to be reduced during magnesium deficiency. Rats that were exposed to chronic suboptimal intake of magnesium had been shown to have greatly reduced plasma magnesium level and bone magnesium content without causing clinical signs of magnesium deficiency [21]. Dietary restriction of magnesium in a group of rats reduced ash weight of magnesium but not calcium in femur tissue [22]. Magnesium-depleted animals exhibited slow growth and greatly reduced concentrations of magnesium in both serum and femur ash [22]. Trabecular bone magnesium content negatively correlated with bone mineral content of the radius measured by x-ray photodensitometry [23]. The increased retention of magnesium in the magnesium load test suggests magnesium deficiency in postmenopausal osteoporosis probably caused by magnesium malabsorption [24].

The Effect of Dietary Magnesium on Bone

Magnesium intake in the diet together with iron and zinc had a positive correlation with forearm bone mineral content in premenopausal women [25, 26]. Higher intakes of magnesium were associated with higher total bone mass [25]. Lumbar spine and femur bone mineral density was higher in women who consumed adequate magnesium in diet [25, 26]. Dietary intake of magnesium is found to be reduced in osteoporotic postmenopausal women and correlated with bone mineral content [27].

The Mechanisms of Bone Loss in Magnesium Deficiency

Histomorphometric analysis demonstrated a reduction in bone mass with a specific loss of trabecular bone volume in the femur [20]. It is postulated that uncoupling of bone formation and resorption is the possible underlining mechanism of bone loss in magnesium-deficient states since increase in osteoclast resorbing surface and osteoclast number but no increase in bone-forming surface or osteoblast number was observed in magnesium-deficient animals [20, 28]. On the other hand, some others reported suppression of bone turnover instead of uncoupling. Moderate to severe dietary restriction of magnesium for 21 days had been shown not only lower the rates of bone formation but also severely reduced bone resorption [29]. Suppression of bone turnover in histological examination of bone from magnesium-deficient animals resulted in progressive subperiosteal hyperplasia, arrest in bone mineralization, and decreased responsiveness of bone to parathyroid hormone [30].

Studies on osteoporosis indicate contribution of cytokines to osteoclastic bone resorption. Circulating cytokine levels that are indicative of a generalized inflammatory state such as IL-1, IL-6, and TNF-α were found to be dramatically elevated during diet-induced magnesium deficiency [31]. The earliest cytokine that was found to be increased within 24 h of magnesium depletion was substance P which was followed by interleukin-1 that began to increase on day 3. Tumor necrosis factor-alpha was a late sequela of magnesium deficiency with a median rise observed after tenth day of depletion [32]. The presence of increased levels of cytokines in cellular elements of the bone marrow microenvironment indicates that osteoclast activation and stimulation of bone resorption is mediated by cytokines [32].

The Effect of Magnesium Deficiency on Bone Mineralization

Mineralization was also disturbed in magnesium deficiency. Osteoid tissue in the metaphyseal bone was found to be unmineralized in an animal study [33]. Short-term magnesium deficiency in the diet of growing rats produced a significant decrease in the mineral content of newly formed metaphysis. Metaphyseal osteocalcin levels were found to be reduced in magnesium-deficient rats [33]. Abnormal mineralization of bones may contribute to fragility.

The Effect of Magnesium Deficiency on Bone Growth and Strength

Bone strength significantly decreased in magnesium-deficient rats. Femur from depleted animals was found to be shorter and weaker in strength as evident from lighter breaking loads [22]. A significant decrease in the three-point bend strength of femurs of magnesium-deficient rats was observed [33]. Growth plate width decreased 33% in young magnesium-deficient mice as a result of decreased chondrocyte numbers and length indicating reduced bone growth in magnesium deficiency [32]. Insufficient dietary magnesium intake significantly decreased chondrocyte density of distal femur articular cartilage and decreased tibial growth plate width in experimental animals compared to controls. The amount of proteoglycans reduced in extracellular matrix of articular cartilage [34].

The Effect of Magnesium Deficiency on Bone Turnover

Experimental Mg deficiency in animal models has resulted in impaired bone turnover in a number of reports. Both formation and resorption of bone are affected by magnesium deficiency. Variability in circulating markers has been reported in magnesium-deficient states. Serum osteocalcin, a marker of bone formation, has been shown to be decreased in dietary-induced magnesium-deprived animals compared to pair-fed normal rats. Serial sampling demonstrated a difference in osteocalcin levels by 2 days of Mg deprivation [35]. Reduced osteocalcin mRNA in calvaria of magnesium-deficient rats indicates that low circulating levels might be due to reduced osteocalcin synthesis in bone [33, 35].

Dietary magnesium reduction to 10% of the nutrient requirement in rats was shown to be associated with a significant fall in both serum alkaline phosphatase and osteocalcin suggesting decreased osteoblast activity [36]. More severe restriction of magnesium in diet, 25–50% of the nutrient requirements, on the other hand, did not create any change in markers of bone formation [37, 38]. But histomorphometric analysis showed an increase in osteoclast and decrease in osteoblast numbers [28, 39]. Alkaline phosphatase activity, another biochemical marker of bone formation, has also been reported to be reduced in bone and serum in severely Mg-deficient rats [40, 41].

Magnesium deficiency has detrimental effects on bone resorption, too. A significant increase in bone resorption markers has been reported in various studies. Significant differences were found in pyridinoline and deoxypyridinoline excretion between women in the lowest and highest quartiles of magnesium intakes. Mean deoxypyridinoline excretion was found to be increased with lower intakes of magnesium and magnesium intake accounted for 12% of the variation in pyrrolidine and deoxypyrrolidine excretion [42]. Moderate to severe restriction of dietary magnesium increased urinary pyridinoline excretion by 34–44% and deoxypyridinoline by 33–40% compared to adequate magnesium-fed rats [29]. Other markers of bone resorption had also been associated with magnesium deficiency. Rats fed on moderate magnesium-restricted diet had significantly higher rates of urinary excretion of C-terminal telopeptide of type 1 collagen compared to animals fed on magnesium-adequate diet [43].

However, contradictory results can also be encountered in the literature. Severe to moderate dietary restriction of magnesium resulted in severely depressed levels of markers of both bone formation and bone resorption. Significant reductions in both serum osteocalcin and urinary pyridinoline and deoxypyridinoline levels were observed during magnesium restriction. Femur osteocalcin mRNA was shown to be reduced 46% and 22% in moderately and severely restricted animals, respectively [29].

Receptor activator of nuclear factor kappa-B ligand (RANKL); its receptor, receptor activator of nuclear factor kappa-B (RANK); and its soluble decoy receptor, osteoprotegerin (OPG), are the cytokines in the final pathway of osteoclastogenesis. Their presence dictates osteoclast activation. Osteoprotegerin was found to be decreased and RANKL increased in low, moderate, and severe magnesium-deficient rats. Increase in the RANKL/OPG ratio indicated an increase in osteoclastogenesis. This may explain the increase in osteoclast number and decrease in bone mass in animal models of magnesium deficiency [44].

The Effect of Magnesium Deficiency on iPTH

Magnesium deficiency impairs mineral homeostasis. Hypocalcemia, impaired parathyroid hormone secretion, and low serum concentrations of 1,25-dihydroxyvitamin D are frequent features of magnesium deficiency. After a 3-week low-magnesium diet, iPTH secretion was found to be impaired in about 77% of healthy subjects and magnesium administration, regardless of the basal concentration, resulted in an immediate rise in serum iPTH. Impaired iPTH secretion in hypomagnesemic patients, in turn, causes a fall in serum 1,25-dihydroxyvitamin D. Lack of response after iv administration of iPTH raises the possibility of resistance to PTH in magnesium-depleted men. These impaired conditions accounts for the hypocalcemia observed in hypomagnesemic states [4, 5].

Treatment

Calcium and vitamin D replacement is the mainstay in standard treatment of osteoporosis. But, other trace minerals are also important and have beneficial effects on osteoporosis. They constitute building components of bone and play important functional roles in bone metabolism and bone turnover.

As being a major storage site of body, magnesium is not a trace element for bone tissue. Although, bone mineral content is lost and osteoporosis develops during magnesium-deficient states, it can easily be replaced by supplementation which may have a restorative effect on bone tissue. (See Table 10.2)

The Effect of Supplementation on Bone Mineral Density

A total dietary program emphasizing magnesium instead of calcium revealed a significant increase in bone mineral density of the calcaneus bone within 1 year in postmenopausal women on hormonal therapy. About 43% of women had bone mineral density rose above the spine fracture threshold which was below before treatment [45]. In a 2-year controlled trial of per oral magnesium in 31 postmenopausal osteoporotic patients, 71% of patients had 1–8% rise in bone density. The mean bone density of all treated patients increased significantly after 1 year and remained unchanged after 2 years [46].

Administration of a supplement constituted of 3.0 g of egg shell powder as a source of bioavailable calcium, 400 IU vitamin D3, and 400 mg of magnesium for 4–8 months increased bone mineral density of the subjects with osteoporosis or osteopenia in the short term. The net increase rate was 5.7% at lumbar spine and 1.8% at total proximal femur [47].

Not only in osteoporotic subjects but also in healthy girls at age between 8 and 14 years old, replacement with magnesium (300 mg elemental Mg per day in two divided doses) given orally for 12 months, provided an increased accrual in hip and spine bone mineral content in a randomized,

Table 10.2 The effects of magnesium supplementation on bone health

Author	Subjects	Type of supplement	Daily dose	Time period	Effect(s)
Abraham et al. [45]	Postmenopausal women	Dietary program	–	12 months	Increase in BMD of calcaneus bone
Stendig-Lindberg et al. [46]	Postmenopausal women	Magnesium hydroxide	250–750 mg	24 months	1–8% rise of bone density
Schaafsma et al. [47]	Women + men	Magnesium + vitamin D + egg shell powder	400 mg	4–8 months	Increase in BMD of lumbar spine and femur
Carpenter et al [48]	8–14 years old healthy girl	Magnesium oxide	300 mg	12 months	Increased accrual in integrated hip and lumbar spine BMC
Wood et al. [49]	Preadolescent girl	Magnesium + calcium + vitamin D + boron + silicon	400 mg	12 months	Gain in trabecular BMD and BMC
Toba et al. [50]	Ovariectomized rats	Magnesium-rich diet	0.15%	1,5 months	Decrease in bone resorption Increase in bone formation Increase in dynamic strength of bone
Sojka et al. [51]	Postmenopausal women	Magnesium hydroxide	–	24 months	Increase in bone density Decrease in fracture rate
Marie et al. [52]	Mice	Magnesium-rich diet	5 or 32 mM	1 month	Increase in bone Mg content Stimulation of bone mineralization
Katsumata et al. [53]	Rats on high-phosphorus diet	Magnesium-rich diet	0.15%	2/3 months	Suppression of bone resorption
Dimai et al. [54]	Healthy men	Magnesium carbonate and oxide	15 mmol	1 month	Reduction in bone formation and resorption Decrease in serum iPTH
Doyle et al. [55]	Healthy female	Magnesium hydroxide	22 mmol	1 month	No effect on bone turnover
Aydin et al. [56]	Postmenopausal osteoporotic women	Magnesium citrate	1,830 mg	1 month	Increase in bone formation Decrease in bone resorption
Bae et al. [57]	Ovariectomized rats	Seaweed calcium extract	0.1%	1 month	Increase in femoral BMD and strength
Bae et al. [58]	Ovariectomized rats	Magnesium-rich diet	0.1%	3 months	Increase in OPG and OPG/RANKL ratio
Day et al. [59]	Postmenopausal women	Magnesium bicarbonate supplemented spring water	120 mg/l	3 months	No effect on bone metabolism

BMD bone mineral density, *BMC* bone mineral content, *OPG* osteoprotegerin, *RANKL* ligand of receptor activator of nuclear factor kappa-B controlled trial [48]. Another double-blind, placebo-controlled trial provided parallel results. One hundred preadolescent girls were given a chewable vitamin-mineral supplement which included 800 mg calcium, 400 mg magnesium, 400 IU vitamin D, 1.33 mg boron, and 9 mg silicon for 12 months that showed a net gain in trabecular bone mineral density of 1.41% and 5.83% in trabecular bone mineral content [49].

Although in most of the studies, magnesium is supplemented together with either calcium or hormonal therapy, magnesium supplementation itself also had beneficial effects on bone metabolism and strength in postmenopausal animal models. Two groups of ovariectomized rats were fed 0.05% or 0.15% magnesium diet for 42 days. Rats on high-magnesium diet had lower urinary deoxypyridinoline excretion and serum PTH levels and higher serum osteocalcin levels. The breaking force and breaking energy of the femur was found to be lower in rats of low-magnesium diet [50].

Besides the effect of magnesium on bone metabolism, in a prospective 2-year study, supplementation with magnesium hydroxide prevented fractures and resulted in a significant increase in bone density in a group of postmenopausal women [51].

Supplementation with 5 or 32 mM magnesium in the normal young mouse for 4 weeks resulted in an increase in serum and urine calcium concentration and bone magnesium content. Bone turnover was stimulated with magnesium supplementation in such a way that both the osteoclastic activity and resorbing surface were increased and osteoblastic surface was reduced without any alteration in PTH secretion. The end result was the stimulation of bone mineralization and increased extracellular mineral availability [52].

The major effect of magnesium supplementation was supposed to be via inhibition of osteoclastogenesis. Magnesium supplementation suppressed bone resorption in rats fed a high-phosphorus diet, which is a well known accelerator of bone resorption by increasing parathyroid hormone secretion [53].

The Effect of Magnesium Supplementation on Bone Turnover

Daily oral supplementation of a moderate dose (15 mmol/day) of magnesium efficiently suppressed bone turnover in healthy, non-magnesium-deficient young adult males. Magnesium supplementation reduced levels of both serum bone formation and resorption biochemical markers [54].

On the other hand, in a double-blind, placebo-controlled, randomized crossover intervention study, a daily dose of 10 mmol magnesium had no effect on biomarkers of bone formation or resorption in 26 healthy young adult females [55].

Since bone turnover increases after menopause and both bone formation and resorption markers are higher compared to the premenopausal period, magnesium supplementation might not have any effect on bone turnover in premenopausal women. But, 30 consecutive days of oral magnesium supplementation at a dose of 1,830 mg/day caused significantly increased serum osteocalcin and decreased urinary deoxypyridinoline levels [56].

References

1. Laires MJ, Moneiro CP, Bicho M. Role of cellular magnesium in health and human disease. Front Biosci. 2004;9:262–76.
2. Webster PO. Magnesium. Am J Clin Nutr. 1987;45:1305–12.
3. Wallach S. Effects of magnesium on skeletal metabolism. Magnes Trace Elem. 1990;9(1):1–14.
4. Fatemi S, Ryzen E, Flores J, et al. Effect of experimental human magnesium depletion on parathyroid hormone secretion and 1,25-dihydroxyvitamin D metabolism. J Clin Endocrinol Metab. 1991;73(5):1067–72.

5. Rude RK, Oldham SB, Sharp Jr CF, et al. Parathyroid hormone secretion in magnesium deficiency. J Clin Endocrinol Metab. 1978;47(4):800–6.
6. Miki H, Maercklein PB, Fitzpatrick LA. Effect of magnesium on parathyroid cells: evidence for two sensing receptors or two intracellular pathways? Am J Physiol. 1997;272(1 Pt 1):E1–6.
7. Suh SM, Tashjian Jr AH, Matsuo N, et al. Pathogenesis of hypocalcemia in primary hypomagnesemia: normal end-organ responsiveness to parathyroid hormone, impaired parathyroid gland function. J Clin Invest. 1973;52(1):153–60.
8. Cholst IN, Steinberg SF, Tropper PJ, et al. The influence of hypermagnesemia on serum calcium and parathyroid hormone levels in human subjects. N Eng J Med. 1984;310(19):1221–5.
9. Marier JR. Magnesium content of the food supply in the modern-day world. Magnesium. 1986;5(1):1–8.
10. Hallfrisch J, Muller DC. Does diet provide adequate amounts of calcium, iron, magnesium, and zinc in a well-educated adult population? Exp Gerontol. 1993;28(4–5):473–83.
11. Barbagallo M, Belvedere M, Dominguez LJ. Magnesium homeostasis and aging. Magnes Res. 2009;22(4):235–46.
12. Alfrey AC, Miller NL, Trow R. Effect of age and magnesium depletion on bone magnesium pools in rats. J Clin Invest. 1974;54(5):1074–81.
13. Cohen L. Recent data on magnesium and osteoporosis. Magnes Res. 1988;1(1–2):85–7.
14. Reginster JY, Strause L, Deroisy R, et al. Preliminary report of decreased serum magnesium in postmenopausal osteoporosis. Magnesium. 1989;8(2):106–9.
15. Ditmar R, Steidl L. The significance of magnesium in orthopedics. V. Magnesium in osteoporosis. Acta Chir Orthop Traumatol Cech. 1989;56(2):143–59.
16. Mutlu M, Argun M, Kilic E, et al. Magnesium, zinc and copper status in osteoporotic, osteopenic and normal postmenopausal women. J Int Med Res. 2007;35(5):692–5.
17. Rude RK, Olerich M. Magnesium deficiency: possible role in osteoporosis associated with gluten-sensitive enteropathy. Osteoporos Int. 1996;6(6):453–61.
18. Abbott L, Nadler J, Rude RK. Magnesium deficiency in alcoholism: possible contribution to osteoporosis and cardiovascular disease in alcoholics. Alcohol Clin Exp Res. 1994;18(5):1076–82.
19. Boncimino K, McMahon DJ, Addesso V, et al. Magnesium deficiency and bone loss after cardiac transplantation. J Bone Miner Res. 1999;14(2):295–303.
20. Rude RK, Kirchen ME, Gruber HE, et al. Magnesium deficiency induces bone loss in the rat. Miner Electrolyte Metab. 1998;24(5):314–20.
21. Héroux O, Peter D, Tanner A. Effect of a chronic suboptimal intake of magnesium on magnesium and calcium content of bone and on bone strength of the rat. Can J Physiol Pharmacol. 1975;53(2):304–10.
22. Kenney MA, McCoy H, Williams L. Effects of magnesium deficiency on strength, mass, and composition of rat femur. Calcif Tissue Int. 1994;54(1):44–9.
23. Manicourt DH, Orloff S, Brauman J, et al. Bone mineral content of the radius: good correlations with physicochemical determinations in iliac crest trabecular bone of normal and osteoporotic subjects. Metabolism. 1981;30(1):57–62.
24. Cohen L, Laor A. Correlation between bone magnesium concentration and magnesium retention in the intravenous magnesium load test. Magnes Res. 1990;3(4):271–4.
25. Angus RM, Sambrook PN, Pocock NA, et al. Dietary intake and bone mineral density. Bone Miner. 1988;4(3):265–77.
26. New SA, Bolton-Smith C, Grubb DA, et al. Nutritional influences on bone mineral density: a cross-sectional study in premenopausal women. Am J Clin Nutr. 1997;65(6):1831–9.
27. Tranquilli AL, Lucino E, Garzetti GG, et al. Calcium, phosphorus and magnesium intakes correlate with bone mineral content in postmenopausal women. Gynecol Endocrinol. 1994;8(1):55–8.
28. Rude RK, Kirchen ME, Gruber HE, et al. Magnesium deficiency-induced osteoporosis in the rat: uncoupling of bone formation and bone resorption. Magnes Res. 1999;12(4):257–67.
29. Creedon A, Flynn A, Cashman K. The effect of moderately and severely restricted dietary magnesium intakes on bone composition and bone metabolism in the rat. Br J Nutr. 1999;82(1):63–71.
30. Jones JE, Schwartz R, Krook L. Calcium homeostasis and bone pathology in magnesium deficient rats. Calcif Tissue Int. 1980;31(3):231–8.
31. Weglicki WB, Phillips TM, Freedman AM, et al. Magnesium-deficiency elevates circulating levels of inflammatory cytokines and endothelin. Mol Cell Biochem. 1992;110(2):169–73.
32. Rude RK, Gruber HE, Wei LY, et al. Magnesium deficiency: effect on bone and mineral metabolism in the mouse. Calcif Tissue Int. 2003;72(1):32–41.
33. Boskey AL, Rimnac CM, Bansal M, et al. Effect of short-term hypomagnesemia on the chemical and mechanical properties of rat bone. J Orthop Res. 1992;10(6):774–83.
34. Gruber HE, Ingram J, Norton HJ, et al. Alterations in growth plate and articular cartilage morphology are associated with reduced SOX9 localization in the magnesium-deficient rat. Biotech Histochem. 2004;79(1):45–52.

35. Carpenter TO, Mackowiak SJ, Troiano N, et al. Osteocalcin and its message: relationship to bone histology in magnesium-deprived rats. Am J Physiol. 1992;263(1 Pt 1):E107–14.
36. Rude RK, Gruber HE, Norton HJ, et al. Bone loss induced by dietary magnesium reduction to 10% of the nutrient requirement in rats is associated with increased release of substance P and tumor necrosis factor-alpha. J Nutr. 2004;134(1):79–85.
37. Rude RK, Gruber HE, Norton HJ, et al. Dietary magnesium reduction to 25% of nutrient requirement disrupts bone and mineral metabolism in the rat. Bone. 2005;37(2):211–9.
38. Rude RK, Gruber HE, Norton HJ, et al. Reduction of dietary magnesium by only 50% in the rat disrupts bone and mineral metabolism. Osteoporos Int. 2006;17(7):1022–32.
39. Rude RK, Singer FR, Gruber HE. Skeletal and hormonal effects of magnesium deficiency. J Am Coll Nutr. 2009;28(2):131–41.
40. Heaton FW. Effect of magnesium deficiency on plasma alkaline phosphatase activity. Nature. 1965;207(5003):1292–3.
41. Lai CC, Singer L, Armstrong WD. Bone composition and phosphatase activity in magnesium deficiency in rats. J Bone Joint Surg Am. 1975;57(4):516–22.
42. New SA, Robins SP, Campbell MK, et al. Dietary influences on bone mass and bone metabolism: further evidence of a positive link between fruit and vegetable consumption and bone health? Am J Clin Nutr. 2000;71(1):142–51.
43. Katsumata SI, Matsuzaki H, Tsuboi R, et al. Moderate magnesium-restricted diet affects bone formation and bone resorption in rats. Magnes Res. 2006;19(1):12–8.
44. Rude RK, Gruber HE, Wei LY, et al. Immunolocalization of RANKL is increased and OPG decreased during dietary magnesium deficiency in the rat. Nutr Metab (Lond). 2005;2(1):24–8.
45. Abraham GE, Grewal H. A total dietary program emphasizing magnesium instead of calcium. Effect on the mineral density of calcaneous bone in postmenopausal women on hormonal therapy. J Reprod Med. 1990;35(5):503–7.
46. Stendig-Lindberg G, Tepper R, Leichter I. Trabecular bone density in a two year controlled trial of peroral magnesium in osteoporosis. Magnes Res. 1993;6(2):155–63.
47. Schaafsma A, Pakan I. Short-term effects of a chicken egg shell powder enriched dairy-based products on bone mineral density in persons with osteoporosis or osteopenia. Bratisl Lek Listy. 1999;100(12):651–6.
48. Carpenter TO, DeLucia MC, Zhang JH, et al. A randomized controlled study of effects of dietary magnesium oxide supplementation on bone mineral content in healthy girls. J Clin Endocrinol Metab. 2006;91(12):4866–72.
49. Wood T, McKinnon T. Calcium-magnesium-vitamin D supplementation improves bone mineralization in preadolescent girls. Salt Lake City: USANA Clinical Research Bulletin, USANA Health Sciences; 2001.
50. Toba Y, Kajita Y, Masuyama R, et al. Dietary magnesium supplementation affects bone metabolism and dynamic strength of bone in ovariectomized rats. J Nutr. 2000;130(2):216–20.
51. Sojka JE, Weaver CM. Magnesium supplementation and osteoporosis. Nutr Rev. 1995;53(3):71–4.
52. Marie PJ, Travers R, Delvin EE. Influence of magnesium supplementation on bone turnover in the normal young mouse. Calcif Tissue Int. 1983;35(6):755–61.
53. Katsumata SI, Matsuzaki H, Uehara M, et al. Effect of dietary magnesium supplementation on bone loss in rats fed a high phosphorus diet. Magnes Res. 2005;18(2):91–6.
54. Dimai HP, Porta S, Wirnsberger G, et al. Daily oral magnesium supplementation suppresses bone turnover in young adult males. J Clin Endocrinol Metab. 1998;83(8):2742–8.
55. Doyle L, Flynn A, Cashman K. The effect of magnesium supplementation on biochemical markers of bone metabolism or blood pressure in healthy young adult females. Eur J Clin Nutr. 1999;53(4):255–61.
56. Aydin H, Deyneli O, Yavuz D, et al. Short-term oral magnesium supplementation suppresses bone turnover in postmenopausal osteoporotic women. Biol Trace Elem Res. 2010;133(2):136–43.
57. Bae YJ, Bu SY, Kim JY, et al. Magnesium supplementation through seaweed calcium extract rather than synthetic magnesium oxide improves femur bone mineral density and strength in ovariectomized rats. Biol Trace Elem Res. 2011;144(1):992–1002.
58. Bae YJ, Kim MH. Calcium and magnesium supplementation improves serum OPG/RANKL in calcium-deficient ovariectomized rats. Calcif Tissue Int. 2010;87(4):365–72.
59. Day RO, Liauw W, Tozer LM, et al. A double-blind, placebo-controlled study of the short term effects of a spring water supplemented with magnesium bicarbonate on acid/base balance, bone metabolism and cardiovascular risk factors in postmenopausal women. BMC Res Notes. 2010;3:180–7.

Chapter 11
Magnesium Salts in a Cancer Patient: Pathobiology and Protective Functions

Gabriel Wcislo and Lubomir Bodnar

Key Points

- Growing population of humans on Earth will be associated with an increasing number of patients with various cancers that will be treated with chemotherapy, radiotherapy, and surgery associated with side effects.
- Magnesium is an essential element in many biological functions; TRPM6, the ion channel protein, regulates magnesium balance in the body by intestinal and renal mechanisms, respectively.
- Anticancer chemotherapy is a systemic method of attacking malignant cells; it can be used as adjuvant or palliative chemotherapy; adverse side effects appear as the result of toxic effects that stem from chemical compounds used as drugs.
- Hypomagnesemia (serum magnesium concentration <1.8 mg/dL) is a common disturbance reported in 12% of hospitalized patients; hypomagnesemia is linked to cisplatin-based chemotherapy and targeted therapy against EGFR with cetuximab.
- Neurotoxicity associated with cancer chemotherapy is mostly observed in patients receiving platinum compounds (cisplatin, carboplatin, oxaliplatin); 85% of patients with colorectal cancer under chemotherapy with oxaliplatin have neurotoxic side effects readily seen as peripheral neuropathy.
- Nephrotoxicity of cisplatin results from the high risk of vulnerable effects of cisplatin on glomerulus and tubal cells. Fifty-five percent of patients require cisplatin dose adjustment due to functional renal abnormalities. Magnesium supplementation seems to have renal protective activity, but more controlled clinical studies need to be performed.

Keywords Cancer treatment • Toxicity • Hypomagnesemia • Magnesium supplementation

Introduction

Demographic projections show how the human population will grow in the next 50 years. The world's population is expected to grow from 6.3 billion today to 8.9 billion in 2050. This projection assumes that family planning will offer good conditions for living at the socioeconomic level. In 2000, about 1.2 billion people lived in more developed regions: Europe, North America, Australia, New Zealand,

G. Wcislo, M.D., Ph.D. (✉) • L. Bodnar, M.D., Ph.D.
Department of Oncology, Military Institute of Medicine, 128 Szaserow Str, Warsaw 04-041, Poland
e-mail: gabrielwcislo@yahoo.pl; lubo@esculap.pl

R.R. Watson et al. (eds.), *Magnesium in Human Health and Disease*, Nutrition and Health,
DOI 10.1007/978-1-62703-044-1_11,

and Japan. On the other hand, 4.9 billion people lived in the economically poor and less developed regions. Slowly growing populations (in more developed countries) have elderly dependency ratio (number of people aged above 65 to the number aged 15–64). At the opposite end, rapidly growing populations have a higher youth dependency ratio (number of people aged 0–14 to the number aged 15–64) [1].

Malignant diseases accompany mankind, and one can expect that in the next 50 years, for the two described models of human population growth, cancer will be of great value at the social level, although this scenario will be made real by different causes. Namely, for the rich and old, cancer will accumulate as a biologically natural event, and for the poor cancer will gnaw due to the lack of money. In 2004, the results of global observation on 26 cancers in 2002 were available with such data as 10.9 million new cases, 6.7 million deaths, and 24.6 million persons alive with cancer within 3 years of diagnosis [2]. The first steps to fight against cancer were made in 1971 when the US President Richard Nixon announced an initiative now known as "war on cancer" [3]. Costly war is capable of giving better results in terms of longer life and higher rate of survivors. After 40 years of the "war on cancer," humanity has reached almost 50% of 5-year survivors. The rate of 5-year survivors for some cancers are as follows: prostate cancer, 78%; breast cancer, 79%; Hodgkin's disease, 81%; testicular cancer, 97%; and for children and youngsters, 80% [4, 5].

The metabolic syndrome, previously known as syndrome X, in cancer survivors is defined as the combination of two large metabolic diseases, i.e., type 2 diabetes and atherosclerosis. Most systems of classification contain main items as dyslipidemia, hypertension, central obesity, and resistance to insulin [6]. So metabolic disturbances can be met exactly in cancer patients during treatment and afterward while living for a long period of 5 years or even more. Among any kind of additional treatments, supplementation with minerals, such as calcium, magnesium, and others, is able to offer clinical benefits for both patients with active malignant disease and cancer survivors. Moreover, there are some data that show the preventive role of supplemented minerals against cancer, e.g., colorectal cancer.

Biological Role of Magnesium

Magnesium (Mg) is an abundant element in the Earth's crust and seawater. It belongs to the alkaline earth metals (group 2 metals) and is present in all known compounds in the +2 oxidation state. The free element is not easy to find because of its chemical reactivity. Therefore, Mg exists predominantly in the form of various compounds. Typically, they are white crystals and are characterized by good solubility in water. Recently, Green et al. [7] reported the reductions of two Mg(II) iodide complexes with potassium metal in toluene, leading to thermally stable Mg(I) compounds. The results of x-ray crystallographic and theoretical studies revealed that central units of Mg_2^{2+} had single, covalent Mg–Mg bonds. So chemically, it is possible to exist in the nature of living organisms, of course, without any functional knowledge. There are many known Mg compounds in nature, but two are of clinical value in medicine, simultaneously having many minerals in nature, namely, magnesium sulfate ($MgSO_4$) and magnesium carbonate ($MgCO_3$)· $MgSO_4$ is commonly used as a saline laxative, together with or without magnesium hydroxide. Epsom salts (heptahydrate $MgSO_4 \cdot 7H_2O$) have practical utility in form of gels for topical application in treating pains. $MgCO_3$ is called magnesite, and this mineral is anhydrous, but other hydrates of $MgCO_3$ have their own names, for instance, barringtonite ($MgCO_3 \cdot 2H_2O$), nesquehonite ($MgCO_3 \cdot 3H_2O$), or lansfordite ($MgCO_3 \cdot 5H_2O$). In medicine, magnesite is used for whitening skulls but probably is best known as food additive E504. Magnesite and $MgSO_4$ are used in laxative drugs alike [8]. Many other Mg compounds are used in clinical medicine during the treatment of various diseases: eclampsia and preeclampsia, cardiac arrhythmias, asthma, dyspepsia, headaches, and constipation. Among these Mg compounds are Mg oxide, Mg hydroxide, Mg citrate, Mg gluconate, Mg chloride, Mg lactate, and Mg aspartate hydrochloride [9].

Mg is an essential element in many biological functions. It has been established that Mg is the fourth most abundant metal in living organisms. This element is distributed in three major compartments in the body: 65% in mineral phase of skeleton, 34% in the intracellular space, and 1% in the extracellular fluid. So it is easy to see that it is really difficult to measure intracellular and bone fraction of Mg. On the other hand, most data indicate that the bone compartment of Mg is able to show alterations of Mg in extracellular fluid, and intracellular Mg levels are stable in spite of wide fluctuations in serum Mg. In plasma, 75–80% of Mg exists in the ionic and complexed form and is filtered in kidneys. The remaining 20–25% of Mg is bound with proteins. Urinary excretion is approximately 3–5% of filtered Mg. In case of functional imbalances producing a rise of Mg levels in serum, the fraction of filtered Mg increases sharply. In case of decrease in the concentration of serum Mg, which can occur as the result of low-Mg diet, Mg quickly disappears from the urine [10, 11].

The inflammatory process is a cofactor in carcinogenesis. Increased risk of malignancy is associated with chronic inflammation which is a complex process where tumor environment is affected by the presence of host leucocytes. Many cell populations with immunocompetence play dubious roles both in carcinogenesis and in reducing tumor growth. The cytokine network of several common tumors is rich in inflammatory cytokines, growth factors, and chemokines. Among the biologically active factors are TNF-α (alpha), IL-1, IL-6, and IL-8 [12]. Mg salts have potential to suppress inflammatory responses by various mechanisms. Experimentally, it turned out that aluminum hydroxide–magnesium combination drug (Maalox) suppressed the expression of the following virulence factors of *Helicobacter pylori*: adhesion, IL-8 inducibility, and expression of extracellular HSP60. Therefore, this drug is a cytoprotective agent with anti-*Helicobacter pylori* activity [13]. Inflammatory responses to LPS (lipopolysaccharide) are blocked when human umbilical vein endothelial cells (HuVECs) were treated with $MgSO_4$. Measured response of HuVECs relied on decreased, statistically significant ($p<0.05$) production of IL-8, expression of ICAM-1 (adhesion molecule), and NF-kappa-B [14].

Iron is a very important factor for practically all microorganisms. In experimental system, Giles and Czuprynski [15] have shown that the yeast Blastomyces dermatitidis can proliferate for short periods of time in the absence of iron but not in the absence of calcium or Mg. Mg as the cation is mainly located inside the cells. Diverse concentrations of the cation have been measured inside the cells what make compartmentalization as the basic interplay of Mg^{2+} among cell organelles and cytosol. Mg^{2+} has intracellular metabolic functions and the action of various effectors on its concentrations together with calcium cations [16].

Mg is the second most abundant element in a cell that plays a pivotal role in many metabolic pathways. This cation is a very important cofactor during DNA replication at the level of proper fidelity with other ions as Co^{2+}, Mn^{2+}, and Ni^{2+}. Mechanisms of this process are regulated by DNA polymerases having right activity in the presence of defined metallic cations. Environmental mutagens have portentous impact on DNA itself. To have controlled low levels of mutations, DNA repair systems watch over how to remove DNA damages. Mg is an essential cofactor in many mechanisms of DNA repair such as the following: NER (nucleotide excision repair) and BER (base excision repair) by keeping many enzymes, for instance, nuclease, N-methylpurine-DNA glycosylase, metal-dependent endonuclease, activated when needed. Mg is a crucial factor for the action of ATPase, prevalent enzyme regulating cell energetics. In a physiological state, Mg together with potassium is bound in polynucleotides that stabilize the double helix of DNA. Moreover, both cations are critical factors for the maintenance of the chromatin structure in the compact state of so-called heterochromatin [17–20].

Mg homeostasis depends mainly on the balance between intestinal absorption and renal excretion. Another factor also crucial for the balanced Mg in the body is the supplementation of this cation in food. Within physiological ranges, decreased Mg intake is balanced by enhanced absorption in the intestine and reduced renal excretion. The main site of Mg absorption is the small intestine with marginal role of colonic absorption. Two different pathways of Mg absorption are known: (1) saturable active transcellular transport and (2) nonsaturable paracellular passive transport. Saturable active transcellular transport is highly involved when there are low intraluminal concentrations of Mg in the small intestine. The molecular mechanism of this active transcellular transport relies on the function of TRPM6 (transient receptor potential melastatin-related ion channel). Paracellular passive transport is regulated by the claudins. Paracellin-1 is a member of the claudin family involved in the formation of tight junction playing a role in the paracellular transport of Mg and calcium ions [21]. In kidneys, approximately 80% of the total Mg is filtered in the glomeruli with more than 95% being reabsorbed along the nephron. This reabsorption is different in various nephron parts. Fifteen to twenty percent is reabsorbed in proximal tube; more than 70% of Mg is reabsorbed in the loop of Henle, especially in the thick ascending limb. Only 5–10% of filtered Mg is reabsorbed in the distant convoluted tube. Three to five percent of the filtered Mg is excreted in the urine [22].

Disturbed Mg balance can be a result of mutations in CASR (Ca^{2+}/Mg^{2+}-sensing receptor) that lead to autosomal-dominant hypoparathyroidism and which is responsible for hypocalcemia and hypocalciuria with hypomagnesemia [23]. An active transcellular transportation of Mg is through cation channels. TRP (transient receptor potential) channels are ion channels named after the primary member TRP protein discovered in Drosophila melanogaster. The TRP superfamily includes many non-voltage cation channels that vary in their selectivity and mode of action. There are three well-characterized subfamilies: TRPC, TRPV, and TRPM. The TRPC (TRP canonical) is highly related to original D. melanogaster TRP protein. The TRPV (TRP vanilloid receptor) was first identified in a search for proteins involved in the neuronal response to pain. They are activated by a vanilloid compound of hot chili pepper, i.e., capsaicin. TRPM (TRP melastatin) channel was discovered in a melanoma cell line characterized by metastatic behavior. The melastatin mRNA is suppressed in highly aggressive melanoma cells [24–26].

TRPM6 and TRPM6/TRPM7 complex are essential ion channel proteins which regulate Mg balance in the body. Mutational disruption of gene encoding TRPM6 causes hypomagnesemia with secondary hypocalcemia. Moreover, TRPM6 heterooligomerizes with TRPM7 and formed complex contributes to epithelial Mg absorption [27, 28].

Mg is an important factor in a living organism at many levels of analysis. A proper balance of this cation is vital to healthy organisms. If so, Mg supplementation should be performed in a right way in order not to lead to Mg deficiency. The latter refers to dietary intake below minimal levels resulting in many symptoms and disturbances. This state can be cured by higher Mg supplementation by using intravenous infusions or needed oral treatment with tablets or capsules. The oral use of Mg provides approximately 30% bioavailability in the body [29]. Hypermagnesemia is a kind of electrolyte disturbance characterized by elevated level of Mg in the blood. This pathologic state appears relatively rare when kidneys function correctly. So the renal insufficiency, with additional Mg-containing drugs, is a primary culprit of abnormal elevation of Mg concentrations, in some cases, leading to life-threatening clinical consequences. On the other hand, there is hypomagnesemia defined by abnormally low level of Mg in blood, typically below 0.7 mmol/L (watch normal ranges in your hospital or clinic). Hypomagnesemia can be present without Mg deficiency. Many pathologic states can result in hypomagnesemia [30]. Special role of the low blood levels of Mg is in cancer patients who can have hypomagnesemia as the result of natural malignant disease course, but of course, more often, it is observed after use of cancer chemotherapy based upon platinum compounds and targeted therapies.

Systemic Cancer Treatment and Its Complications

Tumorigenesis is a process leading to tumor formation. An uncontrolled cell proliferation is the fundamental event providing increased cell numbers that finally are detected as the clinical sign of malignant disease. Cancer is a problem that affects the whole organism. Malignant tumors can originate from almost all types of normal tissues. It is still unknown whether a tumor develops from one or many transformed cells. Although, there are some data giving the notion that malignant tumors develop as the result of transformation of single cells characterized by chromosomal aberrations, malignant tumors are characterized by the presence of distant metastases. The latter are final results of clonal evolution by which some new cells have possibilities to make distant metastatic lesions. Biological characteristics of tumor progression embrace morphological and metabolic changes in particular cells. So any kind of therapy should be focused on mechanisms of cell proliferations (standard chemotherapy) and on specific molecular pathways being readily seen in a given malignant disease (targeted therapies) [31].

Cancer chemotherapy is a systemic method of attacking malignant cells when a tumor has been detected at early stage or when a primary tumor has just been spread in the whole body. When chemotherapy is given at an early stage of a given tumor after surgical removal, it is called adjuvant chemotherapy. On the other hand, palliative chemotherapy is defined when its use is intended to put a curb on cancer cell spread. Both methods of treatment have something in common, i.e., toxicity. Independent of mechanism of action, each chemotherapy agent has toxicity, of course, with different profiles that depend upon a particular drug. Moreover, each form of therapeutic interventions (surgery, radiation, chemotherapy, targeted therapy) has the potential to produce adverse effects on normal host tissues. In case of chemotherapy, there are various toxicities. Some are the result of therapeutic procedure, for instance, extravasation and vascular toxicity or dermatologic toxicity. Some are well-defined as the outcome of systemic action on chemotherapeutic drugs on various organs and systems, for instance, hematologic complications, hypersensitivity reactions, cardiotoxicity, pulmonary toxicity, hepatotoxicity, or renal toxicity. All of them are observed relatively in the short period of time (acute or delayed toxicities) after medication. Finally, some toxicity profiles can be defined as long acting, having serious impact on the formation of secondary malignancies (cisplatin, carboplatin, paclitaxel) [32, 33].

Cisplatin is a commonly used anticancer drug with clinical activity against several malignant diseases. It was discovered in 1965 after the experiment that showed inhibitory effects of an electric field on cell division in bacteria. In the 1970s and 1980s, cisplatin was used in the treatment of lung cancer, bladder cancer, ovarian cancer, cervical cancer, endometrial cancer, germ-cell tumors, testicular cancer, and head and neck cancer with clinically confirmed activity [34]. Cisplatin is the platinum(II) complex with ammonia groups in the *cis* position. An initial aquation reaction relies on replacement of chloride groups by water molecules. The aquated platinum complex can react with different macromolecules such as RNA and DNA, preferentially by binding to N-7 position of guanine and adenine [35]. The N-7 atoms reside in the major groove of the DNA helix. The bridging cisplatin is placed in the minor groove what makes changes in the structure of the double helix of the DNA by reversing to a left-handed form, and the helix is unwound and bent toward the minor groove. The unwinding and the bending of cisplatin interstrand cross-linked DNA may be important in the formation and repair of these cross-links in chromatin [36].

Any kind of toxicity is associated with any activity of anticancer agents and is called adverse side effect. So an oncologist must be familiar with a list of complications and how to manage them to continue anticancer treatment receiving the best clinical response. Cisplatin is a broadly active agent against various tumors and is used very often clinically; therefore, toxicity profile is well known, although mechanisms responsible for their existence are not characterized adequately. There are certain items in a toxicity list such as nausea and vomiting (in 75% of patients), nephrotoxicity (dose limiting in 5–8% of patients), neurotoxicity (dose dependent in 85% of patients), and myelosuppression (in approximately 20–25% of patients) [37]. One of the adverse side effects of cisplatin therapy, very often imperceptible by physicians, is hypomagnesemia which is a frequent complication to this chemotherapy affecting 40–100% of patients [38].

Cancer therapy has extensively been changed during the last decade as the result of better understanding of molecular biology in carcinogenesis. Cancer cell proliferation, apoptosis, angiogenesis, invasion, and metastasis are regulated by many networks of cellular signaling pathways involving extracellular ligands, transmembrane receptors, intracellular protein kinases, and transcriptional factors [39]. New therapies are targeted to inhibit particular pathways responsible for cancer progression. In contrast with most traditional anticancer drugs which interfere with mitosis, DNA synthesis, and repair systems, targeted therapies are involved in blocking some vital pathways in cancer cells. The design rationale of such sophisticated therapies is based on the knowledge of specific biochemical processes and molecular mechanisms within cancer cells and cells of microenvironment with subsequent attempts to tailor treatment with the highest activity and lesser toxicity [40, 41]. Some new therapies are active against many cancers with clinical acceptance, for instance, trastuzumab (anti-HER2 monoclonal antibody in breast cancer and gastric cancer), cetuximab and panitumumab (anti-EGFR [epidermal growth factor receptor] monoclonal antibodies in colorectal cancer), sunitinib (small molecule as multikinase inhibitor of VEGFRs [vascular endothelial growth factor receptors] in renal cell cancer), sorafenib (small molecule as multikinase inhibitor of VEGFRs in renal cell cancer, hepatocellular cancer), gefitinib, and erlotinib (small molecule as kinase inhibitor of EGFR in lung cancer). We live in the decade of fast changes in experimental and clinical medicine that offer a number of clinical trials devoted to many new compounds with the potential to be anticancer drugs [42].

Targeted therapy against various malignant diseases is a powerful method providing clinical benefits in patients. Unfortunately, all targeted drugs have toxicity profile frequently responsible for stopping therapy due to clinical intolerance. Authors of many clinical trials have shown the positive correlation of presence of adverse side effects as predictive factors for a given therapy, for instance, dermatologic toxicity in colorectal cancer patients treated with cetuximab. There is a list of adverse side effects which are associated with targeted therapies against many molecular pathways as EGFR, VEGFR, HGF/c-MET [hepatocyte growth factor/c-MET], IGFR [insulin-like growth factor receptor], mTOR [mammalian target of rapamycin], AKT, MAPK [mitogen-activated protein kinase], and others [42]. The toxicity profiles coincide in many results of clinical trials of various targeted drugs, and the most frequent adverse side effects are hypertension (5–48%), proteinuria (3–40%), cardiotoxicity (3–28%), venous thromboembolic events relevant clinically (approximately 4%), alimentary tract perforation (1–4%), diarrhea (50%), and skin toxicity (25–50%) [43]. Among the adverse side effects of targeted therapy against malignant diseases is hypomagnesemia, which is mostly associated with anti-EGFR therapy.

Hypomagnesemia in Cancer Patients and Magnesium Supplementation

Mg is present in a balanced Western diet at the level of 360 mg/day. From this total supplementation, only 120 mg is absorbed in the intestine where 40 mg of daily Mg is secreted to the lumen, and about 20 mg of this is absorbed in the large bowel. Urinary excretion is approximately 100 mg/day. In healthy humans, typical normal values of Mg in the blood are from 1.8 to 2.3 mg/dL (0.75–0.95 mmol/L). Hypomagnesemia is defined as a serum Mg level of less than 0.74 mmol/L (<1.8 mg/dL) [44].

Hypomagnesemia is a common disturbance seen in up to 12% of hospitalized patients and appears to be of clinical value in 60–65% of patients hospitalized in intensive care unit [45, 46]. Early symptoms of hypomagnesemia are nonspecific and include lethargy and weakness. Symptomatic magnesium depletion is often symptomatic with biochemical abnormalities such as hypokalemia, hypocalcemia, and metabolic alkalosis. Moreover, there are known descriptions, mainly from the past, in which clinical manifestations associated with hypomagnesemia were tetany, positive Chvostek signs, Trousseau signs, and generalized convulsions. Hypomagnesemia has serious consequences for health and plays an important role in many diseases as parathyroid hormone resistance, deficiency of vitamin D, ischemic heart disease, congestive heart failure, cardiac arrhythmia type of torsades de pointes, and others [47].

Moreover, hypomagnesemia, being the result of Mg deficiency, has roles in the pathophysiology of hypertension and diabetes mellitus. Experimental data have shown correlation between Mg-deficient diets and microcirculatory alterations in situ associated with hypertension [48]. In diabetes mellitus, Mg deficiency has a pivotal role in abnormal carbohydrate metabolism by raising the insulin resistance leading to permanent diabetic changes, in turn, readily observed in renal microcirculation and reflected by rapid decline of kidney function [49, 50]. On the other hand, Mg supplementation (MAG21 solution = pure natural concentrated Mg chloride from Lake Deborah) provides clinical benefits in patients with type 2 diabetes by normalizing metabolic changes as triglyceride levels decrease, and patients with hypertension showed significant reduction of systolic, diastolic, and mean blood pressure [51].

Cisplatin is a crucial chemotherapeutic agent used in the treatment of lung cancer, ovarian cancer, germ cell cancer, head and neck cancer, and bladder cancer and is an active compound against other malignant diseases. There are well-known adverse side effects as nephrotoxicity, neurotoxicity, ototoxicity, myelosuppression, and nausea and vomiting, which are often met in patients with malignant diseases who are given cisplatin. Hypomagnesemia caused by renal Mg wasting is a well-characterized adverse side effect of cisplatin chemotherapy. This entity seems to be dose-related, so that six courses of cisplatin, administered alone or in combination with other anticancer drugs, was experienced in almost 100% of the patients [38, 52].

Cisplatin is excreted by kidneys. The presence of this drug in kidneys is responsible for a direct damage of renal tubes. After passing through tubular cells, the appearance of patchy necrosis confined to S3 segmental cells of the proximal tube in the outer strip of the medulla has been reported in rats [53]. In humans, a similar focal necrosis is observed, but morphological changes are predominantly located in the distal convoluted tubules and collecting ducts [54]. Moreover, an interstitial fibrosis is another pathological alteration observed in patients with acute renal failure following repeated courses of cisplatin [55]. Hypomagnesemia after chemotherapy with cisplatin is of renal origin with the occurrence in approximately 40% of cancer patients [56]. The minimal cumulative dose required to induce hypomagnesemia is 300 mg/m^2 of cisplatin [57]. Cisplatin used in cancer patients is responsible for significant Mg and potassium depletion in the majority of patients. Muscle biopsy is not good at measuring intracellular alterations of Mg and potassium; therefore, routine supplementation should be considered in all patients receiving cisplatin [58, 59].

An oral Mg supplementation is well-tolerated; incidentally, Mg can cause gastrointestinal symptoms including nausea, vomiting, and diarrhea. Overdose of Mg can cause thirst, hypotension, drowsiness, muscle weakness, respiratory depression, cardiac arrhythmia, coma, and death. Other drugs such as calcitonin, glucagon, and potassium-sparing diuretics may increase serum Mg levels. On the other hand, the use of Mg supplementation may influence the absorption of fluoroquinolones, aminoglycosides, bisphosphonates, calcium channel blockers, tetracyclines, and skeletal muscle relaxants. Therefore, it is recommended to monitor Mg serum levels in patients administered complex pharmacotherapy. An oral Mg supplementation is safe in adults when used in doses below the upper intake level of 350 mg daily (65 mg/day in children) [9]. Mg from high-Mg-containing food source like almonds is as bioavailable as from soluble Mg acetate. Mg absorption from commercially available enteric-coated Mg chloride is much less than from Mg acetate. Organic Mg salts are probably slightly more available than inorganic Mg salts (used orally). Mg gluconate appears to have the highest bioavailability. Mg supplementation can be performed intravenously or orally. The latter method is very convenient for patients who need permanent supplementation. The second way of faster supplementation is an infusional method [60, 61].

Mg supplementation is a method used in clinical practice which has impact on alleviation of adverse side effects in cancer patients. Its pivotal roles in many functions of very important organs in the body can offer readily seen clinical benefits in particular clinical situations. Controlling the Mg concentrations is a crucial step in cancer patient management, especially in those who are medicated with cisplatin. This anticancer drug is used very often in the treatment of many cancers, but its toxicity profile embraces nephrotoxicity which is of the highest clinical value due to a dose limitation that may interfere with the response rate by dose reduction or even discontinuation of treatment.

Table 11.1 shows the results of seven randomized trials in which cancer patients were given Mg supplementation together with the anticancer treatment. Various methods of supplementation were used with success presented by no statistical decrease of Mg levels after such treatment. Measurements of Mg ions were performed with the use of serum or urine, with one exception where authors [64] took measurements in erythrocytes. From a practical point of view, all authors stated that Mg supplementation should be considered in cancer patients medicated with cisplatin. Moreover, there were no reported clinically relevant adverse side effects of Mg supplementation in cancer patients enrolled in the clinical trials presented in Table 11.1.

Inhibition of EGFR signaling pathway appears to be an integral part in the treatment of various cancers. Cetuximab is a monoclonal antibody against EGFR which is overexpressed in many epithelial cell cancers as colorectal, breast, lung, and head and neck. This antibody has tenfold greater affinity for EGFR than the natural ligand. Clinical trials have demonstrated that cetuximab is synergistic with chemotherapeutic agents like irinotecan which are for patients with colorectal cancer [69].

Patients treated with cetuximab occasionally develop Mg wasting syndrome with intensified urinary excretion. Declining Mg has been responsible for fatigue, symptomatic hypocalcemia, and hypomagnesemia when using cetuximab combined with irinotecan. The group of 154 patients with colorectal cancer treated with cetuximab, among 34 (22%) experienced Mg measurement in serum, and six patients had grade 3 (Mg level <0.9 mg/dL) and two had grade 4 (Mg level <0.7 mg/dL) hypomagnesemia. EGFR is strongly expressed in the kidney; its blockade with cetuximab may be a culprit of hypomagnesemia being a side effect of the immunotherapy. Additionally, symptoms may be rapidly ameliorated with Mg supplementation [70].

The primary results have been confirmed in another paper that 97% (95 out of 98) of colorectal cancer patients treated with cetuximab had decreased Mg concentrations in comparison with baseline measurements. Hypomagnesemia was reported in 38% of patients who were treated with panitumumab (monoclonal antibody against EGFR), with frequency of grade 3 and 4 in 3% of such patients. In some patients, 24-h urinary collections were performed at baseline in 15 patients and in 35 patients on treatment. These results showed inappropriate reabsorption of Mg in kidneys leading to hypomagnesemia. Although this study was not planned to show how Mg supplementation alleviates outcomes of hypomagnesemia, the authors described individual effects of Mg supplementation in patients treated with cetuximab. In such cases, Mg supplementation (both oral or intravenous) is not effective at the level of serum Mg concentration normalization with long-lasting effects. Prospective clinical trials are needed for better understanding how various doses of Mg supplementation may stabilize concentrations of serum Mg [71].

Experimental papers have shown that EGF acts as an autocrine/paracrine hormone regulating the Mg balance. EGF stimulates Mg reabsorption in the renal distal convoluted tubule via engagement of its receptor on the basolateral membrane of these cells and activation of the Mg ion channel TRPM6 in the apical membrane. A point mutation in the gene encoding pro-EGF retains EGF secretion to the apical but not the basolateral membrane that makes disruption in the cascade of normal reabsorption of Mg ions causing renal Mg wasting. This mechanism is probably the most important cause for hypomagnesemia reported in colorectal cancer patients treated with cetuximab [72–75].

Having known the mechanism of the Mg balance disturbance induced by cetuximab, it is to address the problem in relation to the outcome of such therapy which depends on declined Mg serum concentrations. Early results of clinical trial in which 68 metastatic colorectal cancer patients were treated with combination of cetuximab and irinotecan showed at least 20% Mg reduction with 25 patients who had significant difference in response rate (64% vs. 25.6%), time to progression (6.0 months vs.

Table 11.1 Randomized studies of Mg supplementation with Mg cation response in cancer patients receiving cisplatin

Cancer	Number of patients (study vs. control)	Chemotherapy used	Dose of cisplatin	Method of supplementation	Mg measurements	Is there statistical decrease of Mg level after supplementation?	References
Germ cell tumors	16 (8 vs. 8)	Cisplatin + others	120 mg/m^2	Intravenously	Serum	No	[62]
Testicular cancer, ovarian dysgerminoma	17 (8 vs. 9)	Cisplatin + vinblastine + bleomycin	20 mg/m^2/day for 5 days	Intravenously Orally	Serum Urine	No	[63]
Testicular cancer	10 (5 vs. 5)	Cisplatin + vinblastine or bleomycin or etoposide	20 mg/m^2/day for 5 days	Intravenously	Serum Urine Erythrocyte	No	[64]
Head and neck cancer	23 (13 continuous vs. 10 intermittent)	Cisplatin + methotrexate + 5-FU + leucovorin	100 mg/m^2	Orally	Serum	No[a]	[65]
Various cancers	41 (12 i.v. Mg vs. 13 p.o. vs. 14 controls)	Cisplatin	100 mg/m^2	Intravenously Orally	Serum	No	[66]
Upper gastrointestinal tumors	28 (14 vs. 14)	Cisplatin +5-FU + epirubicin	60 mg/m^2	Intravenously Orally	Serum	No	[67]
Ovarian cancer	40 (20 vs. 20)	Cisplatin + paclitaxel	75 mg/m^2	Intravenously Orally	Serum	No	[68]

[a]In this study authors compared intermittent (when serum Mg <1.4 mg/dL) to continuous administration of Mg
i.v. intravenously, *p.o.* orally

3.6 months, $p<0.0001$), and overall survival (10.7 months vs. 8.9 months).The same authors confirmed their previous results in a series of 143 metastatic colorectal cancer patients treated with the third line of combination of cetuximab and irinotecan. Patients with decreased Mg serum after such treatment had better objective response rate (55.8% vs. 16.7%, $p<0.0001$), time to progression (6.3 months vs. 3.6 months, $p<0.0001$), and overall survival (11.0 months vs. 8.1 months, $p<0.002$) [76, 77].

Anemia occurs in more than 50% of cancer patients. There are a set of three main causes of being anemic: (1) age at the time of cancer diagnosis, (2) malignant cells are prone to make an anemic state as the result of interaction with bone marrow, and (3) any kind of treatment like systemic chemotherapy, immunotherapy, radiation, and targeted therapy causes anemia. The impact of anemia on physical and cognitive capacities is various and depends individually, but several studies have shown a linear correlation between hemoglobin levels and relevant quality of life in cancer patients [78, 79]. Cisplatin used in cancer patients is a culprit of cumulative anemia leading to the decrease of production of erythrocytes. The primary etiology of cisplatin-associated anemia is a transient but persisting erythropoietin deficiency state resulting from cisplatin-induced renal tube damage. This renal damage is correlated with hypomagnesemia, hematocrit, and creatinine clearance [80].

The relationship between Mg and carcinogenesis is complex because of the potential of Mg deficiency on tumor incidence, changed distribution of this ion in a body, and the role of Mg supplementation on cancer progression. Accumulating body of evidence has shown that it is impossible to draw a final conclusion because of conflicting results. Some animal studies have provided evidence that Mg-deficient diets impact the increase of incidence of thymic tumors and leukemias [81]. On the other hand, Rubin in the 1970s postulated that Mg was a crucial factor for cell proliferation and coined the term of three M's (membrane, magnesium, mitosis). This theory was based upon an increase of cellular Mg correlated with DNA and protein synthesis, growing cells having more Mg than resting ones, and growth rate correlated with Mg bioavailability [82]. Molecular determinants of growth inhibition by low Mg concentrations have been reported both in normal and malignant cells. In low Mg levels, the cell cycle inhibitory proteins such as p21 and p27 were upregulated, and the cell cycle promoting proteins such as cyclins D and E were downregulated. Mg-deficient mice exhibited approximately 60% reduced growth of primary tumors in comparison with control mice under normal Mg diet. Low Mg availability probably has influence on inhibition of the angiogenic switch, a process that triggers the formation of new vessels within a tumor. But the low degree of angiogenesis has not been in accordance with the fact that far more lung metastases appeared than in controls. One explanation could be made by showing that Mg is an absolute requirement for the activity of NM-23-H1, member of a family of eight gene products with well-known antimetastatic activity [83].

High circulating concentrations of C-peptide are a marker for insulin secretion and have been associated with increased risk of colorectal cancer. Mg supplementation increased insulin sensitivity in healthy humans and patients with type 2 diabetes . Therefore, Larsson et al. [84] conducted a prospective analysis of Mg intake in relation to incidence of colorectal cancer by using data from Swedish Mammography Cohort. Over a mean of 14.8 years (66,651 women representing 74% of the source population), 805 women were diagnosed with colorectal cancer (507, colon cancer; 252, rectal cancer; and 6 patients with both cancers). The participating patients were divided into quintiles of energy-adjusted Mg intake. The authors showed an inverse association of Mg intake with the risk of colorectal cancer (P for trend = 0.006). Compared with women in the lowest quintile of Mg intake, the multivariate rate ratio (RR) was 0.59 (95 CI:0.40–0.87) for those in the highest quintile. Conclusion of the study is that a high Mg intake (>255 mg/day) may reduce the occurrence of colorectal cancer in women. Another population-based study, the Iowa Women's Health Study (cohort of 41,836 women aged 55–69 years), showed that a diet high in Mg may reduce the occurrence of colon cancer among women without protective activity of Mg against rectal cancer [85]. By contrast with these results is the result of the Women's Health Study of 39,876 female health professionals who were 45 years and older. During an average of 11 years of follow-up, 259 women had a confirmed diagnosis of colorectal cancer. The average intake of Mg was 338 mg/day. The authors observed no significant association between intake of total Mg and colorectal cancer incidence. One possible explanation for the different findings among the various cohorts is that a high Mg intake may be related to a reduced risk of colorectal cancer only among populations with relatively low intake levels of Mg [86].

Neurotoxicity and Mg Supplementation

The neurotoxicity profile of chemotherapeutic agents used in cancer patients is a very important issue in clinical practice. Other methods as radiotherapy, immunotherapy, and combination therapy are also complicated by any kind of neurotoxicity. There have been several factors to determine the occurrence of neurotoxicity in treated cancer patients. First, advances in supportive care with growth factors enable oncologists to continue chemotherapy beyond the bone marrow suppression treated exactly by these growth factors such as G-CSF (granulocyte colony-stimulating factor) and GM-CSF (granulocyte-macrophage colony-stimulating factor). Second, the better results of therapy in many cancers allow oncologists to offer patients significantly increased survival mostly without disease progression. Some malignant diseases are successfully controlled by the use of systemic treatment given as maintenance. In such clinical situations, any adverse side effects should be alleviated to continue anticancer treatment with further clinical benefits as the end point. Third, a list of new drugs approved for cancer treatment is permanently growing. Among them, many are characterized by the presence of a neurotoxicity profile. Causes of neurologic symptoms in patients who used chemotherapy are in three categories: (1) direct neurotoxicity from chemotherapy; (2) indirect neurotoxicity from chemotherapy as metabolic encephalopathy, coagulopathy, and myelosuppression with central nervous system; and (3) unrelated to chemotherapy as toxicity of other drugs, neurologic complications of cancer, radiotherapy complications, paraneoplastic syndromes, and others [87].

In experimental rat model, Mg deficiency has been responsible for dramatic increase in serum levels of inflammatory cytokines as IL-1 (interleukin-1), IL-6 (interleukin-6), and TNF-α (alpha) after 3 weeks on a Mg-deficient diet. The authors of this report showed that neuropeptide, substance P, had been elevated after 5 days on this diet. Substance P is known to stimulate production of certain cytokines mentioned previously. Stimulation of the inflammatory cytokines is a crucial step in generating free radicals directly responsible for neural injury [88].

Cisplatin is considered to be one of the most import cytotoxic drugs available to oncologists. Since the 1970s, it has been used successfully in many patients with lung cancer, testicular cancer, ovarian cancer, cervical cancer. On the other hand, cisplatin with its anticancer effects initiated the search for analogues with less toxicity and increased efficacy against malignant diseases. Two of such analogues are now well known in oncology for the treatment of various cancers. Namely, carboplatin and oxaliplatin are clinically useful anticancer drugs having been approved for killing malignant cells in lung cancer, ovarian cancer, head and neck cancer, and colorectal cancer.

Cisplatin has a number of adverse side effects that can limit clinical use in cancer patients: nephrotoxicity, neurotoxicity, nausea and vomiting, ototoxicity, and electrolyte disturbance as hypomagnesemia, hypokalemia, and hypocalcemia. Neurotoxicity in cancer patients medicated with cisplatin is of clinical value and is manifested by peripheral neuropathy, spinal cord toxicity, optic neuropathy, seizures, cortical blindness, and ototoxicity. Neuropathy is the major dose-limiting toxicity of cisplatin. It occurs in about 85% of patients with a cumulative dose greater than 300 mg/m^2 and is irreversible in 30–50% of cases [89].

Peripheral neurotoxicity develops in approximately 50% of patients receiving cisplatin. Signs and symptoms of this complication observed after a cisplatin treatment involve upper and lower extremities with loss of vibration sense, loss of position sense, tingling paresthesia, weakness, tremor, and waste of taste [90]. Tissue cisplatin levels were the highest in the dorsal root ganglia and the lowest in tissue protected by the blood–brain barrier. Moreover, there was a linear relationship between cisplatin levels and cumulative dose in peripheral nerve, dorsal root, and dorsal root ganglia [91]. So the neuropathy predominantly involves the large sensory fibers. Deep tendon reflexes are lost due to the results of toxic effects of cisplatin on the large myelinated sensory fibers which, in turn, are responsible for the afferent arm of the reflex arc. Nevertheless, involvement of motor functions is generally mild but possible to detect especially in cancer patients with already diagnosed severe sensory neuropathy.

Spinal cord toxicity is observed relatively less frequently, but it has been described in several cases. The most important sign of spinal cord neurotoxicity after cisplatin treatment is the well-known Lhermitte's sign, which is an electric shock sensation down the spine or into the extremities provoked by neck flexion (this sign is mainly associated with demyelinating lesions in multiple sclerosis). Cisplatin-induced demyelination is also responsible for the presence of such a sign in cancer patients. Most patients after discontinuation of cisplatin treatment are recovered fully but after several months [92, 93]. Optic neuropathy is also seen very seldom but may have been of clinical value because of prolonged visual loss associated with the optic disk pallor [94]. Cisplatin has another adverse side effect affecting hearing. Tinnitus and hearing loss have been noted in up to 31% of patients treated with cisplatin at a dose of 50 mg/m^2. Transient hearing loss detected by audiometry has been reported in approximately 30% of patients receiving 150 mg/m^2 of cisplatin. Reactive oxygen radicals with reduction of glutathione activity are considered to be the most probable mechanisms leading to hearing disturbances [95–97].

Serum Mg following cisplatin treatment was considerably changed. Analysis of correlation revealed significant correlation between total cisplatin and average Mg ($r=-0.355$) as well as between total cisplatin and Mg nadir ($r=-0.436$). A major adverse side effect of cisplatin treatment relies on alterations of sensory nerve conductance. There was a decrease in ulnar sensory amplitude in 72% and an increase in ulnar sensory latency in 62.2% of patients receiving cisplatin. Unfortunately, the authors have not observed better responses for Mg supplementation in any of the posttherapy conduction parameters, despite the fact that Mg supplementation was effective in raising serum Mg levels [98].

Cisplatin is characterized by the presence of neurotoxicity profile, and there is a great need of searching for chemoprotective agents that could have impact on neurotoxicity limitation to continue cisplatin treatment. Albers et al. [99] have performed search and further analysis focused on a list of chemoprotective agents such as acetylcysteine, amifostine, ACTH, BNP7787, calcium and Mg, diethyldithiocarbamate, glutathione, melanocortin Org 2766, oxycarbamazepine, and vitamin E. Authors' conclusions have revealed that the data used so far are insufficient to conclude that any of the purported chemoprotective agents prevent or limit the neurotoxicity of cisplatin. Most of clinical trials have been powered wrongly; therefore, conclusions are drawn far too early, in spite of encouraging results showing significant protective activity against neurotoxic effect of cisplatin, as it was in the paper dedicated to a protective role of vitamin E during cisplatin treatment [100].

Carboplatin seems to be a drug with less neurotoxic effects than cisplatin. Clinical data have reported that neurological dysfunctions are adverse side effects reported especially in patients receiving high doses of carboplatin or when it is combined with other cytotoxic agents known to be neurotoxic, for instance, paclitaxel. Only 4–6% of patients who received carboplatin may develop peripheral neuropathy. Ototoxicity, after carboplatin therapy, has been reported in a small portion of about 1.1% of cancer patients [101, 102].

Oxaliplatin has become an integral part of chemotherapy schedules in advanced colorectal cancer. Contrary to cisplatin, oxaliplatin produces more neurotoxic effects. Peripheral neuropathy is the most common dose-limiting toxicity of oxaliplatin, and it is one of the major causes of discontinuation of therapy. Sensory peripheral neuropathy caused by the administration of oxaliplatin is divided into two forms: (1) an acute peripheral sensory neuropathy that may appear during the first administration of this drug and (2) a chronic dose-limiting peripheral sensory neuropathy. Acute neurotoxic effects may result from the impairment of voltage-gated sodium channels and occur in about 80–85% of patients receiving oxaliplatin-based chemotherapy due to colorectal cancer. In the literature, this mechanism is called channelopathy. This is based on the similarities that oxaliplatin-induced neuropathy shares with hereditary myotonias and certain toxic exposures [103, 104].

The theory assumes that an oxalate affects the sodium channels by its release intracellularly from oxaliplatin by bicarbonate ions. Unlike the acute neuropathy, the cumulative toxicity of oxaliplatin appears to be related to direct toxicity to the nerves. Morphological changes have been observed in dorsal root ganglia in rats with cumulative doses of oxaliplatin given intraperitoneally. The treated rats had evidence of nuclear, nucleolar, and somatic size reduction of the dorsal root ganglia on microscopic examination [105–107].

Symptoms consist of paresthesias and dysesthesias in the extremities and in the perioral region which can be exacerbated by cold exposure. Other, less frequent, manifestations of the oxaliplatin neurotoxicity are laryngopharyngeal dysesthesia (transient difficulty of breathing without respiratory distress). The risk of acute neuropathy appears to be lower if oxaliplatin is administered in a dose of 85 mg/m^2 every 2 weeks rather than 130 mg/m^2 every 3 weeks [108].

Many clinical trials have examined the efficacy of a number of potential protective agents used against neurotoxicity profile. These agents were administered together with platinum compounds. The use of protective agents aims to reduce the incidence of the neurotoxicity without involvement of the anticancer efficacy of the platinum compounds. There are known several potential neuroprotective agents including amifostine, glutathione, melanocortin Org 2766 (experimentally tested in various platinum compounds), vitamin E (used against cisplatin neurotoxicity), calcium/Mg infusions (clinically tested in oxaliplatin-induced neuropathy in colorectal cancer patients), Mg supplementation (used clinically in cisplatin-treated patients), and anticonvulsants (gabapentin and pregabalin both used in oxaliplatin neurotoxicity), and nonpharmacological approaches to prevent neurotoxicity induced by oxaliplatin embrace the “stop and go” concept. This latter strategy is based on the observation of reversible neurotoxic side effects after discontinuation of oxaliplatin [109–111]. Table 11.2 shows results of some of the clinical trials considering neuroprotective effects of Mg supplementation in cancer patients receiving oxaliplatin or cisplatin combined with other drugs.

Nephrotoxicity and Magnesium Supplementation

Urinary or biliary excretions are the pathways responsible for removal of most chemical compounds taken into the body. Hepatic and renal dysfunction may impair the way how a patient should be treated. Most chemical compounds are metabolized first and then excreted. Renal dysfunction is able to modify a drug metabolism with subsequent excretion by two main mechanisms: (1) a drug transformation in the body may be altered and (2) impairment of a drug metabolite excretion can be of clinical value. So the retention of drug metabolites with pharmacological activities can be responsible for prolonged effects of a given drug which makes accumulated toxic exposures. On the other hand, there is a possibility to induce resistance to the drug which is excreted improperly by kidneys.

Chronic kidney disease has been recognized as a public health problem, and it is recommended to use equations that estimate the glomerular filtration rate (GFR) to facilitate the detection, evaluation, and management of chronic kidney disease. GFR is accepted as the best measure of the kidney excretory function. Normal values (related to age, sex, body size) are 130 mL/min/1.73 m^2 in a young man and 120 mL/1.73 m^2 in a young woman. GRF is measured as the urinary or plasma clearance of an ideal filtration marker, for instance, inulin. Research studies have reported a measurement error of 5–20% variation in the values of GFR on different days in the same patient. Urinary clearance of endogenous filtration of markers such as creatinine or cystatin C can be computed from a 24-h urine collection and blood sampling during the collection period. The accurate diagnosis of chronic kidney

Table 11.2 Neuroprotective effects of Mg supplementation in cancer patients treated with oxaliplatin and cisplatin in combination chemotherapy

Methods of Mg supplementation	No. of patients	Type of cancer	Regimen	Dose of cytostatics	Randomized	Conclusions	References
Magnesium sulfate of 720 mg and calcium gluconate of 850 mg before and after the administration of oxaliplatin or placebo	17 16	Colorectal cancer	FOLFOX6 Oxaliplatin 5-FU Leucovorin	85 mg/m(2) i.v./2 weeks 2,800 mg/m(2) i.v./2 weeks 200 mg/m(2) i.v./2 weeks	Yes	The administration of Ca/Mg: • Is not neuroprotective • Has no any influence on antitumor activity and the blood concentration profile of platinum	[112]
Magnesium sulfate of 1,000 mg and calcium gluconate of 1,000 mg before and after the administration of oxaliplatin or none	96 65	Colorectal cancer	Oxaliplatin 5-FU Leucovorin	85 mg/m(2) i.v./2 weeks or 100 i.v./2 weeks or 130 i.v./3 weeks	No	• Only 4% of patients withdrew for neurotoxicity in the Ca/Mg group versus 31% in the control group (P=0.000003) • The tumor response rate was similar in both groups	[113]
Magnesium sulfate of 1,000 mg and calcium gluconate of 1,000 mg before and after the administration of oxaliplatin or placebo	52 52	Colorectal cancer	Oxaliplatin 5-FU Leucovorin	85 mg/m(2) i.v./2 weeks, 2,800 mg/m(2) i.v./2 weeks 200 mg/m(2) i.v./2 weeks	Yes	Ca/Mg is effective neuroprotectant against oxaliplatin-induced cumulative sensory neurotoxicity	[114]
$MgCl_2$ 4 mmol and calcium gluconate 2.25 mmol before and after oxaliplatin infusion or none	551 181	Colorectal cancer	Oxaliplatin + Capecitabine Bevacizumab or cetuximab	130 mg/m(2) i.v./3 weeks 2,000 mg/m(2) i.v./bid for 14 days/3 weeks 7.5 mg/kg i.v. /3 weeks 400 mg/m(2) i.v. than 250 mg/m(2) i.v./week	No	Ca/Mg significantly reduced all grade oxaliplatin-related neurotoxicity Ca/Mg did not affect the clinical efficiency of treatment	[115]
Magnesium sulfate of 5 g and 20 mg KCl i.v. before administration of cisplatin and 200 mg five times per day p.o. or placebo	19 18	Ovarian (14 pts), cervical (16 pts), endometrial (6 pts) Fallopian tube(1pts)	Cisplatin (17 pts) Cisplatin + Doxorubicin + Cyclophosphamide (20 pts)	70 mg/ m(2) i.v./4 weeks 70 mg/ m(2) i.v./4 weeks NR NR	Yes	Mg supplementation has no significant effect on any posttherapy nerve conduction parameters	[98]

pts patients, *Mg* magnesium, *Ca* calcium, *NR* not reported, *p.o.* orally, bid twice a day, *i.v.* intravenously, *5-FU* 5-fluorouracil disease is the very important step in a cancer patient before making a decision as to what treatment modality should be used because many anticancer drugs have nephrotoxic side effects [116, 117].

A hepatic dysfunction is much commonly reported in cancer patients due to the role of the liver in drug metabolism; renal dysfunction appears to result from preexisting comorbidities or complications of the cancer itself or anticancer treatment modalities with a primary role of systemic drugs used. Kidneys are vulnerable to chemical injury due to their special role in the elimination pathway of many anticancer drugs. The entire anatomic way from glomerulus to distal tubule is at risk to be chemically injured, depending upon the drug involved. The symptoms are various from an asymptomatic rise in serum creatinine or mild proteinuria to acute renal failure with anuria requiring dialysis. In the IRMA study (the Renal Insufficiency and Anticancer Medication), the authors stated that abnormal renal function or renal insufficiency was common in cancer patients reported in more than 50% of cases. Of the patients treated, almost 55% required dose adjustments due to renal insufficiency [118]. Nephrotoxic side effects of many anticancer drugs are well known. Among the anticancer drugs characterized with such clinically relevant shortcomings are cisplatin and carboplatin, mitomycin, methotrexate, streptozocin, carmustine, lomustine, ifosfamide, and other drugs with targeted therapy, e.g., imatinib and bortezomib [119]. Cisplatin has a special place in the list of nephrotoxic drugs and has a substantial role in this toxicity by the mechanism associated with hypomagnesemia.

Cisplatin is a major drug used in many patients diagnosed with cancer. The dose-limiting effect is nephrotoxicity which is reported in about 20% of patients receiving high-dose cisplatin rendering renal dysfunction. The kidney makes cisplatin accumulate to a greater extent than other organs and tissues. The cisplatin concentration in proximal tubular epithelial cells is approximately five times the serum level [120]. Cisplatin is mainly accumulated by peritubular uptake in both the proximal and distal nephrons. The S3 segment of the proximal tubule accumulates the highest concentration of cisplatin, but distal collecting tubule and S1 segment in proximal tubule accumulate less amount of cisplatin as well [121].

Cisplatin is transported in both mechanisms, i.e., active uptake by transport system and passive diffusion through cellular membranes. The first mechanism plays a pivotal role in cisplatin transport in the kidney. OCT2 (organic cation transporter) is the critical transporting mechanism for cisplatin uptake in proximal tubes in both animals and humans. The transport mediated by OCT2 proteins is polyspecific, electrogenic, voltage dependent, bidirectional, pH independent, and regulated by Na^+ cation. There are known three isoforms of the transporters, namely, OCT2 (mostly expressed in the kidney), OCT1 (this form is met in the liver), and OCT3 (the prevalent isoform in a human body with a little expressional skew in the placenta) [122].

Cisplatin activity is regulated and performed by metabolites; some of them are conjugated to glutathione and then metabolized through a GGTP (gamma-glutamyl transpeptidase) and a cysteine S-conjugate beta-lyase-dependent mechanism to form a reactive thiol having nephrotoxic potential [123, 124]. Oxidative stress injury is involved in the process of cisplatin-induced acute kidney impairment. Reactive oxygen species (ROS) together with reactive nitrogen species directly act on cell components including lipids, proteins, and DNA. Cisplatin induces specific gene changes. Genes taking part in drug resistance as following MDR (multidrug resistance), cytoskeleton structure proteins, cell adhesion proteins, and apoptosis regulators (caspases 1, 8, and 9 are initiator caspases that activate caspase 3, which is the primary factor in apoptosis of renal tubule cells) [125–127]. Moreover, cisplatin has induced its renal toxicity by MAPK (mitogen-activated protein kinase) intracellular signaling pathway which is one of the most important regulators of cell proliferation, differentiation, and survival. The three major MAPK pathways are within one set of regulators, i.e., ERK (extracellular regulated kinase), p38, and JNK/SAPK (c-Jun N-terminal kinase/stress-activated protein kinase). Cisplatin appears to be a direct activator of all three MAPKs in the kidney [128].

Inflammation, as shown previously, has been considered as one of the pivotal processes associated with carcinogenesis and further progression in late phase of a malignant disease. Cisplatin induces a series of inflammatory alterations that mediate renal injury both directly and indirectly. This anticancer drug increases degradation of I kappa B (main regulator and controller of biologically active nuclear factor-kappa B = NF-kappa B). NF-kappa B is responsible for enhanced renal expression of TNF-alpha and other cytokines (TGF-beta = transforming growth factor beta; MCP-1 = monocyte chemoattractant protein-1, ICAM = intercellular adhesion molecule) which may regulate infiltration of macrophages and lymphocytes. Among the listed inflammatory factors, TNF-alpha is the primary regulator of all cytokine interactions [129].

Many pathophysiologic mechanisms associated with cisplatin therapy are located in the renal tubules. Cisplatin itself evidently is responsible for damage expressed as necrosis at the sites of tubular reabsorption of Mg which seems to be a main factor for the regulation of Mg homeostasis. Accumulated evidence has shown that cisplatin causes hypomagnesemia dependently of the applied cumulative dose of cisplatin rather than the number of doses given [57, 130].

Cancer patients receiving cisplatin need careful attention to hydration and electrolyte treatment. Clinical recommendations point some simple notions to avoid cisplatin (in doses up to 80 mg/m^2 or higher above 80 mg/m^2, respectively) toxicity with nephrotoxicity by not administrating this drug if serum creatinine level is more than 1.5 mg/dL; not combining cisplatin treatment together with other nephrotoxic drugs, e.g., aminoglycosides; and beginning infusion with 5% dextrose with potassium chloride and Mg sulfate and running 500 mL/h for 1.5–2.0 l or 2.5–3.0 l. After completing 1-h infusion, give 12.5 or 25 g of mannitol by intravenous push; thereafter, immediately start cisplatin infusion for at least 1 h or 2 h with subsequent mannitol 12.5–50 g intravenously to maintain urinary output at the level of 250 mL/h over the duration of hydration [131]. Mannitol diuresis is responsible for amelioration of renal toxicity after Al-Sarraf et al. [132] reported that combination of hydration, cisplatin, and mannitol offered less nephrotoxic complications in comparison with control group without mannitol. Many potential protective compounds have been evaluated in cancer patients with cisplatin chemotherapy, but results are not encouraging enough to use them in clinical practice [133].

From the clinical standpoint, it is very important to show whether Mg salts supplementation, in cancer patients receiving anticancer drugs like EGFR inhibitors or cisplatin, is an effective method to prevent hypomagnesemia. As shown previously in Table 11.1, the data of randomized clinical studies on Mg supplementation are convincing enough to draw a conclusion that cancer patients under Mg supplementation do not experience hypomagnesemia. Nevertheless, there is an undone problem that still needs to be solved – whether Mg supplementation is able to offer clinically relevant response to keep the kidney function intact or almost intact.

So an important item as nephrotoxicity resulting from cancer treatment is not investigated properly at the clinical level, i.e., the level where the rule of Hippocrates *primum non nocere* is legally binding. Table 11.3 shows the results of only two randomized clinical trials carried out on small groups of cancer patients receiving cisplatin-based chemotherapy with or without Mg supplementation. Despite stretched in time, both studies give encouraging results with the hope to perform more accurate clinical studies with the use of Mg salts that are relatively much cheaper than other potential protective compounds existing in a very limited number while taking into account their efficacy.

Conclusion

Cancer is a public health problem. Different modalities of cancer treatment are associated with adverse side effects. One of the side effects is hypomagnesemia, which is associated with anticancer chemotherapy using cisplatin, carboplatin, and oxaliplatin. Mg is an essential element in normal functions of about 300 cellular enzymes. In doing so, metabolic complications that stem from low level of Mg may impair cancer treatment efficacy because of adverse side effects. Platinum compounds are broadly used in oncology with relatively good outcomes, and therefore, all methods used to alleviate side

Table 11.3 Nephroprotection after Mg supplementation assessed in randomized studies carried out in cancer patients receiving cisplatin-based combination chemotherapy

Methods of Mg supplementation	No. of patients	Regimen	Dose of platinum compounds	Randomized	Conclusions	References
Intravenously + orally or none	8	Cisplatin + vinblastine + bleomycin	20 mg/m² i.v./day for 5 days	Yes	Mg supplementation sustains Mg concentrations in serum	[63]
	9				Mg supplementation group has less nephrotoxicity assessed by NAG	
Magnesium sulfate of 5 g and 20 mg KCl i.v. before administration of cisplatin and 200 mg five times per day	20		75 mg i.v./m²/q 3 weeks	Yes	There is a nephroprotective effect of Mg supplementation during chemotherapy with cisplatin	[68]
	20	Cisplatin + paclitaxel	135 mg/m² i.v./q 3 weeks			

i.v. intravenously, *Mg* magnesium, *NAG* N-acetyl-B-D-glucosaminidase effects are welcomed. Mg supplementation may be useful in cancer patients to prevent nephrotoxic effects of cisplatin, but other controlled clinical studies should be performed. Neurotoxicity is rather a more complex problem, and accumulated evidence has so far offered conflicting outcomes. There is a great need to establish the role of Mg supplementation in cancer patients due to the natural character of this element being so important in the normal function of the human body. Mg supplementation could satisfy Linus Pauling's hope of the clinical role of orthomolecular medicine with supplementation of natural substances.

References

1. Cohen JE. Human population: the next half century. Science. 2003;302:1172–5.
2. Ferlay J, Bray F, Pisani P, et al. GLOBOCAN 2002: IARC Cancer Base No 5, Version 2.0. France: IARC-Press; 2004.
3. Beating cancer. Economist. 2004;373:9. Found online at http://www.economist.com/node/3286519
4. Verdecchia A, Francisci S, Brenner H, et al. Recent cancer survival in Europe: a 2000–02 period analysis of EUROCARE-4 data. Lancet Oncol. 2007;8:784–96.
5. Gatta G, Zigon G, Capocaccia R, et al. Survival of European children and young adults with cancer diagnosed 1995–2002. Eur J Cancer. 2009;45:992–1005.
6. De Haas EC, Oosting SF, Lefrand JD, et al. The metabolic syndrome in cancer survivals. Lancet Oncol. 2010;11:193–203.
7. Green SP, Jones C, Stasch A. Stable magnesium(I) compounds with Mg-Mg bonds. Science. 2007;318:1754–7.
8. Pauling L. Magnesium. In: Pauling L, editor. General chemistry. New York: Dover Publications; 1970. p. 626–7.
9. Guerrera MP, Volpe SL, Mao JJ. Therapeutic uses of magnesium. Am Fam Physician. 2009;80:157–62.
10. Levine BS, Coburn JW. Magnesium, the mimic/antagonist of calcium. N Engl J Med. 1984;310:1253–4.
11. Fawcett WJ, Haxby EJ, Male DA. Magnesium: physiology and pharmacology. Br J Anaesth. 1999;83:302–20.
12. Balkwill F, Manovani A. Inflammation and cancer: back to Virchow? Lancet. 2001;357:539–45.
13. Kamiya S, Yamaguchi H, Osaki T, et al. Effect of an aluminium hydroxide-magnesium hydroxide combination drug on adhesion, IL-8 inducibility, and expression of HSP60 by Helicobacter pylori. Scand J Gastroenerol. 1999;34:663–70.
14. Rochelson B, Dowling O, Schwartz N, et al. Magnesium sulfate suppresses inflammatory responses by human umbilical vein endothelial cells (HuVECs) through the NK-kappa pathway. J Reprod Immunol. 2007;73:101–7.
15. Giles SS, Czuprynski CJ. Extracellular calcium and magnesium, but not iron, are needed for optimal growth of Blastomyces dermatitidis yeast form cells. Clin Diag Lab Immunol. 2004;11:426–9.
16. Gunther T. Concentration, compartmentation and metabolic function of intracellular free Mg^{2+}. Magnes Res. 2006;19:225–36.
17. Hartwig A. Role of magnesium in genomic stability. Mut Res. 2001;475:113–21.
18. Adhikari S, Toretsky JA, Yuan L, et al. Magnesium, essential for base excision repair enzymes, inhibits substrate binding of N-methylpurine-DNA glycosylase. J Biol Chem. 2006;281:29525–32.
19. Hibino Y, Kusashoi E, Terakawa T, et al. Enhancement of an Mg^{2+}-dependent nuclease activity in rat liver cells exposed to cisplatin. Biochem Biophys Res Commun. 1994;202:749–56.
20. Mate MJ, Kleanthous C. Structure-based analysis of the metal-dependent mechanism of H-N-H endonucleases. J Biol Chem. 2004;279:34763–9.
21. Colegio OR, Van Itallie CM, McRea H, et al. Claudins create charge-selective channels in the paracellular pathway between epithelial cells. Am J Physiol Cell Physiol. 2002;283:C142–7.
22. Konrad M, Schlingmann KP, Gudermann T. Insights into the molecular of Mg homeostasis. Am J Physiol Renal Physiol. 2004;286:F599–605.
23. Pearce SH, Williamson C, Kifor O, et al. A familial syndrome of hypocalcemia with hypercalciuria due to mutations in the calcium-sensing receptor. N Engl J Med. 1996;335:1115–22.
24. Montell C, Birnbaumer L, Flockerzi V. The TRP channels, a remarkably functional family. Cell. 2002;108:595–8.
25. Wissenbach U, Niemeyer BA, Flockerzi V. TRP channels as potential drug targets. Biol Cell. 2004;96:47–54.
26. Kiselyov K, Soyombo A, Muallem S. TRPpathies. J Physiol. 2007;578(3):641–53.
27. Chubanov V, Waldegger S, Mederos Y, Schnitzler M, et al. Disruption of TRPM6/TRPM7 complex formation by a mutation in the TRPM6 gene causes hypomagnesemia with secondary hypocalcemia. PNAS. 2004;101:2894–9.
28. Voets T, Nilius B, Hoefs S, et al. TRPM6 forms the Mg^{2+} influx channel involved in intestinal and renal Mg^{2+} absorption. J Biol Chem. 2004;279:19–25.

29. Lindberg JS, Zobitz MM, Poindexter JR, et al. Magnesium bioavailability from magnesium citrate and magnesium oxide. J Am Coll Nutr. 1990;9:48–55.
30. Whang R, Hampton EM, Whang DD. Magnesium homeostasis and clinical disorders of magnesium deficiency. Ann Pharmacother. 1994;28:220–6.
31. Wcislo G. Molecular underpinnings of the targeted therapy for cancer. Acta Pol Pharm – Drug Res. 2008;65:633–40.
32. Leonard DGB, Travis LB, Addaye K, et al. p53 mutations in leukemia and myelodysplastic syndrome after ovarian cancer. Clin Cancer Res. 2002;8:973–85.
33. Travis LB, Holowaty EJ, Bergfeldt K, et al. Risk of leukemia after platinum-based chemotherapy for ovarian cancer. N Engl J Med. 1999;340:351–7.
34. Guminski AD, Harnett PR, deFazio A. Scientists and clinicians test their metal-back to the future with platinum compounds. Lancet Oncol. 2002;3:312–8.
35. Pinto AL, Lippard SJ. Binding of the anti-tumor drug cis-diamminedichloroplatinum(II) (cisplatin) to DNA. Biochem Biophys Acta. 1985;780:1345–57.
36. Huang H, Zhu L, Reid BR, et al. Solution of a cisplatin-induced DNA interstrand cross-link. Science. 1995;270:1842–5.
37. Go RS, Adjei AA. Review of the comparative pharmacology and clinical activity of cisplatin and carboplatin. J Clin Oncol. 1999;17:409–22.
38. Buckley JE, Clark VL, Meyer TJ, et al. Hypomagnesenia after cisplatin combination chemotherapy. Arch Intern Med. 1984;144:2347–8.
39. Adjei AA, Hidalgo M. Intracellular signal transduction pathway proteins as targets for cancer therapy. J Clin Oncol. 2005;23:5386–403.
40. Sawyers C. Targeted cancer therapy. Nature. 2004;432:294–7.
41. Ang JE, Kaye SB. Molecular targeted therapies in the treatment of ovarian cancer. Adv Gene Mol Cell Ther. 2007;1:68–79.
42. Ma WW, Adjei AA. Novel agents on the horizon for cancer therapy. CA Cancer J Clin. 2009;59:111–37.
43. Stone RL, Sood AK, Coleman RL. Collateral damage: toxic effects of targeted antiangiogenic therapies in ovarian cancer. Lancet Oncol. 2010;11:465–75.
44. Naderi ASA, Reilly RF. Hereditary etiologies of hypomagnesemia. Nat Clin Prac Nephrol. 2008;4:80–9.
45. Wong ET, Rude RK, Singer FR. A high prevalence of hypomagnesemia in hospitalized patients. Am J Clin Pathol. 1983;79:348–52.
46. Chernov B, Bamberger S, Stoiko M. Hypomagnesemia in patients in postoperative intensive care. Chest. 1989;95:391–7.
47. Agus ZS. Hypomagnesemia. J Am Soc Nephrol. 1999;10:1616–22.
48. Altura BM, Altura BT, Ising H, et al. Magnesium deficiency and hypertension: correlation between magnesium-deficient diets and microcirculatory changes in situ. Science. 1984;223:1315–7.
49. Hans CP, Sialy R, Bansal DD. Magnesium deficiency and diabetes mellitus. Curr Sci. 2002;83:1456–63.
50. Pham PC, Pham PM, Pham PA, et al. Lower serum magnesium levels are associated with more rapid decline of renal function in patients with diabetes mellitus type 2. Clin Nephrol. 2005;63:429–36.
51. Yokota K, Kato M, Lister F, et al. Clinical efficiency of magnesium supplementation in patients with type 2 diabetes. J Am Coll Nutr. 2004;23:506S–9.
52. Markman M, Rothman R, Reichman B, et al. Persistent hypomagnesemia following cisplatin chemotherapy in patients with ovarian cancer. J Cancer Res Clin Oncol. 1991;117:89–90.
53. Mavichak V, Wong NLM, Quamme GA, et al. Studies on the pathogenesis of cisplatin-induced hypomagnesemia in rats. Kidney Int. 1985;28:914–21.
54. Gonzalez-Vitale JC, Hayes DM, Cvitkovic E, et al. The renal pathology in clinical trials of cis-platinum (II) diamminechloride. Cancer. 1977;39:1362–71.
55. Guinee DG, van Zee B, Houghton DC. Clinically silent progressive renal tubulointerstitial disease during cisplatin chemotherapy. Cancer. 1993;71:4050–4.
56. Hashizume N, Mori M. An analysis of hypermagnesemia and hypomagnesemia. Jpn J Med. 1990;29:368–72.
57. Ariceta G, Rodriguez-Soriano J, Vallo A, et al. Acute and chronic effects of cisplatin therapy on renal magnesium homeostasis. Med Pediatr Oncol. 1997;28:35–40.
58. Lajer H, Bundgaard H, Secher NH, et al. Severe intracellular magnesium and potassium depletion in patients after treatment with cisplatin. Br J Cancer. 2003;89:1633–7.
59. Abbasciano V, Mazzotta D, Vacchiatti G, et al. Changes in serum, erythrocyte, and urinary magnesium after a single dose of cisplatin combination chemotherapy. Magnes Res. 1991;4:123–5.
60. Fine KD, Ana CAS, Porter JL, et al. Intestinal absorption of magnesium from food and supplements. J Clin Invest. 1991;88:396–402.
61. Coudray C, Rambeau M, Feillet-Coudray C, et al. Study of magnesium bioavailability from ten organic and inorganic Mg salts in Mg-depleted rats using a stable isotope approach. Magnes Res. 2005;18:215–23.

62. Macaulay VM, Begent RH, Phillips ME, et al. Prophylaxis against hypomagnesemia induced by cis-platinum combination chemotherapy. Cancer Chemother Pharmacol. 1982;9:179–81.
63. Willox JC, McAllister EJ, Sangster G, et al. Effects of magnesium supplementation in testicular cancer patients receiving cisplatin: a randomised trial. Br J Cancer. 1986;54:19–23.
64. Netten PM, de Mulder PH, Theeuwes AG, et al. Intravenous magnesium supplementation during cisdiammine-dichlorplatinum administration prevents hypomagnesemia. Ann Oncol. 1990;1:369–72.
65. Vokes EE, Mick R, Vogelzang NJ, et al. A randomised study comparing intermittent to continuous administration of magnesium aspartate hydrochloride in cisplatin-induced hypomagnesemia. Br J Cancer. 1990;62:1015–7.
66. Mertin M, Diaz-Rubio E, Casado A, et al. Intravenous and oral magnesium supplementation in the prophylaxis of cisplatin hypomagnesemia. Results of a controlled trial. Am J Clin Oncol. 1992;15:348–51.
67. Evans TRJ, Harper CL, Beveridge IG, et al. A randomised study to determine whether routine intravenous magnesium supplements are necessary in patients receiving cisplatin chemotherapy with continuous infusion 5-fluorouracil. Eur J Cancer. 1995;31A:174–8.
68. Bodnar L, Wcislo G, Gasowska-Bodnar A, et al. Renal protection with magnesium subcarbonate and magnesium sulphate in patients with epithelial ovarian cancer after cisplatin and paclitaxel chemotherapy: a randomised phase II study. Eur J Cancer. 2008;44:2608–14.
69. Cunnigham D, Humblet Y, Siena S, et al. Cetuximab monotherapy and cetuximab plus irinotecan in irinotecan-refractory metastatic colorectal cancer. N Engl J Med. 2004;351:337–45.
70. Schrag D, Chung KY, Flombaum C, et al. Cetuximab therapy and symptomatic hypomagnesemia. J Natl Cancer Inst. 2005;97:1221–4.
71. Tejpar S, Piessevaux H, Claes K, et al. Magnesium wasting associated with epidermal-growth-factor receptor-targeting antibodies in colorectal cancer: a prospective study. Lancet Oncol. 2007;8:387–94.
72. Muallem S, Moe OW. When EGF is offside, magnesium is wasted. J Clin Invest. 2007;117:2086–9.
73. Groenestege WMT, Thebault S, van der Wijst J, et al. Impaired basolateral sorting of pro-EGF causes isolated recessive renal hypomagnesemia. J Clin Invest. 2007;117:2260–7.
74. Thebault S, Alexander RT, Groenestege WMT, et al. EGF increases TRPM6 activity and surface expression. J Am Soc Nephrol. 2009;20:78–85.
75. Woudenberg-Vrenken TE, Bindels RJM, Hoenderop JGJ. The role of transient receptor potential channels in kidney disease. Nat Rev Nephrol. 2009;5:441–9.
76. Vincenzi B, Santani D, Galluzzo S, et al. Early magnesium reduction in advanced colorectal cancer patients treated with cetuximab plus irinotecan as predictive factor of efficacy and outcome. Clin Cancer Res. 2008;14:4219–24.
77. Vincenzi B, Gallauzzo S, Santini D, et al. Early magnesium modifications as a surrogate marker of efficacy of cetuximab-based anticancer treatment in KRAS wild-type advanced colorectal cancer patients. Ann Oncol. 2011;22:1141–6.
78. Holzner B, Kemmler G, Greil R, et al. The impact of hemoglobin levels on fatigue and quality of life In cancer patients. Ann Oncol. 2002;13:965–73.
79. Szenajch J, Wcislo G, Jeong J-Y, et al. The role of erythropoietin and its receptor In growth, survival and therapeutic response of human tumor cells. From clinic to bench – a critical review. Biochim Biophys Acta. 2010;1806:82–95.
80. Wood PA, Hrushesky WJM. Cisplatin-associated anemia: an erythropoietin deficiency syndrome. J Clin Invest. 1995;95:1650–9.
81. Bois P. Tumour of the thymus in magnesium-deficient rats. Nature. 1964;204:1316.
82. Rubin H. The logic of the membrane, magnesium, mitosis (MMM) model for the regulation of Animals cell proliferation. Arch Biochem Biophys. 2007;458:16–23.
83. Wolf FI, Cittadini ARM, Maier JAM. Magnesium and tumors: ally or foe? Cancer Treat Rev. 2009;35:378–82.
84. Larsson SC, Bergkvist L, Wolk A. Magnesium intake in relation to risk of colorectal cancer in women. JAMA. 2005;293:86–9.
85. Folsom AR, Hong C-P. Magnesium intake and reduced risk of colon cancer in a prospective study of women. Am J Epidemiol. 2006;163:232–5.
86. Lin J, Cook NR, Lee I-M, et al. Total magnesium intake and colorectal cancer incidence in women. Cancer Epidemiol Biomarkers Prev. 2006;15:2006–9.
87. MacDonald DR. Neurotoxicity of chemotherapy agents. In: Perry MC, editor. The chemotherapy source book. Baltimore: Williams & Wilkins; 1992. p. 666–79.
88. Weglicki WB, Phillips TM. Pathobiology of magnesium deficiency: a cytokine/neurogenic inflammation hypothesis. Am J Physiol Regul Integr Comp Physiol. 1992;263:R734–7.
89. Cersosimo RJ. Cisplatin neurotoxicity. Cancer Treat Rev. 1989;16:195–211.
90. Thompson SW, Davis LE, Kornfeld M, et al. Cisplatin neuropathy. Clinical, electrophysiologic, morphologic, and toxicologic studies. Cancer. 1984;54:1269–75.

91. Gregg RW, Molepo JM, Monpetit VJ, et al. Cisplatin neurotoxicity: the relationship between dosage, time, and platinum concentration in neurologic tissues, and morphologic evidence of toxicity. J Clin Oncol. 1992;10:795–803.
92. Eeles R, Tait DM, Peckham MJ. Lhermitte's sign as a complication of cisplatin-containing chemotherapy for testicular cancer. Cancer Treat Rep. 1986;70:905–7.
93. Wather PJ, Rossith E, Bullard DE. The development of Lhermitte's sign during cisplatin chemotherapy. Cancer. 1987;60:2170–2.
94. Wilding G, Caruso R, Lawrance TS, et al. Retinal toxicity after high-dose cisplatin therapy. J Clin Oncol. 1985;3:1683–9.
95. Hartmann JT, Lipp HP. Toxicity of platinum compounds. Expert Opin Pharmacother. 2003;4:889–901.
96. Laurell G, Beskow C, Frankendal B, et al. Cisplatin administration to gynecologic cancer patients: long term effects on hearing. Cancer. 1996;78:1798–804.
97. Peters U, Preisler-Adams S, Hebeisen A, et al. Glutathione S-transferase genetic polymorphisms and individual sensitivity to the ototoxic effect of cisplatin. Anti-Cancer Drugs. 2000;11:639–43.
98. Ashraf M, Riggs JE, Wearden S, et al. Prospective study of nerve conduction parameters and serum magnesium following cisplatin therapy. Gynecol Oncol. 1990;37:29–33.
99. Albers JW, Chaudrhry V, Cavaletti G, et al. Interventions for preventing neuropathy caused by cisplatin and related compounds. Cochrane Database Syst Rev. 2011;16:CD005228.
100. Pace A, Giannarelli D, Galie E, et al. Vitamin E neuroprotection for cisplatin neuropathy: a randomized, placebo-controlled trial. Neurology. 2010;74:762–6.
101. McWhinney SR, Goldberg RM, McLeod HL, et al. Platinum neurotoxicity pharmacogenetics. Mol Cancer Ther. 2009;8:10–6.
102. Cvitkovic E. Cumulative toxicities from cisplatin therapy and current cytoprotective measures. Cancer Treat Rev. 1998;24:265–81.
103. Grothey A. Oxaliplatin-safety profile: neurotoxicity. Semin Oncol. 2003;30 suppl 15:5–13.
104. Wilson RH, Lehky T, Thomas RR, et al. Acute oxaliplatin-induced peripheral nerve hyperexcitability. J Clin Oncol. 2002;20:1767–74.
105. Adelsberg H, Quasthoff S, Grosskreutz J, et al. The chemotherapeutic oxaliplatin alters voltage-gated Na^+ channel kinetics on rat sensory neurons. Eur J Pharmacol. 2000;406:25–32.
106. Lehky TJ, Leonard GD, Wilson RH, et al. Oxaliplatin-induced neurotoxicity: acute hyperexcitability and chronic neuropathy. Muscle Nerve. 2004;29:387–92.
107. Cavaletti G, Tredici G, Petruccioli MG, et al. Effects of different schedules of oxaliplatin treatment on the peripheral nervous system of the rat. Eur J Cancer. 2001;37:2457–63.
108. Cassidy J, Misset JL. Oxaliplatin-related side effects: characteristics and management. Semin Oncol. 2002;29:11–20.
109. Saif MW, Reardon J. Management of oxaliplatin-induced peripheral neuropathy. Therap Clin Risk Manag. 2005;1:249–58.
110. Amptoulach S, Tsavaris N. Neurotoxicity caused by the treatment with platinum analogues. Chemother Res Prac. Volume 2011, Article ID 843019, 5 pages doi: 10.1155/2011/843019
111. Tournigand C, Cervantes A, Figer A, et al. OPTIMOX1: a randomized study of FOLFOX4 or FOLFOX7 with oxaliplatin in a stop-and-go fashion in advanced colorectal cancer – a GERCOR study. J Clin Oncol. 2006;24:394–400.
112. Ishibasi K, Okada N, Miyazaki T, et al. Effect of calcium and magnesium on neurotoxicity and blood platinum concentrations in patients receiving mFOLFOX6 therapy: a prospective randomized study. Int J Clin Oncol. 2010;15:82–7.
113. Gamelin L, Boisdron-Celle M, Delva R, et al. Prevention of oxaliplatin-related neurotoxicity by calcium and magnesium infusions: a retrospective study of 161 patients receiving oxaliplatin combined with 5-fluorouracil and leucovorin for advanced colorectal cancer. Clin Cancer Res. 2004;10:4055–61.
114. Grothey A, Nikcevich DA, Sloan JA, et al. Intravenous calcium and magnesium for oxaliplatin-induced sensory neurotoxicity in adjuvant colon cancer: NCCTG N04C7. J Clin Oncol. 2011;29:421–7.
115. Knijn N, Tol J, Koopman M, et al. The effect of prophylactic calcium and magnesium infusions on the incidence of neurotoxicity and clinical outcome of oxaliplatin-based systemic treatment in advanced colorectal cancer patients. Eur J Cancer. 2011;47:369–74.
116. Stevens LA, Coresh J, Greene T, et al. Assessing kidney function – measured and estimated glomerular filtration rate. N Engl J Med. 2006;354:2473–83.
117. Levy AS, Coresh J, Balk E, et al. National Kidney Foundation practice guidelines for chronic kidney disease: evaluation, classification, and stratification. Ann Intern Med. 2003;139:137–47.
118. Launay-Vacher V, Oudard S, Janus N, et al. Prevalence of renal insufficiency in cancer patients and implications for anticancer drug management. IRMA Study. Cancer. 2007;110:1376–84.

119. Superfin D, Iannucci AA, Davies AM. Commentary: oncologic drugs in patients with organ dysfunction: a summary. Oncologist. 2007;12:1070–83.
120. Kuhlmann MK, Burkhardt G, Jones RB, et al. Insights into potential cellular mechanisms of cisplatin nephrotoxicity and their clinical application. Nephrol Dial Transplant. 1997;12:2478–80.
121. Kroning R, Lichtenstein AK, Nagami GT. Sulfur-containing amino acids decrease cisplatin cytotoxicity and uptake in renal tubule epithelial cell lines. Cancer Chemother Pharmacol. 2000;45:43–9.
122. Ciarimboli G, Ludwig T, Lang D, et al. Cisplatin nephrotoxicity is critically mediated via the human organic cation transporter 2. Am J Pathol. 2005;167:1477–83.
123. Townsend DM, Deng M, Zhang L, et al. Metabolism of cisplatin to a nephrotoxin in proximal tubule cells. J Am Soc Nephrol. 2003;14:1–10.
124. Townsend DM, Hanigan MH. Inhibition of gamma-glutamyl transpeptidase or cysteine S-conjugate beta-lyase activity blocks the nephrotoxicity of cisplatin in mice. J Pharmacol Exp Ther. 2002;300:142–8.
125. Kawai Y, Nakao T, Kunimura N, et al. Relationship of intracellular calcium and oxygen radicals to cisplatin-related renal injury. J Pharmacol Sci. 2006;100:65–72.
126. Chirino YI, Hernandez-Pando R, Pedraza-Chaverri J. Peroxynitrite decomposition catalyst ameliorates renal damage and protein nitration in cisplatin-induced nephrotoxicity in rats. BMC Pharmacol. 2004;4:20.
127. Huang Q, Dunn IL, Jayadev S, et al. Assessment of cisplatin-induced nephrotoxicity by microarray technology. Toxicol Sci. 2001;63:196–207.
128. Arany I, Megyesi JK, Kaneto H, et al. Cisplatin-induced cell death is EGFR/src/ERK signaling dependent in mouse proximal tubule cells. Am J Physiol Renal Physiol. 2004;287:F543–9.
129. Ramesh G, Reeves WB. TNF-alpha mediates chemokine and cytokine expression and renal injury in cisplatin nephrotoxicity. J Clin Invest. 2002;110:835–42.
130. Schilsky RL, Anderson T. Hypomagnesemia and renal magnesium wasting in patients receiving cisplatin. Ann Intern Med. 1979;90:929–31.
131. Skeel RT. Antineoplastic drugs and biologic response modifiers: classification, use, and toxicity of clinically useful agents. In: Skeel RT, editor. Handbook of cancer chemotherapy. 5th ed. Philadelphia: Lippincott Williams & Wilkins; 1999. p. 89–91.
132. Al.-Sarraf M, Fletcher W, Oishi N, et al. Cisplatin hydration with and without mannitol diuresis in refractory disseminated malignant melanoma: A Southwest Oncology Group Study. Cancer Treat Rep. 1982;66:31–5.
133. Pinzani V, Bressolle F, Haug IJ, et al. Cisplatin-induced renal toxicity and toxicity-modulating strategies: a review. Cancer Chemother Pharmacol. 1994;35:1–9.

Section D
Cardiovascular Disease and Magnesium

Chapter 12
Magnesium and Hypertension

Mark Houston

Key Points

- Magnesium intake of 500–1,000 mg/day will reduce blood pressure by an average of 4/2.5 mmHg.
- The combination of an increased intake of magnesium, potassium, and taurine coupled with reduced sodium intake is more effective in reducing blood pressure than single mineral or nutrient intake. Reducing intracellular sodium and calcium while increasing intracellular magnesium and potassium improves blood pressure response.
- Magnesium will increase the effectiveness of all antihypertensive drug classes.
- Magnesium is most effective in lowering blood pressure in high-renin hypertensive patients with an increased plasma renin activity (PRA).
- Magnesium reduces cardiovascular disease such as coronary heart disease and ischemic stroke, reduces cardiac arrhythmias, and improves insulin resistance, hyperglycemia, diabetes mellitus, left ventricular hypertrophy, and dyslipidemia.
- Genetic defects in magnesium transport are associated with hypertension.
- Magnesium is a natural calcium-channel blocker, increases nitric oxide, improves endothelial dysfunction, induces vasodilation, and reduces blood pressure.

Keywords Hypertension • Magnesium • Cardiovascular disease • Coronary heart disease • Stroke

Introduction

Hypertension remains the leading cause of cardiovascular disease (CVD) affecting approximately one billion individuals worldwide [1]. More than 72 million Americans, or nearly one in three adults, are estimated to have high blood pressure (BP), but only 35% reach goal BP control [2]. Nearly 70 million more adults are at risk of developing prehypertension, and 90% of adults will probably develop

M. Houston, M.D., M.S., FACP, FAHA, FASH (✉)
Department of Medicine, Vanderbilt University School of Medicine, 4230 Harding Road, Suite 400, Nashville 37205, TN, USA

Department of Medicine, Hypertension Institute, Saint Thomas Hospital, 4230 Harding Road, Suite 400, Nashville 37205, TN, USA
e-mail: boohouston@comcast.net

R.R. Watson et al. (eds.), *Magnesium in Human Health and Disease*, Nutrition and Health,
DOI 10.1007/978-1-62703-044-1_12, © Springer Science+Business Media New York 2013

high BP by age 65 [3]. Hypertension is associated with an increased risk of morbidity and mortality from cerebrovascular accidents (CVA), coronary heart disease (CHD), congestive heart failure (CHF), and end-stage renal disease (ESRD). Over the past several years, there has been little change in BP diagnostic thresholds, treatment targets, or in treatment approaches. Due to its high prevalence, hypertension remains the most common reason for visits to physician's offices and the primary reason for prescription drug use.

Diet in the Prevention and Treatment of Hypertension

Several epidemiological studies suggest that diet plays an important role in determining BP [4–6]. Dietary therapies known to lower BP include a reduced sodium intake, increased potassium and magnesium intake, and a diet rich in fruits and vegetables [4–6]. The landmark Dietary Approaches to Stop Hypertension (DASH) trial demonstrated that modification of diet significantly lowered BP in patients with stage 1 hypertension and high-normal BP [7, 8] The DASH diet, which emphasizes fruits, vegetables, high fiber, and low-fat dairy products, also lowers BP in persons with isolated systolic hypertension(ISH) [9]. These BP-lowering effects were seen in 8 weeks and were sustained throughout the study period. The Seventh Report of the Joint National Committee on Prevention, Detection, Evaluation, and Treatment of High Blood Pressure (JNC 7) guidelines as well as the recent American Heart Association (AHA) recommendations for the prevention and management of CVD and numerous other International organizations also recognize the role that various foods, nutrients, and minerals play in lowering BP [10–14].

Effect of Magnesium on BP

Epidemiologic, observational, and clinical trial data show that a diet high in magnesium (at least 500–1,000 mg/day) lowers BP, but the results are inconsistent [15–17]. These varied results may relate to the population studied; duration of the trial; use of concomitant drugs or other nutrients and minerals; type and dose of magnesium administered; inadequate monitoring for adherence using measures of serum magnesium, intracellular magnesium, or 24-h urinary magnesium excretion; as well as evaluation of baseline plasma renin activity, essential fatty acid status, and genetic magnesium transporter status. In most epidemiologic studies, an inverse relationship has been shown between dietary magnesium intake and BP [15–17].

In a study of 60 patients with essential hypertension given magnesium oxide at 20 mmol/day over 8 weeks, significant reductions in ambulatory, home, and office BP were observed [18]. The office BP fell by 3.7/1.7 mmHg, the 24-h ABM was reduced by 2.5/1.4 mmHg, and the home BP was decreased by 2.0/1.4 mmHg. The levels of serum and urinary magnesium correlated with the BP reduction. Those with the highest BP levels at entry had the largest reduction in BP

Witteman et al. [19] demonstrated significant decreases in BP in a double blind placebo-controlled trial of 91 middle-aged to elderly women with mild to moderate hypertension using magnesium aspartate-HCl (20 mmol/day or 485 mg of magnesium) for 6 months. There was a significant decrease in SBP and DBP by 2.7 mmHg ($p<0.18$) and 3.4 mmHg ($P<0.003$), respectively. In addition, BP response was not associated with baseline magnesium level, and those on magnesium had a 50% increase in urinary magnesium excretion. In another study of 48 patients with mild uncomplicated hypertension, those given 600 mg/day of magnesium with lifestyle changes vs. those with lifestyle changes only had significant reductions in 24-h systolic and diastolic BP during daytime and nighttime readings of 5.6/2.8 mmHg versus 1.3/1.8 mmHg ($p<0.001$ and $p<0.002$ respectively), increased serum and intracellular magnesium, increased urinary magnesium excretion, and decreased intracellular calcium and sodium levels [20]. Magnesium supplementation (with calcium and potassium) was administered to 96 patients with hypertension over 6 months by Sacks et al., but no significant reduction in BP was noted [21].

Meta-analysis of magnesium supplementation has also revealed conflicting results. A review of 29 studies of magnesium was inconclusive due to flaws in methodology but suggested that a negative association of BP with magnesium was not present [22]. In contrast, a meta-analysis of 20 randomized clinical trials (RCT's) with a median intake of 15.4 mmol/day of magnesium revealed a dose-dependent BP reduction with magnesium supplementation [15]. In a more recent meta-analysis of 105 trials randomizing 6,805 participants with at least 8 weeks of follow-up, no evidence was found to suggest that magnesium supplements had any important effect on BP [23].

The BP response to magnesium may be dependent in part on the baseline plasma renin activity (PRA) [24]. High-renin hypertension accounts for about 70% of all cases of genetic hypertension [24]. Seventeen inpatients with untreated uncomplicated mild to moderate hypertension and 8 age-matched controls were given 1.0 g/day of magnesium oxide for 2 weeks. The average mean 24-h BP for both daytime and nighttime readings fell from 104.3 to 99.5 mmHg ($p<0.05$) while there was no change in the BP in the control group. The PRA was significantly higher in the responder group than the nonresponder group in those who received the magnesium supplement ($p<0.05$). Magnesium suppresses circulating Na + K + ATPase inhibitory activity to attenuate vascular tone and lower BP. Other studies have shown that oral magnesium improves borderline hypertension [25].

Magnesium may have a more pronounced BP-lowering effect when administered with high potassium intake and low sodium intake [26, 27]. In a double-blind, randomized, placebo-controlled, crossover trial of 32 weeks duration, 37 adults with mild hypertension (DBP< 110 mmHg) were given placebo or potassium 60 mmol/day alone or in combination with magnesium 20 mmol/day in a crossover design. None of the patients were on any other medications or supplements. The potassium/magnesium combination significantly reduced BP ($p<0.001$), but the addition of magnesium to the potassium did not decrease BP further. Other studies suggest that high-potassium, high-magnesium, and low-sodium intake will result in additive reductions in BP [27].

Magnesium given in conjunction with taurine lowers BP, improves insulin resistance, retards atherogenesis, prevents arrhythmias, and stabilizes platelets [28, 29]. The actions may be related to the common mechanism of action of magnesium and taurine to reduce intracellular calcium and sodium levels [28, 29]. In the WHO-CARDIAC study population, those patients with higher 24-h urine magnesium/creatinine and taurine/creatinine levels had significantly lower cardiovascular risks, including CVA, CHD, and MI [29].

Magnesium is also effective in further reducing BP in stage I hypertension, diabetes mellitus, and pregnancy, when coadministered with antihypertensive agents such as angiotensin-converting enzyme inhibitors (ACEI), angiotensin receptor blockers (ARBs), calcium-channel blockers (CCBs) diuretics, beta-blockers (BB), methyldopa, and other pharmacological agents [30–33]. A comprehensive analytical review of 44 human studies of oral magnesium for hypertension showed that Mg supplements enhance BP-lowering effect of antihypertensive medications [31].

Magnesium intake is correlated with reductions in CVA, CVD, arrhythmias, insulin resistance, diabetes mellitus, and left ventricular hypertrophy in most [34–38] but not all studies [39]. In a study of 34,670 women, cerebral infarction was inversely correlated with both magnesium and potassium intake [34]. Left ventricular hypertrophy and LV mass are lower in those with higher magnesium intakes [35]. In the Atherosclerosis Risk in Communities Study (ARIC), higher serum magnesium and magnesium intake were associated with lower prevalence of hypertension, diabetes, and ischemic CVA over 15 years in both men and women [36]. Magnesium intake is also correlated with reductions in serum lipids, hyperglycemia, metabolic syndrome, obesity, insulin resistance, and diabetes mellitus [37, 38].

Mechanisms of Blood Pressure Reduction with Magnesium

The mechanism by which magnesium lowers BP is by acting like a natural calcium-channel blocker (CCB). Specifically, magnesium competes with sodium for binding sites on vascular smooth muscle cells, increases prostaglandin E, binds to potassium in a cooperative manner, induces endothelial-dependent vasodilation, improves endothelial dysfunction in hypertensive and diabetic patients, decreases intracellular calcium and sodium, and reduces BP [28, 29, 40]. Magnesium is more effective in reducing BP when administered as multiple minerals in a natural form as a combination of magnesium, potassium, and calcium than when given alone [41].

Magnesium is also an essential cofactor for the delta-6-desaturase enzyme, which is the rate-limiting step for the conversion of linoleic acid (LA) to gamma-linolenic acid (GLA) [42–44]. GLA in turn elongates to form DGLA (dihomo-gamma-linoleic acid), the precursor for prostaglandin E_1 (PGE_1), which is both a vasodilator and a platelet inhibitor [42–44]. Low magnesium states lead to insufficient amounts of PGE_1, causing vasoconstriction and increased BP [42–44].

In addition to BP, magnesium regulates intracellular calcium, sodium, potassium, and pH as well as left ventricular mass, insulin sensitivity, and arterial compliance [17, 20]. Magnesium also suppresses circulating Na+K+ATPase inhibitory activity that reduces vascular tone [24].

The CCB mimetic effect of magnesium results in production of vasodilator prostacyclins and nitric oxide and alters the vascular responses to vasoactive agonists [44]. These varied biochemical reactions control vascular contraction and dilation, growth and apoptosis, differentiation, and inflammation [44]. Alterations in magnesium transport systems may predispose patients to hypertension and subsequent cardiovascular disease [45–48]. Magnesium efflux and influx transport systems have been well characterized in humans. Magnesium efflux occurs via Na^{++}-dependent and Na^{++}-independent pathways. Magnesium influx is controlled by Mrs2p, SLC41A1, ACDP2, Mag T1, TRPM6, and TRPM7 (melastatin) [45–48]. In particular, increased Mg^{++} efflux through altered regulation of the vascular Na^+/Mg^{++} exchanger and decreased Mg influx due to defective vascular and renal TRPM6/7 expression/activity may be important [45–48]. TRPM6 is found primarily in epithelial cells. TRPM7 is ubiquitously expressed and is implicated as a signaling kinase involved in vascular smooth muscle cell growth, apoptosis, adhesion, contraction, cytoskeletal organization, and migration and is modulated by vasoactive agents, pressure, stretch, and osmotic changes [48]. TRPM7 is altered in hypertension [48].

Research involving new imaging techniques to measure intracellular magnesium, such as P-nuclear magnetic resonance (NMR) and magnesium-specific ion-selective electrodes (ISE), which measure intracellular and extracellular free concentrations of magnesium, and fluorescent probes will further enhance our understanding of the role of magnesium in hypertension [17, 49]. Intracellular magnesium is a more accurate reflection of total body magnesium stores.

The "Ionic Hypothesis" of Resnick and the Role of Magnesium and Other Ions

The "ionic hypothesis" of hypertension and other metabolic disorders by Resnick [49] is characterized by the following:

1. Increased intracellular free calcium and reduced intracellular free magnesium determine the amount of vasoconstriction or vasodilation.
2. An elevated glucose and low-density lipoprotein cholesterol (LDL-C) increase the intracellular calcium and/or lower intracellular magnesium in vascular smooth muscle cells.

3. Hypertension, insulin resistance, and type II diabetes mellitus are associated with an increased intracellular calcium and decreased intracellular magnesium, which all respond to weight loss.
4. Weight loss also decreases intracellular calcium levels.
5. Dietary calcium-suppressible hormones like PTH, 1,25 vitamin D are vasoactive and promote calcium uptake in vascular smooth muscle cells and cardiac muscle.
6. The higher the PTH concentration, the greater the fall in BP, and the greater the reduction in PTH and 1,25 vitamin D, the greater the BP reduction.
7. Individuals with salt-sensitive and calcium-sensitive hypertension have elevated intracellular calcium PTH and 1,25 vitamin D but low intracellular magnesium.
8. Dietary calcium reverses abnormal calcium indices and lowers BP.
9. Dietary potassium reduces urinary calcium excretion and 1,25 vitamin D plasma levels.
10. Magnesium intake reduces tissue calcium accumulation.

Increased intake of magnesium, potassium, and calcium with concomitant reductions in sodium intake will influence the intracellular concentrations of all of the ions and lower BP more effectively than alternating the intake of any single ion. In addition, these intracellular changes modify serum glucose and insulin sensitivity.

Conclusions, Summary, and Recommendations

The overall effect of diet on BP is determined by the net contribution of various nutrients on cytosolic-free minerals such as potassium, calcium, magnesium, and sodium. Measurements of intracellular magnesium are more indicative of body stores and should be used in conjunction with serum magnesium to accurately determine magnesium deficiencies. Consumption of 500–1,000 mg of magnesium will lower BP about 2.7–5.6 mmHg systolic BP(SBP) and 1.7–3.4 mmHg diastolic BP (DBP) as measured by causal office BP readings or by 24-h ambulatory blood pressure monitoring (ABM). Magnesium lowers intracellular sodium and calcium, which enhances BP reduction. Magnesium is natural CCB, blocks sodium attachment to vascular smooth muscle cells, increases vasodilating PGE, binds potassium in a cooperative manner, increases nitric oxide, improves endothelial dysfunction, causes vasodilation, and reduces BP. Patients with high PRA (high-renin hypertension) have more pronounced BP reductions than those with low PRA. Genetic defects in magnesium transport may be causative in hypertension. Combinations of magnesium and potassium with low sodium intakes are more effective in reducing BP than using single minerals. It is recommended that 1,000 mg of magnesium be combined with 4.7 g of potassium and less than 1.5 g of sodium per day through both diet and supplements to maximize BP reduction [27]. Combining magnesium with taurine has additive antihypertensive effects, lowers intracellular sodium and calcium, and reduces CVD. It is recommended that about 1,000–2,000 mg of taurine be added to this regimen. Magnesium has additive antihypertensive effects with all antihypertensive drugs and thus should be routinely administered unless patients have specific contraindications such as renal insufficiency. Chelated forms of magnesium combined to an amino acid are better absorbed and less likely to induce diarrhea with higher magnesium intakes. Magnesium lowers CVD, CHD, and CVA and reduces the incidence of cardiac arrhythmias, insulin resistance, hyperglycemia, diabetes mellitus, LVH, and serum lipids.

Americans consume three to four times the sodium and about one-third of the magnesium and potassium that is recommended by current guidelines. A high intake of potassium, magnesium, and possibly calcium through increased consumption of fruits and vegetables, the DASH diet, and supplements is important for the prevention of hypertension and major public health problems such as CVD, coronary heart disease, and stroke.

References

1. Israili ZH, Hernandez-Hernandez R, Valasco M. The future of antihypertensive treatment. Am J Ther. 2007;14:121–34.
2. Rosamund W. Heart disease and stroke-2008 update. A report from the American Heart Association Statistics Committee and Stroke Statistics Subcommittee. Circulation. 2008;17(4):e25.
3. Svetkey LP, Simons-Morton DG, Proschan MA. Effect of the dietary approaches to stop hypertension diet and reduced sodium intake on blood pressure control. J Clin Hypertens. 2004;6:373–81.
4. Young DB, Lin H, McCabe RD. Potassium's cardiovascular protective mechanisms. Am J Physiol. 1995;268(4 part 2):R825–37.
5. INTERSALT Cooperative Research Group. INTERSALT: an international study of electrolyte excretion and blood pressure. Results for 24 hr urinary sodium and potassium excretion. BMJ. 1988;297:319–28.
6. Efford J, Philips A, Thomsoon AG. Migration and geographic variations in blood pressure in Britain. BMJ. 1990;300:291–5.
7. Appel LJ, Moore TH, Obarzanek E. A clinical trial of the effects of dietary patterns on blood pressure. Research Group. N Engl J Med. 1997;336:1117–24.
8. Sacks FM, Svetkey LP, Vollmer WM, DASH-Sodium Collaborative Research Group. Effects on blood pressure or reduced dietary sodium and the Dietary Approaches to Stop Hypertension (DASH) diet. N Engl J Med. 2001;344:3–10.
9. Moore TJ, Conlin PR, Ard J. DASH (Dietary Approaches to Stop Hypertension) diet is effective treatment for stage 1 isolated systolic hypertension. Hypertension. 2001;38:155–8.
10. Chobanian AV, et al. The seventh report of the joint National Committee on prevention, detection, evaluation and treatment of high blood pressure: the JNC 7 report. JAMA. 2003;289(19):2560–72.
11. Appel LJ, Brands ME, Daniels SR. Dietary approaches to prevent and treat hypertension: a scientific statement from the American Heart Association. Hypertension. 2006;47:296–308.
12. Guidelines Committee. 2003 European Society of Hypertension-European Society of Cardiology guidelines for the management of arterial hypertension. J Hypertens. 2003;21:1011–63.
13. Whitworth JA, World Health Organization. International Society of Hypertension Writing Group. 2003 World Health Organization (WHO) International Society of Hypertension (ISH) statement on management of hypertension. J Hypertens. 2003;21(11):1983–92.
14. Williams B, Poulter NR, Brown MJ. British Hypertension Society guidelines for hypertension management. 2004(BHS -IV): summary. BMJ. 2004;328:634–40.
15. Jee SH, Miller ER, Guallar E. The effect of magnesium supplementation on blood pressure: a meta-analysis of randomized clinical trials. Am J Hypertens. 2002;15:691–6.
16. Touyz RM. Role of magnesium in the pathogenesis of hypertension. Mol Aspects Med. 2003;24:107–36.
17. Resnick LM. Magnesium in the pathophysiology and treatment of hypertension and diabetes mellitus. Where are in 1997? Am J Hypertens. 1997;10:368–70.
18. Kawano Y, Matsuoka H, Takishita S, Omae T. Effects of magnesium supplementation in hypertensive patients: assessment by office, home, and ambulatory blood pressures. Hypertension. 1998;32:260–5.
19. Witteman JC, Grobbee DE, Derkx FH, Bouillon R, de Bruijn AM, Hofman A. Reduction of blood pressure with oral magnesium supplementation in women with mild to moderate hypertension. Am J Clin Nutr. 1994;60(1):129–35.
20. Hatzistavri LS, Sarafidis PA, Georgianos PI, Tziolas IM, Aroditis CP, Zebekakis PE, et al. Oral magnesium supplementation reduces ambulatory blood pressure in patients with mild hypertension. Am J Hypertens. 2001;22(10):1070–5.
21. Sacks FM, Brown LE, Appel L. Combination of potassium, calcium and magnesium supplements in hypertension. Hypertension. 1995;26(6 part 1):950–6.
22. Mizushima S, Cuppauccio FP, Nichols R. Dietary magnesium intake and blood pressure: a qualitative overview of the observational studies. J Hum Hypertens. 1998;12:447–53.
23. Dickinson HO, Nicolson DJ, Campbell F. Potassium supplementation for the management of primary hypertension in adults. Cochrane Database Syst Rev. 2006;3:CD004641.
24. Haga H. Effects of dietary magnesium supplementation on diurnal variations of blood pressure and plasma Na+, K+– ATPase activity in essential hypertension. Jpn Heart J. 1992;33(6):785–800.
25. Kisters K. Oral magnesium supplementation improves borderline hypertension. Magnes Res. 2011;24:17–8.
26. Patki PS, SIngh J, Gokhale SV, Bulakh PM, Shrotri DS, Patwardhan B. Efficacy of potassium and magnesium in essential hypertension: a double-blind placebo controlled, crossover study. BMJ. 1990;301:521–3.
27. Houston MC. The importance of potassium in managing hypertension. Curr Hypertens Rep. 2011;13(4):309–17.
28. McCarty MF. Complementary vascular-protective actions of magnesium and taurine: a rationale for magnesium taurate. Med Hypothesis. 1996;46(2):89–100.

29. Yamori Y, Taquchi T, Mori H, Mori M. Low cardiovascular risks in the middle aged males and females excreting greater 24-hour urinary taurine and magnesium in 41 WHO-CARDIAC study populations in the world. J Biomed Sci. 2010;17 Suppl 1:S 21.
30. Guerrero-Romero F, Rodriquez-Moran M. The effect of lowering blood pressure by magnesium supplementation in diabetic hypertensive adults with low serum magnesium levels: a randomized, double-blind placebo-controlled clinical trial. J Hum Hypertens. 2009;23(4):245–51.
31. Rosanoff A. Magnesium supplements may enhance the effect of antihypertensive medication in stage 1 hypertensive subjects. Magnes Res. 2010;23(1):27–40.
32. Wirell MP, Wester PO, Steqmayr BG. Nutritional dose of magnesium in hypertensive patients on beta blockers lowers systolic blood pressure: a double-blind, cross-over study. J Intern Med. 1994;236(2):189–95.
33. Rudnicki M, Frolich A, Pilagaard K, Nymbereg L, Moller M, Sanchez M, et al. Comparison of magnesium and methyldopa for the control of blood pressure in pregnancies complicated with hypertension. Gynecol Obstet Invest. 2000;49(2):231–5.
34. Larsson SC, Virtamo J, Wolk A. Potassium, calcium and magnesium intakes and risk of stroke in women. Am J Epidemiol. 2011;174(1):35–43. Epub 2011 May 3.
35. Raffelmann T, Dorr M, Ittermann TM, Schwahn C, Volzke H, Ruppert J, et al. Low serum magnesium concentrations predicts increase in left ventricular mass over 5 years independently of common cardiovascular risk factors. Atherosclerosis. 2010;213:563–9.
36. Ohira T, Peacock JM, Iso H, Chambless LE, Rosamond WD, Folsom AR. Serum and dietary magnesium and risk of ischemic stroke: the Atherosclerosis risk in communities study. Am J Epidemiol. 2009;169(12):1437–44.
37. Champagne CM. Magnesium in hypertension, cardiovascular disease, metabolic syndrome and other conditions: a review. Nutr Clin Pract. 2008;23:142–51.
38. Hadjistavri LS, Sarafidis PA, Georgianos PI, Tziolas IM, Aroditis CP, Hitoglou-Makedou A, et al. Beneficial effects of oral magnesium supplementation on insulin sensitivity and serum lipid profile. Med Sci Monit. 2010;16:CR 307–12.
39. Khan AM, Sullivan L, Mccabe E, Levy D, Vasan RS, Want TJ. Lack of association between serum magnesium and the risks of hypertension and cardiovascular disease. Am Heart J. 2010;160:715–20.
40. Barbagallo M, Dominguez LJ, Galioto A, Pineo A, Belvedere M. Oral magnesium supplementation improves vascular function in elderly diabetic patients. Magnes Res. 2010;23:131–7.
41. Preuss HG. Diet, genetics and hypertension. J Am Coll Nutr. 1997;16:296–305.
42. Das UN. Essential fatty acids: biochemistry, physiology and pathology. Biotechnol J. 2006;1(4):420–39.
43. Un D. Nutrients, essential fatty acids and prostaglandins interact to augment immune responses and prevent genetic damage and cancer. Nutrition. 1989;5:106–10.
44. Das UN. Delta 6 desaturase as the target of the beneficial actions of magnesium. Med Sci Monit. 2010;16(8):LE11–2.
45. Sonita B, Touyz RM. Magnesium transport in hypertension. Pathophysiology. 2007;14:205–11.
46. Kisters K, Gremmler B, Hausberg M. Disturbed Mg++ transporters in hypertension. J Hypertens. 2008;26:2450–1.
47. Yogi A, Callera GE, Antuens TT, Tostes RC, Touyz RM. Vascular biology of magnesium and its transporters in hypertension. Magnes Res. 2010;23(4):207–15.
48. Yoga A, Callera GE, Antunes TT, Tostes RC, Touyz RM. Transient receptor potential melastatin 7 (TRPM7) cation channels, magnesium and the vascular system in hypertension. Circ J. 2011;75:237–45.
49. Trapani V, Farruggia G, Marraccini C, Iotti S, Cittadine A, Wolf FL. Intracellular magnesium detection: imaging a brighter future. Analyst. 2010;135:1855–66.

Chapter 13
The Role of Magnesium in the Cardiovascular System

Michael Shechter and Alon Shechter

Key Points

- The rationale for magnesium in heart disease
- Beneficial effects of magnesium on vascular tone
- Anticoagulant and antiplatelet effects of magnesium
- Protective effects of magnesium on vascular endothelial function
- Protective effects of magnesium on myocardial infarct size
- Beneficial effects of magnesium on lipid metabolism
- Beneficial effects of magnesium on cardiac arrhythmias
- The rationale of magnesium supplementation in AMI
- Beneficial effects of magnesium in CHF
- Adverse effects of magnesium
- Reasons for magnesium deficiency

Keywords Magnesium • Diabetes • Nutrition • Endothelium • Myocardial infarction • Heart disease • Hypertension • Platelets • Arrhythmias

Introduction

Prior epidemiological trials from several countries, such as the USA, South Africa, Finland, France, England, Canada, Germany, and the Netherlands [1–17], have demonstrated that water magnesium content is associated with incidence and mortality from coronary artery disease (CAD). Autopsies have demonstrated high concentrations of cardiac muscle magnesium (also called "hard water areas") compared to low magnesium water areas (also called "soft water areas") and vice versa [1, 12, 15–17].

The authors state that no conflict of interest exists regarding the possible publication of this article.

M. Shechter, M.D., M.A. (✉)
Department of Medicine, Tel Aviv University, Tel Hashomer 52621, Israel

Clinical Research Unit, Leviev Heart Center, Chaim Sheba Medical Center, Tel Hashomer 52621, Israel
e-mail: Michael.Shechter@sheba.health.gov.il

A. Shechter
Department of Medicine, Tel Aviv University, Tel Hashomer 52621, Israel
e-mail: alonshechter@gmail.com

R.R. Watson et al. (eds.), *Magnesium in Human Health and Disease*, Nutrition and Health,
DOI 10.1007/978-1-62703-044-1_13, © Springer Science+Business Media New York 2013

After a 4- to 7-year follow-up of the Atherosclerosis Risk in Communities (ARIC) study [18] with 13,922 healthy subjects without CAD on admission, the highest risk for CAD occurred in subjects with the lowest serum magnesium and even after controlling for traditional CAD risk factors. The National Health and Nutrition Examination Survey (NHANES) epidemiologic follow-up study [19] demonstrated an inverse association of serum magnesium and mortality from CAD and all-cause mortality.

The Honolulu Heart Program [20] studied 7,172 men (ages ranging between 45 and 68 years) during the years 1965–1968. After a 30-year follow-up, the consumption of food with low magnesium content was found to increase the incidence of CAD by 2.1 compared to high concentrations of magnesium, even after controlling for traditional CAD risk factors and other food nutrients.

Amighi et al. [21] followed 323 patients with peripheral artery disease and intermittent claudication for 2 years and found that low serum magnesium concentrations were associated with a threefold incidence of cerebrovascular accident compared to those with high serum magnesium levels.

He et al. [22] found that 608 out of 4,637 (11%) young nondiabetic Americans aged 18–30 years without metabolic syndrome had developed the risk factors for cardiovascular disease after a 16-year follow-up. Multivariate analysis demonstrated a significant inverse association between food magnesium content and the incidence of metabolic syndrome.

Prospective epidemiologic studies have reported various associations between magnesium and risk of cardiovascular disease [18–20]. In general, there was a stronger correlation for plasma rather than dietary magnesium. Furthermore, the association between plasma magnesium and CAD risk appears to be stronger for fatal rather than nonfatal events [19], which could be explained if magnesium were protective against fatal ventricular arrhythmias and thus sudden cardiac death (SCD). This hypothesis is further supported by ecologic studies, which have reported inverse associations between regional drinking water hardness and SCD [23] and autopsy studies, where lower myocardial magnesium concentrations were found in victims of SCD compared with death from trauma [24, 25]. In a prospective cohort of 88,375 healthy women (the Nurses' Health Study), Chiuve et al. [26] recently demonstrated that higher plasma concentrations and dietary magnesium intake were associated with a lower risk of SCD over 26 years of follow-up.

Interestingly, while the magnesium content in food products in the USA has fallen over the last two decades, currently standing below the recommended daily allowance, the incidence of CAD is rising.

The Rationale for Magnesium in CAD

It is highly conceivable that the effect of magnesium in cardiovascular disease prevention may be partly related to a decrease in inflammatory response. In animal models, experimental magnesium deficiency induces a clinical inflammatory syndrome characterized by leukocyte and macrophage activation, release of inflammatory cytokines and acute phase proteins, in addition to excessive production of free radicals [27–29]. An increase in extracellular magnesium decreases inflammatory response, while reduction in extracellular magnesium results in phagocyte and endothelial cell activation. Inflammation occurring in experimental magnesium deficiency is the mechanism that induces hypertriglyceridemia and proatherogenic changes in lipoprotein profile. Endothelial cell actively contributes to inflammation in magnesium deficiency states. Magnesium intake has been shown to be inversely associated with markers of systemic inflammation and endothelial dysfunction in healthy [30] and postmenopausal women [31].

The available data suggest that a combination of mechanisms may act additively or even synergistically to protect myocytes and constitute the rationale of magnesium supplementation in patients with heart disease [1, 3, 32–34] (Table 13.1). Exogenic administration of magnesium prevents intracellular depletion of magnesium, potassium, and high-energy phosphates; improves myocardial metabolism; prevents intramitochondrial calcium accumulation; and reduces vulnerability to oxygen-derived free

Table 13.1 The protective effects of magnesium in heart disease

Antiplatelet/anticoagulant
Coronary and systemic vasodilation
Enhanced angiogenesis
Improved lipid profile
Improved exercise duration time and cardiac performance
Improved quality of life
Improved vascular endothelial function
Calcium influx inhibition
Inhibition of vulnerability to oxygen free radicals
Inhibition of reperfusion injury
Catecholamine inhibition
Mild reduction of blood pressure
Reduced cardiac arrhythmias
Reduced systemic vascular resistance

radicals. Magnesium can impact on vascular tone, platelet aggregation and the coagulation system, endothelial function, infarct (scar) size, lipid metabolism, cardiac arrhythmias, myocardial infarction, and heart failure.

Beneficial Effects of Magnesium on Vascular Tone

Magnesium is considered to be nature's physiologic calcium blocker [35]. It reduces the release of calcium from and into the sarcoplasmic reticulum and protects cells against calcium overload under ischemic conditions [35–48]. Furthermore, it reduces systemic and pulmonary vascular resistance, with a concomitant decrease in blood pressure and a slight increase in cardiac index [35–37]. Elevation of extracellular magnesium levels reduces arteriolar tone and tension in a wide variety of arteries [38–40] and potentiates the dilatory action of some endogenous (adenosine, potassium, and some prostaglandins) and exogenous (isoproterenol and nitroprusside) vasodilators [38, 39, 41, 42]. As a result, magnesium has a mild reducible effect on systolic and diastolic blood pressure [49] and may act as afterload reduction and thus unload the ischemic ventricle. Kugiyama et al. [48] demonstrated that exercise-induced angina is suppressed by intravenous magnesium in patients with variant angina, most probably as a result of improved regional myocardial blood flow by suppression of coronary artery spasm. Altura and Altura [43, 44] found in an experimental vascular smooth muscle model that magnesium deficiency, through potentiation of increased cellular calcium activity, may be responsible for the arterial hypertension that accompanies toxemia of pregnancy. The proven effectiveness of parenteral magnesium therapy in toxemia of pregnancy [39, 50] is most likely the result of its calcium antagonist action.

Shechter et al. [51] found that intralymphocytic magnesium levels in stable CAD patients after myocardial infarction and/or coronary artery bypass grafting were highly correlated to exercise duration time and cardiac performance and inversely correlated to the peak exercise double product (heart rate x systolic blood pressure). Thereafter, Shechter et al. [52] demonstrated that a 6-month oral magnesium supplementation significantly improved exercise tolerance, exercise duration time, ischemic threshold, and quality of life in stable CAD patients in Austria, Israel, and the USA.

Pokan et al. [53] reinforced the findings of Shechter et al. by demonstrating that a 6-month oral magnesium supplementation significantly improved intracellular magnesium levels, VO_{2max}, and left ventricular ejection fraction while reducing exercise-induced heart rate.

Anticoagulant and Antiplatelet Effects of Magnesium

In 1943, Greville and Lehmann [54] found that a small amount of magnesium added to fresh unclotted human plasma prolonged the clotting time. In Germany, during and shortly after the World War II, magnesium sulfate was widely used as a muscle relaxant, and it was seen that the blood of patients examined post-mortem after such treatment was unclotted [55]. In 1959, Anstall et al. [56] demonstrated that magnesium inhibits human blood coagulation.

Adams and Mitchel [57] found that magnesium, both topically and parenterally, suppressed thrombus formation and increased ADP concentration needed to initiate thrombus production at minor human injury sites. Some experimental studies have demonstrated the antiplatelet effects of magnesium, which may prevent the propagation of coronary artery thrombi or reocclusion of the infarct-related coronary artery after spontaneous or fibrinolysis-induced recanalization [57–60]. Recently, some studies have demonstrated that magnesium reduces platelet aggregation in healthy volunteers [58]. High magnesium levels inhibit blood coagulation [56] and thrombus formation in vivo [57], diminish platelet aggregation [59–61], reduce synthesis of platelet agonist thromboxane A_2 [59], and inhibit thrombin-stimulated calcium influx [59].

Platelet activation is a key element in acute vascular thrombosis, an important factor in the pathogenesis of acute myocardial infarction (AMI) and complications of coronary balloon angioplasty and stenting. Studies have demonstrated that magnesium can suppress platelet activation either by inhibiting platelet-stimulating factors, such as thromboxane A_2, or by stimulating synthesis of platelet inhibitory factors, such as prostacyclin (PGI_2) [58–64]. Intravenous administration of magnesium to healthy volunteers inhibited both ADP-induced platelet aggregation by 40% and the binding of fibrinogen or surface expression of glycoprotein IIb-IIIa complex GMP-140 by 30% [61–63]. Thus, pharmacological concentrations of magnesium effectively inhibit platelet function in vitro and ex vivo.

Using an ex vivo perfusion (Badimon) chamber [64], Shechter et al. [65] recently demonstrated that platelet-dependent thrombosis was significantly increased in stable CAD patients who, despite receiving antiplatelet treatment with aspirin, had low mononuclear intracellular levels of magnesium.

Furthermore, in a randomized, prospective, double-blind, crossover, placebo-controlled trial, Shechter et al. [66] found that 3 months of magnesium oxide tablets (800–1,200 mg/day) significantly reduced median platelet-dependent thrombosis by 35% compared to placebo in stable CAD patients who were on aspirin therapy. Despite the 100% use of aspirin therapy, the antithrombotic effect of magnesium treatment was observed.

Gawaz et al. [61, 63] demonstrated that platelet aggregation, fibrinogen binding, and expression of P-selectin on the platelet surface are all effectively inhibited by intravenous magnesium supplementation. Since glycoprotein IIb-IIIa is the only glycoprotein on the platelet surface that binds fibrinogen, Gawaz et al. speculated that magnesium supplementation directly impaired fibrinogen interaction with glycoprotein IIb-IIIa complex. Since fibrinogen binding to the platelet membrane and surface expression of P-selectin require previous cellular activation, the inhibitory effect of magnesium might be a consequence of direct interference of the cation with the agonist-receptor interaction or with the intracellular signal transduction event. Fibrinogen-glycoprotein IIb-IIIa interaction is regulated by divalent cations, and at the pharmacological level, magnesium may inhibit the binding of fibrinogen to glycoprotein IIb-IIIa by altering receptor conformation. This process might be caused by the competition of magnesium with calcium ions for calcium-binding sites in the glycoprotein IIb subunit.

Rukshin et al. [67] recently demonstrated that treatment with intravenous magnesium sulfate produced a time-dependent inhibition of acute stent thrombosis under high-shear flow conditions without any hemostatic or significant hemodynamic complications in an ex vivo porcine arteriovenous shunt model of high-shear blood flow, suggesting that magnesium inhibits acute stent thrombosis in an animal model. Thereafter, the same group [68] demonstrated that intravenous magnesium sulfate is a safe agent in acute coronary syndrome patients undergoing nonacute percutaneous coronary intervention with stent implantation, while magnesium therapy significantly inhibited platelet activation [69].

Protective Effects of Magnesium on Vascular Endothelial Function

The vascular endothelium is an active paracrine, endocrine, and autocrine organ, which plays a critical role in vascular homeostasis by secreting several mediators regulating vessel tone and diameter, coagulation factors, vascular inflammation, cell proliferation and migration, platelet and leukocyte interaction and activity, and thrombus formation [70–77]. Endothelial dysfunction is therefore recognized as a major factor in the development of atherosclerosis, hypertension, and heart failure. Vascular endothelial dysfunction is an independent risk factor for cardiovascular events and provides important prognostic data in addition to the classic cardiovascular risk factors and may be a "crystal ball prediction for enhanced cardiovascular risk" [78].

Shechter et al. [79] recently demonstrated that endothelial function is significantly correlated to intracellular magnesium levels, measured in sublingual epithelial cells, in CAD patients, while 30 mmol/day oral magnesium (total magnesium 730 mg/day) for 6 months significantly increased intracellular magnesium compared to placebo. In addition, magnesium therapy resulted in a significant improvement in endothelial function, which was associated with improved exercise duration, exercise-induced chest pain, and exercised-induced cardiac arrhythmias. Pearson et al. [80] demonstrated that hypomagnesemia selectively impaired the release of nitric oxide (NO) from coronary endothelium in a canine model. In a model of hypomagnesemia, Paravicini et al. [81] demonstrated a significant increase in blood pressure in low intracellular magnesium levels, compared with normal to high intracellular magnesium levels. These low levels were associated with impaired endothelial function and decreased plasma nitrate levels and endothelial NO synthase expression, when compared with normal to high intracellular magnesium levels. Since NO is a potent endogenous nitrovasodilator and inhibitor of platelet aggregation and adhesion, hypomagnesemia may promote vasoconstriction and coronary thrombosis in hypomagnesemic states.

Endothelial cells actively contribute to inflammation in magnesium deficiency states. Song et al. [30] and Chacko et al. [31] observed that magnesium intake was inversely associated with markers of systemic inflammation and endothelial dysfunction in healthy and postmenopausal women.

Protective Effects of Magnesium on Myocardial Infarct Size

Hypomagnesemia may increase coronary and systemic vasoconstriction and afterload, leading to increased myocardial oxygen depth [3, 32, 33]. Low concentrations of magnesium in laboratory animals seem to potentiate catecholamine-induced myocardial necrosis and cardiomyopathy [82]. Magnesium deficiency may adversely influence the healing and reendothelialization of vascular injuries and the healing of myocardial infarction and may also result in delayed or inadequate angiogenesis [83, 84]. Such effects could potentially lead to inadequate collateral development and infarct expansion. Magnesium reduces vulnerability to oxygen-derived free radicals [85], reperfusion injury, and stunning of the myocardium.

Beneficial Effects of Magnesium on Lipid Metabolism

While magnesium plays an interesting role in lipid regulation, its mechanism is not yet fully understood [86–91]. Magnesium is an important cofactor of two enzymes that are essential in lipid metabolism: lecithin-cholesterol acyltransferase and lipoprotein lipase. In an animal model, a rabbit was fed either a normal diet or a high-cholesterol diet, supplemented with varying amounts of magnesium, which achieved a dose-dependent reduction not only in the area of the aortic lesions but also in the cholesterol content of the aortas [89]. The 1% cholesterol diet significantly increased plasma cholesterol and triglyceride concentrations, while it decreased the high-density lipoprotein (HDL) cholesterol concentration. Additional magnesium had no further effect on cholesterol and HDL cholesterol concentrations, but it slightly decreased the rise in triglyceride concentrations [89]. In contrast, rats placed on severely depleted magnesium diets developed adverse lipid changes [90]. In one rat model, magnesium-deficient diets demonstrated an elevated plasma cholesterol level, low-density lipoprotein (LDL), and triglycerides with a proportionate reduction in HDL [87]. Rassmussen et al. [86] administered 15 mmol of magnesium hydroxide daily to humans and found a 27% reduction in triglycerides and very-low-density lipoprotein (VLDL) after 3 months of therapy, as well as a reduction in apoprotein B and HDL elevation. Davis et al. [87] demonstrated significant improvement in the ratio of HDL to LDL plus VLDL by giving 18 mmol of magnesium per day in a 4-month clinical trial.

Niemela et al. [88] showed that there was a significant inverse correlation in platelet intracellular magnesium levels with serum total cholesterol ($r=-0.52$, $p<0.02$), LDL ($r=-0.54$, $p<0.009$), and apolipoprotein B ($r=-0.42$, $p<0.04$) in men, but not in women. These investigators also speculated that decreased platelet intracellular magnesium levels are a possible marker for platelet membrane alterations that may affect platelet involvement in thrombosis and atherogenesis [88].

Beneficial Effects of Magnesium on Cardiac Arrhythmias

Magnesium deficiency is associated with intracellular hypopotassemia, hypernatremia, and augmentation of cell excitability [92]. Magnesium has the following modest electrophysiologic effects: it prolongs actual and corrected sinus node recovery time; it prolongs atrioventricular nodal function, as well as relative and effective refractory periods; and it slightly increases QRS duration during ventricular pacing at cycle lengths of 250 and 500 milliseconds and increases the atrial-His interval and atrial-paced cycle length causing atrioventricular nodal Wenckebach conduction [93]. In 1935, Zwillinger [94] was the first to recognize the arrhythmic effect of magnesium, when it was used to convert paroxysmal tachycardia to normal sinus rhythm. Later on, it was successfully used in resistant ventricular tachycardias [95], ventricular arrhythmias induced by digitalis toxicity [96], and episodes of torsades de pointes, a life-threatening ventricular arrhythmia [96, 97].

Magnesium was also found to be effective in the termination of episodes of supraventricular arrhythmia, such as multifocal atrial tachycardia [98], while it increased the susceptibility of atrial tachycardia to pharmacological conversion with digoxin [86].

In the Nurses' Health Study [26], the NHANES Epidemiologic Follow-up Study [19], and the multiethnic ARIC Study [18] population, higher plasma magnesium was associated with lower risk of SCD.

Extracellular magnesium influences cardiac ion channel properties [99] and regulates potassium homeostasis through activation of sodium potassium ATPase [100]. Magnesium administration suppresses early after depolarization and dispersion of repolarization [101, 102], whereas magnesium deficiency results in polymorphic ventricular tachycardia and SCD in animal models [103]. In clinical studies, magnesium therapy is efficacious in the treatment of arrhythmias secondary to acquired torsades de pointes [97] or hypomagnesemia [104]. Apart from antiarrhythmic actions, magnesium may also influence SCD risk through other pathways, including improvement in vascular tone, lipid metabolism, endothelial function, inflammation, blood pressure, diabetes, and inhibition of platelet function.

Magnesium has recently been recommended by the American Heart Association as the third drug of choice (after amiodarone and lidocaine) in the resuscitation of patients with pulseless ventricular tachycardias or ventricular fibrillation [57].

Magnesium therapy may correct resistant hypokalemia since it is a cofactor of the ATP molecule [105].

The Rationale of Magnesium Supplementation in AMI

Some relatively small prospective, randomized, double-blind, and controlled trials comparing intravenous magnesium to placebo in AMI patients have been reported throughout the last two decades [106–115]. In 1984, Morton et al. [106] published a pioneer study which showed that magnesium reduced infarct size by 20% in patients in Killip class I, as well as in-hospital mortality in AMI patients.

The second Leicester Intravenous Magnesium Intervention Trial (LIMIT-2) [116, 117] was the first large clinical trial, where 30% of the 2,316 patients received thrombolytic therapy. Intravenous magnesium reduced congestive heart failure (CHF) by 25% and all-cause mortality by 24% at 28 days [116, 117] with a 20% reduction in ischemic heart disease-related mortality over a mean follow-up of 4.5 years [113].

In the mid-1990s, Shechter et al. [118] demonstrated that 22 g (92 mmol) of intravenous magnesium sulfate administered over a 48-h period in 215 AMI patients considered unsuitable for reperfusion reduced in-hospital mortality by almost 50% and the incidence of arrhythmias and CHF by 33% in elderly patients above the age of 70 years.

At the same time, the Fourth International Study of Infarct Survival (ISIS-4) and Magnesium in Coronaries studies [119] were conducted with approximately 58,000 AMI patients, of whom almost 70% received thrombolytic therapy, and showed no survival benefit from intravenous magnesium sulfate over placebo at 35 days and at 1 year. The magnesium dose was almost identical to that of the LIMIT-2 study but with an open control. However, the time from onset of symptoms to randomization was substantially longer (median of 8 h rather than 3). The 30% of patients not receiving thrombolytic therapy were randomized at a median of 12 h after symptom onset. The low mortality rate in the ISIS-4 control group; the late enrollment of patients, particularly those who did not receive thrombolytic treatment; and the fact that magnesium infusions were delayed by 1–2 h after thrombolytic therapy suggest the possibility that the majority of patients in the ISIS-4 study were at low mortality risk and that an elevated magnesium blood level was not reached until well beyond the narrow time frame for salvage of the myocardium or prevention of reperfusion injury suggested by the experimental data [83, 84].

Shortly thereafter, Shechter et al. [120] showed a significant long-term (mean follow-up of 4.5 years) mortality reduction of 40% in 194 AMI patients, considered unsuitable candidates for reperfusion therapy at the time of enrollment, who received intravenous magnesium compared to placebo for 48 h. Rest left ventricular ejection fraction, measured in all patients who survived the last year of follow-up, was significantly higher in patients who received magnesium versus placebo. Thus, the favorable effects of intravenous magnesium therapy can last for several years after acute treatment, probably due to preserved left ventricular ejection fraction.

The Magnesium in Coronaries (MAGIC) trial [121], published in 2002, randomized 6,213 patients ≥ 65 years, of whom an unexpected high percentage (45%) were female with acute ST elevation AMI of < 6 h. They were either eligible for reperfusion therapy (median age 73 years) [stratum 1] or were patients of any age not eligible for reperfusion therapy (median age 67 years) [stratum 2]. All of them received either a 2-g intravenous bolus of magnesium sulfate, administered over 15 min, followed by a 17-g infusion of magnesium sulfate over 24 h (n = 3,113) or matching placebo (n = 3,100). The "magnesium community" was very disappointed by the results which demonstrated the null effects of magnesium on 30-day mortality or heart failure. Compared to the MAGIC trial, the study of Shechter et al. [120] comprised thrombolysis-ineligible AMI patients, of whom one third were > 75 years and therefore were similar to the MAGIC stratum 2 patients but differed from them in two aspects: (a) they received a higher dose of intravenous magnesium sulfate (22 vs. 19 g) and (b) administration was for a longer period of time (48 vs. 24 h). Furthermore, a significantly higher proportion of the MAGIC study patients received aspirin, β-blockers, and angiotensin-converting enzyme inhibitors than those in the Shechter et al. trial. As a result, the postulated cardioprotective effects of magnesium could have been superseded by the effects of these medical regimens.

Recently published random-effect meta-analyses have demonstrated a significant reduction in early mortality when comparing magnesium with placebo (OR 0.66, 95% CI 0.53–0.82), especially in patients not treated with thrombolysis (OR 0.73, 95% CI 0.56–0.94) and in those treated with < 75 mmol of magnesium (OR 0.59, 95% CI 0.49–0.70) [122].

Following the data from the ISIS-4 and MAGIC studies, the current guideline recommendation is that magnesium should not be routinely administered to all AMI patients. However, it should be an adjunct therapy option in selected cases of high-risk AMI patients, such as in the elderly; in those with left ventricular dysfunction and/or CHF; and/or in those patients not suitable for reperfusion therapy [33].

Beneficial Effects of Magnesium in CHF

Patients with CHF are magnesium deficient, based on the fact that the activation of the renin-angiotensin-aldosterone system and the use of diuretics are associated with depletion of potassium and magnesium in CHF [1, 3, 32, 123]. Magnesium deficiency stimulates aldosterone production and secretion, while magnesium infusion decreases aldosterone production by inhibiting cellular calcium influx [124]. Adamopoulos et al. [125] recently found in a long-term follow-up of 36 months that CHF in patients [mainly those in New York Heart Association (NYHA) class II-II] with low serum magnesium (≤2 mEq/L) was associated with increased cardiovascular mortality (albeit not cardiovascular hospitalization), compared to those with higher serum magnesium levels (>2 mEq/L), suggesting that most of these deaths were likely sudden (arrhythmic) in nature.

Furthermore, Stepura and Martynow [126] demonstrated that oral magnesium orotate used as adjuvant therapy in severe NYHA class IV CHF patients increased 1-year survival rate and improved clinical symptoms as well patients' quality of life, compared to placebo.

Adverse Effects of Magnesium

Magnesium supplementation is relatively safe [3, 32–34]. In all previous randomized controlled clinical trials, only a few adverse effects have been reported. In the ISIS-4 trial [113] comprising 58,000 patients with suspected AMI, no overall increase in the incidence of second- or third-degree heart block was observed, although there was a slight but not statistically significant increase during or just after magnesium infusion. There was no evidence of adverse effects in the LIMIT 2 trial [116, 117] with 1,500 patients and in the MAGIC trial [121] with 6,200 AMI patients. However, nonclinical significant sinus bradycardia was observed in some but not all of the randomized clinical trials. Since magnesium is a physiological calcium competitor, an intravenous bolus dose of 1 g over 5 min is recommended due to the fact that rapid intravenous (bolus) administration can reduce blood pressure and is therefore prohibited [97].

A patient with normal kidney function filters approximately 2.5 g of magnesium and reclaims 95%, excreting some 100 mg/dl into the urine to maintain homeostasis. Approximately 25–30% is reclaimed in the proximal tube through a passive transport system that depends on sodium reabsorption and tubular fluid flow. Usually, as serum magnesium concentration increases, there is a linear increase in urinary magnesium excretion, paralleling that of insulin. With normal kidney function, hypermagnesemia or magnesium intoxication does not usually develop, even during high intravenous magnesium infusion [3, 32–34]. Furthermore, oral magnesium supplementation may cause diarrhea, soft stools, gastrointestinal irritation, weakness, nausea, vomiting, and abdominal pain.

Reasons for Magnesium Deficiency

The prevalence of hypomagnesemia in hospitalized patients ranges from 8% to 30% [1, 3]. Elderly patients, particularly those with CAD and/or CHF, can have low body magnesium levels, the mechanisms of which are likely to be multifactorial. Evidence suggests that the occidental "American-type diet" is relatively deficient in magnesium [1, 3, 10, 11], while the "oriental diet," characterized by a greater intake of fruits and vegetables, is richer in magnesium [4]. It has also been observed that CAD patients absorb more magnesium during magnesium loading tests than non-CAD patients, suggesting that CAD is associated with excessive magnesium loss and a relative magnesium-deficient condition [13].

Magnesium deficiency may usually be reflected in low-magnesium diet, blood loss, excessive sweating, drug and/or alcohol abuse or due to certain medication intake (such as loop diuretics and thiazides, cytotoxic drugs, aminoglycosides, digoxin, steroids), or some physiological condition, such as in pregnancy or infancy growth when overutilization of magnesium is present. Mental stress can also lead to magnesiuresis due to high serum adrenalin [127, 128]. Diabetes mellitus is also associated with magnesium deficiency mainly due to urinary magnesium loss [1]. Other diseases associated with magnesium deficiency are liver cirrhosis, diseases of the thyroid and parathyroid glands, and renal diseases. Moreover, diets rich in animal foods but low in vegetables can induce acidosis as well as an increase in magnesium urinary excretion.

Pure magnesium deficiency is characterized by a number of clinical features including muscular tremor, vertigo, ataxia, tetany, convulsions, and organic brain syndrome.

Conclusions

Magnesium plays a vital role in many cellular processes. Since it is associated with a variety of enzymes which control carbohydrates, fats, proteins, and electrolyte metabolism, magnesium is essential for a number of metabolic activities. Several hundreds of enzymes are directly or indirectly dependent on magnesium. Most important among these enzymes are those which hydrolyze and transfer phosphate groups, including enzymes that are concerned with reactions involving energy production and ATP. Magnesium deficiency or reduction in dietary intake plays an important role in the etiology of diabetes and numerous cardiovascular diseases, including thrombosis, atherosclerosis, ischemic heart disease, myocardial infarction, hypertension, cardiac arrhythmias, and CHF in humans.

Magnesium deficiency may lead to reduced energetic metabolite production and the sense of fatigue and/or "chronic fatigue syndrome." Modern lifestyle and Western industrial diet enhance the reduction of magnesium in our food which directly contributes to marginal or absolute magnesium deficiency. This is particularly marked in the elderly population, in those with myocardial infarction and/or CHF, diabetics, patients with chronic airway obstruction, pre- or toxemia of pregnancy, in post-transplantation patients (especially in heart transplantation), patients with malignancies who receive cytotoxic chemical therapy, in competitive athletes, and in metabolic syndrome patients.

It should be noted that magnesium deficiency can be easily treated by magnesium supplementation on the condition that we are aware of the situation. The best recommendation is to increase consumption of magnesium-rich food. However, since this is difficult to do in practice, it is recommended to take magnesium supplements which safely increase the amount of magnesium in the body and correct the deficit.

There are theoretical potential benefits of magnesium supplements as a cardioprotective agent in CAD patients, as well as promising results from previous studies in animal and humans. Magnesium is an essential element in treating CAD patients, especially high-risk groups such as CAD patients with heart failure, the elderly, and hospitalized patients with hypomagnesemia. Furthermore, magnesium therapy is indicated in life-threatening ventricular arrhythmias such as torsades de pointes and intractable ventricular tachycardia.

While the screening of serum magnesium levels need not be routinely advocated, the treating physician should be highly suspicious of magnesium deficiency, unless proved otherwise.

It should be remembered that magnesium is neither a "panacea" nor a "wonder drug" to be aggressively pushed by the pharmaceutical industry. On balance, it is a relatively simple nutrient, inexpensive and easy to administer with few adverse effects, and may be described as a "nutrient which is the sparkle of life" and a dedicated caretaker.

References

1. Seelig MS, Rosanoff A. In: Seelig MS, Rosanoff A, editors. The magnesium factor. New York: Avery Publishers; 2003.
2. Wacker WEC, Parisi AF. Magnesium metabolism. N Engl J Med. 1968;278:658–63.
3. Shechter M, Kaplinsky E, Rabinowitz B. The rationale of magnesium supplementation in acute myocardial infarction. A review of the literature. Arch Intern Med. 1992;152:2189–96.
4. Whang R, Flink E, Dyckner T, et al. Magnesium depletion as a cause of refractory potassium repletion. Arch Intern Med. 1985;145:1686–9.
5. Ryzen E, Elkayam U, Rude RK. Low blood mononuclear cell magnesium content in intensive cardiac care unit patients. Am Heart J. 1986;111:475–80.
6. Reinhart RA. Magnesium metabolism. Arch Intern Med. 1988;148:2415–20.
7. Elin RJ. Status of the determination of magnesium in mononuclear blood cells in humans. Magnesium. 1988;7:300–5.
8. Haigney MCP, Silver B, Tanglao E, et al. Noninvasive measurement of tissue magnesium and correlation with cardiac levels. Circulation. 1995;92:2190–7.
9. Cohen L. Physiologic assessment of magnesium status in humans: a combination of load retention and renal excretion. IMAJ. 2000;2:938–9.
10. Seelig MS. The requirement of magnesium by the normal adult. Am J Clin Nutr. 1964;6:342–90.
11. Centers for Disease Control and Prevention. Dietary intake of vitamins, minerals, and fiber of persons ages 2 months and over in the United States: Third National Health and Nutrition Examination Survey, Phase I, 1988–91, Advance data from vital and health statistics, vol. 258. Hyattsville: National Center for Health Statistics; 1994. p. 1–28.
12. Lowenstein FW, Stanton MF. Serum magnesium levels in the United States, 1971–1974. J Am Coll Nutr. 1986;5(4):399–414.
13. Seelig MS. Cardiovascular consequences of magnesium deficiency and loss: pathogenesis, prevalence and manifestations- magnesium and chloride loss in refractory potassium repletion. Am J Cardiol. 1989;63:4G–21.
14. Lichton IJ. Dietary intake levels of requirements of Mg and Ca for different segments of the U.S. population. Magnesium. 1989;8:117–23.
15. Peterson DR, Thompson DJ, Nam JM. Water hardness, arteriosclerotic heart disease and sudden death. Am J Epidemiol. 1970;92:90–3.
16. Shaper AG. Soft water, heart attacks, and stroke. J Am Med Assoc. 1974;230:130–1.

17. Anderson TW, Neri LC, Schreiber GB, et al. Ischemic heart disease, water hardness and myocardial magnesium. CMA J. 1975;113:199–203.
18. Liao F, Folsom AR, Brancati FL. Is low magnesium concentration a risk factor for coronary heart disease? The Atherosclerosis Risk in Communities (ARIC) Study. Am Heart J. 1998;136:480–90.
19. Ford ES. Serum magnesium and ischemic heart disease: findings from national sample of US adults. Int J Epidemiol. 1999;28:645–51.
20. Abbott RD, Ando F, Masaki KH, et al. Dietary magnesium intake and the future risk of coronary heart disease (The Honolulu Heart Program). Am J Cardiol. 2003;92:665–9.
21. Amighi J, Sabeti S, Schlager O, et al. Low serum magnesium predicts neurological events in patients with advanced atherosclerosis. Stroke. 2004;35:22–7.
22. He K, Liu K, Daviglus ML, Morris SJ, et al. Magnesium intake and incidence of metabolic syndrome among young adults. Circulation. 2006;113:1675–82.
23. Anderson TW, Le Riche WH, MacKay JS. Sudden death and ischemic heart disease. Correlation with hardness of local water supply. N Engl J Med. 1969;280:805–7.
24. Johnson CJ, Peterson DR, Smith EK. Myocardial tissue concentrations of magnesium and potassium in men dying suddenly from ischemic heart disease. Am J Clin Nutr. 1979;32:967–70.
25. Peacock JM, Ohira T, Post W, et al. Serum magnesium and risk of sudden cardiac death in the Atherosclerosis Risk in Communities (ARIC) Study. Am Heart J. 2010;160:464–70.
26. Chiuve SE, Korngold EC, Januzzi Jr JL, et al. Plasma and dietary magnesium and risk of sudden cardiac death in women. Am J Clin Nutr. 2011;93:253–60.
27. Pachikian BD, Neyrinck AM, Deldicque L, et al. Changes in intestinal bifidobacteria levels are associated with the inflammatory response in magnesium-deficient mice. J Nutr. 2010;140:509–14.
28. Lin CY, Tsai PS, Hung YC, et al. L-type calcium channels are involved in mediating the anti-inflammatory effects of magnesium sulphate. Br J Anaesth. 2010;104:44–51.
29. King DE. Inflammation and elevation of C-reactive protein: does magnesium play a key role? Magnes Res. 2009;22:57–9.
30. Song Y, Li TY, van Dam RM, et al. Magnesium intake and plasma concentrations of markers of systemic inflammation and endothelial dysfunction in women. Am J Clin Nutr. 2007;85:1068–74.
31. Chacko SA, Song Y, Nathan L, et al. Relations of dietary magnesium intake to biomarkers of inflammation and endothelial dysfunction in an ethnically diverse cohort of postmenopausal women. Diabetes Care. 2010;33:304–10.
32. Shechter M, Kaplinsky E, Rabinowitz B. Review of clinical evidence – is there a role for supplemental magnesium in acute myocardial infarction in high-risk populations (patients ineligible for thrombolysis and the elderly)? Coron Artery Dis. 1996;7:352–8.
33. Shechter M. Does magnesium have a role in the treatment of patients with coronary artery disease? Am J Cardiovasc Drugs. 2003;3:231–9.
34. Shechter M, Shechter A. Magnesium and myocardial infarction. Clin Calcium. 2005;11:111–5.
35. Iseri LT, French JH. Magnesium: nature's physiologic calcium blocker. Am Heart J. 1984;108:188–93.
36. Holroyde MJ, Robertson SP, Johnson JD, et al. The calcium and magnesium binding sites on cardiac troponin and their role in the regulation of myofibrillar adenosine triphosphatase. J Biol Chem. 1980;255:11688–91.
37. Sordahl LA. Effects of magnesium on initial rates of calcium uptake and release of heart mitochondria. Arch Biochem Biophys. 1975;167:104–7.
38. Mroczek WJ, Lee WR, Davidov ME. Effect of magnesium sulfate on cardiovascular hemodynamics. Angiology. 1977;28:720–4.
39. Cotton DB, Gonik B, Dorman KF. Cardiovascular alterations in severe pregnancy-induced hypertension: acute effects of intravenous magnesium sulfate. Am J Obstet Gynecol. 1984;148:162–5.
40. Rasmussen HS, Larsen OG, Meier K, et al. Hemodynamic effects of intravenous administered magnesium in patients with ischemic heart disease. Int J Cardiol. 1988;11:824–8.
41. Altura BM, Altura BT. New perspectives on the role of magnesium in the pathophysiology of the cardiovascular system. Magnesium. 1985;4:245–71.
42. Altura BM. Magnesium neurohypophyseal hormone interactions in contraction of vascular smooth muscle. Am J Physiol. 1975;228:1615–20.
43. Altura BM, Altura BT, Corella A. Magnesium deficiency-induced spasms of umbilical vessels: relation to preeclampsia, hypertension, growth retardation. Science. 1983;221:376–8.
44. Altura BM, Altura BT. Vascular smooth muscle and prostaglandins. Fed Proc. 1976;35:2360–6.
45. Askar AD, Mustafa SJ. Role of magnesium in the relaxation of coronary arteries by adenosine. Magnesium. 1983;2:17–25.
46. Whelton PK, Klay MJ. Magnesium and blood pressure: review of the epidemiologic and clinical trial experience. Am J Cardiol. 1989;63:26G–30.

47. Mizushima S, Cappuccio FP, Nichols R, et al. Dietary magnesium intake and blood pressure: a qualitative overview of the observational studies. J Hum Hypertens. 1998;12(7):447–53.
48. Kugiyama K, Yasue H, Okumara K, et al. Suppression of exercise-induced angina by magnesium sulfate in patients with variant angina. J Am Coll Cardiol. 1988;12:1177–83.
49. Jee HS, Miller 3rd ER, Guallar E, et al. The effect of magnesium supplementation on blood pressure. A meta analysis of clinical randomized trials. Am J Hypertens. 2002;15:691–6.
50. Lucas MJ, Leveno KJ, Cunningham FG. A comparison of magnesium sulfate with phenytoin for the prevention of eclampsia. N Engl J Med. 1995;333:201–5.
51. Shechter M, Paul-Labrador M, Rude RK, et al. Intracellular magnesium predicts functional capacity in patients with coronary artery disease. Cardiology. 1998;90:168–72.
52. Shechter M, Bairey Merz CN, Stuehlinger HG, et al. Oral magnesium supplementation improves exercise duration and quality of life in patients with coronary artery disease. Am J Cardiol. 2003;91:517–21.
53. Pokan R, Hofmann P, von Duvillard SP, et al. Oral magnesium therapy, exercise heart rate, exercise tolerance, and myocardial function in coronary artery disease patients. Br J Sports Med. 2006;40:773–8.
54. Greville GD, Lehmann H. Cation antagonism in blood coagulation. J Physiol. 1943;103:175–84.
55. Schnitzler B. Thromboseprophylaxe mit Magnesium. Munch med Wschr. 1957;99:81–4.
56. Anstall HB, Huntsman RG, Lehmann H, et al. The effect of magnesium on blood coagulation in human subjects. Lancet. 1959;1:814–5.
57. Adams JH, Mitchel JRA. The effect of agents which modify platelet behavior and of magnesium ions on thrombus formation in vivo. Thromb Haemost. 1979;42:603–10.
58. Frandsen NJ, Winther K, Pedersen F, et al. Magnesium and platelet function: in vivo influence on aggregation and alpha-granule release in healthy volunteers. Magnesium Bull. 1995;17:37–40.
59. Hwang DL, Yen CF, Nadler JL. Effect of extracellular magnesium on platelet activation and intracellular calcium mobilization. Am J Hypertens. 1992;5:700–6.
60. Born GVR, Cross GP. Effect of inorganic ions and plasma proteins on the aggregation of blood platelets by adenosine diphosphate. J Physiol. 1964;170:397–414.
61. Gawaz M, Ott I, Reininger AJ, et al. Effects of magnesium on platelet aggregation and adhesion. Magnesium modulates surface expression of glycoproteins on platelets in vitro and ex vivo. Thromb Haemost. 1994;72:912–8.
62. Nadler JL, Goodson S, Rude RK. Evidence that prostacyclin mediates the vascular action of magnesium in humans. Hypertension. 1987;9:379–83.
63. Gawaz M. Effects of intravenous magnesium on platelet function and platelet-leukocyte adhesion in symptomatic coronary heart disease. Thromb Res. 1996;83:341–9.
64. Badimon L, Badimon JJ, Galvez A, et al. Influence of arterial damage and wall shear rate on platelet formation: ex vivo study in a swine model. Arteriosclerosis. 1986;6:312–20.
65. Shechter M, Bairey Merz CN, Rude RK, et al. Low intracellular magnesium levels promote platelet-dependent thrombus formation in patients with coronary artery disease. Am Heart J. 2000;140:212–8.
66. Shechter M, Bairey Merz CN, Paul-Labrador M, et al. Oral magnesium supplementation inhibits platelet-dependent thrombosis in patients with coronary artery disease. Am J Cardiol. 1999;84:152–6.
67. Rukshin V, Azarbal B, Shah PK, et al. Intravenous magnesium in experimental stent thrombosis in swine. Arterioscler Thromb Vasc Biol. 2001;21:1544–9.
68. Rukshin V, Shah PK, Cercek B, et al. Comparative antithrombotic effects of magnesium sulfate and platelet glycoprotein IIb/IIIa inhibitors tirofiban and eptifibatide in a canine model of stent thrombosis. Circulation. 2002;105:1970–5.
69. Rukshin V, Santos R, Gheorghiu M, et al. A prospective, nonrandomized, open-labeled pilot study investigating the use of magnesium in patients undergoing nonacute percutaneous coronary intervention with stent implantation. J Cardiovasc Pharmacol Ther. 2003;8:193–200.
70. Bonetti PO, Lerman LO, Lerman A. Endothelial dysfunction. A marker of atherosclerotic risk. Arteriosc Thromb Vasc Biol. 2003;23:168–75.
71. Corretti MC, Anderson TJ, Benjamin EJ, International Brachial Artery Reactivity Task Force, et al. Guidelines for the ultrasound assessment of endothelial-dependent flow-mediated vasodilatation of the brachial artery. J Am Coll Cardiol. 2002;39:257–65.
72. Vogel RA. Coronary risk factors, endothelial function, and atherosclerosis: a review. Clin Cradiol. 1997;20:426–32.
73. McLenachan JM, Williams JK, Fish RD, et al. Loss of flow-mediated endothelium-dependent dilation occurs early in the development of atherosclerosis. Circulation. 1991;84:1272–7.
74. Rizzoni D. Endothelial function in hypertension: fact or fantasy? J Hypertens. 2002;20:1479–81.
75. Drexler H, Hayoz D, Münzel T, et al. Endothelial function in chronic congestive heart failure. Am J Cardiol. 1992;69:1596–601.

76. Widlansky ME, Gokce N, Keaney Jr JF, et al. The clinical implication of endothelial dysfunction. J Am Coll Cardiol. 2003;42:1149–60.
77. Lerman A, Zeiher AM. Endothelial function cardiac events. Circulation. 2005;111:363–8.
78. Shechter M, Sherer Y. Endothelial dysfunction: a crystal ball prediction for enhanced cardiovascular risk? Isr Med Assoc J. 2003;5:736–8.
79. Shechter M, Sharir M, Labrador MJ, et al. Oral magnesium therapy improves endothelial function in patients with coronary artery disease. Circulation. 2000;102:2353–8.
80. Pearson PJ, Evora PR, Seccombe JF, et al. Hypomagnesemia inhibits nitric oxide release from coronary endothelium: protective role of magnesium infusion after cardiac operations. Ann Thor Surg. 1998;65:967–72.
81. Paravicini TM, Yogi A, Mazur A, et al. Dysregulation of vascular TRPM7 and annexin-1 is associated with endothelial dysfunction in inherited hypomagnesemia. Hypertension. 2009;53:423–9.
82. Vormann J, Fiscer G, Classen HG, et al. Influence of decreased and increased magnesium supply on cardiotoxic effects of epinephrine in rats. Arzneimittelforschung. 1983;33:205–10.
83. Banai S, Haggroth L, Epstein SE, et al. Influence of extracellular magnesium on capillary endothelial cell proliferation and migration. Circ Res. 1990;67:645–50.
84. Weisman HF, Bush DE, Mannisi JA, et al. Cellular mechanisms of myocardial infarct expansion. Circulation. 1988;78:186–201.
85. Dickens BF, Weglicki WB, Li YS, et al. Magnesium deficiency in vitro enhances free radical-induced intracellular oxidation and cytotoxicity in endothelial cells. FEBS. 1992;311:187–91.
86. Rasmussen HS, Aurup P, Goldstein K, et al. Influence of magnesium substitution therapy on blood lipid composition in patients with ischemic heart disease. Arch Intern Med. 1989;149:1050–3.
87. Davis WH, Leary WP, Reyes AH. Monotherapy with magnesium increases abnormally low high density lipoprotein cholesterol: a clinical assay. Curr Ther Res. 1984;36:341–4.
88. Niemela JE, Csako G, Bui MN, et al. Gender-specific correlation of platelet ionized magnesium and serum low-density cholesterol concentrations in apparently healthy subjects. J Lab Clin Med. 1997;129:89–96.
89. Ouchi Y, Tabata RE, Stergiopoulos K, et al. Effect of dietary magnesium on development of atherosclerosis in cholesterol fed rabbits. Arteriosclerosis. 1990;10:732–7.
90. Rayssiguier Y, Guex E, Weiser D. Effect of magnesium deficiency on lipid metabolism in rats fed a high carbohydrate diet. J Nutr. 1981;111:1876–83.
91. Luthringer C, Rayssiguier Y, Gueux E, et al. Effect of moderate magnesium deficiency on serum lipids, blood pressure and cardiovascular reactive in normotensive rats. Br J Nut. 1988;59:243–50.
92. Beller GA, Hood Jr WB, Abelmann WH, et al. Correlation of serum magnesium levels and cardiac digitalis intoxication. Am J Cardiol. 1974;33:225–9.
93. Arsenian MA. Magnesium and cardiovascular disease. Prog Cardiovasc Dis. 1993;35:271–310.
94. Zwillinger L. Uber die Magnesiumwirkung auf das Herz. Klin Wochenschr. 1935;14:1329–433.
95. Chadda KD, Lichstein E, Gapta P. Hypomagnesemia and refractory cardiac arrhythmia in a nondigitalized patient. Am J Cardiol. 1973;31:98–100.
96. Tzivoni D, Keren A. Suppression of ventricular arrhythmias by magnesium. Am J Cardiol. 1990;65:1397–9.
97. Tzivoni D, Keren A, Cohen AM, et al. Magnesium therapy for torsade de pointes. Am J Cardiol. 1984;53:528–30.
98. Iseri LT, Fairshter RD, Hardemann JL, et al. Magnesium and potassium therapy in multifocal atrial tachycardia. Am Heart J. 1985;110:789–94.
99. Agus ZS, Morad M. Modulation of cardiac ion channels by magnesium. Annu Rev Physiol. 1991;53:299–307.
100. Saris NE, Mervaala E, Karppanen H, et al. Magnesium: an update on physiological, clinical and analytical aspects. Clin Chim Acta. 2000;294:1–26.
101. Verduyn SC, Vos MA, van der Zande J, et al. Role of interventricular dispersion of repolarization in acquired torsade de-pointes arrhythmias: reversal by magnesium. Cardiovasc Res. 1997;34:453–63.
102. Davidenko JM, Cohen L, Goodrow R, et al. Quinidine induced action potential prolongation, early after depolarizations, and triggered activity in canine Purkinje fibers. Effects of stimulation rate, potassium, and magnesium. Circulation. 1989;79:674–86.
103. Fiset C, Kargacin ME, Kondo CS, et al. Hypomagnesemia: characterization of a model of sudden cardiac death. J Am Coll Cardiol. 1996;27:1771–6.
104. Ceremuzynski L, Gebalska J, Wolk R, et al. Hypomagnesemia in heart failure with ventricular arrhythmias. Beneficial effects of magnesium supplementation. J Intern Med. 2000;247:78–86.
105. Horner SM. Efficacy of intravenous magnesium in acute myocardial infarction in reducing arrhythmias and mortality. Circulation. 1992;86:774–9.
106. Morton BC, Nair RC, Smith FM, McKibbon TG, Poznanski WJ. Magnesium therapy in acute myocardial infarction. Magnesium. 1984;3:346–52.
107. Rasmussen HS, McNair P, Norregard P, et al. Magnesium infusion in acute myocardial infarction. Lancet. 1986;1:234–6.

108. Rasmussen HS, Grønbaek M, Cintin C, et al. One-year rate in 270 patients with suspected acute myocardial infarction, initially treated with intravenous magnesium or placebo. Clin Cardiol. 1988;11:377–81.
109. Smith LF, Heagerty AM, Bing RF, et al. Intravenous infusion of magnesium sulphate after acute myocardial infarction: effects on arrhythmias and mortality. Int J Cardiol. 1986;12:175–80.
110. Abraham AS, Rosenman D, Meshulam Z, et al. Magnesium in the prevention of lethal arrhythmias in acute myocardial infarction. Arch Intern Med. 1987;147:753–5.
111. Ceremuzynski L, Jurgiel R, Kulakowski P, et al. Threatening arrhythmias in acute myocardial infarction are prevented by intravenous magnesium sulfate. Am Heart J. 1989;118:1333–4.
112. Shechter M, Hod H, Marks N, et al. Beneficial effect of magnesium sulfate in acute myocardial infarction. Am J Cardiol. 1990;66:271–4.
113. Shechter M, Hod H. Magnesium therapy in aged patients with acute myocardial infarction. Magnesium Bull. 1991;13:7–9.
114. Feldstedt M, Boesgaard S, Bouchelouche P, et al. Magnesium substitution in acute ischaemic heart syndromes. Eur Heart J. 1991;12:1215–8.
115. Teo KK, Yusuf S, Collins R, et al. Effects of intravenous magnesium in suspected acute myocardial infarction: overview of randomized trials. Br Med J. 1991;303:1499–503.
116. Woods KL, Fletcher S, Roffe C, et al. Intravenous magnesium sulphate in suspected acute myocardial infarction: results of the Second Leicester Intravenous Magnesium Intervention Trial (LIMIT-2). Lancet. 1992;339:1553–8.
117. Woods KL, Fletcher S. Long term outcome after intravenous magnesium sulphate in suspected acute myocardial infarction: the Second Leicester Intravenous Magnesium Intervention Trial (LIMIT-2). Lancet. 1994;343:816–9.
118. Shechter M, Hod H, Chouraqui P, et al. Magnesium therapy in acute myocardial infarction when patients are not candidates for thrombolytic therapy. Am J Cardiol. 1995;75:321–3.
119. ISIS-4. A randomised factorial trial assessing early oral captopril, oral mononitrate, and intravenous magnesium sulphate in 58050 patients with suspected acute myocardial infarction. Lancet. 1995;345:669–85.
120. Shechter M, Hod H, Rabinowitz B, et al. Long-term outcome in thrombolysis-ineligible acute myocardial infarction patients treated with intravenous magnesium. Cardiology. 2003;99:203–10.
121. Magnesium in Coronaries (MAGIC) trial investigators. Early administration of intravenous magnesium to high-risk patients with acute myocardial infarction in the coronaries (MAGIC) trial: a randomized controlled trial. Lancet. 2002;360:1189–96.
122. Li J, Zhang Q, Zhang M, et al. Intravenous magnesium for acute myocardial infarction. Cochrane Database Syst Rev. 2007; 2. Art. No.: CD002755. doi: 10.1002/14651858.CD002755.pub2
123. Cohen N, Almoznino-Sarfian D, Zaidenstein R, et al. Serum magnesium aberrations in furosemide treated patients with chronic congestive heart failure: pathophysiological correlates and prognostic evaluation. Heart. 2003;89:411–6.
124. Fakunding JL, Chow R, Catt KJ. The role of calcium in the stimulation of aldosterone production by adrenocorticotropin, angiotensin II, and potassium in isolated glomerulosa cells. Endocrinology. 1979;105:327–33.
125. Adamopoulos C, Pitt B, Sui X, et al. Low serum magnesium and cardiovascular mortality in chronic heart failure: a propensity-matched study. Int J Cardiol. 2009;136:270–7.
126. Stepura OB, Martynow AI. Magnesium orotate in severe congestive heart failure (MACH). Int J Cardiol. 2009;131:293–5.
127. Seelig MS. Possible roles of magnesium in disorders of the aged. In: Regelson W, Sinex FM, editors. Intervention in the aging process, Part A: quantitation, epidemiology, clinical research. New York: AR Liss, Inc; 1983. p. 279–305.
128. Johansson G. Magnesium metabolism: studies in health, primary hyperparathyroidism and renal stone disease. Scand J Urol Nephrol. 1979;51:1–47.

Chapter 14
Vascular Biology of Magnesium: Implications in Cardiovascular Disease

Tayze T. Antunes, Glaucia Callera, and Rhian M. Touyz

Key Points

- Alterations in magnesium (Mg^{2+}) status have been implicated in impaired endothelial function and altered vascular function and structure, processes that play an important role in vascular remodeling and inflammation associated with cardiovascular disease, such as atherosclerosis, hypertension, stroke, cardiac failure, and chronic kidney disease.
- Unlike our knowledge of other major cations, mechanisms regulating cellular Mg^{2+} handling are poorly understood. Until recently, little was known about protein transporters controlling transmembrane Mg^{2+} influx.
- However, new research has uncovered a number of genes and proteins identified as transmembrane Mg^{2+} transporters, particularly transient receptor potential melastatin (TRPM) cation channels, TRPM6 and TRPM7, and magnesium transporter subtype 1 (MagT1).
- Whereas TRPM6 is found primarily in epithelial cells, TRPM7 and MagT1, are ubiquitously expressed. Vascular TRPM7 has been implicated as a signaling kinase involved in vascular smooth muscle cell (VSMC) growth, apoptosis, adhesion, contraction, cytoskeletal organization, and migration and is modulated by vasoactive agents, pressure, stretch, and osmotic changes.
- Emerging evidence suggests that vascular TRPM7 function may be altered in hypertension. This chapter discusses the importance of Mg^{2+} in vascular biology and implications in cardiovascular health and disease.

Keywords TRPM7 • Magnesium • Cardiovascular disease

T.T. Antunes, Ph.D. • G. Callera, Ph.D.
Kidney Research Center, Ottawa Hospital Research Institute, University of Ottawa, Ottawa, Canada

R.M. Touyz, M.D., Ph.D. (✉)
Kidney Research Center, Ottawa Hospital Research Institute, University of Ottawa, Ottawa, Canada

Institute of Cardiovascular and Medical Sciences, BHF Glasgow Cardiovascular Research Centre, University of Glasgow, Glasgow, Scotland, United Kingdom
e-mail: Rhian.Touyz@glasgow.ac.uk

R.R. Watson et al. (eds.), *Magnesium in Human Health and Disease*, Nutrition and Health,
DOI 10.1007/978-1-62703-044-1_14, © Springer Science+Business Media New York 2013

Introduction

Magnesium (Mg^{2+}), an abundant intracellular divalent cation, plays an essential role in numerous cellular functions. It is an allosteric modulator of several proteins; controls nucleotide and protein synthesis; regulates Na^+, K^+, and Ca^{2+} channels; and is critical for many enzymatic reactions, particularly those that are ATP-dependent [1–3]. Mg^{2+} is stored mainly in bone and muscle with less than 1% of total body Mg^{2+} circulating in the blood [4, 5]. Mg^{2+} homeostasis depends on the balance between intestinal uptake and renal excretion. The kidneys play an important role in maintaining Mg^{2+} balance, and bone acts as a buffer providing a rapidly exchangeable pool to protect against acute changes in the plasma [6]. Mg^{2+} deficiency results from reduced dietary intake, intestinal malabsorption, or renal loss. Mammalian cells control Mg^{2+} levels within a narrow range (0.70–1.1 mmol/l) by many mechanisms, including intraorganelle compartmentalization, intracellular Mg^{2+} buffering, and transmembrane transport through ion channels and transporters.

Epidemiological studies indicate that low plasma and cellular Mg^{2+} levels are associated with various cardiovascular diseases including hypertension, stroke, cardiac disease, atherosclerosis, and diabetes/metabolic syndrome [6–10]. In the cardiovascular system, Mg^{2+} regulates vascular tone, endothelial function, cardiac rhythm, myocardial contractility, platelet function, and vascular cell growth [11, 12]. Increased intracellular free Mg^{2+} concentration ($[Mg^{2+}]_i$) causes vasodilation and attenuates agonist-induced vasoconstriction, whereas reduced $[Mg^{2+}]_i$ has opposite effects, leading to hypercontractility and impaired vasorelaxation [13–16]. Low $[Mg^{2+}]_i$ also results in abnormal vascular cell growth, inflammation, and increased intracellular Ca^{2+} accumulation in endothelial and vascular smooth muscle cells, which may be important in the regulation of endothelial function, vascular contraction, and tone [17, 18].

The understanding of Mg^{2+} homeostasis has been greatly enhanced since 2000; however, there is still a paucity of information regarding molecular mechanisms whereby cells regulate intracellular Mg^{2+} status. Transporters and exchangers implicated in transmembrane Mg^{2+} transport include, among others, the Na^+/Mg^{2+} exchanger, Mg^{2+}/Ca^{2+} exchanger, Mg^{2+} transporter subtype 1 (MagT1), and the transient receptor potential melastatin 6 and 7 cation channels (TRPM6, TRPM7) [19–23]. This chapter provides an update on the molecular biology and cellular regulation of Mg^{2+}, focusing specifically on the vascular system and discusses the role of vascular Mg^{2+} in cardiovascular health and disease (Fig. 14.1).

Cellular Mg Homeostasis

Unlike our knowledge of other major cations, mechanisms regulating cellular Mg^{2+} handling are still poorly understood. More than 95% of Mg^{2+} is sequestrated by chelators or bound to other biomolecules, including phospholipids, ribosomes, and phosphonucleotides (ATP, ADP) [24]. Intracellular Mg^{2+} is maintained below the concentration predicted from the transmembrane electrochemical potential.

Mg^{2+} efflux occurs against the electrochemical gradient; therefore, an energy-coupled mechanism for its extrusion must be present. Mg^{2+} efflux appears to be regulated by at least two pathways: the Na^+/Mg^{2+} exchange driven by the Na^+ gradient and the Na^+-independent "passive" Mg^{2+} transport via Mg^{2+}-permeable channels [25, 26]. Na^+-dependent Mg^{2+} transport occurs mainly via the Na^+/Mg^{2+} exchanger and has been demonstrated in many cell types, including VSMC and cardiomyocytes [27, 28]. On the other hand, Na^+-independent transport, demonstrated mainly in erythrocytes and hepatic cells, involves Ca^{2+} (Ca^{2+}/Mg^{2+} exchanger), Mn^{2+} (Mn^{2+}/Mg^{2+} antiporter), and Cl^- (Cl^-/Mg^{2+} co-transporter)-dependent mechanisms [29–31]. These exchangers have been identified at the functional level but not yet at the protein level.

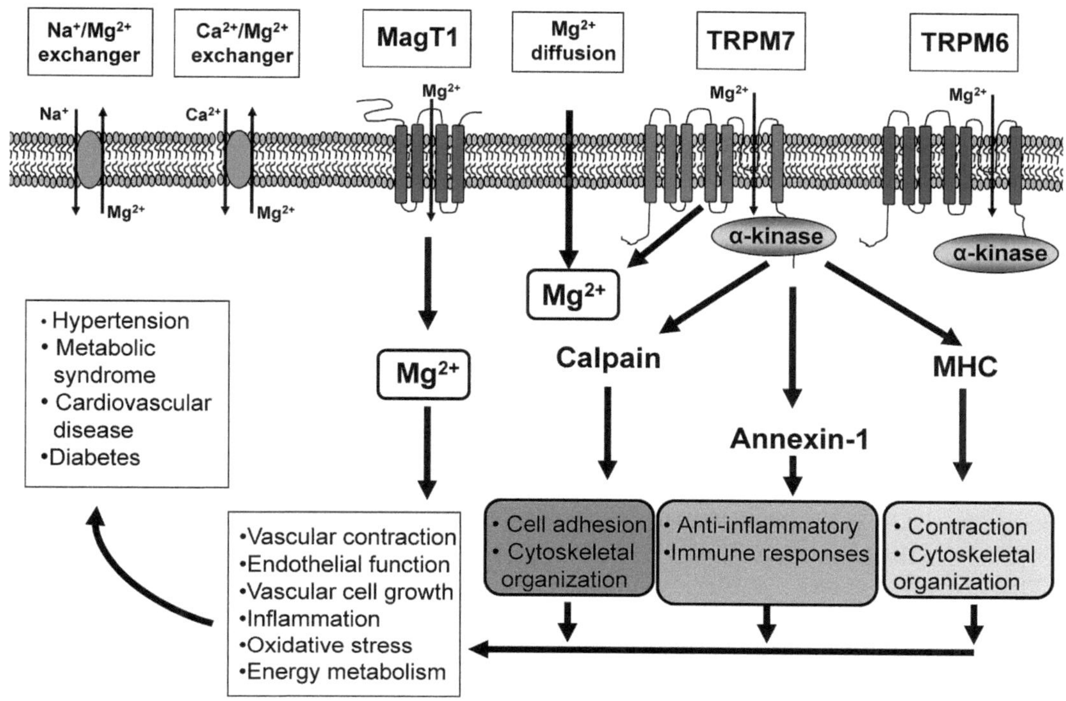

Fig. 14.1 Mechanisms regulating magnesium transport across the cell membrane and magnesium-dependent cellular responses that may impact on vascular function. Magnesium influx occurs through unique Mg^{2+} transporters including MagT1, TRPM6, and TRPM7 and efflux occurs through numerous Mg^{2+} exchangers. In vascular cells, changes in intracellular Mg^{2+} levels influence contraction, dilation, growth, inflammation, which are processes that contribute to vascular remodeling and altered tone in cardiovascular disease. Signaling through TRPM7 kinase targets, including calpain, annexin-1, and MHC, also influences cellular function. Whereas TRPM6 and TRPM7 have been identified in vascular cells, presence of MagT1 has yet to be demonstrated

Advances in the field have now identified specific transmembrane Mg^{2+} transporters. The first mammalian Mg^{2+} transporter to be identified at the molecular level was Mrs2 (mitochondrial RNA splicing2), responsible for mitochondrial Mg^{2+} uptake [32, 33]. Other proteins shown to regulate Mg^{2+} homeostasis include MagT1, the solute carrier (SLC) family 41 subtypes 1 and 2 (SLC41A1, SLAC41A2, respectively), and the ancient conserved domain protein 2 (ACDP2) [34–37]. Microarray analysis showed that the NIPA1 and 2 genes, named for "nonimprinted in Prader-Willi/Angelman," membrane Mg^{2+} transporters 1 and 2 (MMgT1 and 2, respectively), and Huntington interacting protein genes, HIP14 and HIP14L also encode an Mg^{2+} transporter [38–40].

Analysis of different forms of human disorders characterized by low serum Mg^{2+} levels due to defective intestinal absorption and/or renal Mg^{2+} wasting led to the identification of a paracellular (between cells) Mg^{2+} transporter, paracellin-1 (claudin-16), a member of the claudin family of tight-junction proteins [41]. Paracellin-1 mutations are associated with a hereditary disease, hypomagnesemia with hypercalciuria and nephrocalcinosis (FHHNC), characterized by massive renal Mg^{2+} and Ca^{2+} wasting leading to end-stage renal disease.

TRPM6 and TRPM7 were identified as Mg^{2+} transporters through genetic analyses of patients with primary hypomagnesemia and secondary hypocalcemia (HSH), another Mg^{2+}-wasting disorder [42, 43]. TRPM6 and TRPM7 are the only known examples of ion channels covalently linked to a kinase domain, making them unique channels. The ubiquitously expressed TRPM7 was characterized functionally as a constitutively active ion channel permeable for divalent cations including Mg^{2+} and regulated by intracellular concentrations of Mg^{2+}, magnesium-nucleotide complexes, and humoral factors [44, 45]. Whereas TRPM6 and TRPM7 are primarily Mg^{2+} transporters, they can also transport other ions, such as Ca^{2+}. Currently, the only known selective plasma membrane Mg^{2+} transporter is MagT1 [46].

TRPM6/7 Cation Channels, Mg^{2+}, and Vascular (Patho)Biology

TRPM proteins, members of the transient receptor potential (TRP) superfamily [47], have been implicated in cardiovascular pathologies [48, 49]. The founding member, TRPM1 (melastatin), was detected in malignant melanoma cells [50]. Members of the TRPM family are divided into four groups [50]: TRPM1/3, TRPM2/8, TRPM4/5, and TRPM6/7. TRPM channels exhibit highly varying cation permeability, from Ca^{2+} impermeable (TRPM4/5) to highly Ca^{2+} and Mg^{2+} permeable (TRPM6 and TRPM7). TRPM6 and TRPM7 share ~50% sequence homology, are constitutively active, and contain protein kinase domains in their COOH termini, which bear sequence similarity to elongation factor 2 (eEF-2) serine/threonine kinases [45].

TRPM6 mRNA shows a restricted expression pattern, with the highest levels along the intestine and the distal convoluted tubule (DCT) of the kidney [51–53]. The critical role of TRPM6 for epithelial Mg^{2+} transport became evident when positional cloning showed that HSH, an autosomal-recessive disorder and characterized by low serum Mg^{2+} levels due to diminished intestinal Mg^{2+} absorption and decreased renal Mg^{2+} reabsorption, is caused by mutations in the TRPM6 gene [53]. Factors implicated in TRPM6 regulation remain unclear, but 17beta-estradiol and Mg^{2+} are important [54]. TRPM6 currents are inhibited by millimolar concentrations of intracellular Mg^{2+} [55, 56]. Epidermal growth factor (EGF) has been identified as an autocrine/paracrine magnesiotropic hormone [57, 58]. EGF binds to its EGFR at the basolateral membrane of the DCT and stimulates endomembrane trafficking and increased number of TRPM6 channels at the plasma membrane [58, 59]. The signaling pathway downstream of EGFR activation, which stimulates TRPM6-mediated Mg^{2+} influx, is dependent on Src, ERK/MEK, PI3K, and AP-1 [58, 59]. The role of EGFR signaling in Mg^{2+} homeostasis is further supported by the clinical observation that patients treated with monoclonal antibodies directed against the EGFR, e.g., cetuximab, develop hypomagnesemia [60–63].

TRPM7 has a widespread distribution [64, 65]. In physiological conditions, TRPM7 preferentially transports Mg^{2+} and to a lesser extent Ca^{2+} and other divalent cations. Since Mg^{2+} plays such an important role in cell function, it seems unlikely that TRPM7 is the only transporter involved in maintaining intracellular Mg^{2+} concentration. In support of this, a recent study using primary mammalian cells found that deletion of TRPM7 did not alter cellular Mg^{2+} concentration [66]. In addition, cells that are deficient in TRPM7−/− exhibit increased expression of MagT1, possibly as a compensation to prevent cellular Mg^{2+} deficiency [67]. Several other studies demonstrated that targeted disruption of TRPM7 gene in cell lines is lethal, underpinning a salient and nonredundant role of this channel in cell physiology and cell survival. TRPM7 is constitutively active and like TRPM6 is negatively regulated by changes in cytosolic Mg^{2+} or Mg^{2+}-ATP [64, 65].

The dual ability of TRPM7 to act as a channel and at the same time as a kinase suggests that this protein is involved in regulating both cellular Mg^{2+} status and intracellular signaling [68]. The role of TRPM7 kinase domain for the channel function is still controversial. While some studies reported that TRPM7 mutants displayed reduced sensitivity for intracellular Mg^{2+} and reduced current amplitude upon G protein-coupled receptor stimulation or were even not functional, others found no changes in Mg^{2+} sensitivity of TRPM7 channels lacking the α-kinase domain versus wild-type channel [68, 69].

Since the α-kinase is adjacent to the conducting pore of TRPM7, changes in free Mg^{2+} levels may activate the associated kinase domain. The kinase in turn would signal via other phosphorylated substrates and/or feedback to modulate TRPM7 channel gating. Thus, TRPM7 could function as an "Mg^{2+} sensor" [70, 71].

A number of TRPM7 kinase substrates, all of which are important in regulating vascular cell function, have been identified. These include annexin-1, myosin IIA heavy chain, and calpain [72–74]. Annexin-1 is an endogenous mediator of the anti-inflammatory actions of glucocorticoids and is implicated in cell growth and apoptosis [72]. Myosin II heavy chain is involved in cell migration, growth, apoptosis, contraction, and cytoskeletal organization. TRPM7 also activates calpain and initiates the disassembly or turnover of peripheral adhesion complexes, which is involved in cell adhesion, contraction, motility, metastasis, apoptosis, and ischemic cell death. Other possible functions of the kinase domain include a role in phospholipase signaling [75]. The physiological significance of TRPM7-regulated downstream targets still remains unclear, since studies on these substrates were performed mainly in cell lines.

TRPM6 and TRPM7 have been implicated in a number of vascular pathologies, including endothelial dysfunction and hypertension [76, 77]. We and others demonstrated that vascular smooth muscle and endothelial cells from rodents and humans possess TRPM6 and TRPM7 cation channels and that TRPM7 is critically involved in regulating vascular cell Mg^{2+} influx, viability, proliferation, and contraction/dilation in vascular cells [71, 78]. Fluid flow and fluid-induced shear stress stimulates cytosol-to-membrane translocation of TRPM7 and increases TRPM7-like current in vascular cells [79]. These observations suggest that TRPM7 may act as a mechanotransducer, which could be important in pathological responses to vessel wall injury, particularly in the context of endothelial damage.

Angiotensin II and aldosterone, important vasoactive agents, regulate vascular TRPM7 acutely by inducing phosphorylation and chronically by increasing expression at the mRNA and protein levels [71, 78]. We identified a novel signaling pathway whereby bradykinin, through TRPM7, influences proinflammatory responses in vascular cells [79]. Downregulation of vascular TRPM7 by siRNA reduced basal $[Mg^{2+}]_i$ and angiotensin II-stimulated $[Mg^{2+}]_i$ transients and attenuated vascular smooth muscle cell growth. These findings confirmed that TRPM7 is a key regulator of vascular cell Mg^{2+} homeostasis and that it plays a major role in vascular cell function [80–82].

Vascular smooth muscle cells from spontaneously hypertensive rats exhibit downregulation of vascular TRPM7 and reduced Mg^{2+} influx, which may contribute to low $[Mg^{2+}]_i$ [71]. In experimental models of hypertension and in patients with hypertensive intracerebral hemorrhage, TRPM7 is downregulated, and this is associated with altered vascular cells Mg^{2+} transport, decreased $[Mg^{2+}]_i$, and impaired vasodilation [83, 84]. These data suggest that aberrant cellular Mg^{2+} homeostasis and abnormal vascular smooth muscle cell function in hypertension may be related to defective TRPM7 expression/activity which could contribute to proliferation, inflammation, fibrosis, and contraction, important in processes associated with vascular remodeling in cardiovascular diseases, such as hypertension and atherosclerosis.

Magnesium and Magnesium Transporter Subtype 1 (MagT1)

MagT1 is essential for Mg^{2+} influx, cell growth and development [46, 85]. Knockdown of MagT1 leads to early developmental arrest in zebrafish embryos, an effect that could be rescued by high supplemental Mg^{2+} [86]. In conditions where TRPM7 is downregulated or deficient, expression of MagT1 is increased indicating a close interaction between these transporters, possibly through changes in cellular Mg^{2+} concentrations. Hence, when cells are exposed to suboptimal cellular Mg^{2+} levels, they respond by transcriptional upregulation of MagT1 to increase Mg^{2+} influx, possibly as a protective mechanism to prevent intracellular Mg^{2+} deficiency [67, 87].

MagT1 is an important regulator of Mg^{2+} influx in T lymphocytes. Patients with MagT1 mutations exhibit immunodeficiency characterized by CD4 lymphopenia, viral infections, and defective T lymphocyte activation due to abnormal Mg^{2+} signaling [88, 89]. As such, MagT1 has been suggested to be a putative target for novel therapeutics in patients with defective T lymphocyte function [79].

Although MagT1 has a widespread distribution, nothing is known about this Mg^{2+} transporter in the cardiovascular system. However, considering the importance of Mg^{2+} in the regulation of cardiac and vascular function, future studies may identify MagT1 as a major Mg^{2+} transporter in the cardiovascular system.

Magnesium and Vascular Function

Mg^{2+}influences vascular tone by regulating endothelial function and vascular smooth muscle cell contraction [90–93]. Mg^{2+} stimulates production of vasodilators, such as prostacyclin and nitric oxide (NO), and promotes endothelium-dependent and endothelium-independent vasodilation [94, 95]. Isolated vessels exposed to reduced levels of Mg^{2+} display a transient vasorelaxation followed by sustained constriction. In the presence of endothelial damage, low Mg^{2+} induces a sustained contraction without the transient vasorelaxation phase [96, 97], suggesting that Mg^{2+} could have a dual effect in the regulation of vascular reactivity, depending on the integrity of the endothelium. An intact endothelium prevents against the unfavorable effects of hypomagnesemia, whereas in the presence of endothelial damage, as is the case in many cardiovascular diseases, the compensatory vasodilatory effect is absent and low Mg^{2+} promotes constriction [98, 99].

Mg^{2+} also modulates vascular tone and reactivity by altering responses to vasoconstrictor and vasodilator agents. Increased extracellular Mg^{2+}concentration blunts vasoconstrictor actions and potentiates vasorelaxant properties of vasoactive agents [100–103]. These effects may be related to altered binding of agonists to their specific cell membrane receptors and/or to production of vasoactive agents such as endothelin-1 (ET-1), angiotensin II (Ang II), and prostacyclin (PGI_2) [104, 105]. In Mg^{2+}-deficient rats, plasma ET-1 levels are elevated, whereas in Mg^{2+}-supplemented rats, plasma ET-1 levels are reduced [106, 107]. Increased Mg^{2+} attenuates ET-1-induced contraction, and reduced Mg^{2+} levels augment ET-1-stimulated contraction [107, 108]. Mg^{2+} stimulates endothelial release of vasodilator PGI_2 from human umbilical arteries and cultured umbilical vein endothelial cells [109]. These effects may be particularly relevant in $MgSO_4$ treatment of eclampsia and preeclampsia [109, 110].

Another possible mechanism whereby Mg^{2+} could have an impact on vascular function is via its antioxidant, anti-inflammatory, and growth-regulatory properties [111, 112]. Mg^{2+} has antioxidant/anti-inflammatory properties scavenging oxygen radicals possibly by the spontaneous dismutation of superoxide anions ($\bullet O_2^-$). Mg^{2+} is also an essential requirement for the synthesis of some important natural antioxidants that could attenuate damaging actions of oxidative stress and inflammation in the vasculature [113]. In fact, Mg^{2+} deficiency enhanced oxidative damage, increased plasma levels of lipoproteins, and induced oxidative damage to erythrocytes, neuronal, and vascular cells [71]. Increased production of $\bullet O_2^-$ and hydrogen peroxide (H_2O_2) provoked by reduced Mg^{2+} levels is associated not only with contraction, hypertrophy, and hyperplasia of VSMC but also with inflammatory responses and subsequent tissue injury [114, 115]. There is a growing interest in the role of Mg^{2+} in inflammation and cardiovascular disease [116, 117].

Magnesium and Cardiovascular Disease

Epidemiological studies have linked the prevalence of cardiovascular diseases, including hypertension, ischemic heart disease, arrhythmias, and stroke, with "soft water," low in Mg^{2+}, and protection against cardiovascular disease with "hard water," high in Mg^{2+} [116–118].

1. Hypertension

   The best epidemiological evidence linking Mg^{2+} and blood pressure comes from the Honolulu Heart study, in which the relationships of dietary variables with blood pressure were examined [116–119]. Of all the nutrients, dietary Mg^{2+} intake had the strongest (inverse) relationship with blood pressure. Numerous subsequent epidemiological and clinical investigations further supported the view that increased Mg^{2+} intake contributes to prevention of hypertension and cardiovascular disease [120, 121].

   Some clinical studies have shown hypomagnesemia (serum and/or tissue) in hypertensive patients, with significant negative correlations between Mg^{2+} concentration and blood pressure [122, 123]. A relationship has also been described between the renin-angiotensin system, Mg^{2+}, and blood pressure. High rennin hypertensive patients have lower serum Mg^{2+} levels than normotensive subjects [124, 125] and serum Mg^{2+} is inversely associated with plasma renin activity. Recent studies have also reported a negative dependency between $[Mg^{2+}]_i$ and arterial compliance in humans, the lower the $[Mg^{2+}]_i$, the stiffer the blood vessels and the greater the blood pressure [126]. In earlier investigations, total Mg^{2+} levels were determined. Today, with the availability of selective fluorescent Mg^{2+} probes and Mg^{2+}-specific ion-selective electrodes, which measure $[Mg^{2+}]_i$ in living cells, experimental evidence indicates that many cell types from hypertensive patients have significantly lower $[Mg^{2+}]_i$ than cells from normotensive subjects, even if total Mg^{2+} levels are within the normal range [127]. Underlying causes for altered Mg^{2+} metabolism in hypertension are unclear, but genetic, dietary, hormonal factors, or drugs may play a role.

   Not all clinical investigations have reported Mg^{2+} deficiency in hypertension. Some studies found no differences in serum Mg^{2+} levels or in $[Mg^{2+}]_i$ in hypertensive patients, while others reported increased erythrocyte $[Mg^{2+}]_i$ in patients with essential hypertension [128–130]. A few epidemiological studies also failed to show associations between Mg^{2+} intake and blood pressure [130, 131]. It is evident that not all hypertensive patients are hypomagnesemic and not all patients with Mg^{2+} deficiency are hypertensive. Despite the inconsistencies in the literature regarding Mg^{2+} status in hypertension, there are subgroups of hypertensive patients who consistently demonstrate altered Mg^{2+} metabolism, including the elderly and African-Americans [132, 133]. Clinically, there are a number of well-defined conditions where high blood pressure is associated with hypomagnesemia, including hyperaldosteronism, use of diuretics, calcineurin inhibitors, and anti-epidermal growth factor receptor (EGFR) monoclonal antibodies used to treat colorectal cancer [134–136].

   The therapeutic value of Mg^{2+} in the treatment of clinical hypertension was suggested in 1925 when Mg^{2+} infusion was found to improve malignant hypertension [137]. Since then, many investigations have supported a putative role for Mg^{2+} in the treatment of hypertension. In general, data from clinical trials of Mg^{2+} therapy in hypertension have been disappointing. Some studies reported significant blood pressure-lowering effects of oral or intravenous Mg^{2+} treatment [138, 139], and Mg^{2+} supplementation to patients already receiving diuretics or other antihypertensive agents appears to reduce blood pressure further [140, 141]. However, other trials have failed to demonstrate any hypotensive action of Mg^{2+} supplementation [142, 143], and results from the Trial of Hypertension Prevention (TOHP) study showed no benefit of Mg^{2+} therapy in 698 patients followed for 6 months [144]. The inconsistency in results may be due to the number of studies that were small or of short duration, differing treatment protocols, variable forms of Mg^{2+} salts used, different concentrations of Mg^{2+} supplemented, and the heterogeneity of the population of hypertensives investigated. Nonetheless, studies have consistently shown a beneficial effect of Mg^{2+}

treatment in African-American patients, those with established hypomagnesemia, those with diuretic-associated hypertension, and in patients where Mg^{2+} was supplemented long term [132, 133]. In addition, Mg^{2+} is therapeutically effective in lowering blood pressure in secondary hypertension and in preeclampsia [145, 146]. Taken together, these data suggest Mg^{2+} is at best weakly hypotensive. Thus, although Mg^{2+} may not be a universally effective antihypertensive agent, it does seem to benefit a subgroup of hypertensive patients.

2. Magnesium, Preeclampsia, and Eclampsia
Preeclampsia, defined as hypertension after 20 weeks of gestation with proteinuria [147], has been treated with Mg^{2+} salts since the early 1900s. During preeclampsia, both cardiac output and plasma volume are reduced, whereas systemic vascular resistance is increased [148]. These changes result in reduced perfusion of the placenta, kidney, liver, and brain, leading to maternal and fetal morbidity and mortality [148]. Mg^{2+} has been shown to improve endothelial function in preeclampsia [149]. This may be due to the direct vasodilatory properties of Mg^{2+} and/or to the ability of Mg^{2+} to stimulate release of the endothelial vasodilator prostacyclin, which induces vasodilation and inhibits platelet adherence and aggregation [150].
Acute Mg^{2+} sulfate administration elicits a rapid fall in systemic vascular resistance, a rise in the cardiac index, and a transient decrease in blood pressure [151]. Mg^{2+} sulfate infusion also increases renal blood flow and stimulates production and release of prostacyclin in preeclampsia but not in preterm labor [152, 153]. In healthy nonpregnant women, these effects are inhibited by indomethacin, a cyclooxygenase inhibitor, suggesting that the fall in blood pressure is mediated by Mg^{2+}-induced prostacyclin release [154].
Data relating to Mg^{2+} concentrations in preeclampsia are conflicting. Some studies demonstrate no differences between preeclamptic versus uncomplicated pregnancies. Others report decreased serum and intracellular Mg^{2+} in preeclamptic women [155, 156]. Although the exact role of Mg^{2+} in the pathogenesis of preeclampsias is still unclear, it has been suggested that Mg^{2+} can be used as a predictive tool for preeclampsia. Standley et al. [157] found that Mg^{2+} decreases in both preeclamptic and uncomplicated pregnancies but that the Mg^{2+} concentration was lowered earlier in women with preeclampsia. This difference has been proposed as a marker of severity of the condition. It has also been suggested that alterations in the Na^+/Mg^{2+} exchanger in trophoblast cells may be important [158].
Mg^{2+} sulfate remains the most frequently used treatment in the management of preeclampsia and eclampsia [159]. The Collaborative Eclampsia Trial provides level I evidence (high level of evidence) of the superiority of Mg^{2+} sulfate for the treatment of eclampsia. Mg^{2+} sulfate had a 52% lower risk of recurrent convulsions versus diazepam and a 67% lower risk of recurrent convulsions versus phenytoin [160]. Use of Mg^{2+} sulfate for prophylactic treatment in women with preeclampsia is more controversial. A recent large trial among severe preeclamptic women compared Mg^{2+} sulfate with placebo. The trial was terminated prematurely after finding a significant reduction in the development of eclampsia with Mg^{2+} sulfate (0.3% vs. 3.2%) [161]. The largest clinical trial comparing Mg^{2+} sulfate with phenytoin in hypertensive pregnancies also reported that eclampsia was significantly reduced in women taking Mg^{2+} sulfate compared to those on phenytoin [162]. The Magpie trial, which involved 10,141 women with preeclampsia in 175 hospitals in 33 countries recently showed that Mg^{2+} sulfate significantly reduces the risks of eclampsia among women with preeclampsia [163]. These data clearly demonstrate that Mg^{2+} sulfate has a very important role in preventing as well as controlling eclampsia, and the available evidence suggests that adverse effects are minimal. Although the use of Mg^{2+} sulfate for eclampsia is well substantiated, there is little evidence supporting the routine use of Mg^{2+} sulfate in pregnancy-induced hypertension (gestational hypertension).

3. Acute Myocardial Infarction
Results from autopsy studies demonstrated lower myocardial and muscle total Mg^{2+} in subjects who died from ischemic heart disease as compared to those who died from non-cardiac causes [164].

During myocardial ischemia, free ionized intracellular Mg^{2+} increases, while total intracellular Mg^{2+} decreases [165]. The reduction in free Mg^{2+} in ischemia has been attributed, in part, to decreased ATP content [166]. Furthermore, ischemia leads to intracellular calcium overload, which is exacerbated during reperfusion. Mg^{2+} administration may confer cellular protection during ischemia by acting as a calcium antagonist, thereby reducing calcium overload, by conserving cellular ATP as the Mg^{2+} salt and thereby preserving energy-dependent cellular activity, by improving myocardial contractility, and by limiting infarct size [167]. Furthermore, Mg^{2+} may prevent arrhythmias associated with ischemia, decrease catecholamine release, and protect against oxidative stress-induced cardiac damage [168].

Given the above Mg^{2+} effects, it is not surprising that Mg^{2+} therapy has been studied extensively in the context of acute myocardial infarction. Numerous small clinical trials reported that Mg^{2+} administration is a safe and effective method of reducing arrhythmias and mortality in acute myocardial infarction. The second Leicester Intravenous Magnesium Intervention Trial (LIMIT2) was the first randomized, double-blind, placebo-controlled study demonstrating that intravenous Mg^{2+} therapy has a protective effect during the treatment of acute myocardial infarction [169]. 2,316 patients with acute myocardial infarction were allocated randomly to receive Mg^{2+} or placebo. Treatment consisted of Mg^{2+} 8 mmol over 5 min before thrombolytic therapy, followed by 65 mmol as an infusion over the following 24 h. There was a 24% relative reduction in mortality after 28 days and a 25% lower incidence of left ventricular failure. The reduction in left ventricular failure was associated with a corresponding reduction in mortality from ischemic heart disease over a mean follow-up period of 2.7 years. The conclusion from LIMIT 2 was that early intravenous Mg^{2+} is a useful addition to standard therapy in acute myocardial infarction.

However, findings from two recent mega-trials, the Fourth International Study of Infarct Survival (ISIS 4) [170] and the Magnesium in Coronaries (MAGIC) trial [171], failed to demonstrate a beneficial effect of Mg^{2+} therapy in acute myocardial infarction. In ISIS 4, effects of early intervention with oral captopril, isosorbide-5-mononitrate, or intravenous $MgSO_4$ were assessed in 58,050 patients with suspected acute myocardial infarction. Results showed a trend toward increased mortality at 35 days with an excess incidence of cardiogenic shock and cardiac failure in the Mg^{2+} group, although there was a significant reduction in the early occurrence of ventricular fibrillation. No benefit was observed in the treatment group across all major subgroups, whether they were treated early or late and whether or not they received thrombolysis. The aim of the MAGIC trial was to compare short-term mortality in patients with ST-elevation myocardial infarction who received either intravenous MgSO4 or placebo. 6,213 patients were randomly assigned to a 2 g intravenous bolus of $MgSO_4$ (8 mmol elemental Mg^{2+}) administered over 15 min, followed by a 17 g infusion of $MgSO_4$ (68 mmol elemental Mg^{2+}) over 24 h ($n = 3,113$) or matching placebo ($n = 3,100$). At 30 days, 15.3% patients in the Mg^{2+} group and 15.2% in the placebo group had died. No benefit or harm of Mg^{2+} was observed.

Differences in results between LIMIT 2, ISIS 4, and MAGIC have led to confusion as to whether or not Mg^{2+} should be administered routinely as first-line therapy during the acute phase of myocardial infarction. Findings from a meta-analysis of all randomized controlled studies of Mg^{2+} in acute myocardial infarction (68,684 patients) demonstrated that patients at low risk of mortality from acute myocardial infarction and who benefit from thrombolysis and aspirin probably gain little benefit from Mg^{2+} therapy [172]. This was further supported by an Italian study, which demonstrated that intravenous Mg^{2+}, delivered before, during, and after reperfusion, did not decrease myocardial damage and did not improve short-term clinical outcomes in patients with acute myocardial infarction treated with direct angioplasty [173]. In high-risk patients who may not be suitable for thrombolysis, Mg^{2+} appears to be useful [174]. Overall, there is no indication for the routine administration of intravenous Mg^{2+} in patients with acute myocardial infarction. However, Mg^{2+} is well tolerated, and there is no apparent harm from its use. Hence, Mg^{2+} can be continued to be administered for repletion of documented electrolyte deficits and for life-threatening arrhythmias.

Conclusions

Inadequate Mg^{2+} intake and hypomagnesemia may contribute to the pathophysiology of cardiovascular diseases. Mg^{2+} normally regulates vascular tone and reactivity by modulating Na^+, K^+, and Ca^{2+} and by influencing activity of multiple enzymes. Several studies have demonstrated that Mg^{2+} plays an important role in the maintenance of vascular integrity, where decreased Mg^{2+} content is associated with endothelial dysfunction, increased reactivity, enhanced contractility, vascular remodeling and inflammation, and elevated blood pressure. On the other hand, increased vascular Mg^{2+} levels are associated with vasodilation, anti-inflammatory responses, and reduced blood pressure. Unlike our knowledge of other major cations, cellular Mg^{2+} handling is poorly understood, and until recently, little was known about pathways regulating Mg^{2+} transport. To date, a number of Mg^{2+} efflux pathways, including the Na^+/Mg^{2+} antiporter and the Ca^{2+}/Mg^{2+} exchanger, have been demonstrated at a functional level. Recently, TRPM6 and TRPM7 have been identified as unique Mg^{2+} transporters regulating Mg^{2+} influx. Vascular TRPM7 expression/activity appears to be altered in experimental models of hypertension. Such aberrations may contribute to Mg^{2+} dysregulation and vascular dysfunction associated with high blood pressure. There is still much to be discovered regarding the function and regulation of TRPM6, TRPM7, and TRPM6/7 and of the importance of these channels in cardiovascular biology in health and disease.

Acknowledgement Work from the author's laboratory was supported by grants from the Canadian Institutes of Health Research (CIHR) and the Herat and Stroke Foundation of Canada. RMT is supported through a Canada Research Chair/ Canadian Foundation for Innovation award. CAN is supported through a fellowship from the CIHR.

References

1. Wolf FI, Cittadini A. Magnesium in cell proliferation and differentiation. Front Biosci. 1999;4:D607–17.
2. Rubin H. The logic of the Membrane, Magnesium, Mitosis (MMM) model for the regulation of animal cell proliferation. Arch Biochem Biophys. 2007;458(1):16–23.
3. Fox C, Ramsoomair D, Carter C. Magnesium: its proven and potential clinical significance. South Med J. 2001;94:1195–201.
4. Alexander RT, Hoenderop JG, Bindels RJ. Molecular determinants of magnesium homeostasis: insights from human disease. J Am Soc Nephrol. 2008;19(8):1451.
5. Romani AM. Cellular magnesium homeostasis. Arch Biochem Biophys. 2011;512(1):1–23.
6. Spiegel DM. Magnesium in chronic kidney disease: unanswered questions. Blood Purif. 2011;31(1–3):172–6.
7. Guerrero-Romero F, Rodriguez-Moran M. Low serum magnesium levels and metabolic syndrome. Acta Diabetol. 2002;39:209–13.
8. Swaminathan R. Magnesium metabolism and its disorders. Clin Biochem Rev. 2003;24:47–66.
9. Barbagallo M, Dominguez LJ. Magnesium metabolism in type 2 diabetes mellitus, metabolic syndrome and insulin resistance. Arch Biochem Biophys. 2007;458:40–7.
10. Resnick LM, Laragh JH, Sealey JE, Alderman MH. Divalent cations in essential hypertension. Relations between serum ionized calcium, magnesium, and plasma renin activity. N Engl J Med. 1983;309(15):888–91.
11. Sontia B, Touyz RM. Magnesium transport in hypertension. Pathophysiology. 2007;14:205–11.
12. Shechter M, Merz CN, Paul-Labrador M, Meisel SR, Rude RK, Molloy MD, et al. Oral magnesium supplementation inhibits platelet-dependent thrombosis in patients with coronary artery disease. Am J Cardiol. 1999;84: 152–6.
13. Touyz RM, Laurant P, Schiffrin EL. Effect of magnesium on calcium responses to vasopressin in vascular smooth muscle cells of spontaneously hypertensive rats. J Pharmacol Exp Ther. 1998;284:998–1005.
14. Yang ZW, Wang J, Zheng T, Altura BT, Altura BM. Low $[Mg^{2+}]_o$ induces contraction and $[Ca^{2+}]_i$ rises in cerebral arteries: roles of Ca^{2+}, PKC, and PI3. Am J Physiol Heart Circ Physiol. 2000;279:H2898–907.
15. Northcott CA, Watts SW. Low $[Mg^{2+}]_e$ enhances arterial spontaneous tone via phosphatidylinositol 3-kinase in DOCA-salt hypertension. Hypertension. 2004;43:125–9.
16. Resnick LM, Gupta RK, Bhargava KK, Gruenspan H, Alderman MH, Laragh JH. Cellular ions in hypertension, diabetes, and obesity. A nuclear magnetic resonance spectroscopic study. Hypertension. 1991;17(6 Pt 2):951–7.

17. Yoshimura M, Oshima T, Matsuura H, Ishida T, Kambe M, Kajiyama G. Extracellular Mg^{2+} inhibits capacitance Ca^{2+} entry in vascular smooth muscle cells. Circulation. 1997;95(11):2567–72.
18. Resnick LM. Cellular calcium and magnesium metabolism in the pathophysiology and treatment of hypertension and related metabolic disorders. Am J Med. 1992;93(2A):115–205.
19. Schmitz C, Perraud AL, Johnson CO, Inabe K, Smith MK, Penner R, et al. Regulation of vertebrate cellular Mg2+ homeostasis by TRPM7. Cell. 2003;114(2):191–200.
20. Zoller MK, Hermosura MC, Nadler MJ, Scharenberg AM, Penner R, Fleig A. TRPM7 provides an ion channel mechanism for cellular entry of trace metal ions. J Gen Physiol. 2003;121(1):49–60.
21. Schmitz C, Dorovkov MV, Zhao X, Davenport BJ, Ryazanov AG, Perraud AL. The channel kinases TRPM6 and TRPM7 are functionally nonredundant. J Biol Chem. 2005;280(45):37763–71.
22. Romani AM. Regulation of magnesium homeostasis and transport in mammalian cells. Arch Biochem Biophys. 2007;458(1):90–102.
23. Schlingmann KP, Waldegger S, Konrad M, Chubanov V, Gudermann T. TRPM6 and TRPM7-gatekeepers of human magnesium metabolism. Biochim Biophys Acta. 2007;1772:81–821.
24. Romani AM, Maguire ME. Hormonal regulation of Mg2+ transport and homeostasis in eukaryotic cells. Biometals. 2002;15(3):271–83.
25. Flatman PW, Smith LM. Magnesium transport in ferret red cells. J Physiol. 1990;431:11–25.
26. Handy RD, Gow IF, Ellis D, Flatman PW. Na^+-dependent regulation of intracellular free magnesium concentration in isolated rat ventricular myocytes. J Mol Cell Cardiol. 1996;28:1641–51.
27. Okada K, Ishikawa S, Saito T. Cellular mechanisms of vasopressin and endothelin to mobilize [Mg2+]i in vascular smooth muscle cells. Am J Physiol. 1992;263(4 Pt 1):C873–8.
28. Fagan TE, Romani A. Activation of Na(+)- and Ca(2+)-dependent Mg(2+) extrusion by alpha(1)- and beta-adrenergic agonists in rat liver cells. Am J Physiol Gastrointest Liver Physiol. 2000;279(5):G943–50.
29. Almulla HA, Bush PG, Steele MG, Ellis D, Flatman PW. Loading rat heart myocytes with Mg2+ using low-[Na+] solutions. J Physiol. 2006;575(Pt 2):443–54.
30. Gunther T. Mechanisms, regulation and pathologic significance of Mg2+ efflux from erythrocytes. Magnes Res. 2006;19(3):190–8.
31. Cefaratti C, Romani AM. Functional characterization of two distinct Mg(2+) extrusion mechanisms in cardiac sarcolemmal vesicles. Mol Cell Biochem. 2007;303(1–2):63–72.
32. Bui DM, Gregan J, Jarosch E, Ragnini A, Schweyen RJ. The bacterial magnesium transporter CorA can functionally substitute for its putative homologue Mrs2p in the yeast inner mitochondrial membrane. J Biol Chem. 1999;274:20438–43.
33. Kolisek M, Zsurka G, Samaj J, Weghuber J, Schweyen RJ, Schweigel M. Mrs2p is an essential component of the major electrophoretic Mg2+ influx system in mitochondria. EMBO J. 2003;22:1235–44.
34. Goytain A, Quamme GA. Identification and characterization of a novel mammalian Mg2+ transporter with channel-like properties. BMC Genomics. 2005;6(1):48.
35. Goytain A, Quamme GA. Functional characterization of human SLC41A1, a Mg2+ transporter with similarity to prokaryotic MgtE Mg2+ transporters. Physiol Genomics. 2005;21(3):337–42.
36. Goytain A, Quamme GA. Functional characterization of ACDP2 (ancient conserved domain protein), a divalent metal transporter. Physiol Genomics. 2005;22:382–9.
37. Sahni J, Nelson B, Scharenberg AM. SLC41A2 encodes a plasma-membrane Mg2+ transporter. Biochem J. 2007;401(2):505–13.
38. Goytain A, Hines RM, El-Husseini A, Quamme GA. NIPA1 (SPG6), the basis for autosomal dominant form of hereditary spastic paraplegia encodes a functional Mg2 transporter. J Biol Chem. 2007;282:8060–8.
39. Goytain A, Quamme GA. Identification and characterization of a novel family of magnesium transporters, MMgT1 and MMgT2. Am J Physiol Cell Physiol. 2008;294:C495–502.
40. Goytain A, Hines RM, Quamme GA. Huntingtin-interacting proteins, HIP14 and HIP14L, mediate dual functions: palmitoyl acyltransferase and Mg2+ transport. J Biol Chem. 2008;283:33365–74.
41. Simon DB, Lu Y, Choate KA, Velazquez H, Al-Sabban E, Praga M, et al. Paracellin-1, a renal tight junction protein required for paracellular Mg2+ resorption. Science. 1999;285(5424):103–6.
42. Schlingmann KP, Weber S, Peters M, Niemann Nejsum L, Vitzthum H, Klingel K, et al. Hypomagnesemia with secondary hypocalcemia is caused by mutations in TRPM6, a new member of the TRPM gene family. Nat Genet. 2002;31:166–70.
43. Walder RY, Landau D, Meyer P, Shalev H, Tsolia M, Borochowitz Z, et al. Mutation of TRPM6 causes familial hypomagnesemia with secondary hypocalcemia. Nat Genet. 2002;31(2):171–4.
44. Nadler MJ, Hermosura MC, Inabe K, Perraud AL, Zhu Q, Stokes AJ, et al. LTRPC7 is a Mg ATP-regulated divalent cation channel required for cell viability. Nature. 2001;411:590–5.
45. Runnels LW, Yue L, Clapham DE. TRP-PLIK, a bifunctional protein with kinase and ion channel activities. Science. 2001;291:1043–7.
46. Quamme GA. Molecular identification of ancient and modern mammalian magnesium transporters. Am J Physiol Cell Physiol. 2010;298(3):C407–29.

47. Kraft R, Harteneck C. The mammalian melastatin-related transient receptor potential cation channels: an overview. Pflugers Arch. 2005;451(1):204–11.
48. Watanabe H, Murakami M, Ohba T, Ono K, Ito H. The pathological role of transient receptor potential channels in heart disease. Circ J. 2009;73:419–27.
49. Ando J, Yamamoto K. Vascular mechanobiology. Circ J. 2009;73:1983–92.
50. Wolf FI, Trapani V. MagT1: a highly specific magnesium channel with important roles beyond cellular magnesium homeostasis. Magnes Res. 2011;24(3):S86–91.
51. Chubanov V, Gudermann T, Schlingmann KP. Essential role for TRPM6 in epithelial magnesium transport and body magnesium homeostasis. Pflugers Arch. 2005;451(1):228–34.
52. Chubanov V, Schlingmann KP, Waring J, Heinzinger J, Kaske S, Waldegger S, et al. Hypomagnesemia with secondary hypocalcemia due to a missense mutation in the putative pore-forming region of TRPM6. J Biol Chem. 2007;282(10):7656–67.
53. Voets T, Nilius B, Hoefs S, van der Kemp AW, Droogmans G, Bindels RJ, et al. TRPM6 forms the Mg2+ influx channel involved in intestinal and renal Mg2+absorption. J Biol Chem. 2004;279(1):19–25.
54. Groenestege WM, Hoenderop JG, van den Heuvel L, Knoers N, Bindels RJ. The epithelial Mg^{2+} channel transient receptor potential melastatin 6 is regulated by dietary Mg^{2+} content and estrogens. J Am Soc Nephrol. 2006;17(4):1035–43.
55. Ferrè S, Hoenderop JG, Bindels RJ. Insight into renal Mg^{2+} transporters. Curr Opin Nephrol Hypertens. 2011;20(2):169–76.
56. Runnels LW. TRPM6 and TRPM7: a Mul-TRP-PLIK-cation of channel functions. Curr Pharm Biotechnol. 2011;12(1):42–53.
57. Groenestege WM, Thebault S, van der Wijst J, van den Berg D, Janssen R, Tejpar S, et al. Impaired basolateral sorting of pro-EGF causes isolated recessive renal hypomagnesemia. J Clin Invest. 2007;117:2260–7.
58. Thebault S, Alexander RT, Tiel Groenestege WM, Hoenderop JG, Bindels RJ. EGF increases TRPM6 activity and surface expression. J Am Soc Nephrol. 2009;20(1):78–85.
59. Ikari A, Sanada A, Okude C, Sawada H, Yamazaki Y, Sugatani J, et al. Up-regulation of TRPM6 transcriptional activity by AP-1 in renal epithelial cells. J Cell Physiol. 2010;222(3):481–7.
60. Schrag D, Chung KY, Flombaum C, Saltz L. Cetuximab therapy and symptomatic hypomagnesemia. J Natl Cancer Inst. 2005;97:1221–4.
61. Tejpar S, Piessevaux H, Claes K, Piront P, Hoenderop JG, Verslype C, et al. Magnesium wasting associated with epidermal growth- factor receptor-targeting antibodies in colorectal cancer: a prospective study. Lancet Oncol. 2007;8:387–94.
62. Costa A, Tejpar S, Prenen H, Van Cutsem E. Hypomagnesaemia and targeted anti-epidermal growth factor receptor (EGFR) agents. Target Oncol. 2011;6(4):227–33.
63. Melichar B, Králíčková P, Hyšpler R, Kalábová H, Cerman Jr J, Holečková P, et al. Hypomagnesaemia in patients with metastatic colorectal carcinoma treated with cetuximab. Hepatogastroenterology. 2012;59(114):366–71.
64. Holzer P. Transient receptor potential (TRP) channels as drug targets for diseases of the digestive system. Pharmacol Ther. 2011;131(1):142–70.
65. Rychkov GY, Barritt GJ. Expression and function of TRP channels in liver cells. Adv Exp Med Biol. 2011;704:667–86.
66. Jin J, Desai BN, Navarro B, Donovan A, Andrews NC, Clapham DE. Deletion of Trpm7 disrupts embryonic development and thymopoiesis without altering Mg^{2+} homeostasis. Science. 2008;322:756–60.
67. Deason-Towne F, Perraud AL, Schmitz C. The Mg^{2+} transporter MagT1 partially rescues cell growth and Mg^{2+} uptake in cells lacking the channel-kinase TRPM7. FEBS Lett. 2011;585(14):2275–8.
68. Demeuse P, Penner R, Fleig A. TRPM7 channel is regulated by magnesium nucleotides via its kinase domain. J Gen Physiol. 2006;127(4):421–34.
69. Yamaguchi H, Matsushita M, Nairn AC, Kuriyan J. Crystal structure of the atypical protein kinase domain of a TRP channel with phosphotransferase activity. Mol Cell. 2001;7(5):1047–57.
70. Matsushita M, Kozak JA, Shimizu Y, McLachlin DT, Yamaguchi H, Wei FY, et al. Channel function is dissociated from the intrinsic kinase activity and autophosphorylation of TRPM7/ChaK1. J Biol Chem. 2005;280:20793–803.
71. Sontia B, Montezano AC, Paravicini T, Tabet F, Touyz RM. Downregulation of renal TRPM7 and increased inflammation and fibrosis in aldosterone-infused mice: effects of magnesium. Hypertension. 2008;51(4): 915–21.
72. Dorovkov MV, Ryazanov AG. Phosphorylation of annexin I by TRPM7 channel-kinase. J Biol Chem. 2004;279(49):50643–6.
73. Clark K, Langeslag M, van Leeuwen B, Ran L, Ryazanov AG, Figdor CG, et al. TRPM7, a novel regulator of actomyosin contractility and cell adhesion. EMBO J. 2006;25(2):290–301.
74. Su LT, Agapito MA, Li M, Simonson WT, Huttenlocher A, Habas R, et al. TRPM7 regulates cell adhesion by controlling the calcium-dependent protease calpain. J Biol Chem. 2006;281(16):11260–70.

75. Sahni J, Scharenberg AM. TRPM7 ion channels are required for sustained phosphoinositide 3-kinase signaling in lymphocytes. Cell Metab. 2008;8:84–93.
76. Paravicini TM, Yogi A, Mazur A, Touyz RM. Dysregulation of vascular TRPM7 and annexin-1 is associated with endothelial dysfunction in inherited hypomagnesemia. Hypertension. 2009;53:423–9.
77. He Y, Yao G, Savoia C, Touyz RM. Transient receptor potential melastatin 7 ion channels regulate magnesium homeostasis in vascular smooth muscle cells: role of angiotensin II. Circ Res. 2005;96(2):207–15.
78. Touyz RM, He Y, Montezano AC, Yao G, Chubanov V, Gudermann T, et al. Differential regulation of transient receptor potential melastatin 6 and 7 cation channels by Ang II in vascular smooth muscle cells from spontaneously hypertensive rats. Am J Physiol Regul Integr Comp Physiol. 2006;290(1):R73–8.
79. Oancea E, Wolfe JT, Clapham DE. Functional TRPM7 channels accumulate at the plasma membrane in response to fluid flow. Circ Res. 2006;98(2):245–53.
80. Yogi A, Callera GE, Tostes R, Touyz RM. Bradykinin regulates calpain and proinflammatory signaling through TRPM7-sensitive pathways in vascular smooth muscle cells. Am J Physiol Regul Integr Comp Physiol. 2009;296(2):R201–7.
81. Zholos A, Johnson C, Burdyga T, Melanaphy D. TRPM channels in the vasculature. Adv Exp Med Biol. 2011;704:707–29.
82. Bates-Withers C, Sah R, Clapham DE. TRPM7, the Mg(2+) inhibited channel and kinase. Adv Exp Med Biol. 2011;704:173–83.
83. Thilo F, Suess O, Liu Y, Tepel M. Decreased expression of transient receptor potential channels in cerebral vascular tissue from patients after hypertensive intracerebral hemorrhage. Clin Exp Hypertens. 2011;33(8):533–7.
84. Yogi A, Callera GE, O'Connor SE, He Y, Correa JW, Tostes RC, et al. Dysregulation of renal transient receptor potential melastatin 6/7 but not paracellin-1 in aldosterone-induced hypertension and kidney damage in a model of hereditary hypomagnesemia. J Hypertens. 2011;29(7):1400–10.
85. Zhou H, Clapham DE. Mammalian MagT1 and TUSC3 are required for cellular magnesium uptake and vertebrate embryonic development. Proc Natl Acad Sci U S A. 2009;106(37):15750.
86. Schmitz C, Deason F, Perraud AL. Molecular components of vertebrate Mg^{2+}–homeostasis regulation. Magnes Res. 2007;20(1):6–11.
87. Wolf FI, Trapani V, Simonacci M, Mastrototaro L, Cittadini A, Schweigel M. Modulation of TRPM6 and Na(+)/Mg(2+) exchange in mammary epithelial cells in response to variations of magnesium availability. J Cell Physiol. 2010;222(2):374–81.
88. Li FY, Chaigne-Delalande B, Kanellopoulou C, Davis JC, Matthews HF, Douek DC, et al. Second messenger role for Mg^{2+} revealed by human T-cell immunodeficiency. Nature. 2011;475(7357):471–6.
89. Wu N, Veillette A. Immunology: magnesium in a signalling role. Nature. 2011;475(7357):462–3.
90. Laurant P, Berthelot A. Influence of endothelium in the in vitro vasorelaxant effect of magnesium on aortic basal tension in DOCA-salt hypertensive rat. Magnes Res. 1992;5(4):255–60.
91. Laurant P, Berthelot A. Influence of endothelium on Mg^{2+}-induced relaxation in noradrenaline-contracted aorta from DOCA-salt hypertensive rat. Eur J Pharmacol. 1994;258(3):167–72.
92. Teragawa H, Matsuura H, Chayama K, Oshima T. Mechanisms responsible for vasodilation upon magnesium infusion in vivo: clinical evidence. Magnes Res. 2002;15(3–4):241–6.
93. Maier JA, Bernardini D, Rayssiguier Y, Mazur A. High concentrations of magnesium modulate vascular endothelial cell behaviour in vitro. Biochim Biophys Acta. 2004;1689(1):6–12.
94. Gold ME, Buga GM, Wood KS, Byrns RE, Chadhuri G, Ignarro LJ. Antagonistic modulatory roles of magnesium and calcium on release of endothelium-derived relaxing factor and smooth muscle tone. Circ Res. 1990;66:355–66.
95. Pearson PJ, Evora PR, Seccombe JF, Schaff HV. Hypomagnesemia inhibits nitric oxide release from coronary endothelium: protective role of magnesium infusion after cardiac operations. Ann Thorac Surg. 1998;68:967–72.
96. Yogi A, Callera GE, Antunes TT, Tostes RC. Touyz RM Transient receptor potential melastatin 7 (TRPM7) cation channels, magnesium and the vascular system in hypertension. Circ J. 2011;75(2):237–45.
97. Maier JA. Endothelial cells and magnesium: implications in atherosclerosis. Clin Sci (Lond). 2012;122(9):397–407.
98. Soltani N, Keshavarz M, Sohanaki H, Zahedi Asl S, Dehpour AR. Relaxatory effect of magnesium on mesenteric vascular beds differs from normal and streptozotocin induced diabetic rats. Eur J Pharmacol. 2005;508(1–3):177–81.
99. Laurant P, Touyz RM. Physiological and pathophysiological role of magnesium in the cardiovascular system: implications in hypertension. J Hypertens. 2000;18(9):1177–91.
100. Ko EA, Park WS, Earm YE. Extracellular Mg(2+) blocks endothelin-1-induced contraction through the inhibition of non-selective cation channels in coronary smooth muscle. Pflugers Arch. 2004;449(2):195–204.
101. McHugh D, Beech DJ. Modulation of Ca^{2+} channel activity by ATP metabolism and internal Mg^{2+} in guinea-pig basilar artery smooth muscle cells. J Physiol. 1996;492(pt 2):359–76.

102. Zhang HF, Chen XQ, Hu GY, Wang YP. Magnesium lithospermate B dilates mesenteric arteries by activating BKCa currents and contracts arteries by inhibiting K(V) currents. Acta Pharmacol Sin. 2010;31(6):665–70.
103. Shimosawa T, Takano K, Ando K, Fujita T. Magnesium inhibits norepinephrine release by blocking N-type calcium channels at peripheral sympathetic nerve endings. Hypertension. 2004;44(6):897–902.
104. Satake K, Lee JD, Shimizu H, Uzui H, Mitsuke Y, Yue H, et al. Effects of magnesium on prostacyclin synthesis and intracellular free calcium concentration in vascular cells. Magnes Res. 2004;17(1):20–7.
105. Wells IC, Agrawal DK. Abnormal magnesium metabolism in two rat models of genetic hypertension. Can J Physiol Pharmacol. 1992;70:1225–9.
106. Laurant P, Gaillard E, Kantelip JP, Berthelot A. Lack of magnesium supplementation effects on blood pressure and vascular responsiveness in aged spontaneously hypertensive rats. Magnesium Bull. 1996;18:38–43.
107. Laurant P, Berthelot A. Endothelin-1-induced contraction in isolated aortae from normotensive and DOCA-salt hypertensive rats: effect of magnesium. Br J Pharmacol. 1996;119:1367–74.
108. Laurant P, Touyz RM, Schiffrin EL. Effect of magnesium on vascular tone and reactivity in pressurized mesenteric arteries from SHR. Can J Physiol Pharmacol. 1997;5:293–300.
109. Laurant P, Hayoz D, Brunner HR, Berthelot A. Effect of magnesium deficiency on blood pressure and mechanical properties of rat carotid artery. Hypertension. 1999;33:1105–10.
110. Briel RC, Lippert TH, Zahradnik HP. Action of magnesium sulfate on platelet interaction and prostacyclin of blood vessels. Am J Obstet Gynecol. 1985;153:232–4.
111. Mak IT, Chmielinska JJ, Kramer JH, Weglicki WB. AZT-induced oxidative cardiovascular toxicity: attenuation by Mg-supplementation. Cardiovasc Toxicol. 2009;9(2):78–85.
112. Hur KY, Kim SH, Choi MA, Williams DR, Lee YH, Kang SW, et al. Protective effects of magnesium lithospermate B against diabetic atherosclerosis via Nrf2-ARE-NQO1 transcriptional pathway. Atherosclerosis. 2010;211(1):69–76.
113. Qu J, Ren X, Hou RY, Dai XP, Zhao YC, Xu XJ, et al. The protective effect of magnesium lithospermate B against glucose-induced intracellular oxidative damage. Biochem Biophys Res Commun. 2011;411(1):32–9.
114. Weglicki WB, Chmielinska JJ, Kramer JH, Mak IT. Cardiovascular and intestinal responses to oxidative and nitrosative stress during prolonged magnesium deficiency. Am J Med Sci. 2011;342(2):125–9.
115. Nielsen FH. Magnesium, inflammation, and obesity in chronic disease. Nutr Rev. 2010;68(6):333–40.
116. Elwood PC, Pickering J. Magnesium and cardiovascular disease: a review of epidemiological evidence. J Clin Basic Cardiol. 2002;5:61–6.
117. Joffres MR, Reed DM, Yano K. Relation of magnesium intake and other dietary factors to blood pressure the Honolulu Heart Study. Am J Clin Nutr. 1987;45:469–75.
118. Whelton PK, Klag MJ. Magnesium and blood pressure: review of the epidemiologic and clinical trial experience. Am J Cardiol. 1989;63:26G–30.
119. Abbott RD, Ando F, Masaki KH, Tung KH, Rodriguez BL, Petrovitch H, et al. Dietary magnesium intake and the future risk of coronary heart disease (the Honolulu Heart Program). Am J Cardiol. 2003;92(6):665–9.
120. Stuehlinger HG. The wider use of magnesium. Eur Heart J. 2001;22:713–4.
121. Ascherio A, Rimm EB, Giovannucci EL, Colditz GA, Rosner B, Willett WC, et al. A prospective study of nutritional factors and hypertension among US men. Circulation. 1992;86:1475–84.
122. Touyz RM, Milne FJ, Reinach SG. Intracellular Mg2+, Ca2+, Na+and K+in platelets and erythrocytes of essential hypertensive patients: relation to blood pressure. Clin Exp Hypertens. 1992;14(6):1189–209.
123. Kesteloot H, Joossens JV. Relationship of dietary sodium, potassium, calcium, and magnesium with blood pressure. Belgian Interuniversity Research on Nutrition and Health. Circulation. 1988;12:594–9.
124. Resnick LM, Oparil S, Chait A, Haynes RB, Kris-Etherton P, Stern JS, et al. Factors affecting blood pressure responses to diet: the Vanguard study. Am J Hypertens. 2000;13(9):956–65.
125. Resnick LM, Gupta RK, Sosa RE, Corbett ML, Laragh JH. Intracellular free magnesium in erythrocytes of essential hypertension. Proc Natl Acad Sci U S A. 1987;84:7663–7.
126. Resnick LM, Militianu D, Cunnings AJ, Pipe JG, Evelhoch JL, Soulen RL. Direct magnetic resonance determination of aortic distensibility in essential hypertension. Relation to age, abdominal visceral fat, and in situ intracellular free magnesium. Hypertension. 1997;30:654–9.
127. Touyz RM, Milne FJ. Alterations in intracellular cations and cell membrane ATPase activity in patients with malignant hypertension. J Hypertens. 1995;13:867–74.
128. Ferrara LA, Iannuzzi R, Castaldo A, Iannuzzi A, Dello Russo A, Mancini M. Long-term magnesium supplementation in essential hypertension. Cardiology. 1992;81:25–33.
129. Hiraga H, Oshima T, Yoshimura M, Matsuura H, Kajiyama G. Abnormal platelet Ca^{2+} handling accompanied by increased cytosolic free Mg^{2+} in essential hypertension. Am J Physiol. 1998;275:R574–9.
130. Reffelmann T, Ittermann T, Dörr M, Völzke H, Reinthaler M, Petersmann A, et al. Low serum magnesium concentrations predict cardiovascular and all-cause mortality. Atherosclerosis. 2011;219(1):280–4.
131. Rowe WJ. Correcting magnesium deficiencies may prolong life. Clin Interv Aging. 2012;7:51–4.

132. Barbagallo M, Belvedere M, Dominguez LJ. Magnesium homeostasis and aging. Magnes Res. 2009;22(4):235–46.
133. Fox CH, Mahoney MC, Ramsoomair D, Carter CA. Magnesium deficiency in African-Americans: does it contribute to increased cardiovascular risk factors? J Natl Med Assoc. 2003;95(4):257–62.
134. Sarafidis PA, Georgianos PI, Lasaridis AN. Diuretics in clinical practice. Part II: electrolyte and acid–base disorders complicating diuretic therapy. Expert Opin Drug Saf. 2010;9(2):259–73.
135. Zia AA, Kamalov G, Newman KP, McGee JE, Bhattacharya SK, Ahokas RA, et al. From aldosteronism to oxidative stress: the role of excessive intracellular calcium accumulation. Hypertens Res. 2010;33(11):1091–101.
136. Sánchez-Fructuoso AI, Santín Cantero JM, Pérez Flores I, Valero San Cecilio R, Calvo Romero N, Vilalta Casas R. Changes in magnesium and potassium homeostasis after conversion from a calcineurin inhibitor regimen to an mTOR inhibitor-based regimen. Transplant Proc. 2010;42(8):3047–9.
137. Blackfan KD, Hamilton B. Uremia in acute glomerular nephritis: the cause and treatment in children. Boston Med Surg J. 1925;193:617–21.
138. Kawasaki T, Itoh K, Kawasaki M. Reduction in blood pressure with a sodium-reduced potassium-and magnesium-enriched salt in subjects with mild essential hypertension. Hypertens Res. 1998;21(4):235–43.
139. Itoh K, Kawasaka T, Nakamura M. The effects of high oral magnesium supplementation on blood pressure, serum lipids and related variables in apparently healthy Japanese subjects. Br J Nutr. 1997;78(5):737–50.
140. Katz A, Rosenthal T, Maoz C, Peleg E, Zeidenstein R, Levi Y. Effect of a mineral salt diet on 24 hr blood pressure monitoring in elderly hypertensive patients. J Hum Hypertens. 1999;13(11):777–80.
141. Pastori C, Delva P, Degan M, Lechi A. Preliminary communication on intralymphocyte ionized magnesium in hypertensive patients under treatment with beta-blockers. Magnes Res. 1999;1291:49–55.
142. Resnick LM. Ionic basis of hypertension, insulin resistance, vascular disease and related disorders: mechanisms of Syndrome X. Am J Hypertens. 1993;6:123S–34.
143. Dominguez LJ, Barbagallo M, Sowers JR, Resnick LM. Magnesium responsiveness to insulin and insulin-like growth factor 1 in erythrocytes from normotensive and hypertensive subjects. J Clin Endocr Metab. 1998;83:4402–7.
144. Yamamoto ME, Applegate WB, Klag MJ, Borhani NO, Cohen JD, Kirchner KA, et al. Lack of blood pressure effect with calcium and magnesium supplementation in adults with high-normal blood pressure. Results from phase 1 of the Trials of Hypertension Prevention (TOHP). Ann Epid. 1995;5(2):96–107.
145. Jurcovicova J, Krueger KS, Nandy I, Lewis DF, Brooks GG, Brown EG. Expression of platelet-derived growth factor-A mRNA in human placenta: effect of magnesium infusion in pre-eclampsia. Placenta. 1998;19(5–6):423–7.
146. Mason BA, Standley CA, Whitty JE, Cotton DB. Fetal ionized magnesium levels parallel maternal levels during magnesium sulfate therapy for preeclampsia. Am J Obstet Gynecol. 1996;175(1):213–7.
147. Zwilleinger L. Uber die magnesiumwirkung auf das Herz. Klin Wochenscher. 1935;14:1429–33.
148. Robert JM, Redman CW. Pre-eclampsia: more than pregnancy-induced hypertension. Lancet. 1993;341:1447–51 [Erratum, Lancet 1993, 342: 504].
149. Roberts JM, Taylor RN, Musci MK. Preeclampsia: an endothelial cell disorder. Am J Obstet Gynecol. 1989;161:1200–4.
150. Nadler JL, Goodson S, Rude RK. Evidence that prostacyclin mediates the vascular action of magnesium in humans. Hypertension. 1987;9:379–83.
151. Kisters K, Barenbrock M, Louwen F, Hausberg M, Rahn KH, Kosch M. Membrane, intracellular, and plasma magnesium and calcium concentrations in preeclampsia. Am J Hypertens. 2000;13(7):765–9.
152. Hibbard JU. Hypertensive disease and pregnancy. J Hypertens. 2002;20:S29–33.
153. Scardo JA, Hogg BB, Newman RB. Favorable hemodynamic effects of magnesium sulfate in preeclampsia. Am J Obstet Gynecol. 1995;173:1249–53.
154. Handwerker SM, Altura BT, Altura BM. Ionized serum magnesium and potassium levels in pregnant women with preeclampsia and eclampsia. J Reprod Med. 1995;40:201–8.
155. Sanders R, Konijnenberg A, Huijgen HJ, Wolf H, Boer K, Sanders GTB. Intracellular and extracellular ionized and total magnesium in pre-eclampsia and uncomplicated pregnancy. Clin Chem Lab Med. 1998;37:55–9.
156. Frenkel Y, Weiss M, Shefi M, Lusky A, Mashiach S, Dolev E. Mononuclear cell magnesium content remains unchanged in various cell hypertensive disorders of pregnancy. Gynecol Obstet Inves. 1994;38:220–2.
157. Standley CA, Whitty JE, Mason BA, Cotton DB. Serum ionized levels in normal and preeclamptic gestation. Obstet Gynecol. 1997;89:24–7.
158. Standley PR, Standley CA. Identification of a functional Na^{+}/Mg^{2+} exchanger in human trophoblast cells. Am J Hypertens. 2002;15:565–70.
159. Malapaka SV, Ballal PK. Low-dose magnesium sulfate versus Pritchard regimen for the treatment of eclampsia imminent eclampsia. Int J Gynaecol Obstet. 2011;115(1):70–2.
160. Lucas MJ, Leveno KJ, Cunningham FG. A comparison of magnesium sulfate with phenytoin for the prevention of eclampsia. N Engl J Med. 1995;333:201–5.

161. The Eclampsia Collaborative Group. Which anticonvulsant for women with eclampsia? Evidence from the Collaborative Trial. Lancet. 1995;345:1455–63.
162. Duley L, Gülmezoglu AM, Henderson-Smart DJ, Chou D. Magnesium sulphate and other anticonvulsants for women with pre-eclampsia. Cochrane Database Syst Rev. 2010;12:CD000127.
163. The Magpie trial collaborative group. Do women with pre-eclampsia, and their babies, benefit from magnesium sulphate? The Magpie trial: a randomised placebo-controlled trial. Lancet. 2002;359:1877–90.
164. Johnson C, Peterson D, Smith E. Myocardial tissue concentrations of magnesium and potassium in men dying suddenly from ischemic heart disease. Am J Clin Nutr. 1979;32:967–70.
165. Gasser RNA. Free intracellular magnesium during myocardial ischemia: the state of the art. In: Smetena R, editor. Advances in magnesium research. London: John Libbey and Company Ltd. 1996. pp. 103–10.
166. Yoshimura M, Oshima T, Hiraga H, Nakano Y, Matsuura H, Yamagata T, et al. Increased cytosolic free Mg^{2+} and Ca^{2+} in platelets of patients with vasospastic angina. Am J Physiol. 1989;274:R548–54.
167. Antman EM. Magnesium in acute myocardial infarction: overview of available evidence. Am Heart J. 1996;132: 487–95.
168. Tejero-Taldo MI, Kramer JH, Mak IuT, Komarov AM, Weglicki WB. The nerve-heart connection in the pro-oxidant response to Mg-deficiency. Heart Fail Rev. 2006;11(1):35–44.
169. Woods KL, Fletcher S, Roffe C, Haider Y. Intravenous magnesium sulphate in suspected acute myocardial infarction: results of the second Leicester Intravenous Magnesium Intervention Trial (LIMIT-2). Lancet. 1992;339:1553–8.
170. ISIS-4: a randomized factorial trial assessing early oral captopril, oral mononitrate and intravenous magnesium sulphate in 58,050 patients with suspected acute myocardial infarction. ISIS-4 (Fourth International Study of Infarct Survival) Collaborative Group. Lancet. 1995;345(8951)669–85.
171. The Magnesium in Coronaries (MAGIC) Trial Investigators. Early administration of intravenous magnesium to high-risk patients with acute myocardial infarction in the Magnesium in Coronaries (MAGIC) Trial: a randomized controlled trial. Lancet. 2002;360:1189–96.
172. Baxter G, Sumeray M, Walker J. Infarct size and magnesium insights into LIMIT 2 and ISIS 4 from experimental studies. Lancet. 1996;348:1424–6.
173. Santoro GM, Antoniucci D, Bolognese L, Valenti R, Buonamici P, Trapani M, et al. A randomized study of intravenous magnesium in acute myocardial infarction treated with direct coronary angioplasty. Am Heart J. 2000; 140(6):891–7.
174. Prasad A, Reeder G. Modern adjunctive pharmacotherapy of myocardial infarction. Expert Opin Pharmacother. 2000;1(3):405–18.

Chapter 15
Intravenous Magnesium for Cardiac Arrhythmias in Humans: A Role?

Kwok M. Ho

Key Points

- Intravenous magnesium has a high therapeutic to toxic ratio making it a relatively safe antiarrhythmic drug to use. Symptoms of tingling, flushing, and dizziness are the main side effects. Hypotension and bradycardia are less common compared to intravenous amiodarone and calcium channel blockers.
- Although intravenous magnesium has been used for a variety of supraventricular and ventricular arrhythmias, the evidence to support its antiarrhythmic roles is strongest in the prevention of atrial fibrillation after cardiac surgery, reduction of ventricular rate in acute-onset atrial fibrillation, and prevention and treatment of polymorphic ventricular tachycardia.
- Publication bias favoring intravenous magnesium in small randomized controlled trials and case series exists. Furthermore, none of the studies has, so far, demonstrated a significant improvement in patient-centered outcomes after using intravenous magnesium as an antiarrhythmic agent.
- Until the benefits of magnesium as an antiarrhythmic agent are confirmed by adequately powered randomized controlled trials, maintaining serum magnesium concentrations above the upper limit of normal by intravenous magnesium may present the most reasonable approach in the prevention and treatment of cardiac arrhythmias that are possibly related to magnesium deficiency.

Keywords Arrhythmias • Evidence • Magnesium • Outcomes • Side effects

Introduction

Magnesium is an important intracellular ion and a cofactor for many enzymes involved in cell function of many body systems, including cardiovascular, metabolic, and nervous system. Magnesium has a wide range of physiological effects on the heart and circulatory system including calcium antagonism at both L- and T-type calcium channels, endothelial cell function and secretion, and regulation of energy transfer, and in pharmacological doses, magnesium decreases outward potassium current intensity resulting in membrane stabilization and acts as an indirect antagonist of digitalis at the sarcolemma Na-K-ATPase pump [1–3].

K.M. Ho, MPH, Ph.D., FRCP, FANZCA, FCICM (✉)
Department of Intensive Care Medicine, Royal Perth Hospital, Perth, Australia

School of Population Health, University of Western Australia, Perth, Australia
e-mail:kwok.ho@health.wa.gov.au

R.R. Watson et al. (eds.), *Magnesium in Human Health and Disease*, Nutrition and Health,
DOI 10.1007/978-1-62703-044-1_15, © Springer Science+Business Media New York 2013

As a potential antiarrhythmic agent, the mechanism of its action may include reducing atrial automaticity [4], increasing sinus node recovery time [5], reducing atrioventricular nodal conduction [6], blocking the antegrade and retrograde conduction over an accessory pathway [7–9] including a dominant effect on the slow atrioventricular nodal pathway in patients with dual atrioventricular nodal physiology [10], prolonging His-ventricular conduction [11], and homogenizing transmural ventricular repolarization without affecting autonomic nervous activity [12–15]. Homogenization of ventricular repolarization explains the antiarrhythmic effect of magnesium in torsade-de-pointes polymorphic ventricular tachycardia [12]. Magnesium appears to have no direct effect on excitable gap arrhythmias associated with a fixed anatomic substrate as in monomorphic ventricular tachycardia [14, 16]. In addition to its direct cardiovascular effects, magnesium is also involved in the process of glucose utilization, fatty acid synthesis, and energy metabolism, potentially contributing to its long-term beneficial effects on the health of the cardiovascular system.

Intravenous Magnesium as an Antiarrhythmic Agent

Magnesium deficiency is very common in hospitalized patients, especially if they have cardiovascular diseases. Magnesium deficiency is commonly associated with potassium deficiency and is associated with an increased risk of cardiac arrhythmias, especially for patients who undergo cardiac surgery [2]. Due to its diverse electrophysiological effects on the heart [17], magnesium has a potential to be an excellent drug in the prevention and treatment of different types of cardiac arrhythmia. Intravenous magnesium has been reported to be very useful in the prevention of atrial fibrillation and ventricular arrhythmias after cardiac and thoracic surgery [18, 19], reducing rapid ventricular response to less than 100 beats per minute in recent-onset atrial fibrillation [20] including patients with Wolff-Parkinson-White syndrome [9, 21], improving success rate of achieving rhythm control with ibutilide [22, 23], and treating digitalis-induced supraventricular and ventricular arrhythmias [24], multifocal atrial tachycardia [25], drug-induced and hypomagnesemia-induced polymorphic ventricular tachycardia or torsade de pointes [26–30] or severe sinus tachycardia, and hypertension for patients with pheochromocytoma or tetanus [31, 32]. Intravenous magnesium has also been used as a second-line drug in the treatment of supraventricular tachycardia when adenosine is not effective [33]. However, intravenous magnesium is not effective in converting shock-resistant ventricular fibrillation [34, 35] and monomorphic ventricular tachycardia into sinus rhythm compared to placebo [14, 16]. The possible antiarrhythmic roles of intravenous magnesium are described in Table 15.1.

Intravenous magnesium has minimal negative inotropic effect and a high therapeutic to toxic ratio. When compared to intravenous calcium channel blockers, such as diltiazem or verapamil and amiodarone, intravenous magnesium appears to cause less significant detrimental hemodynamic side effects such as hypotension and bradycardia. Minor symptoms of flushing, tingling, and dizziness are,

Table 15.1 Reported indications of intravenous magnesium as an antiarrhythmic agent

1	Prevention of atrial fibrillation after cardiac and thoracic surgery
2	Reducing ventricular response in acute-onset atrial fibrillation
3	Multifocal atrial tachycardia
4	Paroxysmal supraventricular tachycardia
5	Ventricular ectopics
6	Ventricular tachycardia from digoxin toxicity or class IC antiarrhythmic agent toxicity
7	Polymorphic ventricular tachycardia
8	Ventricular fibrillation from amitriptyline toxicity
9	Sinus tachycardia and hypertension in severe tetanus or pheochromocytoma

however, common after intravenous magnesium (17%) [20]. Because magnesium has multiple pharmacological actions other than calcium antagonism on the atrioventricular node, some of these minor side effects may be due to the other actions of magnesium, including its antagonism on the N-methyl-D-aspartate (NMDA) receptors and peripheral vasodilatation [2].

Evidence to Support Intravenous Magnesium as a Prophylaxis to Prevent Cardiac Arrhythmias

Because magnesium is predominantly an intracellular ion [2] and less than 1% of the total body magnesium is, in fact, in the blood, serum magnesium concentrations can be relatively normal despite having significant deficiency in the total body magnesium [36, 37]. Evidence also suggests that serum magnesium concentrations correlate poorly with myocyte magnesium concentrations [38], making the diagnosis of magnesium deficiency as a cause or contributing factor for cardiac arrhythmia extremely difficult. As such, some authorities have suggested that it is entirely reasonable to give and keep serum magnesium concentrations at the upper end of the normal limits (>0.85 mmol.L^{-1}) to avoid magnesium deficiency and its consequences [37].

Both magnesium deficiency and atrial fibrillation are very common after cardiac surgery [2, 18]. Because magnesium has diverse electrophysiological effects on the sinus node and atrial conduction function of the heart, it is possible that magnesium deficiency may contribute to the high incidence of atrial fibrillation after cardiac surgery [2]. As serum magnesium concentration is a poor predictor of total body magnesium deficiency, it is reasonable to provide magnesium supplementation or prophylaxis for patients who are likely to have magnesium deficiency to reduce their risk of atrial fibrillation. This hypothesis has been tested by many randomized controlled trials in the last two decades.

The first major meta-analysis involving 2,490 patients from 20 randomized controlled trials was published in 2005 and suggested that magnesium supplementation was effective in preventing atrial fibrillation in cardiac surgery compared to placebo (18% vs. 28%, odds ratio 0.54; 95% confidence interval 0.38–0.75)[18]. The results of this meta-analysis were subsequently challenged by a large randomized controlled trial involving a total of 927 patients [39]. The difference in the conclusions of the meta-analysis and this large randomized controlled trial requires careful consideration. First, there is a fundamental design problem in the large randomized controlled trial. In this study, patients randomized into the control group were given 2 g (4 mmol) of intravenous magnesium when the serum magnesium concentration was <1.2 mmol.L^{-1} and 5 g (20 mmol) of intravenous magnesium when the serum magnesium concentration was <0.8 mmol.L^{-1}, resulting in substantial number of patients in the control group who received supplementary intravenous magnesium after surgery. This trial became a comparison of on-demand intravenous magnesium supplement (control group) and predefined magnesium supplementation (treatment group). Second, there were significantly more patients with heart failure in the magnesium group than in the control group (76% vs. 49%). These two factors may have, at least in part, contributed to the negative results of this otherwise well-designed study.

By including this large randomized controlled trial [39] and also other trials that have been published since the last meta-analysis [18], we identified a total of 5,069 patients from 32 randomized controlled trials assessing the effects of intravenous magnesium in cardiac surgery as of 25 August 2011. When these 32 trials were pooled, the beneficial effect of intravenous magnesium on risk of atrial fibrillation in cardiac surgery remained highly significant (relative risk 0.69, 95% confidence interval 0.57–0.83) (Fig. 15.1). Nevertheless, meta-analysis is prone to bias including publication bias. Using atrial fibrillation as an endpoint, the shape of the funnel plot suggests that publication bias favoring the publication of small positive studies exists (Fig. 15.2). Furthermore, significant heterogeneity between the results of the pooled 32 studies also exists, making any definitive conclusion

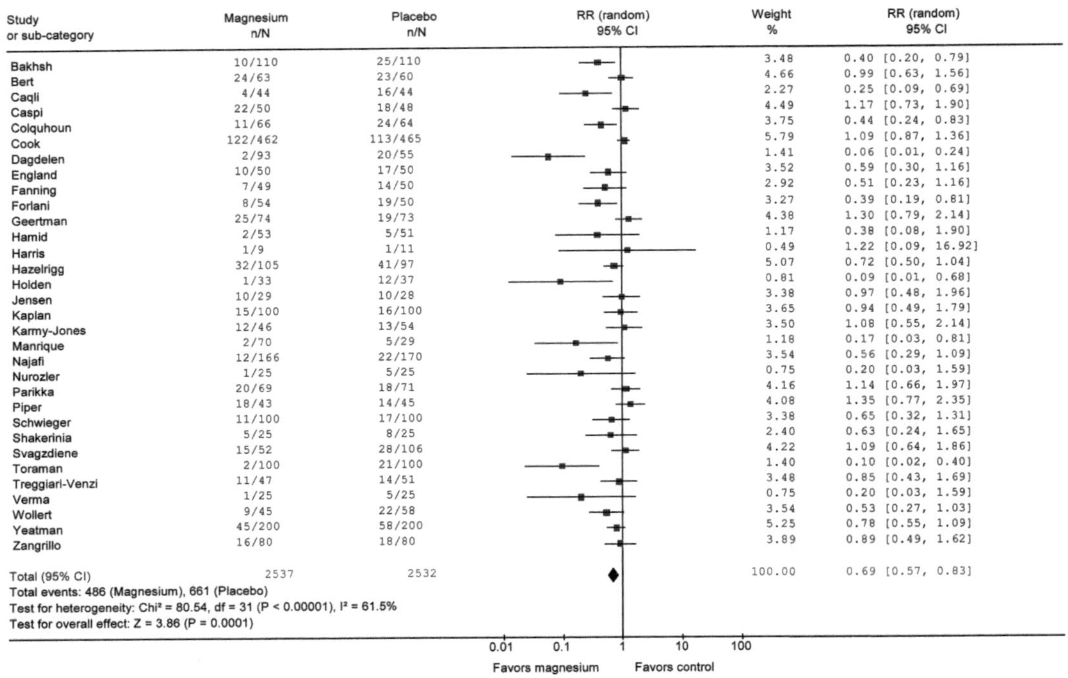

Fig. 15.1 Forest plot showing the effect of magnesium prophylaxis on risk of atrial fibrillation after cardiac surgery compared to placebo

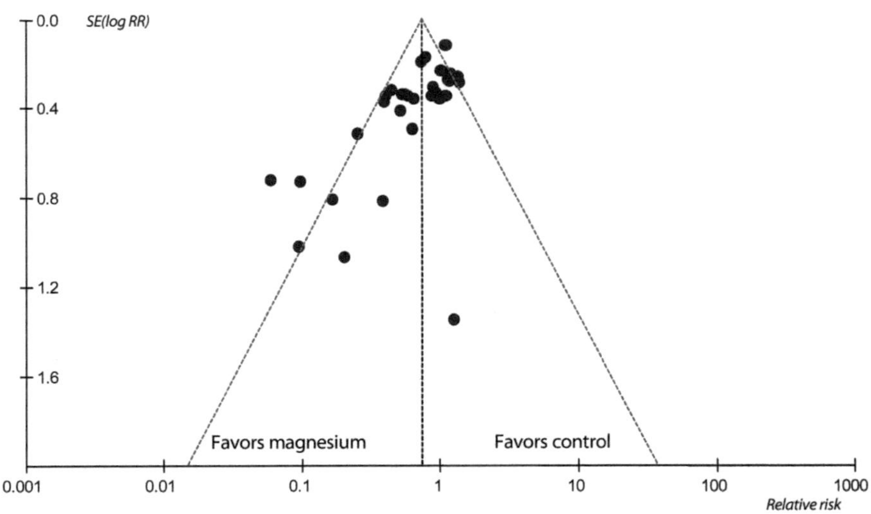

Fig. 15.2 Funnel plot showing the possibility of publication bias in trials on benefits of magnesium prophylaxis on risk of atrial fibrillation after cardiac surgery

about the benefits of intravenous magnesium on risk of atrial fibrillation in cardiac surgery impossible.

Evidence to Support Intravenous Magnesium as a Therapeutic Antiarrhythmic Agent to Treat Recent-Onset Atrial Fibrillation

There is a suggestion that patients who have both magnesium deficiency and atrial fibrillation are more likely to convert to sinus rhythm when given intravenous magnesium [40, 41]. Because magnesium deficiency is difficult to diagnose, intravenous magnesium has been used as a therapeutic drug at a dose between 3 g (12 mmol) and 10 g (40 mmol) in patients with recent-onset atrial fibrillation aiming at converting it to sinus rhythm [20]. Our previous meta-analysis showed that intravenous magnesium was not effective in converting recent-onset atrial fibrillation to sinus rhythm, but it was effective in reducing ventricular rate to below 100 beats per minute compared to placebo (58.5% vs. 32.6%, odds ratio 3.2, 95% confidence interval 1.93–5.42). We have updated this meta-analysis by including two studies that were published after 2007. The results remained unchanged; there was insufficient evidence to suggest that intravenous magnesium was effective in converting recent-onset atrial fibrillation to sinus rhythm either compared to placebo (relative risk 1.12, 95% confidence interval 0.79–1.60) or an alternative antiarrhythmic agent such as calcium channel blocker (relative risk 1.76, 95% confidence interval 0.79–3.91) (Fig. 15.3).

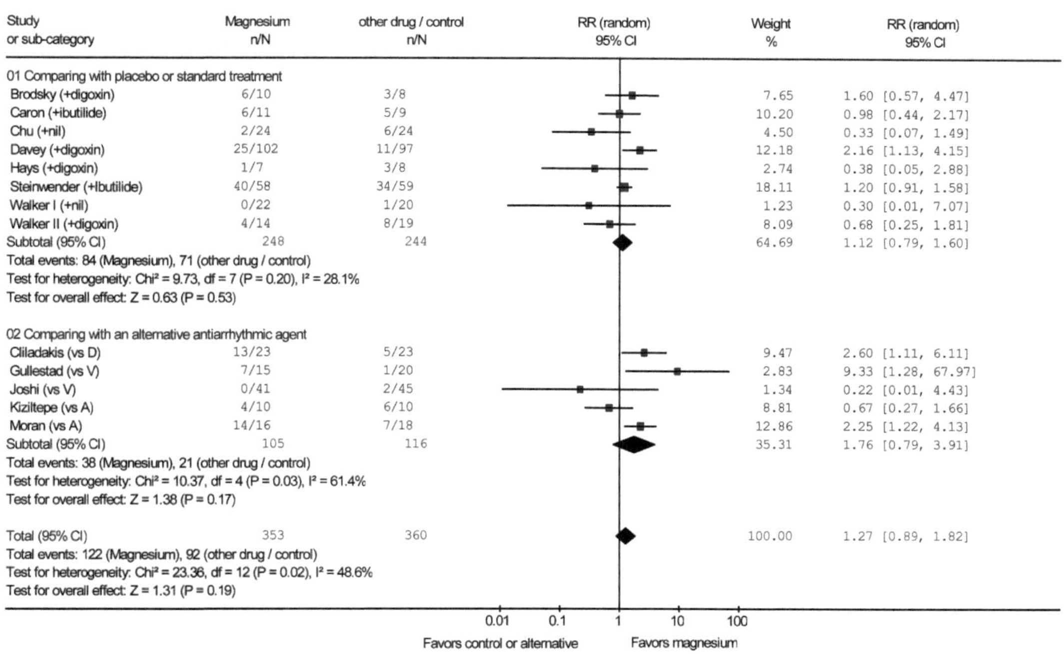

Fig. 15.3 Forest plot showing the effect of intravenous magnesium on conversion of acute-onset atrial fibrillation or flutter to sinus rhythm compared to placebo or an alternative antiarrhythmic drug. *D* diltiazem, *A* amiodarone, *V* verapamil

Future of Intravenous Magnesium for Cardiac Arrhythmias in Humans

Although many successful reports of using intravenous magnesium as an antiarrhythmic agent come from case reports, animal studies, and randomized controlled studies using physiological endpoints only [42], this does not imply that intravenous magnesium is not that useful for cardiac arrhythmias. Absence of evidence is not the same as evidence of absence, and in this regard, the current evidence is insufficient to exclude the full antiarrhythmic potential of intravenous magnesium.

First, magnesium supplementation is most likely to have therapeutic benefits against cardiac arrhythmias in patients who are deficient in total body magnesium. As such, we need a simple and more reliable bedside test than serum magnesium concentration to diagnose magnesium deficiency. Availability of a simple reliable test of magnesium deficiency will greatly improve the design of clinical trials on therapeutic effects of intravenous magnesium. Second, the optimal dose of intravenous magnesium to prevent and treat different types of cardiac arrhythmias remains uncertain. Even with the classic indication of intravenous magnesium for eclampsia, a sufficient dose of intravenous magnesium (1–2 g per hour infusion) is needed to achieve its therapeutic effect [43]. A significant difficulty in interpreting trials on intravenous magnesium supplementation comes from the different doses used in different trials, and negative results may, in part, be due to insufficient amount of magnesium being used [44]. Dose–response studies on intravenous magnesium and whether magnesium may have synergistic antiarrhythmic effects with other commonly used drugs in humans are sparse in the literature [45].

Once we have worked out how to diagnose magnesium deficiency reliably and the optimal dose response relationship between magnesium and cardiac arrhythmias is established, we will be in a position to design adequately powered clinical studies to confirm the precise roles of intravenous magnesium in different types of cardiac arrhythmias in humans. Before then, maintaining serum magnesium concentrations above the upper limit of normal (>1.0 mmol.L^{-1}) by intravenous magnesium may represent the most reasonable approach in the prevention and treatment of cardiac arrhythmias that are possibly related to magnesium deficiency.

References

1. Wu JY, Lipsius SL. Effects of extracellular Mg^{2+} on T- and L-type Ca^{2+} currents in single atrial myocytes. Am J Physiol. 1990;259:H1842–50.
2. Fawcett WJ, Haxby EJ, Male DA. Magnesium: physiology and pharmacology. Br J Anaesth. 1999;83:302–20.
3. Ingemansson MP, Smideberg B, Olsson SB. Intravenous MgSO4 alone and in combination with glucose, insulin and potassium (GIK) prolong the atrial cycle length in chronic atrial fibrillation. Europace. 2000;2:106–14.
4. Iseri LT, Allen BJ, Ginkel ML, Brodsky MA. Ionic biology and ionic medicine in cardiac arrhythmias with particular reference to magnesium. Am Heart J. 1992;123:1404–9.
5. DiCarlo Jr LA, Morady F, de Buitleir M, Krol RB, Schurig L, Annesley TM. Effects of magnesium sulfate on cardiac conduction and refractoriness in humans. J Am Coll Cardiol. 1986;7:1356–62.
6. Rasmussen HS, Larsen OG, Meier K, Larsen J. Hemodynamic effects of intravenously administered magnesium on patients with ischemic heart disease. Clin Cardiol. 1988;11:824–8.
7. Christiansen EH, Frost L, Andreasen F, Mortensen P, Thomsen PE, Pedersen AK. Dose-related cardiac electrophysiological effects of intravenous magnesium. A double-blind placebo-controlled dose–response study in patients with paroxysmal supraventricular tachycardia. Europace. 2000;2:320–6.
8. Viskin S, Belhassen B, Sheps D, Laniado S. Clinical and electrophysiologic effects of magnesium sulfate on paroxysmal supraventricular tachycardia and comparison with adenosine triphosphate. Am J Cardiol. 1992;70:879–85.
9. Sideris AM, Galiatsu E, Filippatos GS, Kappos K, Anthopoulos LP. Effects of magnesium and potassium on Wolff-Parkinson-White syndrome. J Electrocardiol. 1996;29:11–5.
10. Stiles MK, Sanders P, Disney P, Brooks A, John B, Lau DH, et al. Differential effects of intravenous magnesium on atrioventricular node conduction in supraventricular tachycardia. Am J Cardiol. 2007;100:1249–53.
11. Satoh Y, Sugiyama A, Tamura K, Hashimoto K. Effect of magnesium sulfate on the haloperidol-induced QT prolongation assessed in the canine in vivo model under the monitoring of monophasic action potential. Jpn Circ J. 2000;64:445–51.

12. Chinushi M, Sugiura H, Komura S, Hirono T, Izumi D, Tagawa M, et al. Effects of intravenous magnesium in a prolonged QT interval model of polymorphic ventricular tachycardia focus on transmural ventricular repolarization. Pacing Clin Electrophysiol. 2005;28:844–50.
13. Ince C, Schulman SP, Quigley JF, Berger RD, Kolasa M, Ferguson R, et al. Usefulness of magnesium sulfate in stabilizing cardiac repolarization in heart failure secondary to ischemic cardiomyopathy. Am J Cardiol. 2001;88:224–9.
14. Haigney MC, Berger R, Schulman S, Gerstenblith G, Tunin C, Silver B, et al. Tissue magnesium levels and the arrhythmic substrate in humans. J Cardiovasc Electrophysiol. 1997;8:980–6.
15. Chinushi M, Izumi D, Komura S, Ahara S, Satoh A, Furushima H, et al. Role of autonomic nervous activity in the antiarrhythmic effects of magnesium sulfate in a canine model of polymorphic ventricular tachyarrhythmia associated with prolonged QT interval. J Cardiovasc Pharmacol. 2006;48:121–7.
16. Farouque HM, Sanders P, Young GD. Intravenous magnesium sulfate for acute termination of sustained monomorphic ventricular tachycardia associated with coronary artery disease. Am J Cardiol. 2000;86:1270–2.
17. Crippa G, Sverzellati E, Giorgi-Pierfranceschi M, Carrara GC. Magnesium and cardiovascular drugs: interactions and therapeutic role. Ann Ital Med Int. 1999;14:40–5.
18. Miller S, Crystal E, Garfinkle M, Lau C, Lashevsky I, Connolly SJ. Effects of magnesium on atrial fibrillation after cardiac surgery: a meta-analysis. Heart. 2005;91:618–23.
19. Saran T, Perkins GD, Javed MA, Annam V, Leong L, Gao F, et al. Does the prophylactic administration of magnesium sulphate to patients undergoing thoracotomy prevent postoperative supraventricular arrhythmias? A randomized controlled trial. Br J Anaesth. 2011;106:785–91.
20. Ho KM, Sheridan DJ, Paterson T. Use of intravenous magnesium to treat acute onset atrial fibrillation: a meta-analysis. Heart. 2007;93:1433–40.
21. Merrill JJ, DeWeese G, Wharton JM. Magnesium reversal of digoxin-facilitated ventricular rate during atrial fibrillation in the Wolff-Parkinson-White syndrome. Am J Med. 1994;97:25–8.
22. Kalus JS, Spencer AP, Tsikouris JP, Chung JO, Kenyon KW, Ziska M, et al. Impact of prophylactic i.v. magnesium on the efficacy of ibutilide for conversion of atrial fibrillation or flutter. Am J Health Syst Pharm. 2003;60:2308–12.
23. Tercius AJ, Kluger J, Coleman CI, White CM. Intravenous magnesium sulfate enhances the ability of intravenous ibutilide to successfully convert atrial fibrillation or flutter. Pacing Clin Electrophysiol. 2007;30:1331–5.
24. Kinlay S, Buckley NA. Magnesium sulfate in the treatment of ventricular arrhythmias due to digoxin toxicity. J Toxicol Clin Toxicol. 1995;33:55–9.
25. McCord JK, Borzak S, Davis T, Gheorghiade M. Usefulness of intravenous magnesium for multifocal atrial tachycardia in patients with chronic obstructive pulmonary disease. Am J Cardiol. 1998;81:91–3.
26. Gupta A, Lawrence AT, Krishnan K, Kavinsky CJ, Trohman RG. Current concepts in the mechanisms and management of drug-induced QT prolongation and torsade de pointes. Am Heart J. 2007;153:891–9.
27. Winters SL, Sachs RG, Curwin JH. Nonsustained polymorphous ventricular tachycardia during amiodarone therapy for atrial fibrillation complicating cardiomyopathy. Management with intravenous magnesium sulfate. Chest. 1997;111:1454–7.
28. Sarisoy O, Babaoglu K, Tugay S, Barn E, Gokalp AS. Efficacy of magnesium sulfate for treatment of ventricular tachycardia in amitriptyline intoxication. Pediatr Emerg Care. 2007;23:646–8.
29. Knudsen K, Abrahamsson J. Magnesium sulphate in the treatment of ventricular fibrillation in amitriptyline poisoning. Eur Heart J. 1997;18:881–2.
30. Meikle A, Milne B. Management of prolonged QT interval during a massive transfusion: calcium, magnesium or both? Can J Anaesth. 2000;47:792–5.
31. van der Heide K, de Haes A, Wietasch GJ, Wiesfeld AC, Hendriks HG. Torsades de pointes during laparoscopic adrenalectomy of a pheochromocytoma: a case report. J Med Case Reports. 2011;5:368.
32. Thwaites CL, Yen LM, Loan HT, Thuy TT, Thwaites GE, Stepniewska K, et al. Magnesium sulphate for treatment of severe tetanus: a randomised controlled trial. Lancet. 2006;368:1436–43.
33. LeDuc TJ, Carr JD. Magnesium sulfate for conversion of supraventricular tachycardia refractory to intravenous adenosine. Ann Emerg Med. 1996;27:375–8.
34. Allegra J, Lavery R, Cody R, Birnbaum G, Brennan J, Hartman A, et al. Magnesium sulfate in the treatment of refractory ventricular fibrillation in the prehospital setting. Resuscitation. 2001;49:245–9.
35. Hassan TB, Jagger C, Barnett DB. A randomised trial to investigate the efficacy of magnesium sulphate for refractory ventricular fibrillation. Emerg Med J. 2002;19:57–62.
36. Ulger Z, Ariogul S, Cankurtaran M, Halil M, Yavuz BB, Orhan B, et al. Intra-erythrocyte magnesium levels and their clinical implications in geriatric outpatients. J Nutr Health Aging. 2010;14:810–4.
37. Elin RJ. Assessment of magnesium status for diagnosis and therapy. Magnes Res. 2010;23:S194–8.
38. Ralston MA, Murnane MR, Kelley RE, Altschuld RA, Unverferth DV, Leier CV. Magnesium content of serum, circulating mononuclear cells, skeletal muscle, and myocardium in congestive heart failure. Circulation. 1989;80:573–80.

39. Cook RC, Humphries KH, Gin K, Janusz MT, Slavik RS, Bernstein V, et al. Prophylactic intravenous magnesium sulphate in addition to oral {beta}-blockade does not prevent atrial arrhythmias after coronary artery or valvular heart surgery: a randomized, controlled trial. Circulation. 2009;120(11 Suppl):S163–9.
40. Kiziltepe U, Eyileten ZB, Sirlak M, Tasoz R, Aral A, Eren NT, et al. Antiarrhythmic effect of magnesium sulfate after open heart surgery: effect of blood levels. Int J Cardiol. 2003;89:153–8.
41. Cybulski J, Budaj A, Danielewicz H, Maciejewicz J, Ceremuzy ski L. A new-onset atrial fibrillation: the incidence of potassium and magnesium deficiency. The efficacy of intravenous potassium/magnesium supplementation in cardioversion to sinus rhythm. Kardiol Pol. 2004;60:578–81.
42. Ho KM. Intravenous magnesium for cardiac arrhythmias: jack of all trades. Magnes Res. 2008;21:65–8.
43. Lowe SA, Brown MA, Dekker GA, Gatt S, McLintock CK, McMahon LP, et al. Society of obstetric medicine of Australia and New Zealand. Guidelines for the management of hypertensive disorders of pregnancy 2008. Aust N Z J Obstet Gynaecol. 2009;49:242–6.
44. Ho KM. Atrial tachyarrhythmia after cardiac surgery: role of magnesium infusion. Intensive Care Med. 1999;25:243.
45. Ho KM, Lewis JP. Prevention of atrial fibrillation in cardiac surgery: time to consider a multimodality pharmacological approach. Cardiovasc Ther. 2010;28:59–65.

Section E
Magnesium and Neurological Function

Chapter 16
Magnesium in Inflammation-Associated Fetal Brain Injury

Christopher Wayock, Elisabeth Nigrini, Ernest Graham, Michael V. Johnston, and Irina Burd

Key Points

- Injury to the fetal and perinatal brain is one of the leading causes of lifetime neurodevelopmental morbidity and mortality in children.
- Based on recent clinical studies, antenatal magnesium sulfate for the prevention of cerebral palsy in pregnancies at risk for preterm birth appears to be protective.
- The basic science mechanisms for the neuroprophylaxis are not completely understood.
- While the exact mechanism of action that allows magnesium to exert its neuroprotective effects is unknown, magnesium has been shown to act via multiple mechanisms.
- Further research is urgently needed to elucidate magnesium's actions on preterm fetal brain to facilitate broader application of this tried and true ion.

Keywords Magnesium • Preterm birth • Mechanisms • Fetal brain injury • Perinatal brain injury • Cerebral palsy

Introduction

Injury to the fetal and perinatal brain is one of the leading causes of lifetime neurodevelopmental morbidity and mortality in children and affects more than 6,000 children born in the United States each year [1, 2]. The neonatal brain is vulnerable due to its rapid rate of growth and development and immature immunological mechanisms [3]. While the etiology of injury differs across the developmental spectrum with regard to gestational age, infection and inflammation play an important role in brain injury, especially in the very preterm neonate [4]. We now know that the most prevalent mechanism responsible for preterm birth and especially preterm birth <28 weeks is mediated by inflammation [5].

C. Wayock, M.D. • E. Nigrini, M.D. • E. Graham, M.D. • M.V. Johnston, M.D.
Department of Gynecology and Obstetrics, Division of Maternal Fetal Medicine,
Johns Hopkins University, Baltimore, MD, USA

I. Burd, M.D., Ph.D. (✉)
Department of Gynecology and Obstetrics, Division of Maternal Fetal Medicine,
Johns Hopkins University, 600 North Wolfe Street, Phipps 212, Baltimore, MD 21287, USA

Neuroscience Laboratory, Kennedy Krieger Institute, Baltimore, MD, USA
e-mail: iburd1@jhmi.edu

R.R. Watson et al. (eds.), *Magnesium in Human Health and Disease*, Nutrition and Health,
DOI 10.1007/978-1-62703-044-1_16, © Springer Science+Business Media New York 2013

The neonates born in the setting of infection or inflammation are at significantly increased risk for adverse neurodevelopmental outcomes [6, 7]. Clinical and epidemiological studies demonstrate the association between maternal infection, neonatal brain damage, and increased levels of proinflammatory cytokines in the fetal brain, umbilical cord, and amniotic fluid [8, 9]. Two meta-analyses conclude that clinical chorioamnionitis is a risk factor for both cerebral palsy and cystic PVL [10, 11]. There is also suggestion that less conspicuous infection as evidenced by histological chorioamnionitis is associated with brain injury and cerebral palsy [11]. Up to 50% of infants who survive very preterm birth will suffer cognitive, behavioral, attention, and socialization deficits [12, 13]. Basic science studies using animal models (rodents, rabbits, cats, and sheep) have induced PWMD by inducing inflammation and infectious insults [14, 15]. Recently, a meta-analysis of six randomized controlled trials of antenatal magnesium sulfate for the prevention of cerebral palsy and other neurological abnormalities has concluded that magnesium appears to be protective in neonates at risk for delivery less than 34 weeks. The basic science mechanisms for the neuroprophylaxis are not completely understood. This chapter will concentrate on best possible evidence for mechanisms by which magnesium sulfate appears to act as neuroprotectant in the premature brain at risk for preterm birth.

What Are the Mechanisms?

Most cases of intrauterine inflammation/infection likely arise from ascending infection. Microorganisms enter the amniotic cavity and gain access to the fetus by ascending through the cervix from the vagina [16, 17]. Proinflammatory cytokines are produced by an innate immune response and cause inflammation of the chorionic and amniotic membranes [16]. Antenatal inflammation and infection are linked to brain damage as evidenced by the increased rate of cerebral palsy seen in premature infants born in the setting of clinical or histological chorioamnionitis with an odds ratio of 9.3 [18]. Several studies have examined the neurotoxic effects of inflammatory cytokines and the development of white matter injury [8, 19, 20, 21]. Inflammation also likely plays a role in enhancing the susceptibility of the brain to hypoxic–ischemic (HI) events [22].

The permeability of the blood–brain barrier is altered as is the hemodynamic stability of the fetus, and thus, inflammatory-mediated injury is likely synergistic with fetal hypoxemia [23]. Circulating inflammatory cytokines have also been shown to result in an unstable circulation which can contribute to subsequent white matter injury [3]. Some authors hypothesize regarding a "multiple-hit" mechanism such that there are antenatal, perinatal, and postnatal factors that induce or modulate brain lesions in human preterm neonates. Inflammation and infection may play a role in sensitizing neuronal cells thus predisposing them to injury [14]. Other mechanisms involve glutamate receptor activation and cell-mediated apoptosis. These injuries primarily target the developing brain of premature fetuses <32 weeks and are demonstrated clinically as periventricular leukomalacia (PVL).

PVL is primarily confined to preterm infants and occurs as the result of injury to preoligodendrocytes in the premature brain. Preoligodendrocytes are precursors of myelinating oligodendrocytes (major glial population in the white matter). Damage to these preoligodendrocytes by cytokines leads to poor myelination later and subsequent reduction in white matter volume diagnosed by three-dimensional MRI [3]. In addition to the reduction in white matter, there are also reductions in gray matter volumes and subsequent cortical development. This may explain why the cognitive ability of preterm infants <32 weeks is impaired as evidenced by lower standard intelligence test scores even excluding major neurological deficits [3].

Intraventricular hemorrhage (IVH) is an important and not uncommon cause of brain injury in the preterm neonate. Very low birth weight neonates born at less than 32 weeks have up to a 40% incidence of developing IVH [24]. Those neonates who are exposed to an infected environment in utero, as evidenced by clinical chorioamnionitis, have greater than threefold higher risk of developing IVH

[25, 26]. A more recent prospective large cohort study confirms the findings of increased risk of severe IVH (grade III or IV) in the setting of clinical chorioamnionitis [27]. Even histological inflammation alone increases the risk of IVH 2.6-fold in VLBW neonates [28]. After accounting for confounders, the aforementioned studies demonstrate a relationship between infection and IVH, and an explanation may lie in hypoxia–ischemia in the setting of inflammation.

Eklind et al. in 2001 and Duncan et al. in 2006 demonstrated that LPS administered to animal models (rat and ovine, respectively) caused hypoxic–ischemic injury when administered acutely [29, 30]. However, fetal hypoxemia and the resulting brain damage were not as severe when LPS was administered to the ovine fetus via prolonged intravenous infusion or intra-amniotically [30, 31]. There appears to be a potentiation of hypoxic–ischemic damage that occurs in the setting of inflammation/infection. While the exact etiology of inflammation-associated fetal brain injury is unknown, it is likely multifactorial in nature.

Given the wealth of information in both animal and human studies regarding the role of inflammation-associated brain injury, our focus must shift towards protecting the developing brain during the antepartum, intrapartum, and neonatal periods. This is one of the most significant challenges in perinatal medicine [32]. Currently, hypothermia and magnesium are two of the few rigorously tested and implemented clinically significant therapeutic agents used to prevent adverse neurodevelopmental outcomes.

Hypothermia, unfortunately, is only being used in neonates born at ≥36 weeks secondary to the lack of studies conducted in preterm neonate and the suspected risk of systemic hypothermia on the fragile preterm neonate. The risks associated with hypothermia include thrombocytopenia, hypotension, increased oxygen consumption, decreased surfactant production, and concern for adversely affecting coagulation and increasing the risk of IVH in a high-risk population [33]. While much work has been done to help identify early markers of neonatal brain injury, therapies are still scant [34].

Magnesium sulfate has been considered as a drug for fetal neuroprophylaxis in cases of mothers "at risk" for preterm birth since data were first published by Nelson and Grether in 1995 showing that VLBW neonates exposed to magnesium had an OR of developing cerebral palsy of 0.18 after controlling for confounders [35]. Given that magnesium has the benefit of years of research and a favorable safety profile in the antepartum period, prospective randomized controlled trials were conducted to assess the neuroprotective effect of magnesium sulfate. Five landmark randomized controlled trials (MagNET, ACTOMgSO4, MAGPIE, PREMAG, BEAM) and two meta-analyses of these trials concluded that magnesium sulfate demonstrated significantly decreased risk of moderate or severe cerebral palsy and substantial gross motor dysfunction without a significant effect on the risk of total pediatric mortality and thus should be used to protect the developing brain in this high-risk population [36–43].

The Cochrane review concluded that magnesium sulfate should be implemented for neuroprophylaxis of fetuses exposed to preterm birth [44]. However, the mechanism by which magnesium sulfate appears to have an effect is unknown. In rodent models, aimed to simulate the most common clinical scenario associated with inflammation-associated fetal brain injury, Burd et al. and Tam Tam et al. used lipopolysaccharide to induce an inflammatory response in fetal brains of mice and rats, respectively [45, 46]. Those animals exposed to magnesium demonstrated less neuronal injury in the setting of inflammation (Fig. 16.1).

Basic science studies support the anti-inflammatory role of magnesium. In vitro studies by Rochelson et al. demonstrated the attenuation of cytokine production by magnesium in cultured endothelial cells [47]. In vivo studies by Malpuech-Brugere et al. demonstrated the anti-inflammatory effects of magnesium in the presence and absence of lipopolysaccharide in magnesium-deficient animals [48, 49]. In a rat model of maternal infection, maternal administration of magnesium reduced proinflammatory mediators in maternal and fetal compartments, including the fetal brain [46].

As the previous chapter discussed, magnesium is successfully being used to mitigate the risk of cerebral palsy in preterm neonates. This chapter will focus on inflammation-associated brain injury

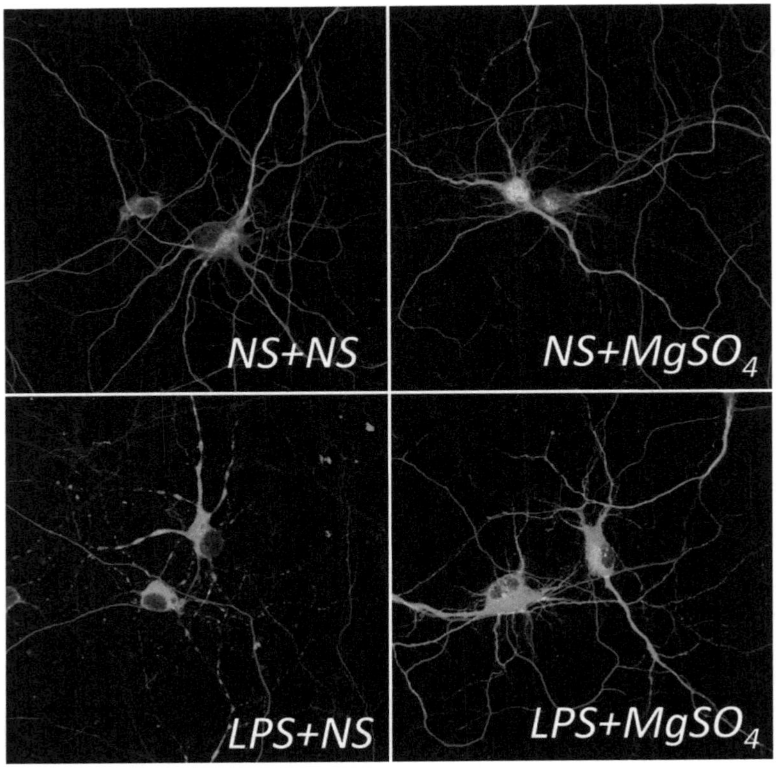

Fig. 16.1 Representative images from confocal microscopy, evaluating four treatment groups. Microtubule-associated protein 2 (MAP-2) staining is seen in *green*, and NF200 staining is seen in *red*. *NS + NS*, normal MAP2 staining and neuronal aggregation in the normal saline solution (control group); *NS + MgSO4*, normal MAP2 staining and neuronal aggregation in the NS + MgSO4 group. Lipopolysaccharide-exposed neurons that were treated with NS (*LPS + NS*) demonstrated decreased MAP2 staining, fragility, decreased aggregation, and a reduced number of dendritic processes. Lipopolysaccharide-exposed neurons that were treated with MgSO4 (*LPS + MgSO4*) had an appearance that was similar to the control group, with a normal pattern of MAP2 staining and aggregation

and the role that magnesium plays in therapy and prevention. While we do not know the exact pathway by which magnesium appears to mitigate its effects, literature supports at least three possible mechanisms: (1) nonselective antagonist at N-methyl-D-aspartate (NMDA) receptor, (2) blood–brain barrier stabilizer, and possibly as (3) an anti-apoptotic agent. These mechanisms will be explored below.

Inflammation, Magnesium, N-Methyl-D-Aspartate (NMDA) Receptors

Following hypoxic–ischemic injury and energy failure in the developing human brain, extracellular glutamate increases. Glutamate acts to depolarize neurons via N-methyl-D-aspartate (NMDA) receptors, Alpha-amino-3-hydroxy-5-methyl-4-isox-azole propionate (AMPA) receptors, and kainate receptors. With excitation of neurons, there is an influx of sodium into the cells. Increased intracellular sodium leads to an influx of chloride and, consequently, water within the cell, edema, cell lysis, and cell death [50, 51, 52].

Sanders et al. propose that excitation of neurons by glutamate following hypoxic brain injury leads to release of inflammatory mediators which activate microglia, leukocytes, and mast cells as well as stimulates apoptosis and delayed cell injury [52]. In neonates, intrapartum fever, infection, and

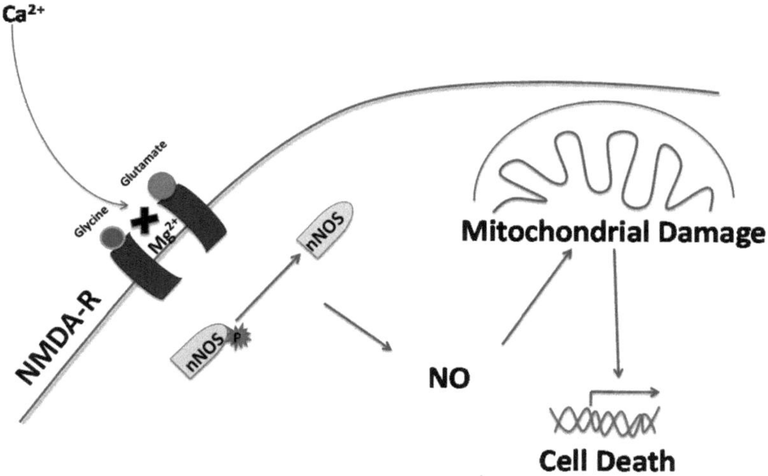

Fig. 16.2 Magnesium may act as a NMDA antagonist by blocking the influx of calcium and serving to block the NMDA channel directly

inflammation have been associated with worsened brain injury [53]. Similar to hypoxic–ischemic injury in neonates, we hypothesize that inflammation is associated with an excess of glutamate release and consequently opens NMDA receptors. This activation could lead to cellular injury and death via both immediate and delayed processes.

Magnesium may act to decrease inflammation-mediated injury at the NMDA receptor. Magnesium can act on the NMDA receptor in one of three ways: (1) It may increase the affinity of glycine for the receptor as shown by Wang and McDonald in vitro in fetal mice [54]. (2) Magnesium may act independently when NMDA receptors are saturated with glycine [55]. (3) Magnesium may block the voltage-dependent NMDA receptor by binding directly [56, 57, 58] (Fig. 16.2). Gilland et al. showed that MK-801, an NMDA receptor antagonist, decreased energy utilization [59]. In addition, increased levels of magnesium in perinatal rats reduced the neurotoxic effects mediated by NMDA receptors in a dose-dependent manner, suggesting magnesium is acting as an antagonist of the NMDA receptor [60]. These findings have lead to ongoing research about the therapeutic effects of magnesium in traumatic brain injury in adults, hypoxic–ischemic brain injury in neonates, and fetal injury following exposure to intrauterine inflammation.

NMDA receptors are involved in signaling throughout the nervous system. Specifically, these receptors are expressed in developing oligodendrocyte processes in mice, which are responsible for myelinating neuronal axons [50]. Decreased myelination of the oligodendrocytes is associated with periventricular leukomalacia (PVL) and brain injury in neonates [50]. This injury may be mediated via NMDA receptors as they have a low resistance to oxidative stress, are calcium-gated, and are susceptible to energy failure [61]. It has also been shown that oligodendrocytes are selectively injured during brain ischemia in the neonatal mouse. After inducing ischemia in mice, NMDA receptor activation resulted in disintegration of oligodendrocytes. This injury was ameliorated by NMDA-selective antagonists [62].

Volpe et al. proposed that ischemia in human neonates is potentiated by inflammation via cerebral white matter accumulation of microglial cells [63]. Lipopolysaccharide (LPS) exposure has been used to study the role of acute inflammation in neonatal animal models. There is a synergistic effect between LPS-induced inflammation and perinatal brain injury [64]. Notably, magnesium has been shown to prevent neuronal injury and to preserve dendritic processes in fetal mice when administrated to mothers [45]. We hypothesize that in the setting of inflammation, magnesium may act via NMDA receptors to prevent neonatal brain injury.

During brain development, the NMDA receptors open more easily, have a larger flux of calcium, and are less easily blocked by magnesium than in a mature brain [51]. There is an influx of calcium into the cell after excess excitation in an ischemic event. Excess activation of NMDA receptors with neuronal insult may lead to upregulation of NMDA receptors. Johnston et al. in 2002 proposed that the delayed clinical injury of PVL and cerebral palsy (CP) may be related to mitochondrial failure from calcium flooding, excessive nitric oxide production, and release of free radicals [51]. Gilland et al. in 1998 showed that delayed mitochondrial injury in the rat brain can be alleviated by NMDA receptor antagonists, suggesting a significant role for the NMDA receptor in cell injury. Magnesium can act as a voltage-dependent NMDA antagonist or can block the NMDA channel directly [60]. It has been proposed that administration of magnesium in labor or prior to preterm labor may help to alleviate PVL or CP in the neonatal brain.

Blood–Brain Barrier

The blood–brain barrier (BBB) serves an important function to prevent the entry of microbes, leukocytes, and macromolecules into the delicate central nervous system. The BBB is made up of endothelial cells held together by tight junction, pericytes, basal lamina, and astrocytes. The BBB is reliant on tight junctions between the endothelial cells lining the blood vessels in the brain to form the separation necessary between the parenchymal and vascular compartments. Tight junction permeability is altered, increasing the risk of neurological injury, by inflammatory mediators such as tumor necrosis factor-alpha (TNF-α) in response to focal cerebral ischemia or sepsis in animal models [23, 65]. Magnesium sulfate has been shown to stabilize these tight junctions which may serve as a mechanism of magnesium's neuroprotection.

In a fetal lamb model, Goni-de-Cerio et al. in 2009 induced cerebral hypoxia–ischemia through partial umbilical cord clamping and compared groups with or without magnesium sulfate exposure [66]. In the magnesium sulfate-exposed group, the percentage of S-100 protein, an astroglial protein that represents an early and sensitive biochemical marker of neurological injury, was similar to control and suggestive of neuroprotection. In a rat model, Esen et al. demonstrated the neuroprotective effects of magnesium after induced sepsis [65]. BBB permeability to Evans blue dye was significantly lower in the magnesium-treated septic rats compared to control. Magnesium likely serves to stabilize endothelial cell tight junctions by altering the calcium-dependent cadherins present between endothe-

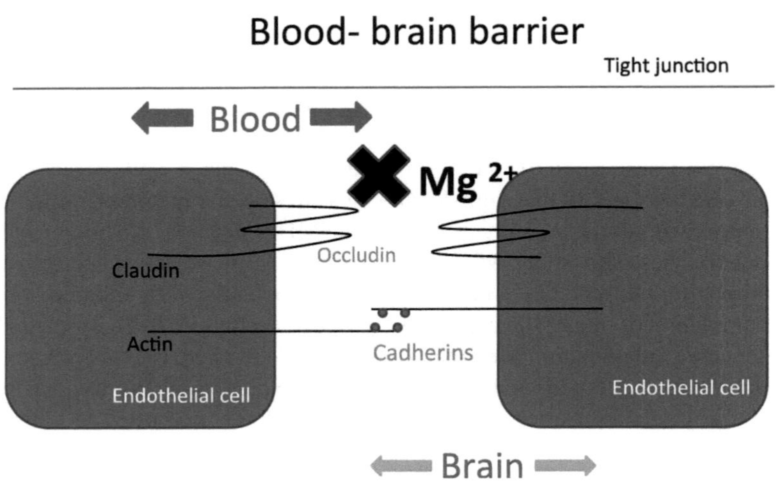

Fig. 16.3 Magnesium is shown stabilizing the tight junctions between endothelial cells of the blood–brain barrier

lial cells and decrease permeability in the setting of inflammation and thus exert a neuroprotective effect for those at risk of injury (Fig. 16.3).

Inflammation, Magnesium, and Apoptosis

Lastly, there is a considerable controversy in the literature in regard to the relationship between inflammation, magnesium, and apoptosis. Some studies suggest magnesium is protective against apoptosis, and others suggest magnesium potentiates cell death.

Following a hypoxic–ischemic event, excitation of neurons is followed by release of free radicals and cell death, or apoptosis. Sanders et al. proposed that apoptosis may occur due to loss of synaptic connectivity, trophic factors, inflammatory activation, and mitochondrial impairment [52]. Edwards and Mehmet studied induced ischemic injury in piglets and found that the proportion of apoptotic cells correlated with the degree of energy failure in hypoxic–ischemic injury [67]. As previously discussed, we believe that inflammation potentiates the effects of neuronal ischemic injury.

Fas molecule binding on the B cell lymphocyte has been shown to lead to apoptotic cell death. Chien et al. showed that the higher concentration of anti-Fas antibody (indicating cell death), the higher the percentage of cells in vitro mobilizing magnesium, fragmenting DNA, and externalizing phosphatidylserine to signal macrophages. They concluded that magnesium is required for apoptosis [68].

Several studies have been done to evaluate the role of magnesium on apoptosis in hypoxic–ischemic brain injury in neonates using animal models. Turkyilmaz et al. evaluated rat neonates following hypoxic–ischemic injury [69]. In rats pretreated with magnesium prior to insult, there was a significant decrease in apoptosis, specifically in the hippocampus. Enomoto et al. looked further at magnesium pretreatment and traumatic brain injury to determine if magnesium administration could protect rat neonates from neuronal loss and cognitive dysfunction [70]. They found that magnesium pretreatment prevented neuronal loss and cognitive dysfunction in the radial arm maze test.

In the above studies, magnesium appears to protect cells from apoptosis (Fig. 16.4). In contrast, Dribben et al. assessed the effects of magnesium exposure on the neonatal mouse brain to determine if high doses of magnesium could cause cell death [71]. They found that high doses of magnesium sulfate exposure induced apoptotic cell death in developing mouse brain. Evaluation of mouse brains at postnatal day of life 7 (corresponding to the third trimester in humans) showed widespread neuro-

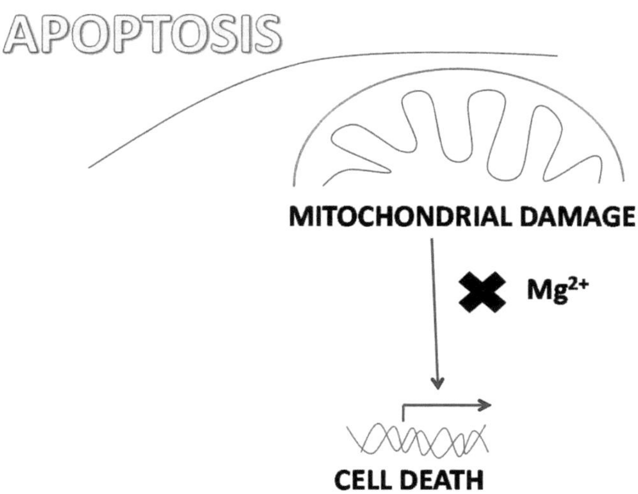

Fig. 16.4 Magnesium may interact with damaged cells and mitigate apoptosis, thus rescuing the cell from "programmed death"

degeneration in the brain from apoptotic processes in the mice exposed to magnesium. In vitro studies by Black et al. demonstrated apoptosis at physiologic levels of magnesium. This suggests that magnesium is necessary for apoptosis in human tissue [72]. The effect of magnesium was mitigated by antioxidants such as vitamin C, vitamin E, and acetylcysteine, which decrease free radicals, apoptosis, and cell death.

Conclusion

Inflammation plays an important role in neuronal injury and perinatal brain damage, as it has been demonstrated in both basic science and clinical studies. The morbidity associated with preterm birth, especially inflammation-associated preterm birth, is astoundingly costly. The impact neuronal injury has on developing life in terms of potential and resources is impossible to quantify and calculate. There is nothing more devastating in perinatal medicine. Preventing injury early in development is paramount to impacting the morbidity of inflammation-associated brain injury.

Magnesium has been used safely and effectively in medicine for decades. In vivo studies confirm many in vitro findings of the neuroprotective effects of magnesium sulfate. While the exact mechanism of action that allows magnesium to exert its neuroprotective effects is unknown, magnesium has been shown to act via multiple mechanisms. Since inflammation-associated brain injury likely occurs via various mechanisms, magnesium, as a therapeutic agent, has the opportunity to impact the deleterious effects of inflammation at multiple levels. We still have a lot to learn about the mechanisms of action of magnesium with respect to inflammation-associated brain injury. Further research is urgently needed to elucidate magnesium's actions to facilitate broader application of this tried and true ion.

References

1. Mathews TJ, MacDorman MF. Infant mortality statistics from the period linked birth/infant death data set. Natl Vital Stat Rep. 2006;58(17):1–31.
2. Msall ME, Tremont MR. Measuring functional outcomes after prematurity: developmental impact of very low birth weight and extremely low birth weight status on childhood disability. Ment Retard Dev Disabil Res Rev. 2002;8(4):258–72.
3. Cooke RWI, Abernethy LJ. The sequelae of neonatal brain injury. Paediatr Child Health. 2010;20(8):374–81.
4. Andrews WW, et al. Early preterm birth: association between in utero exposure to acute inflammation and severe neurodevelopmental disability at 6 years of age. Am J Obstet Gynecol. 2008;198(4):466e1–11.
5. Goldenberg RL, Hauth JC, Andrews WW. Intrauterine infection and preterm delivery. N Engl J Med. 2000;342(20):1500–7.
6. Wu YW, Colford Jr JM. Chorioamnionitis as a risk factor for cerebral palsy: a meta-analysis. JAMA. 2000;284(11):1417–24.
7. Romero R, et al. The role of infection in preterm labour and delivery. Paediatr Perinat Epidemiol. 2001;15 Suppl 2:41–56.
8. Yoon BH, et al. Interleukin-6 concentrations in umbilical cord plasma are elevated in neonates with white matter lesions associated with periventricular leukomalacia. Am J Obstet Gynecol. 1996;174(5):1433–40.
9. Nelson KB, et al. Neonatal cytokines and coagulation factors in children with cerebral palsy. Ann Neurol. 1998;44(4):665–75.
10. Shatrov JG, et al. Chorioamnionitis and cerebral palsy: a meta-analysis. Obstet Gynecol. 2010;116 (2 Pt 1):387–92.
11. Wu YW. Systematic review of chorioamnionitis and cerebral palsy. Ment Retard Dev Disabil Res Rev. 2002;8(1):25–9.
12. Allin M, et al. Cognitive maturation in preterm and term born adolescents. J Neurol Neurosurg Psychiatry. 2008;79(4):381–6.
13. Larroque B, et al. Neurodevelopmental disabilities and special care of 5-year-old children born before 33 weeks of gestation (the EPIPAGE study): a longitudinal cohort study. Lancet. 2008;371(9615):813–20.

14. Degos V, et al. Neuroprotective strategies for the neonatal brain. Anesth Analg. 2008;106(6):1670–80.
15. Woodward LJ, et al. Object working memory deficits predicted by early brain injury and development in the preterm infant. Brain. 2005;128(Pt 11):2578–87.
16. Romero R, et al. Infection and labor. V. Prevalence, microbiology, and clinical significance of intraamniotic infection in women with preterm labor and intact membranes. Am J Obstet Gynecol. 1989;161(3):817–24.
17. Romero R, et al. Infection and labor. VI. Prevalence, microbiology, and clinical significance of intraamniotic infection in twin gestations with preterm labor. Am J Obstet Gynecol. 1990;163(3):757–61.
18. Grether JK, Nelson KB. Maternal infection and cerebral palsy in infants of normal birth weight. JAMA. 1997;278(3):207–11.
19. Leviton A, et al. Maternal infection, fetal inflammatory response, and brain damage in very low birth weight infants. Developmental Epidemiology Network Investigators. Pediatr Res. 1999;46(5):566–75.
20. Wheater M, Rennie JM. Perinatal infection is an important risk factor for cerebral palsy in very-low-birthweight infants. Dev Med Child Neurol. 2000;42(6):364–7.
21. Faix RG, Donn SM. Association of septic shock caused by early-onset group B streptococcal sepsis and periventricular leukomalacia in the preterm infant. Pediatrics. 1985;76(3):415–9.
22. Coumans AB, et al. Intracisternal application of endotoxin enhances the susceptibility to subsequent hypoxic-ischemic brain damage in neonatal rats. Pediatr Res. 2003;53(5):770–5.
23. Yang GY, et al. Tumor necrosis factor alpha expression produces increased blood–brain barrier permeability following temporary focal cerebral ischemia in mice. Brain Res Mol Brain Res. 1999;69(1):135–43.
24. Paneth N, et al. Incidence and timing of germinal matrix/intraventricular hemorrhage in low birth weight infants. Am J Epidemiol. 1993;137(11):1167–76.
25. Morales WJ. The effect of chorioamnionitis on the developmental outcome of preterm infants at one year. Obstet Gynecol. 1987;70(2):183–6.
26. Salafia CM, et al. Maternal, placental, and neonatal associations with early germinal matrix/intraventricular hemorrhage in infants born before 32 weeks' gestation. Am J Perinatol. 1995;12(6):429–36.
27. Soraisham AS, et al. A multicenter study on the clinical outcome of chorioamnionitis in preterm infants. Am J Obstet Gynecol. 2009;200(4):372.e1–6.
28. Dammann O, Leviton A. Maternal intrauterine infection, cytokines, and brain damage in the preterm newborn. Pediatr Res. 1997;42(1):1–8.
29. Eklind S, et al. Bacterial endotoxin sensitizes the immature brain to hypoxic–ischaemic injury. Eur J Neurosci. 2001;13(6):1101–6.
30. Duncan JR, et al. Chronic endotoxin exposure causes brain injury in the ovine fetus in the absence of hypoxemia. J Soc Gynecol Investig. 2006;13(2):87–96.
31. Nitsos I, et al. Chronic exposure to intra-amniotic lipopolysaccharide affects the ovine fetal brain. J Soc Gynecol Investig. 2006;13(4):239–47.
32. Rees S, Harding R, Walker D. An adverse intrauterine environment: implications for injury and altered development of the brain. Int J Dev Neurosci. 2008;26(1):3–11.
33. Gunn AJ, Bennet L. Brain cooling for preterm infants. Clin Perinatol. 2008;35(4):735–48. Vi–vii.
34. Donohue PK, Graham EM. Earlier markers for cerebral palsy and clinical research in premature infants. J Perinatol. 2007;27(5):259–61.
35. Nelson KB, Grether JK. Can magnesium sulfate reduce the risk of cerebral palsy in very low birthweight infants? Pediatrics. 1995;95(2):263–9.
36. Rouse DJ, et al. A randomized, controlled trial of magnesium sulfate for the prevention of cerebral palsy. N Engl J Med. 2008;359(9):895–905.
37. The Magpie Trial: a randomised trial comparing magnesium sulphate with placebo for pre-eclampsia. Outcome for children at 18 months. BJOG. 2007;114(3):289–99.
38. Crowther CA, et al. Effect of magnesium sulfate given for neuroprotection before preterm birth: a randomized controlled trial. JAMA. 2003;290(20):2669–76.
39. Doyle LW, et al. Antenatal magnesium sulfate and neurologic outcome in preterm infants: a systematic review. Obstet Gynecol. 2009;113(6):1327–33.
40. Mittendorf R, et al. Is tocolytic magnesium sulphate associated with increased total paediatric mortality? Lancet. 1997;350(9090):1517–8.
41. Mittendorf R, et al. Association between the use of antenatal magnesium sulfate in preterm labor and adverse health outcomes in infants. Am J Obstet Gynecol. 2002;186(6):1111–8.
42. Conde-Agudelo A, Romero R. Antenatal magnesium sulfate for the prevention of cerebral palsy in preterm infants less than 34 weeks' gestation: a systematic review and metaanalysis. Am J Obstet Gynecol. 2009;200(6): 595–609.
43. Marret S, et al. Magnesium sulphate given before very-preterm birth to protect infant brain: the randomised controlled PREMAG trial*. BJOG. 2007;114(3):310–8.

44. Doyle LW, et al. Magnesium sulphate for women at risk of preterm birth for neuroprotection of the fetus. Cochrane Database Syst Rev. 2009;(1):CD004661.
45. Burd I, et al. Magnesium sulfate reduces inflammation-associated brain injury in fetal mice. Am J Obstet Gynecol. 2010;202(3):292. e1–9.
46. Tam Tam HB, et al. Magnesium sulfate ameliorates maternal and fetal inflammation in a rat model of maternal infection. Am J Obstet Gynecol. 2011;204(4):364.e1–8.
47. Rochelson B, et al. Magnesium sulfate suppresses inflammatory responses by human umbilical vein endothelial cells (HuVECs) through the NFkappaB pathway. J Reprod Immunol. 2007;73(2):101–7.
48. Malpuech-Brugere C, et al. Enhanced tumor necrosis factor-alpha production following endotoxin challenge in rats is an early event during magnesium deficiency. Biochim Biophys Acta. 1999;1453(1):35–40.
49. Malpuech-Brugere C, et al. Inflammatory response following acute magnesium deficiency in the rat. Biochim Biophys Acta. 2000;1501(2–3):91–8.
50. Jensen FE. Role of glutamate receptors in periventricular leukomalacia. J Child Neurol. 2005;20(12):950–9.
51. Johnston MV, Nakajima W, Hagberg H. Mechanisms of hypoxic neurodegeneration in the developing brain. Neuroscientist. 2002;8(3):212–20.
52. Sanders RD, et al. Preconditioning and postinsult therapies for perinatal hypoxic-ischemic injury at term. Anesthesiology. 2010;113(1):233–49.
53. Edwards AD, Tan S. Perinatal infections, prematurity and brain injury. Curr Opin Pediatr. 2006;18(2):119–24.
54. Wang LY, MacDonald JF. Modulation by magnesium of the affinity of NMDA receptors for glycine in murine hippocampal neurones. J Physiol. 1995;486(Pt 1):83–95.
55. Paoletti P, Neyton J, Ascher P. Glycine-independent and subunit-specific potentiation of NMDA responses by extracellular Mg2+. Neuron. 1995;15(5):1109–20.
56. Antonov SM, Johnson JW. Permeant ion regulation of N-methyl-D-aspartate receptor channel block by Mg(2+). Proc Natl Acad Sci U S A. 1999;96(25):14571–6.
57. Zhu Y, Auerbach A. Na(+) occupancy and Mg(2+) block of the n-methyl-d-aspartate receptor channel. J Gen Physiol. 2001;117(3):275–86.
58. Qian A, Antonov SM, Johnson JW. Modulation by permeant ions of Mg(2+) inhibition of NMDA-activated whole-cell currents in rat cortical neurons. J Physiol. 2002;538(Pt 1):65–77.
59. Gilland E, et al. Mitochondrial function and energy metabolism after hypoxia-ischemia in the immature rat brain: involvement of NMDA-receptors. J Cereb Blood Flow Metab. 1998;18(3):297–304.
60. McDonald JW, Silverstein FS, Johnston MV. Magnesium reduces N-methyl-D-aspartate (NMDA)-mediated brain injury in perinatal rats. Neurosci Lett. 1990;109(1–2):234–8.
61. Wong R. NMDA receptors expressed in oligodendrocytes. Bioessays. 2006;28(5):460–4.
62. Salter MG, Fern R. NMDA receptors are expressed in developing oligodendrocyte processes and mediate injury. Nature. 2005;438(7071):1167–71.
63. Volpe JJ, et al. The developing oligodendrocyte: key cellular target in brain injury in the premature infant. Int J Dev Neurosci. 2011;29(4):423–40.
64. Wang X, et al. Lipopolysaccharide-induced inflammation and perinatal brain injury. Semin Fetal Neonatal Med. 2006;11(5):343–53.
65. Esen F, et al. Effect of magnesium sulfate administration on blood–brain barrier in a rat model of intraperitoneal sepsis: a randomized controlled experimental study. Crit Care. 2005;9(1):R18–23.
66. Goni-de-Cerio F, et al. MgSO4 treatment preserves the ischemia-induced reduction in S-100 protein without modification of the expression of endothelial tight junction molecules. Histol Histopathol. 2009;24(9):1129–38.
67. Edwards AD, Mehmet H. Apoptosis in perinatal hypoxic-ischaemic cerebral damage. Neuropathol Appl Neurobiol. 1996;22(6):494–8.
68. Chien MM, et al. Fas-induced B cell apoptosis requires an increase in free cytosolic magnesium as an early event. J Biol Chem. 1999;274(11):7059–66.
69. Turkyilmaz C, et al. Magnesium pre-treatment reduces neuronal apoptosis in newborn rats in hypoxia-ischemia. Brain Res. 2002;955(1–2):133–7.
70. Enomoto T, et al. Pre-Injury magnesium treatment prevents traumatic brain injury-induced hippocampal ERK activation, neuronal loss, and cognitive dysfunction in the radial-arm maze test. J Neurotrauma. 2005;22(7):783–92.
71. Dribben WH, et al. High dose magnesium sulfate exposure induces apoptotic cell death in the developing neonatal mouse brain. Neonatology. 2009;96(1):23–32.
72. Black S, et al. Physiologic concentrations of magnesium and placental apoptosis: prevention by antioxidants. Obstet Gynecol. 2001;98(2):319–24.

Chapter 17
Magnesium and Its Interdependency with Other Cations in Acute and Chronic Stressor States

Babatunde O. Komolafe, M. Usman Khan, Rami N. Khouzam, Dwight A. Dishmon, Kevin P. Newman, Jesse E. McGee, Syamal K. Bhattacharya, and Karl T. Weber

Key Points

- Mg^{2+}, an abundant intracellular cation, has a number of interdependent reactions with other cations, including K^+ and Ca^{2+}.
- The interplay among Mg^{2+}, K^+, and Ca^{2+} is particularly evident during acute and chronic stressor states unmasked by neurohormonal activation.
- Acute elevations in circulating catecholamines and chronic aldosteronism readily unmask this interconnection.
- The interdependency between Mg^{2+}, K^+, and Ca^{2+} includes the activity of Mg^{2+}-dependent Na/K ATPase and Mg^{2+}-mediated secretion of parathyroid hormone.
- Restoration of K^+ and Ca^{2+} dyshomeostasis during stressor states demands that Mg^{2+} dyshomeostasis be resolved first.

Keywords Magnesium • Catecholamines • Potassium • Aldosteronism • Calcium • Stressor states

Abbreviations

CHF	Congestive heart failure
PTH	Parathyroid hormone
SHPT	Secondary hyperparathyroidism

This work was supported, in part, by NIH grants R01-HL73043 and R01-HL90867 (KTW). Its contents are solely the responsibility of the authors and do not necessarily represent the official views of the NIH. Authors have no conflicts of interest to disclose.

B.O. Komolafe, M.D. • M.U. Khan, M.D. • R.N. Khouzam, M.D. • D.A. Dishmon, M.D. • K.P. Newman, M.D.
S.K. Bhattacharya, Ph.D. • K.T. Weber, M.D. (✉)
Division of Cardiovascular Diseases, University of Tennessee Health Science Center, Coleman College of Medicine Bldg., Suite A312, 956 Court Avenue, Memphis, TN 38163, USA
e-mail: bkomolaf@uthsc.edu; mkhan10@uthsc.edu; rkhouzam@uthsc.edu; ddishmon@uthsc.edu; knewman@uthsc.edu; sbhattachary@uthsc.edu; ktweber@uthsc.edu

J.E. McGee, M.D.
Division of Cardiovascular Diseases, University of Tennessee Health Science Center and Veterans Affairs Medical Center, Memphis, TN, USA
e-mail: jesse.mcgee@va.gov

R.R. Watson et al. (eds.), *Magnesium in Human Health and Disease*, Nutrition and Health,
DOI 10.1007/978-1-62703-044-1_17, © Springer Science+Business Media New York 2013

Introduction

Mg^{2+} and Ca^{2+} are major divalent intracellular cations. Mg^{2+} is integral to a myriad of enzymatic reactions and physiologic responses, especially in tissues such as the heart, where metabolic activity and ATP consumption are high [1]. In addition, there are intricate interdependencies between Mg^{2+} and Ca^{2+} and other cations. Numerous examples of the interplay that exists between Mg^{2+}, Ca^{2+}, and K^+ have been well recognized. Herein, we focus specifically on several inseparable interconnections involving Mg^{2+} and those cations which are clinically relevant to the heart during acute and chronic stressor states, wherein neurohormonal activation involving the adrenergic and renin-angiotensin-aldosterone systems unmasks their interdependency.

Acute Stressor States

Pathophysiologically, acute stressor states can be referred to as acute bodily injury. They include thermal or electrical burns, head or musculoskeletal trauma, subarachnoid hemorrhage or intracerebral bleed, acute myocardial infarction, and major cardiac or noncardiac surgery. Sepsis and diabetic ketoacidosis are examples of stressor states involving acute systemic inflammatory responses. Irrespective of their etiologic origins, acute stressor states are accompanied by neurohormonal activation. This involves the adrenergic nervous and renin-angiotensin-aldosterone systems, whose effector hormones are integral to stressor state-mediated adoptive homeostatic responses that can beget dyshomeostasis (see Fig. 17.1).

Catecholamine-Driven K^+ and Mg^{2+} Dyshomeostasis

Mg^{2+}-Dependent Na/K ATPase

Na/K ATPase is a sarcolemmal energy-dependent pump whose activity contributes to the regulation of intracellular K^+. It is an obligatory Mg^{2+}-dependent pump [2]. It is regulated by catecholamines, whether derived from endogenous sources or when given as a pharmacologic agent. A large number of Na/K ATPase pumps are present in skeletal muscle [3]. Increments in plasma epinephrine and norepinephrine that accompany acute stressor states will activate these pumps leading to marked K^+ uptake by muscle and the rapid induction of hypokalemia. Reductions in myocardial K^+ are accompanied by delayed repolarization and prolongation of the QTc interval of the electrocardiogram—a pathophysiologic state that favors an increased propensity for supra- and ventricular arrhythmias.

Catecholamines are therefore responsible for the acute hypokalemia that appears with acute stressor states. This response was demonstrated in normal human volunteers who were given intravenous epinephrine. A prompt and marked fall in serum K^+ from 4.0 to 3.2 mEq/L (0.8 ± 0.19 mEq/L) occurred and could be prevented by either a β_2-adrenergic receptor or Ca^{2+} channel blocker [4–7]. In addition to hypokalemia, a contemporaneous fall in serum Mg^{2+} and Ca^{2+} was also found. The intravenous infusion of epinephrine in normal human volunteers is accompanied by the rapid appearance of hypomagnesemia [8, 9]. When acute bodily injury is accompanied by hemorrhagic shock, the release of endogenous catecholamines and rise in their plasma concentrations will be marked. These levels rise even further when pharmacologic doses of norepinephrine are given to promote arteriolar vasoconstriction and to restore arterial pressure. Under these conditions, the reductions in serum K^+ (<3.0 mEq/dL) and Mg^{2+} (<1.5 mg/dL) can be profound and may be accompanied by either atrial fibrillation or ventricular tachycardia [10]. The hypomagnesemia which accompanies hyperadrenergic states includes catecholamine-induced lipolysis with Mg^{2+} bound to free fatty acids and its sequestration in adipocytes.

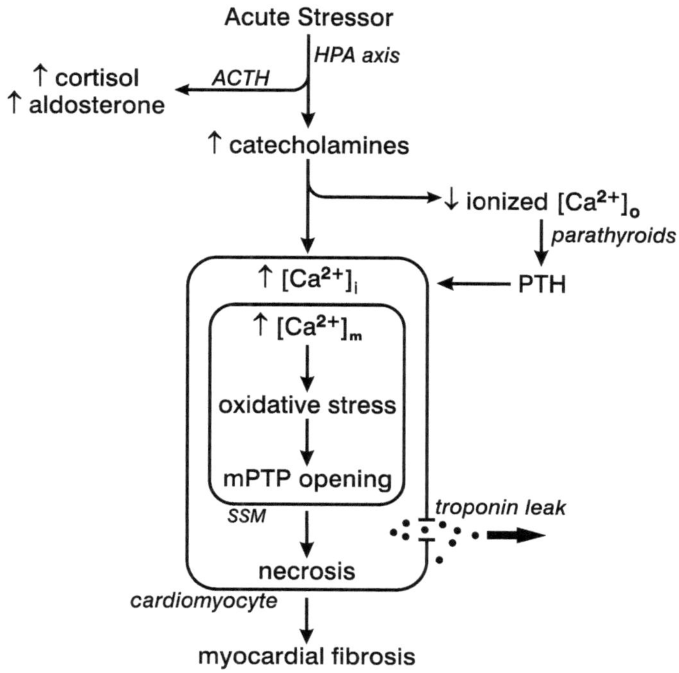

Fig. 17.1 An acute stressor state, such as bodily injury, is accompanied by an activation of the hypothalamic-pituitary-adrenal (*HPA*) axis with elevated circulating catecholamines and adrenocorticotropin stimulating the release of cortisol and aldosterone from the adrenal glands. Epinephrine and norepinephrine are responsible for intracellular Ca^{2+} overloading with a subsequent fall in plasma ionized $[Ca^{2+}]_o$, which, in turn, provokes the parathyroid glands to release parathyroid hormone (*PTH*). It too is responsible for intracellular Ca^{2+} overloading. In cardiomyocytes, this is accompanied by the induction of oxidative stress that leads to the opening of the mitochondrial permeability transition pore (*mPTP*) and osmotic injury of these organelles. The necrosis of a small number of cardiomyocytes follows accompanied by the leak of troponins into the interstitial space and ultimate rise in plasma troponins, albeit a lesser rise than seen with acute myocardial infarction due to a critical reduction in blood flow to a segment of myocardium. Lost to necrosis, cardiac myocytes are replaced by fibrosis, or scarring, to preserve the structural integrity of myocardium (Adapted with permission from Whitted AD, et al. Am J Med Sci. 2010;340:48–53)

The underlying K^+ balance that exists prior to bodily injury is a determinant of the severity of the ensuing hypokalemia that occurs during an acute stressor state. This was again demonstrated in normal volunteers who were given a thiazide diuretic prior to the epinephrine infusion. The diuretic-induced loss of K^+ predisposed to marked hypokalemia in response to the catecholamine [11]. Spironolactone, an aldosterone antagonist and K^+-sparing diuretic, on the other hand, was protective against hypokalemia [12]. Patients with arterial hypertension or congestive heart failure (CHF), who have been receiving long-term thiazide or loop diuretic treatment, respectively, may have marginal K^+ and Mg^{2+} reserves, which are then further compromised by the presence of a hyperadrenergic state (e.g., motor vehicle accident) that quickly leads to marked hypokalemia and hypomagnesemia with QTc prolongation and a propensity for arrhythmias. Albuterol, a catecholamine with bronchodilator properties, also predisposes to hypokalemia and hypomagnesemia in normal volunteers and even more so in those receiving a diuretic [5, 13]. Chronic excessive use of these β_2-receptor agonists may lead to marked hypokalemia and a greater propensity for arrhythmias. It must also be recognized that drug-induced prolongation of myocardial repolarization and QTc interval occurs in association with certain antibiotics (e.g., levofloxacin, ciprofloxacin, erythromycin), antidepressants (e.g., amitriptyline, imipramine), and antipsychotics (e.g., ziprasidone) [14–16].

Elevations in plasma catecholamines also are accompanied by hypomagnesemia [17]. This is related to a cyclic AMP-mediated rise in intracellular Mg^{2+}, together with increased lipolysis and Mg^{2+} binding to free fatty acids and increased urinary Mg^{2+} losses [18]. Hypomagnesemia is common in both critically ill children and adults. Predisposing risk factors include hypokalemia, hypocalcemia, thiazide and loop diuretic use, and sepsis. The hypomagnesemia which is already present at the time of admission in these patients will become more severe during prolonged hospital stay due to ongoing excretory losses and reduced Mg^{2+} intake [19–21].

Atrial and ventricular arrhythmias appear when hypomagnesemia is of moderate to marked severity (1.68 ± 0.27 mg/dL) [22–24]. Mg^{2+}-dependent Na/K ATPase activity will be reduced with hypomagnesemia and will further prolong the QTc interval and raise the propensity for arrhythmias. Moreover, the correction of impaired K^+ balance and hypokalemia will prove difficult unless Mg^{2+} is first replaced. The QTc interval and its abnormal prolongation (>440 ms) identifies a deficiency of myocardial K^+ and Mg^{2+}. Daily monitoring of the QTc interval and its normalization during Mg^{2+} and K^+ supplementation can be used to gauge the adequacy of their intracellular replacement. The attainment of normal QTc with these supplements may require several days more than needed for the more rapid return to normal for their serum levels and the correction of hypokalemia and hypomagnesemia. Moreover, less than 1% of Mg^{2+} is extracellular, and hence, serum Mg^{2+} is not an accurate indicator of intracellular Mg^{2+} stores and why the QTc interval is a valuable surrogate.

Concurrent hypokalemia and hypomagnesemia are common in critically ill patients, where the interactions between K^+ and Mg^{2+} are diverse and complex, including the importance of Mg^{2+} deficiency that interferes with K^+ retention while urinary K^+ excretion is increased [25, 26]. Mg^{2+} deficiency contemporaneously begets K^+ deficiency. The effective clinical resolution of hypokalemia mandates the simultaneous reversal of hypomagnesemia [27–29].

In the absence of gastrointestinal losses or diuretic usage, hypomagnesemia and hypokalemia due to impaired renal tubular reabsorption and presenting as urinary K^+ and Mg^{2+} wasting must be considered. Diagnostic evaluation of inheritable renal tubular disorders associated with such excessive urinary K^+ and Mg^{2+} losses should be considered (e.g., the Gitelman syndrome in adults and Bartter syndrome in children). This is especially the case when the resolution of these cations, using oral Mg^{2+} and K^+ supplements, proves difficult to achieve.

Mg^{2+} Efflux from Cardiomyocytes

Mg^{2+} is an endogenous antagonist to Ca^{2+} entry in cardiomyocytes and their mitochondria and vice versa [30–32]. Catecholamines promote the efflux of Mg^{2+} from cardiomyocytes, which, in turn, augments Ca^{2+} entry and the potential for excessive intracellular Ca^{2+} accumulation [33]. A β_1-adrenergic receptor antagonist prevents catecholamine-induced Mg^{2+} loss. Reduced myocardial Mg^{2+} content slows repolarization and prolongs QTc interval to favor arrhythmias. The efficacy of β-receptor blockade in critically ill patients, including those with myocardial infarction, therefore includes their favorable impact on catecholamine-driven dyshomeostasis of K^+ and Mg^{2+} that accompanies hyperadrenergic states.

Catecholamine-Driven Ca^{2+} and Mg^{2+} Dyshomeostasis

Intracellular Ca^{2+} Overloading and Oxidative Stress

Reductions in plasma ionized $[Ca^{2+}]_o$, or ionized hypocalcemia, are commonly found in children or adults presenting to the emergency department or admitted to intensive care units with an acute stressor state [34–44]. The fall in $[Ca^{2+}]_o$ correlates with the degree of the hyperadrenergic response and, in turn, the severity of illness (see Fig. 17.1). Ionized hypocalcemia serves as an in-hospital marker of survival. Hypoalbuminemia can contribute to reduced total Ca^{2+} concentration as a lesser amount of serum proteins are available to bind with Ca^{2+}.

The appearance of acute ionized hypocalcemia in critically ill patients is based on a shift in circulating Ca^{2+} into the intracellular compartment of various tissues that includes the heart, skeletal muscle, and peripheral blood mononuclear cells. This hypocalcemia occurs in response to catecholamine-induced intracellular Ca^{2+} overloading and is then followed by parathyroid hormone (PTH)-mediated excessive Ca^{2+} entry that appears in response to hypocalcemia (see Fig. 17.1). Thus, catecholamine- and PTH-facilitated intracellular Ca^{2+} overloading of cardiomyocytes leads to mitochondrial Ca^{2+} overloading where these organelles account for the induction of oxidative stress, when the rate of reactive oxygen and nitrogen species overwhelm their rate of detoxification by endogenous antioxidant defenses. The ensuing necrotic death of cardiomyocytes is followed by tissue repair that includes inflammatory cells and myofibroblasts to eventuate in a replacement fibrosis, or scarring. This fibrous tissue response preserves the structural integrity of the myocardium. It is a morphologic footprint of previous necrotic cell death. Fibrosis, however, has adverse consequences that include compromised ventricular function in diastole and systole and its serving as substrate for reentrant arrhythmias. In contrast to necrosis, apoptosis is a sterile form of cell death which does not involve inflammatory cells or fibroblasts, and accordingly, fibrosis does not appear.

The catecholamine-induced necrosis of cardiomyocytes is accompanied by the release of troponins, an intracellular enzyme that plays a crucial role in discerning myocardial injury. Catecholamine-induced cardiomyocyte necrosis with increased plasma troponin levels occurs in critically ill patients, including those having sepsis, hemorrhagic shock, subarachnoid hemorrhage, trauma, gastrointestinal bleeding, or pulmonary embolus [45–48]. The levels to which plasma troponins rise in such patients do not reach the more marked elevations seen with a segmental loss of myocardium that accompanies an acute myocardial infarction due to reduced coronary blood flow with a thrombosed coronary artery.

At the same time, catecholamines promote intracellular Ca^{2+} overloading; they induce an efflux of Mg^{2+} from these cells, where Mg^{2+} served as an endogenous Ca^{2+} channel blocker. Hence, the propensity for Ca^{2+} overloading and oxidative stress is enhanced during stressor states, while the loss of $[Mg^{2+}]_i$ favors QTc prolongation and propensity for arrhythmias.

Mg^{2+}-Dependent Parathyroid Hormone Secretion

In response to catecholamine-driven hypocalcemia, the Ca^{2+}-sensing receptor of the parathyroid glands provokes the increased secretion of PTH. In turn, secondary hyperparathyroidism (SHPT) with increased plasma PTH seeks to restore extracellular Ca^{2+} homeostasis by promoting the resorption of bone Ca^{2+} and increased Ca^{2+} absorption from the gut and kidneys by $25(OH)_2D_3$ which is synthesized by the kidneys in response to PTH. When hypocalcemia is associated with hypomagnesemia, PTH secretion will be impaired and corrected by reversing hypomagnesemia [49–51].

Chronic Stressor States

Congestive heart failure (CHF) is a chronic stressor state with neurohormonal activation that involves the ANS and RAAS, which are integral pathophysiologic features, irrespective of its etiologic origins or patient age. Elevated plasma levels of cortisol, renin activity, angiotensin II, aldosterone, norepinephrine, and endothelin-1 are present in CHF [52–56].

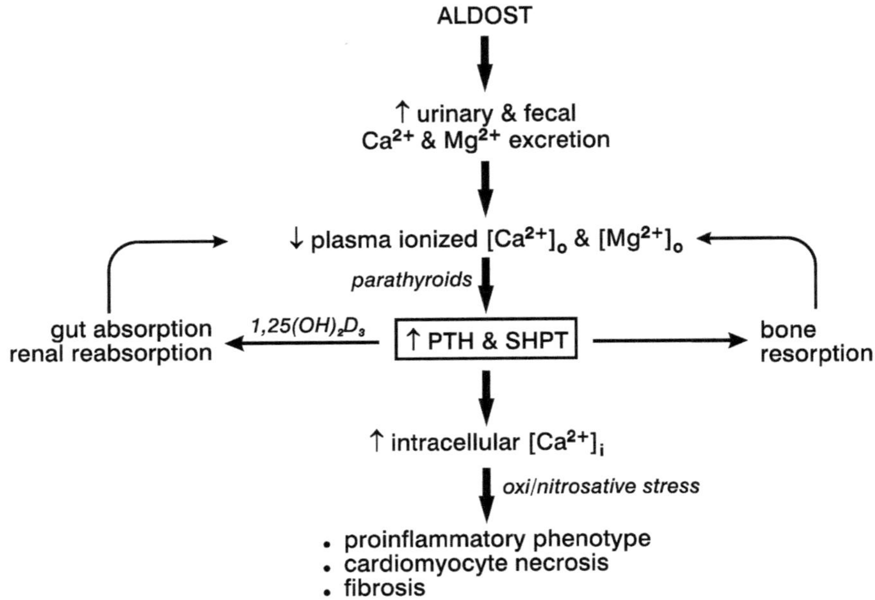

Fig. 17.2 Inappropriate (relative to dietary Na^+) elevations in plasma aldosterone comparable to those found in human CHF accompany aldosterone/salt treatment (*ALDOST*) in rats. This chronic aldosteronism is responsible for increased urinary and fecal excretion of Ca^{2+} and Mg^{2+}. This leads to plasma ionized hypocalcemia and hypomagnesemia. Secondary hyperparathyroidism (*SHPT*) is invoked to restore extracellular Ca^{2+} and Mg^{2+} homeostasis via parathyroid hormone (*PTH*)-mediated bone resorption and increased absorption and reabsorption of these cations from the gut and kidneys, respectively, mediated by the steroid hormone $1{,}25(OH)_2D_3$ synthesized by the kidneys. Paradoxically, PTH is responsible for intracellular Ca^{2+} overloading, which leads to oxi-/nitrosative stress and a proinflammatory phenotype with cardiomyocyte necrosis (Reprinted with permission from Kamalov G, et al. J Cardiovasc Pharmacol. 2010;56:320–8)

Aldosteronism-Related Ca^{2+} and Mg^{2+} Dyshomeostasis

Increased Excretory Losses and Secondary Hyperparathyroidism

Renal tubular Mg^{2+} reabsorption regulates Mg^{2+} balance, particularly involving Mg^{2+} reabsorbed in the ascending limb of Henle and proximal tubule. Mg^{2+} reabsorption depends on Na^+ reabsorption. RAAS activation in patients with heart failure leads to a salt-avid state with Na^+ and water retention that eventuates in the appearance of CHF. Splanchnic congestion with gut edema impairs Mg^{2+} absorption. Urinary and fecal excretion of K^+ and Mg^{2+} are increased during CHF based on inappropriate (relative to dietary Na^+) elevations and endocrine properties of circulating aldosterone acting at these sites, where high-density aldosterone receptor binding occurs (see Fig. 17.2) [57, 58]. The loss of these cations is accentuated by a loop diuretic commonly used in the management of CHF [27, 59]. Chronic hypomagnesemia is frequently associated with hypokalemia and hypocalcemia and has an adverse prognosis [60]. Loop as well as thiazide diuretics promote excessive urinary loss of K^+ and Mg^{2+} that can lead to both hypokalemia and hypomagnesemia. Combining either of these diuretics with spironolactone preserves K^+ and Mg^{2+} homeostasis, provided renal function is not markedly impaired (serum creatinine <2.0 mg/dL) and K^+ supplements are discontinued. Mg^{2+} is a physiologic regulator of adrenal aldosterone production with Mg^{2+} deficiency associated with elevated plasma aldosterone levels [61–63].

The importance of hypokalemia on patient mortality has been well documented. The Digitalis Investigative Group (DIG) trial database involving more than 7,700 patients revealed that in ambulatory patients having either systolic or diastolic heart failure, serum K^+ <4.0 mEq/L, and Mg^{2+} <2.0 mg/dL were associated with increased mortality [64, 65]. The same was true in patients with heart failure having associated chronic kidney disease [66]. This database also demonstrated the adverse impact of loop diuretics on death, cardiovascular mortality, and heart failure-related hospitalization in ambulatory patients, including the elderly [67, 68]. Diuretics and digoxin can each intensify existing Mg^{2+} deficiency in the elderly [69]. Mg^{2+} deficiency can raise arrhythmogenicity due to digoxin, a Na/K ATPase inhibitor [25]. This raises the prospect that prolonged routine use of a potent loop diuretic, in the absence of symptoms and signs of salt and water retention, can be quite deleterious and should be discontinued [70]. The agent can be reinstituted if and when the patient is again avidly retaining Na^+ and water.

In the Studies of Left Ventricular Dysfunction (SOLVD) trial with a cohort of more than 6,700 patients, the aforementioned adverse events were not seen with potassium-sparing diuretics. Indeed, such agents may be associated with reduced risk of all-cause mortality or death from or hospitalization for progressive heart failure [71–73]. In the Randomized Aldactone Evaluation (RALES) trial, the efficacy and safety of spironolactone, which conserves both K^+ and Mg^{2+}, was shown when it was combined with an ACE inhibitor or angiotensin receptor blocker and a loop diuretic. This included a 30% risk reduction for all-cause and cardiovascular-related mortality and sudden cardiac death and cardiovascular morbidities [73].

Mg^{2+}-Dependent Parathyroid Hormone Secretion

The secondary aldosteronism of CHF and increased fecal and urinary Ca^{2+} excretion account for consequent ionized hypocalcemia and, in turn, SHPT with elevated plasma PTH levels (see Fig. 17.2) [74–78]. However, PTH secretion is Mg^{2+}-dependent, and therefore, if the extent and duration of hypomagnesemia is marked, then the increased secretion and rise in circulating levels of PTH needed to promote bone resorption may prove difficult.

Elevated plasma levels of PTH and SHPT are also found in patients with pulmonary hypertension or obstructive airway disease [79, 80]. Reductions in systemic blood flow, including renal perfusion, account for RAAS activation in this setting. SHPT is also found in patients with primary aldosteronism [81–84], where aberrations in serum ionized and total Ca^{2+}, together with elevated PTH, are normalized by either spironolactone or adrenal surgery [83, 84]. Furthermore, elevated PTH is a known stimulus to adrenal aldosterone production and can account for elevated plasma aldosterone levels. In patients with primary hyperparathyroidism, preoperative PTH levels in excess of 100 ng/mL are independent predictors of abnormally elevated plasma aldosterone levels [85]. The impact of chronic aldosteronism on the increased incidence of adverse cardiovascular outcomes in patients with primary hyperparathyroidism [86] remains uncertain. Our experimental findings would underscore the importance of PTH-mediated intracellular Ca^{2+} overloading and induction of oxidative stress as accounting for cardiomyocyte necrosis with fibrosis as contrasted to elevations in circulating aldosterone per se [58, 87, 88].

Abnormal elevations in serum PTH (>65 pg/mL), the mediator of excessive intracellular Ca^{2+} accumulation in cardiac myocytes and mitochondria during aldosteronism [58, 89, 90], are found in patients hospitalized with decompensated heart failure and those awaiting cardiac transplantation [74, 78, 91, 92]. In outpatients having heart failure, elevated serum PTH levels are found, where it serves as an independent predictor of CHF and the need for hospitalization [93–95]. Plasma PTH levels are also an independent risk factor for mortality and cardiovascular events in patients undergoing coronary angiography [96], and they predict increased risk for cardiovascular mortality and the risk of heart failure in a community-based cohort of elderly men followed longitudinally [97, 98]. SHPT is especially prevalent in African-Americans (AA) with protracted decompensated biventricular failure, where chronic elevations in plasma aldosterone account for symptoms and signs of CHF [78]. SHPT is also related to the prevalence of hypovitaminosis D in AA with CHF [78]. The increased melanin content of darker skin in AA serves as a natural sunscreen. Accordingly, the prevalence of hypovitaminosis D, often of marked severity (<10 ng/mL), compromises Ca^{2+} homeostasis predisposing AA to hypocalcemia and consequent SHPT [78, 99, 100].

Other factors which may relate to compromised Ca^{2+} stores and contribute to the appearance of SHPT, especially in AA with CHF, include reduced dietary Ca^{2+} intake because of lactose intolerance and an active avoidance of dairy products rich in Ca^{2+} [101] and a preference for a high-Na^+ diet that enhances urinary Ca^{2+} excretion. A high-salt diet and consequential calciuria is well-known for predisposing a patient to ionized hypocalcemia and SHPT with a resorption of bone which is invoked to restore extracellular Ca^{2+} homeostasis. Over time, osteopenia and osteoporosis appear as an adverse outcome to SHPT predisposing to atraumatic bone fractures [102, 103]. Patients with heart failure have low bone density, which is related to SHPT and vitamin D deficiency coupled with reduced physical activity that may be part of their effort intolerance due to symptomatic failure [74, 91, 104–108]. The risk of such fractures is increased in elderly patients with heart failure [109], where SHPT may be contributory and which appears to be preventable when spironolactone is combined with today's standard of care [110].

Elevations in serum troponins, biomarkers of cardiomyocyte necrosis, but not due to acute MI or renal failure, are found in patients hospitalized because of their decompensated heart failure and are associated with increased in-hospital and overall cardiac mortality [111–120]. The role of intracellular Ca^{2+} overloading and oxidative stress, induced by neurohormonal activation that includes calcitropic hormones, catecholamines, and PTH, in promoting nonischemic cardiomyocyte necrosis must be considered. An ongoing loss of cardiomyocytes to both necrotic and apoptotic cell death pathways inevitably contributes to the progressive nature of heart failure.

Mg^{2+} Deficiency

Hypomagnesemia and Mg^{2+} deficiency can occur independent of each other; they are not synonymous. Mg^{2+} deficiency is a common disorder, especially in patients receiving prolonged diuretic therapy with loop or thiazide diuretics. Chemotherapeutics and immunosuppressive agents are also associated with enhanced excretory losses of Mg^{2+} and the consequent appearance of hypomagnesemia. Reduced dietary Mg^{2+} intake and impaired intestinal absorption (e.g., diarrhea) are other sources of Mg^{2+} deficiency. Chronic stressor states with prolonged hyperadrenergic activity will raise Mg^{2+} requirements. Reductions in serum Mg^{2+}, together with compromised Mg^{2+} body stores, accentuate the interdependence between Mg^{2+}, K^+, and Ca^{2+} and the appearance of a prooxidant/proinflammatory cardiac phenotype, in which pathologic remodeling of the right and left atria and ventricles is observed. The prooxidant phenotype is based on several pathophysiologic events that accompany Mg^{2+} deficiency, each of which leads to intracellular Ca^{2+} overloading. Glutathione, a component of the endogenous antioxidant defenses that is consumed in response to oxidative stress, also has a dependency on Mg^{2+} for its synthesis. A reduction in this antioxidant will accompany Mg^{2+} deficiency. Weglicki et al. have systematically addressed pathophysiologic responses associated with dietary Mg^{2+} deficiency, including the role of a neurogenic peptide, substance P. The interested reader should consult their recent reviews [121, 122].

Summary and Conclusions

Homeostasis, when invoked inappropriately and/or persistently by an acute or chronic stressor state with its inextricable coupling to neurohormonal activation, can go awry and beget dyshomeostasis (see Fig. 17.3). Acute and chronic stressor state-induced iterations in Mg^{2+}, K^+, and Ca^{2+} are numerous

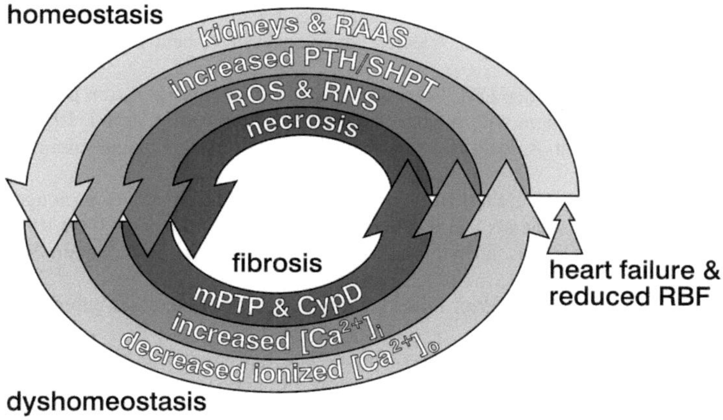

Fig. 17.3 A progressive downward spiral, where homeostasis begets dyshomeostasis at organ, cellular, and molecular levels leading to cardiomyocyte necrosis. The cycle begins with heart failure and reduced renal blood flow (*RBF*) leading to the homeostatic activation of the renin-angiotensin-aldosterone system (*RAAS*). Ionized hypocalcemia $[Ca^{2+}]_o$ appears due to an accompanying increased excretory loss of Ca^{2+}. In turn, this dyshomeostatic reaction accounts for the subsequent homeostatic response which is initiated by the appearance of secondary hyperparathyroidism (*SHPT*) with increased circulating PTH. The dyshomeostatic response to SHPT is PTH-mediated intracellular $[Ca^{2+}]_i$ overloading wherein induction of oxidative stress follows with the generation of reactive oxygen and nitrogen species (*ROS/RNS*). Together, intracellular Ca^{2+} overloading and oxidative stress contribute to the pathologic opening of the mPTP pore and activation of cyclophilin D with the ensuing osmotic injury to mitochondria and ultimately necrotic cell death (Reprinted with permission from Kamalov G, et al. J Cardiovasc Pharmacol. 2010;56:320–8)

and draw attention to the interdependence between Mg^{2+} and these cations, including the importance of Mg^{2+}-dependent Na/K ATPase. Effector hormones that accompany activation orchestrate the concordant appearance of hypokalemia, ionized hypocalcemia, and hypomagnesemia. Contemporaneously, there follows a coordinated translocation of cations to tissues, but at the expense of serum cation concentrations. Ionized hypocalcemia provokes heightened Mg^{2+}-dependent secretion of PTH with SHPT which may prove compromised during Mg^{2+} deficiency. PTH-mediated intracellular Ca^{2+} overloading, together with the induction of oxidative stress and mPTP opening, eventuates in cardiomyocyte necrosis with the release of cardiac troponins. Intracellular cationic shifts adaptively regulate the pathophysiologic equilibrium between prooxidants and antioxidants, a critical determinant of cell survival. Systematic and serial surveillance of Mg^{2+}, together with K^+ and ionized Ca^{2+}, in patients with acute and chronic stressor states will be advantageous by enabling their preservation and, when needed, correction to avoid adverse pathophysiologic consequences. Maintenance of serum K^+ at $\geq$4.0 mEq/L and Mg^{2+} at $\geq$2.0 mg/dL is considered normal and safe.

References

1. Rude RK, Shils ME. Magnesium. In: Shils ME, Shike M, Ross AC, Caballero B, Cousins RJ, editors. Modern nutrition in health and disease. 10th ed. Philadelphia: Lippincott Williams & Wilkins; 2006. p. 223–47.
2. Grubbs RD, Maguire ME. Magnesium as a regulatory cation: criteria and evaluation. Magnesium. 1987;6:113–27.
3. Kjeldsen K. Hypokalemia and sudden cardiac death. Exp Clin Cardiol. 2010;15:e96–9.
4. Brown MJ, Brown DC, Murphy MB. Hypokalemia from beta2-receptor stimulation by circulating epinephrine. N Engl J Med. 1983;309:1414–9.
5. Reid JL, Whyte KF, Struthers AD. Epinephrine-induced hypokalemia: the role of beta adrenoceptors. Am J Cardiol. 1986;57:23F–7.

6. Hansen O, Johansson BW, Nilsson-Ehle P. Metabolic, electrocardiographic, and hemodynamic responses to increased circulating adrenaline: effects of selective and nonselective beta adrenoceptor blockade. Angiology. 1990;41:175–88.
7. Darbar D, Smith M, Mörike K, Roden DM. Epinephrine-induced changes in serum potassium and cardiac repolarization and effects of pretreatment with propranolol and diltiazem. Am J Cardiol. 1996;77:1351–5.
8. Whyte KF, Addis GJ, Whitesmith R, Reid JL. Adrenergic control of plasma magnesium in man. Clin Sci (Lond). 1987;72:135–8.
9. Joborn H, Hjemdahl P, Larsson PT, Lithell H, Olsson G, Wide L, et al. Effects of prolonged adrenaline infusion and of mental stress on plasma minerals and parathyroid hormone. Clin Physiol. 1990;10:37–53.
10. Tarditi DJ, Hollenberg SM. Cardiac arrhythmias in the intensive care unit. Semin Respir Crit Care Med. 2006;27:221–9.
11. Struthers AD, Whitesmith R, Reid JL. Prior thiazide diuretic treatment increases adrenaline-induced hypokalaemia. Lancet. 1983;1(8338):1358–61.
12. Lipworth BJ, McDevitt DG, Struthers AD. Hypokalemic and ECG sequelae of combined beta-agonist/diuretic therapy. Protection by conventional doses of spironolactone but not triamterene. Chest. 1990;98:811–5.
13. Ahrens RC, Smith GD. Albuterol: an adrenergic agent for use in the treatment of asthma pharmacology, pharmacokinetics and clinical use. Pharmacotherapy. 1984;4:105–21.
14. Gupta A, Lawrence AT, Krishnan K, Kavinsky CJ, Trohman RG. Current concepts in the mechanisms and management of drug-induced QT prolongation and torsade de pointes. Am Heart J. 2007;153:891–9.
15. Buckley MS, Leblanc JM, Cawley MJ. Electrolyte disturbances associated with commonly prescribed medications in the intensive care unit. Crit Care Med. 2010;38:S253–64.
16. Curtis LH, Østbye T, Sendersky V, Hutchison S, Allen LaPointe NM, Al-Khatib SM, et al. Prescription of QT-prolonging drugs in a cohort of about 5 million outpatients. Am J Med. 2003;114:135–41.
17. Ebel H, Günther T. Role of magnesium in cardiac disease. J Clin Chem Clin Biochem. 1983;21:249–65.
18. Rayssiguier Y. Hypomagnesemia resulting from adrenaline infusion in ewes: its relation to lipolysis. Horm Metab Res. 1977;9:309–14.
19. Escuela MP, Guerra M, Añon JM, Martínez-Vizcaino V, Zapatero MD, Garcia-Jalón A, et al. Total and ionized serum magnesium in critically ill patients. Intensive Care Med. 2005;31:151–6.
20. Ryzen E. Magnesium homeostasis in critically ill patients. Magnesium. 1989;8:201–12.
21. Ueshima K, Tachibana H, Suzuki T, Hiramori K. Factors affecting the blood concentration of ionized magnesium in patients in the acute phase of myocardial infarction. Heart Vessels. 2004;19:267–70.
22. Ceremuzynski L, Van Hao N. Ventricular arrhythmias late after myocardial infarction are related to hypomagnesemia and magnesium loss: preliminary trial of corrective therapy. Clin Cardiol. 1993;16:493–6.
23. Saleem AF, Haque A. On admission hypomagnesemia in critically ill children: risk factors and outcome. Indian J Pediatr. 2009;76:1227–30.
24. Safavi M, Honarmand A. Admission hypomagnesemia–impact on mortality or morbidity in critically ill patients. Middle East J Anesthesiol. 2007;19:645–60.
25. Sheehan JP, Seelig MS. Interactions of magnesium and potassium in the pathogenesis of cardiovascular disease. Magnesium. 1984;3:301–14.
26. Solomon R. The relationship between disorders of K^+ and Mg^+ homeostasis. Semin Nephrol. 1987;7:253–62.
27. Leier CV, Dei Cas L, Metra M. Clinical relevance and management of the major electrolyte abnormalities in congestive heart failure: hyponatremia, hypokalemia, and hypomagnesemia. Am Heart J. 1994;128:564–74.
28. Milionis HJ, Alexandrides GE, Liberopoulos EN, Bairaktari ET, Goudevenos J, Elisaf MS. Hypomagnesemia and concurrent acid-base and electrolyte abnormalities in patients with congestive heart failure. Eur J Heart Fail. 2002;4:167–73.
29. Kraft MD, Btaiche IF, Sacks GS, Kudsk KA. Treatment of electrolyte disorders in adult patients in the intensive care unit. Am J Health Syst Pharm. 2005;62:1663–82.
30. Fedelesová M, Ziegelhöffer A, Luknárová O, Kostolansky S. Prevention by K^+, Mg^{2+}-aspartate of isoproterenol-induced metabolic changes in the myocardium. Recent Adv Stud Cardiac Struct Metab. 1975;6:59–73.
31. Hermes-Lima M, Castilho RF, Valle VG, Bechara EJ, Vercesi AE. Calcium-dependent mitochondrial oxidative damage promoted by 5-aminolevulinic acid. Biochim Biophys Acta. 1992;1180:201–6.
32. Szanda G, Rajki A, Gallego-Sandin S, Garcia-Sancho J, Spät A. Effect of cytosolic Mg^{2+} on mitochondrial Ca^{2+} signaling. Pflugers Arch. 2009;457:941–54.
33. Romani A, Marfella C, Scarpa A. Regulation of magnesium uptake and release in the heart and in isolated ventricular myocytes. Circ Res. 1993;72:1139–48.
34. Carlstedt F, Lind L, Joachimsson PO, Rastad J, Wide L, Ljunghall S. Circulating ionized calcium and parathyroid hormone levels following coronary artery by-pass surgery. Scand J Clin Lab Invest. 1999;59:47–53.
35. Carlstedt F, Lind L, Rastad J, Stjernstrom H, Wide L, Ljunghall S. Parathyroid hormone and ionized calcium levels are related to the severity of illness and survival in critically ill patients. Eur J Clin Invest. 1998;28:898–903.

36. Carlstedt F, Lind L, Wide L, Lindahl B, Hänni A, Rastad J, et al. Serum levels of parathyroid hormone are related to the mortality and severity of illness in patients in the emergency department. Eur J Clin Invest. 1997;27:977–81.
37. Hästbacka J, Pettilä V. Prevalence and predictive value of ionized hypocalcemia among critically ill patients. Acta Anaesthesiol Scand. 2003;47:1264–9.
38. Choi YC, Hwang SY. The value of initial ionized calcium as a predictor of mortality and triage tool in adult trauma patients. J Korean Med Sci. 2008;23:700–5.
39. Cherry RA, Bradburn E, Carney DE, Shaffer ML, Gabbay RA, Cooney RN. Do early ionized calcium levels really matter in trauma patients? J Trauma. 2006;61:774–9.
40. Dickerson RN, Henry NY, Miller PL, Minard G, Brown RO. Low serum total calcium concentration as a marker of low serum ionized calcium concentration in critically ill patients receiving specialized nutrition support. Nutr Clin Pract. 2007;22:323–8.
41. Burchard KW, Simms HH, Robinson A, DiAmico R, Gann DS. Hypocalcemia during sepsis. Relationship to resuscitation and hemodynamics. Arch Surg. 1992;127:265–72.
42. Joborn H, Hjemdahl P, Larsson PT, Lithell H, Lundin L, Wide L, et al. Platelet and plasma catecholamines in relation to plasma minerals and parathyroid hormone following acute myocardial infarction. Chest. 1990;97:1098–105.
43. Karlsberg RP, Cryer PE, Roberts R. Serial plasma catecholamine response early in the course of clinical acute myocardial infarction: relationship to infarct extent and mortality. Am Heart J. 1981;102:24–9.
44. Klein GL, Nicolai M, Langman CB, Cuneo BF, Sailer DE, Herndon DN. Dysregulation of calcium homeostasis after severe burn injury in children: possible role of magnesium depletion. J Pediatr. 1997;131:246–51.
45. Gunnewiek JM, Van Der Hoeven JG. Cardiac troponin elevations among critically ill patients. Curr Opin Crit Care. 2004;10:342–6.
46. Jeremias A, Gibson CM. Narrative review: alternative causes for elevated cardiac troponin levels when acute coronary syndromes are excluded. Ann Intern Med. 2005;142:786–91.
47. Maeder M, Fehr T, Rickli H, Ammann P. Sepsis-associated myocardial dysfunction: diagnostic and prognostic impact of cardiac troponins and natriuretic peptides. Chest. 2006;129:1349–66.
48. Vasile VC, Babuin L, Rio Perez JA, Alegria JR, Song LM, Chai HS, et al. Long-term prognostic significance of elevated cardiac troponin levels in critically ill patients with acute gastrointestinal bleeding. Crit Care Med. 2009;37:140–7.
49. Anast CS, Winnacker JL, Forte LR, Burns TW. Impaired release of parathyroid hormone in magnesium deficiency. J Clin Endocrinol Metab. 1976;42:707–17.
50. Rude RK, Oldham SB, Singer FR. Functional hypoparathyroidism and parathyroid hormone end-organ resistance in human magnesium deficiency. Clin Endocrinol (Oxf). 1976;5:209–24.
51. Iwasaki Y, Asai M, Yoshida M, Oiso Y, Hashimoto K. Impaired parathyroid hormone response to hypocalcemic stimuli in a patient with hypomagnesemic hypocalcemia. J Endocrinol Invest. 2007;30:513–6.
52. Anker SD, Chua TP, Ponikowski P, Harrington D, Swan JW, Kox WJ, et al. Hormonal changes and catabolic/anabolic imbalance in chronic heart failure and their importance for cardiac cachexia. Circulation. 1997;96:526–34.
53. Dutka DP, Olivotto I, Ward S, Nihoyannopoulos P, al-Subaili M, Oakley CM, et al. Plasma neuro-endocrine activity in very elderly subjects and patients with and without heart failure. Eur Heart J. 1995;16:1223–30.
54. Bolger AP, Sharma R, Li W, Leenarts M, Kalra PR, Kemp M, et al. Neurohormonal activation and the chronic heart failure syndrome in adults with congenital heart disease. Circulation. 2002;106:92–9.
55. Emdin M, Passino C, Prontera C, Iervasi A, Ripoli A, Masini S, et al. Cardiac natriuretic hormones, neuro-hormones, thyroid hormones and cytokines in normal subjects and patients with heart failure. Clin Chem Lab Med. 2004;42:627–36.
56. Buchhorn R, Hammersen A, Bartmus D, Bürsch J. The pathogenesis of heart failure in infants with congenital heart disease. Cardiol Young. 2001;11:498–504.
57. Wester PO, Dyckner T. Intracellular electrolytes in cardiac failure. Acta Med Scand Suppl. 1986;707:33–6.
58. Chhokar VS, Sun Y, Bhattacharya SK, Ahokas RA, Myers LK, Xing Z, et al. Hyperparathyroidism and the calcium paradox of aldosteronism. Circulation. 2005;111:871–8.
59. Law PH, Sun Y, Bhattacharya SK, Chhokar VS, Weber KT. Diuretics and bone loss in rats with aldosteronism. J Am Coll Cardiol. 2005;46:142–6.
60. Soliman HM, Mercan D, Lobo SS, Melot C, Vincent JL. Development of ionized hypomagnesemia is associated with higher mortality rates. Crit Care Med. 2003;31:1082–7.
61. Ginn HE, Cade R, McCallum T, Fregley M. Aldosterone secretion in magnesium-deficient rats. Endocrinology. 1967;80:969–71.
62. Solounias BM, Schwartz R. The effect of magnesium deficiency on serum aldosterone in rats fed two levels of sodium. Life Sci. 1975;17:1211–7.

63. Atarashi K, Matsuoka H, Takagi M, Sugimoto T. Magnesium ion: a possible physiological regulator of aldosterone production. Life Sci. 1989;44:1483–9.
64. Ahmed A, Zannad F, Love TE, Tallaj J, Gheorghiade M, Ekundayo OJ, et al. A propensity-matched study of the association of low serum potassium levels and mortality in chronic heart failure. Eur Heart J. 2007;28: 1334–43.
65. Adamopoulos C, Pitt B, Sui X, Love TE, Zannad F, Ahmed A. Low serum magnesium and cardiovascular mortality in chronic heart failure: a propensity-matched study. Int J Cardiol. 2009;136:270–7.
66. Bowling CB, Pitt B, Ahmed MI, Aban IB, Sanders PW, Mujib M, et al. Hypokalemia and outcomes in patients with chronic heart failure and chronic kidney disease: findings from propensity-matched studies. Circ Heart Fail. 2010;3:253–60.
67. Domanski M, Tian X, Haigney M, Pitt B. Diuretic use, progressive heart failure, and death in patients in the DIG study. J Card Fail. 2006;12:327–32.
68. Ahmed A, Husain A, Love TE, Gambassi G, Dell'Italia LJ, Francis GS, et al. Heart failure, chronic diuretic use, and increase in mortality and hospitalization: an observational study using propensity score methods. Eur Heart J. 2006;27:1431–9.
69. Seelig M. Cardiovascular consequences of magnesium deficiency and loss: pathogenesis, prevalence and manifestations–magnesium and chloride loss in refractory potassium repletion. Am J Cardiol. 1989;63:4G–21.
70. Weber KT. Furosemide in the long-term management of heart failure. The good, the bad and the uncertain. J Am Coll Cardiol. 2004;44:1308–10.
71. Cooper HA, Dries DL, Davis CE, Shen YL, Domanski MJ. Diuretics and risk of arrhythmic death in patients with left ventricular dysfunction. Circulation. 1999;100:1311–5.
72. Domanski M, Norman J, Pitt B, Haigney M, Hanlon S, Peyster E. Diuretic use, progressive heart failure, and death in patients in the Studies Of Left Ventricular Dysfunction (SOLVD). J Am Coll Cardiol. 2003;42:705–8.
73. Pitt B, Zannad F, Remme WJ, Cody R, Castaigne A, Perez A, et al. The effect of spironolactone on morbidity and mortality in patients with severe heart failure. Randomized Aldactone Evaluation Study Investigators. N Engl J Med. 1999;341:709–17.
74. Shane E, Mancini D, Aaronson K, Silverberg SJ, Seibel MJ, Addesso V, et al. Bone mass, vitamin D deficiency, and hyperparathyroidism in congestive heart failure. Am J Med. 1997;103:197–207.
75. Khouzam RN, Dishmon DA, Farah V, Flax SD, Carbone LD, Weber KT. Secondary hyperparathyroidism in patients with untreated and treated congestive heart failure. Am J Med Sci. 2006;331:30–4.
76. LaGuardia SP, Dockery BK, Bhattacharya SK, Nelson MD, Carbone LD, Weber KT. Secondary hyperparathyroidism and hypovitaminosis D in African-Americans with decompensated heart failure. Am J Med Sci. 2006;332: 112–8.
77. Arroyo M, LaGuardia SP, Bhattacharya SK, Nelson MD, Johnson PL, Carbone LD, et al. Micronutrients in African-Americans with decompensated and compensated heart failure. Transl Res. 2006;148:301–8.
78. Alsafwah S, LaGuardia SP, Nelson MD, Battin DL, Newman KP, Carbone LD, et al. Hypovitaminosis D in African Americans residing in Memphis, Tennessee with and without heart failure. Am J Med Sci. 2008;335:292–7.
79. Ulrich S, Hersberger M, Fischler M, Huber LC, Senn O, Treder U, et al. Bone mineral density and secondary hyperparathyroidism in pulmonary hypertension. Open Respir Med J. 2009;3:53–60.
80. Franco CB, Paz-Filho G, Gomes PE, Nascimento VB, Kulak CA, Boguszewski CL, et al. Chronic obstructive pulmonary disease is associated with osteoporosis and low levels of vitamin D. Osteoporos Int. 2009;20:1881–7.
81. Fertig A, Webley M, Lynn JA. Primary hyperparathyroidism in a patient with Conn's syndrome. Postgrad Med J. 1980;56:45–7.
82. Hellman DE, Kartchner M, Komar N, Mayes D, Pitt M. Hyperaldosteronism, hyperparathyroidism, medullary sponge kidneys, and hypertension. JAMA. 1980;244:1351–3.
83. Resnick LM, Laragh JH. Calcium metabolism and parathyroid function in primary aldosteronism. Am J Med. 1985;78:385–90.
84. Rossi E, Sani C, Perazzoli F, Casoli MC, Negro A, Dotti C. Alterations of calcium metabolism and of parathyroid function in primary aldosteronism, and their reversal by spironolactone or by surgical removal of aldosterone-producing adenomas. Am J Hypertens. 1995;8:884–93.
85. Brunaud L, Germain A, Zarnegar R, Rancier M, Alrasheedi S, Caillard C, et al. Serum aldosterone is correlated positively to parathyroid hormone (PTH) levels in patients with primary hyperparathyroidism. Surgery. 2009;146:1035–41.
86. Andersson P, Rydberg E, Willenheimer R. Primary hyperparathyroidism and heart disease–a review. Eur Heart J. 2004;25:1776–87.
87. Vidal A, Sun Y, Bhattacharya SK, Ahokas RA, Gerling IC, Weber KT. Calcium paradox of aldosteronism and the role of the parathyroid glands. Am J Physiol Heart Circ Physiol. 2006;290:H286–94.
88. Selektor Y, Ahokas RA, Bhattacharya SK, Sun Y, Gerling IC, Weber KT. Cinacalcet and the prevention of secondary hyperparathyroidism in rats with aldosteronism. Am J Med Sci. 2008;335:105–10.

89. Gandhi MS, Deshmukh PA, Kamalov G, Zhao T, Zhao W, Whaley JT, et al. Causes and consequences of zinc dyshomeostasis in rats with chronic aldosteronism. J Cardiovasc Pharmacol. 2008;52:245–52.
90. Kamalov G, Deshmukh PA, Baburyan NY, Gandhi MS, Johnson PL, Ahokas RA, et al. Coupled calcium and zinc dyshomeostasis and oxidative stress in cardiac myocytes and mitochondria of rats with chronic aldosteronism. J Cardiovasc Pharmacol. 2009;53:414–23.
91. Lee AH, Mull RL, Keenan GF, Callegari PE, Dalinka MK, Eisen HJ, et al. Osteoporosis and bone morbidity in cardiac transplant recipients. Am J Med. 1994;96:35–41.
92. Schmid C, Kiowski W. Hyperparathyroidism in congestive heart failure. Am J Med. 1998;104:508–9.
93. Ogino K, Ogura K, Kinugasa Y, Furuse Y, Uchida K, Shimoyama M, et al. Parathyroid hormone-related protein is produced in the myocardium and increased in patients with congestive heart failure. J Clin Endocrinol Metab. 2002;87:4722–7.
94. Zittermann A, Schleithoff SS, Tenderich G, Berthold HK, Korfer R, Stehle P. Low vitamin D status: a contributing factor in the pathogenesis of congestive heart failure? J Am Coll Cardiol. 2003;41:105–12.
95. Sugimoto T, Tanigawa T, Onishi K, Fujimoto N, Matsuda A, Nakamori S, et al. Serum intact parathyroid hormone levels predict hospitalisation for heart failure. Heart. 2009;95:395–8.
96. Pilz S, Tomaschitz A, Drechsler C, Ritz E, Boehm BO, Grammer TB, et al. Parathyroid hormone level is associated with mortality and cardiovascular events in patients undergoing coronary angiography. Eur Heart J. 2010;31:1591–8.
97. Hagström E, Hellman P, Larsson TE, Ingelsson E, Berglund L, Sundström J, et al. Plasma parathyroid hormone and the risk of cardiovascular mortality in the community. Circulation. 2009;119:2765–71.
98. Hagström E, Ingelsson E, Sundström J, Hellman P, Larsson TE, Berglund L, et al. Plasma parathyroid hormone and risk of congestive heart failure in the community. Eur J Heart Fail. 2010;12:1186–92.
99. Bell NH, Greene A, Epstein S, Oexmann MJ, Shaw S, Shary J. Evidence for alteration of the vitamin D-endocrine system in blacks. J Clin Invest. 1985;76:470–3.
100. Sawaya BP, Monier-Faugere MC, Ratanapanichkich P, Butros R, Wedlund PJ, Fanti P. Racial differences in parathyroid hormone levels in patients with secondary hyperparathyroidism. Clin Nephrol. 2002;57:51–5.
101. Jarvis JK, Miller GD. Overcoming the barrier of lactose intolerance to reduce health disparities. J Natl Med Assoc. 2002;94:55–66.
102. Cohen AJ, Roe FJ. Review of risk factors for osteoporosis with particular reference to a possible aetiological role of dietary salt. Food Chem Toxicol. 2000;38:237–53.
103. Teucher B, Dainty JR, Spinks CA, Majsak-Newman G, Berry DJ, Hoogewerff JA, et al. Sodium and bone health: impact of moderately high and low salt intakes on calcium metabolism in postmenopausal women. J Bone Miner Res. 2008;23:1477–85.
104. Kerschan-Schindl K, Strametz-Juranek J, Heinze G, Grampp S, Bieglmayer C, Pacher R, et al. Pathogenesis of bone loss in heart transplant candidates and recipients. J Heart Lung Transplant. 2003;22:843–50.
105. Nishio K, Mukae S, Aoki S, Itoh S, Konno N, Ozawa K, et al. Congestive heart failure is associated with the rate of bone loss. J Intern Med. 2003;253:439–46.
106. Kenny AM, Boxer R, Walsh S, Hager WD, Raisz LG. Femoral bone mineral density in patients with heart failure. Osteoporos Int. 2006;17:1420–7.
107. Frost RJ, Sonne C, Wehr U, Stempfle HU. Effects of calcium supplementation on bone loss and fractures in congestive heart failure. Eur J Endocrinol. 2007;156:309–14.
108. Abou-Raya S, Abou-Raya A. Osteoporosis and congestive heart failure (CHF) in the elderly patient: double disease burden. Arch Gerontol Geriatr. 2009;49:250–4.
109. van Diepen S, Majumdar SR, Bakal JA, McAlister FA, Ezekowitz JA. Heart failure is a risk factor for orthopedic fracture: a population-based analysis of 16,294 patients. Circulation. 2008;118:1946–52.
110. Carbone LD, Cross JD, Raza SH, Bush AJ, Sepanski RJ, Dhawan S, et al. Fracture risk in men with congestive heart failure. Risk reduction with spironolactone. J Am Coll Cardiol. 2008;52:135–8.
111. Ishii J, Nomura M, Nakamura Y, Naruse H, Mori Y, Ishikawa T, et al. Risk stratification using a combination of cardiac troponin T and brain natriuretic peptide in patients hospitalized for worsening chronic heart failure. Am J Cardiol. 2002;89:691–5.
112. Kuwabara Y, Sato Y, Miyamoto T, Taniguchi R, Matsuoka T, Isoda K, et al. Persistently increased serum concentrations of cardiac troponin in patients with acutely decompensated heart failure are predictive of adverse outcomes. Circ J. 2007;71:1047–51.
113. Peacock 4th WF, De Marco T, Fonarow GC, Diercks D, Wynne J, Apple FS, et al. Cardiac troponin and outcome in acute heart failure. N Engl J Med. 2008;358:2117–26.
114. Zairis MN, Tsiaousis GZ, Georgilas AT, Makrygiannis SS, Adamopoulou EN, Handanis SM, et al. Multimarker strategy for the prediction of 31 days cardiac death in patients with acutely decompensated chronic heart failure. Int J Cardiol. 2009;141:284–90.
115. Löwbeer C, Gustafsson SA, Seeberger A, Bouvier F, Hulting J. Serum cardiac troponin T in patients hospitalized with heart failure is associated with left ventricular hypertrophy and systolic dysfunction. Scand J Clin Lab Invest. 2004;64:667–76.

116. Horwich TB, Patel J, MacLellan WR, Fonarow GC. Cardiac troponin I is associated with impaired hemodynamics, progressive left ventricular dysfunction, and increased mortality rates in advanced heart failure. Circulation. 2003;108:833–8.
117. Sukova J, Ostadal P, Widimsky P. Profile of patients with acute heart failure and elevated troponin I levels. Exp Clin Cardiol. 2007;12:153–6.
118. Ilva T, Lassus J, Siirilä-Waris K, Melin J, Peuhkurinen K, Pulkki K, et al. Clinical significance of cardiac troponins I and T in acute heart failure. Eur J Heart Fail. 2008;10:772–9.
119. Sato Y, Nishi K, Taniguchi R, Miyamoto T, Fukuhara R, Yamane K, et al. In patients with heart failure and non-ischemic heart disease, cardiac troponin T is a reliable predictor of long-term echocardiographic changes and adverse cardiac events. J Cardiol. 2009;54:221–30.
120. Miller WL, Hartman KA, Burritt MF, Grill DE, Jaffe AS. Profiles of serial changes in cardiac troponin T concentrations and outcome in ambulatory patients with chronic heart failure. J Am Coll Cardiol. 2009;54:1715–21.
121. Kramer JH, Spurney C, Iantorno M, Tziros C, Mak IT, Tejero-Taldo MI, et al. Neurogenic inflammation and cardiac dysfunction due to hypomagnesemia. Am J Med Sci. 2009;338:22–7.
122. Weglicki WB, Chmielinska JJ, Kramer JH, Mak IT. Cardiovascular and intestinal responses to oxidative and nitrosative stress during prolonged magnesium deficiency. Am J Med Sci. 2011;342:125–8.

Chapter 18
Magnesium and Traumatic Brain Injury

Renée J. Turner and Robert Vink

Key Points

- Intracellular free magnesium concentration declines after acute injury to the CNS.
- The decline is highly significant (~50%) and persists for at least 5 days.
- Low magnesium levels facilitate secondary injury processes including inflammation, excitotoxicity, mitochondrial dysfunction, energy failure, edema formation, free radical production, and apoptosis, among others.
- Post-traumatic administration of magnesium to restore normal magnesium homeostasis reduces neuronal cell death and improves functional outcome in experimental studies.
- Facilitating the entry of magnesium across the blood-brain barrier and into the CNS will increase the likelihood of successful clinical translation.

Keywords Traumatic brain injury • Magnesium • Secondary injury • Central nervous system

Introduction

Magnesium is one of the most important ions in the body and is present in high concentrations within all cells. It is indispensable in terms of maintenance and regulation of general cellular metabolism and function due to the central roles it plays in nearly every aspect of cell function, including energy metabolism and maintenance of ionic gradients. Given that magnesium is so essential for normal cellular function, disruption of magnesium homeostasis has deleterious consequences. Indeed, the detrimental effects of the disruption of magnesium homeostasis are clearly observed following trauma to the CNS, leading to serious biochemical changes. Accordingly, the aim of this chapter is to review both the role of magnesium in secondary injury following traumatic brain injury (TBI) and also the efficacy of the experimental and clinical administration of magnesium as a novel therapeutic for the treatment of TBI.

R.J. Turner, Ph.D. (✉) • R. Vink
Adelaide Centre for Neuroscience Research, School of Medical Sciences, University of Adelaide, North Terrace, Adelaide, SA 5005, Australia
e-mail: renee.turner@adelaide.edu.au; robert.vink@adelaide.edu.au

R.R. Watson et al. (eds.), *Magnesium in Human Health and Disease*, Nutrition and Health,
DOI 10.1007/978-1-62703-044-1_18, © Springer Science+Business Media New York 2013

Traumatic Brain Injury

TBI is the leading cause of death in individuals under the age of 45 years, with an estimated incidence of death reported as 20–30 per 100,000 [1]. The majority of TBI cases can be attributed to motor vehicle accidents, motorcycle accidents, bicycle accidents, and pedestrian injuries [2]. Survivors often are left with debilitating neurological deficits after injury [3], and putting aside the enormous personal burden to victims and their families, the financial impact for the community in terms of hospitalization, treatment, rehabilitation, and specialized care runs into the billions of dollars annually. Despite this devastating impact, there is currently no approved therapy for the treatment of head trauma, largely because the mechanisms associated with neuronal cell death are poorly understood. TBI results from acceleration/deceleration forces that produce rapid movement of the brain within the skull or from the head impacting with an object [4]. The type and severity of the resultant injury is dependent upon the nature of the initiating force, in addition to the site, direction, and magnitude of the impact [5].

Primary and Secondary Injury

There are two mechanisms by which injury can occur in TBI, designated as primary injury and secondary injury. Primary injury is irreversible and occurs at the time of impact. It encompasses the mechanical forces at the time of injury that damage blood vessels, axons, neurons, and glia through shearing, tearing, and stretching [6]. It also includes surface contusions and lacerations, diffuse axonal injury, and hemorrhage [7]. The shearing forces applied to neurons in response to injury cause massive ion fluxes across neuronal membranes, resulting in the widespread loss of membrane potential and the excessive release of neurotransmitters [8]. Such cellular events are part of an evolving sequence of cellular, neurochemical, and metabolic alterations termed secondary injury, which is initiated by the initial traumatic events and ensues in the minutes to days following injury. Secondary injury has profound effects on ion channels, membranes, intracellular biochemical events, and second messenger systems including changes in neurotransmitter release, ion homeostasis, blood flow and cellular bioenergetic state, along with oxidative stress and lactoacidosis [9]. Unlike primary injury, such secondary injury is potentially reversible, thereby providing a therapeutic window for pharmacological intervention. The aim of such therapy is to reduce injury and improve both outcome and survival. However, despite the large number of experimental studies successfully targeting individual injury factors, none have resulted in an effective therapy that can be used clinically. Given the diverse nature of secondary injury and the failure to date to find a "magic bullet" targeting an individual injury factor, it is likely that agents that target multiple aspects of the injury cascade will be the most efficacious. Indeed, magnesium salts have been implicated in a number of secondary injury cascades and, accordingly, have been extensively investigated in both experimental and clinical TBI as a potential neuroprotective agent for the treatment of head injury.

Magnesium in Secondary Injury

The association between magnesium, its decline, and outcome following TBI has been extensively studied and its central role in a number of secondary injury mechanisms identified (Fig. 18.1). Magnesium's role in each of these mechanisms is briefly discussed below.

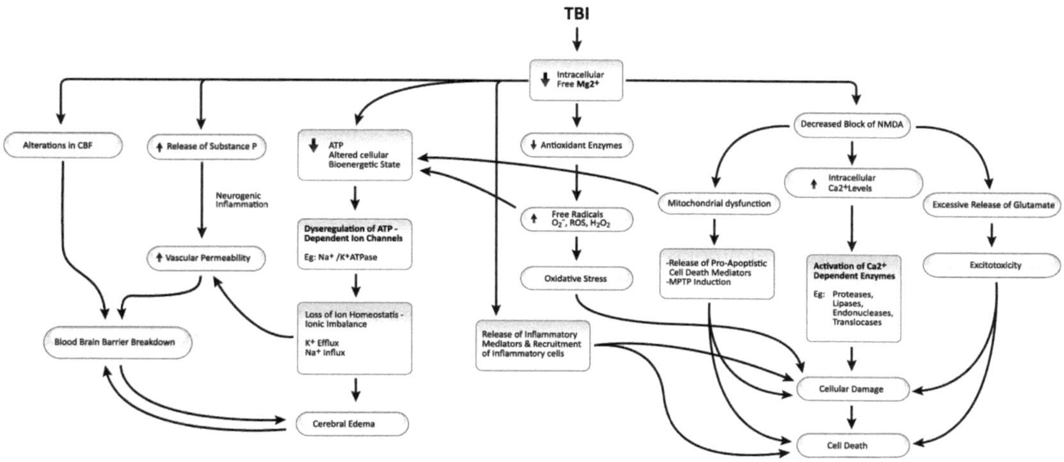

Fig. 18.1 Secondary injury pathways initiated by intracellular free magnesium decline following trauma

Energy and Enzyme Regulation

Probably the most important of magnesium's roles is as an essential cofactor for energy production [10]. Many of the enzymes of ATP production require magnesium for their action, including those involved in glycolysis, oxidative phosphorylation, and cellular respiration. Furthermore, all reactions that consume or synthesize energy require magnesium as a cofactor [11]. Magnesium is also required for the functioning of other enzymes such as those involved in DNA replication, DNA transcription, lipid metabolism [10], and protein synthesis [12]. In addition, magnesium plays a key role at the level of the cell membrane, regulating membrane permeability and enzyme activity while also ensuring that the stability, integrity, and normal functioning of the membrane is maintained [13].

Such functions of magnesium are particularly important following trauma in light of the significant alterations in oxygen and glucose consumption and tissue concentrations of phosphocreatine, lactate/pyruvate, and ATP that occur, producing marked changes in cerebral metabolism and cerebral energy state [14, 15]. Furthermore, changes in cerebral blood flow and mitochondrial function lead to impaired electron transport and reduced ATP production results [16, 17]. Hence, when magnesium levels are significantly decreased after trauma, profound changes in cellular bioenergetic state occur. Cells are less capable of providing sufficient energy for repair and to restore the disrupted ion gradients, events that may result in cell death [18]. Excess calcium contributes to reduced capacity of cells to produce energy as increased intracellular calcium levels directly impair mitochondrial function. The decreased magnesium levels, as well as the increase in inorganic phosphate and calcium, serve to induce the mitochondrial permeability transition pore [19]. Induction of the mitochondrial permeability transition pore allows the indiscriminate passage of ions in and out of the mitochondria and increased mitochondrial uptake of calcium, causing mitochondrial swelling. All of these factors mean that there is a subsequent depletion of the ATP-producing capacity by the mitochondria and energy failure ensues. The markedly reduced ATP levels have devastating consequences for cells. Indeed, there is a correlation between the intracellular free magnesium decline concentration and the cytosolic phosphorylation potential, suggesting that reduced magnesium levels following trauma greatly impair the cells' ability to produce sufficient energy for basic functions such as the maintenance of ionic gradients and cell repair [18].

Ionic Gradients

Magnesium plays a critical role in the maintenance of normal ionic gradients. The sodium/potassium ATPase co-transporter is responsible for restoring concentration gradients disrupted by action potentials [20] and has an absolute requirement for magnesium for normal functioning [13]. Being an ATPase, it is dependent on the supply of ATP, whose hydrolysis also has an absolute requirement for magnesium. When ATP levels decline as a result of the decreased magnesium levels, the result is decreased activity of the sodium/potassium ATPase and a disruption of ion homeostasis. Such an ionic destabilization places enormous energy demands upon cells to restore ionic gradients through the activation of pumping mechanisms, energy demands which are not met due to the cellular bioenergetic crisis that occurs following injury [21]. Cerebral edema is a consequence of the ionic imbalances [11, 19, 22] and is characterized by an increase in the water content of the brain either within the interstitial space, neurons, or glia [23].

Calcium Homeostasis and Excitotoxicity

Intracellular calcium is a common signal and transduction element that must be tightly regulated by cells [24]. Within minutes of injury, there is a rapid and massive increase in free intracellular calcium levels [8, 21]. Calcium can enter cells via many different channels, opened by either voltage-dependent opening induced by the mechanical deformation of the membrane or by agonist-dependent opening mediated by the release of neurotransmitters into the extracellular space [21]. The significant increase in calcium levels has widespread and detrimental effects on cell function and has been linked to much of the damage associated with secondary injury [8]. In fact, the observed changes in calcium homeostasis have been found to underlie delayed neuronal death and degeneration following injury [7].

Magnesium plays a regulatory role as an endogenous blocker of voltage and ligand-gated calcium channels [25], an example of which is the glutamate N-methyl-D-aspartate (NMDA) receptor/channel. This channel is permeable to calcium and is present in the highest density within the hippocampus and cerebral cortex [26]. Under normal conditions, voltage-gated block of the NMDA channel by magnesium regulates intracellular calcium concentrations and prevents excitotoxicity [27–30]. The mechanism whereby magnesium blocks the NMDA channel is thought to be by changing the conformation of one of the receptor components, such that the receptor is disabled [26]. After TBI, the increased glutamate release [31] facilitates entry of calcium into the cell, which may also, in part, be attributed to decreased block of the NMDA receptors by magnesium or by a receptor conformational change [32]. With magnesium no longer blocking NMDA, calcium can pass freely into cells.

Such a massive influx of calcium allows the activation numerous processes, pathways, and substrates that require calcium. The activation of calcium-dependent enzymes, including lipases, proteases, and endonucleases, occurs, and once activated, their activity may trigger harmful cascades that affect cellular metabolism and lead to the production of toxic products such as free radicals [7]. The expression of the immediate early genes c-fos, c-jun, heat shock proteins, calbindin, and glucose-regulated proteins is also increased post-trauma. These genes are activated by the increased intracellular calcium levels [7]. The rise in calcium increases release of glutamate, which in addition to activating the NMDA receptors, stimulates cell swelling, vacuolization, and neuronal death via excitotoxicity [7, 33]. Indeed, through the regulation of glutamate release, magnesium is able to protect cells from glutamate-induced excitotoxicity that leads to extensive cell damage and cell death [31].

Oxidative Stress

Under normal conditions, the mitochondria produce low levels of free radicals including reactive oxygen species (ROS), superoxide, and hydrogen peroxide. These are produced as normal by-products of cellular respiration and are normally scavenged by the antioxidant enzymes superoxide dismutase, glutathione peroxidase, and catalase and converted to less dangerous metabolites [34]. However, following TBI, the production of these harmful reactive species is markedly increased [35] through contributions from a number of secondary injury factors including excitotoxicity, the arachidonic acid cascade, increased leakage from the electron transport chain, and activation of neutrophils [36, 37]. Such an increase overwhelms the endogenous antioxidant enzyme defenses, leading to oxidative stress which in turn causes devastating damage to many cellular components/structures. The ROS generated by the secondary injury cascade are extremely damaging due to the unpaired electron in their outer shell, making them highly reactive and able to cause damage by peroxidation. Because free radicals can react with almost all cellular structures, they have the potential to cause extensive damage and even cell death. This results in injury to cellular and vascular structures, protein oxidation, cleavage of DNA, and inhibition of the mitochondrial electron transport chain [37]. The increased calcium levels further exacerbate such injury, which stimulates lipid membrane breakdown, resulting in the accumulation of cytotoxic free radicals. The partial inactivation of protective enzyme systems such as peroxidase and xanthine oxidase further contributes to the oxidative stress [37].

Lipid peroxidation of the cell membrane also occurs as a result of the decreased magnesium levels, a process that leads to increased membrane turnover and ultimately free radical formation [37]. One of the products of lipid peroxidation is 4-hydroxylnonenal. This breakdown product has several adverse effects on cell function, such as impairment of sodium/potassium ATPase function, leading to membrane depolarization, increased vulnerability to excitotoxicity, and disruption of homeostasis [38]. It is also impairs glucose transport within neurons through direct conjugation with the glucose transporter GLUT-3 [39]. This impairment of glucose transport leads to impairment of mitochondrial function and subsequent ATP depletion, further placing neurons at risk of excitotoxic damage [38, 39].

Inflammation

Alterations in magnesium levels affect both classical inflammatory pathways and neurally elicited inflammatory pathways. Magnesium deficiency significantly influences the inflammatory state by stimulating the release of proinflammatory factors such as cytokines and substance P (SP). Indeed, 1 week of magnesium deficiency has been shown to result in an elevation of plasma SP levels [40], corresponding with the development of neurogenic inflammation. Furthermore, neurogenic inflammation has been recognized as contributing to the secondary injury that occurs following trauma [41]. Neurogenic inflammation is a neurally elicited response having features of classical inflammation, such as vasodilation and increased microvascular permeability, but characterized by the release of neuropeptides such as SP and calcitonin gene-related peptide from sensory nerves [42, 43]. Recent experimental studies have demonstrated that SP release is increased following acute injury to the brain and is correlated with the degree of blood-brain barrier dysfunction, vasogenic edema, and functional deficits [41, 44–48].

Classical inflammation is also influenced by magnesium level. Under conditions of low magnesium levels, such as following TBI, intracellular calcium levels are elevated within immune cells, rendering them more reactive [49, 50], initiating a robust inflammatory response including the activation of neurons and glia as well as the infiltration of monocytes and lymphocytes into the brain. This in turn leads to the release of more inflammatory mediators such as interleukin (IL)-1B, IL-1, IL-6 and tumor necrosis factor alpha (TNF-α). Increased levels of these cytokines have been detected in the

CSF and brain parenchyma of both clinical patients and experimental animals following trauma [51–56]. These cytokines are key mediators of the inflammatory response and can stimulate the release of other mediators and further propagate the inflammatory response. In particular, IL-1 promotes the release of damaging mediators such as prostaglandins, reactive oxygen species, and proteases [55, 56], whereas TNF-α promotes the release of proteolytic enzymes involved in blood-brain barrier dysfunction and subsequent cerebral edema formation [57].

Blood Flow

Calcium entry into neurons causes phospholipase A2 (PLA2) activation, which in turn activates the synthesis of platelet-aggregating factor (PAF) [7, 58]. Once synthesized, PAF may cause ischemia, inflammatory changes, impairment of cerebral blood flow through profound vasoconstriction, and increased blood-brain barrier (BBB) permeability and may even be directly neurotoxic [7]. It is hypothesized that disruption of the BBB may lead to neuronal loss by causing inflammatory reactions, vasogenic edema, and impairment of nutrient and metabolic by-product transport [59]. Changes in vascular tone and subsequent alterations in cerebral blood flow may also occur directly as a result of the decreased magnesium levels [13] as magnesium is an essential element for circulatory regulation [60]. Phospholipase C (PLC) levels are increased following trauma. It has been suggested that PLC-induced products may also be able to chelate magnesium, thereby providing a possible mechanism by which magnesium levels decrease following trauma [61].

Apoptotic Cell Death

Apoptotic cell death is a normal part of growth and development and cell turnover. However, following trauma, the signals that induce and inhibit apoptosis are altered by a number of factors, leading to an increase in apoptotic cell death. Under normal conditions, magnesium may directly act to inhibit cell death by attenuating the apoptotic pathway [62]. Specifically, magnesium blocks the induction of the mitochondrial permeability transition pore, which allows the passage of ions and cytochrome c out of the mitochondria, thereby preventing the fluxes of components that provide a stimulus for cell death [19, 63]. Magnesium also prevents cell death through inhibition of DNA fragmentation factor and p53 gene expression [64]. p53 is able to act as a transcription factor for the pro-apoptotic protein Bax and repress the promoter of the anti-apoptotic protein Bcl-2, thereby promoting cell death [62].

The expression of the pro-apoptotic protein Bax is stimulated by calcium, and this alters the Bax/Bcl-2 ratio, providing a positive stimulus for apoptosis [65]. The Bax protein itself is harmful as it forms ion-conducting pores in the membrane, leading to nuclear envelope breakdown and further increases in calcium, which in turn leads to cell death [62]. In addition to the adverse effects of the Bax protein, it has been observed that the levels of Bcl-2 mRNA are decreased following trauma [65, 66]. This would thereby provide an imbalance in the normal Bax/Bcl-2 ratio and stimulate the release of cytochrome c, which activates destructive caspases that lead to cell death [66]. Magnesium has been observed to restore the normal Bax/Bcl-2 ratio after hypoxia [62].

Disruption of Magnesium Homeostasis Following TBI

Experimental TBI studies were the first to demonstrate that changes in magnesium levels occur following acute injury to the brain [67, 68]. Specifically, they reported a decline in both brain total and free magnesium levels following trauma. The levels of free magnesium were found to correlate with the degree of cellular damage following injury, as evidenced by poor functional outcome. Numerous laboratories have since investigated the decline in magnesium levels with trauma and other CNS pathologies [69]. It is now known that the decline in free magnesium is in the order of 40–60% and that magnesium decline is a post-traumatic phenomenon that occurs across all species and trauma models [70–73]. Given that alterations in magnesium levels are observed irrespective of model or species suggests that this is a universal feature of acute injury to the brain, one which may have devastating consequences in terms of normal cellular function and neurological functioning. Subsequent studies have shown that magnesium decline also occurs in other CNS pathologies such as stroke, spinal cord injury, mitochondrial myopathies, neurodegeneration, and even psychiatric conditions [67, 74], although its rapid decline seems to only occur in the acute conditions.

This rapid decline in magnesium is a widely documented post-traumatic phenomenon and has been detected within 20 min of injury. Moreover, it only occurs at sites of tissue damage, making it an accurate marker of CNS injury. The decreased magnesium levels have been found to persist for anywhere from 4 to 7 days following trauma [71, 75], and it appears that it is the length of time for which magnesium is decreased, rather than the magnitude of the decline, that is most important in terms of neurological outcome [18]. Despite this widely recognized phenomenon, the mechanism(s) by which intracellular free magnesium levels decrease post-trauma remain unknown. While earlier studies identified that changes in the binding capacity of the cell membranes alter intracellular free magnesium concentration [18], recent studies have suggested that the newly identified transient receptor potential melastatin (TRPM) channels that have the potential to transport magnesium may play an important role [76].

Not only do brain magnesium levels decrease by up to 40–60% following trauma [67] but blood levels also decrease [77]. However, it appears that the decrease in brain magnesium levels is the most important pathophysiologically due to the key role that magnesium plays in normal cell function [18, 78]. Furthermore, it has been observed that magnesium levels do not drop below 0.2 mM, suggesting that this may be the threshold beyond which magnesium levels cannot fall physiologically. The decreased magnesium levels are associated with neuronal death and functional impairments in neurological motor and cognitive outcome post-trauma [68] and indeed correlate with such deficits [78–80].

Magnesium Therapy: Experimental Studies

In view of the deleterious effects associated with a decline in magnesium post-trauma, an attempt at restoring magnesium homeostasis may presumably have a positive effect. Indeed, the beneficial effects of magnesium administration following experimental trauma to the CNS have now been well documented [77–83]. Numerous experimental studies have found that the administration of magnesium post-trauma significantly improves neurological outcome [77–83]. Early studies showed that administration of $MgSO_4$ at 30 min post-trauma significantly improved motor outcome [15], while $MgCl_2$ at 15 min following trauma significantly increased memory retention and cognitive function [84]. Subsequent studies have shown that magnesium therapy appears to be effective up to 24 h following the traumatic event, with the best outcome being achieved with magnesium administration within the first 12 h post-trauma [78]. The critical period in magnesium homeostasis therefore appears to occur within the first 24 h post-trauma. Any magnesium deficiency induced prior to trauma, and presumably persisting after trauma, exacerbated the neurological dysfunction [74].

Both $MgCl_2$ and $MgSO_4$ have been tested in the laboratory for their neuroprotective effects, with each showing the same success. Clearly, the magnesium ion itself is protective irrespective of the salt administered [85]. Both these salts have been shown to significantly improve neurological outcome, irrespective of whether they are administered as a single dose or in multiple doses [77, 82]. The salts have been shown to be effective in a dose-dependent manner, with small doses of $MgCl_2$ having a small effect on neurological function, whereas a large dose of $MgCl_2$ was shown to have a significantly greater effect on neurological function [83]. There is, however, some debate over the chemical properties of each of these salts. It has been suggested that $MgSO_4$ may act marginally quicker than $MgCl_2$ [77], but there are safety concerns with use of $MgSO_4$ in humans despite the fact that it has been shown to have favorable effects on blood flow, acting to decrease vascular resistance and cause vascular dilation [60]. It has also been suggested that $MgCl_2$ exerts better ionic exchange through the cellular membranes [86]. Current opinion is that the chloride salt may be the preferred choice. The route of administration, whether it be intramuscular or intravenous, does not appear to affect the efficacy of these salts in improving neurological outcome [85]. Administered magnesium (intravenous or intramuscular) has been shown to penetrate the BBB and enter the cerebrospinal fluid (CSF) following experimental trauma, leading to an increase in the brain intracellular free magnesium concentration [26, 78].

Magnesium: Possible Mechanisms of Neuroprotection

Despite magnesium therapy having been shown to be protective following a variety of CNS pathologies including experimental TBI [83, 84, 87, 88], the exact mechanism whereby magnesium is protective remains to be elucidated. The possibilities are outlined below.

Restoration of Magnesium Homeostasis

In experimental studies, magnesium administered post-trauma is able to penetrate the BBB and enter the brain, where it acts to increase the extracellular concentration of magnesium [26]. Magnesium is then able to move into the cell down its concentration gradient and restore all the cellular processes that depend on magnesium back to normal. For example, post-traumatic magnesium administration has been shown to inhibit NMDA-induced cell death after brain injury [26]. By restoring the block of the NMDA receptor, magnesium reduces the activation of the NMDA receptor and the influx of calcium that results from such activation [26]. It has even been suggested that increased magnesium concentrations may cause a downregulation of the NMDA receptor population [26]. Magnesium administration also has positive effects on cellular metabolism as restoration of magnesium homeostasis allows all the energetic processes that depend on magnesium to return to normal, thus facilitating energy production, restoring ionic gradients, attenuating regional cerebral edema [89], and permitting cell repair mechanisms to proceed.

Increased Neuronal Cell Survival

Magnesium treatment has widely been shown to attenuate cell death caused by experimental brain injury [78, 86, 89] (Saatman et al. 2001). One mechanism may be by attenuating the acute cytoskeletal alterations, thereby maintaining both neuronal cytoskeletal integrity and cell viability. Notably, the attenuation of cytoskeletal damage only occurred in cortical neurons and not the hippocampus [89]. Similar regional results were obtained by Muir et al. (1999) who found that the post-traumatic attenuation of p53 expression by $MgCl_2$ only occurred in the cortex and not the hippocampus or thalamus [64]. The above studies provide evidence for a molecular basis of neuroprotection by magnesium and suggest that the mechanisms of magnesium's neuroprotective action may be regionally distinct. Magnesium therapy has also been found to decrease the expression of the pro-apoptotic Bax protein and increase the expression of the anti-apoptotic protein Bcl-2, thus favoring cell survival. Bcl-2 is able to prevent induction of the MPTP [62]. Magnesium may also act to increase cell survival by preventing the induction of the MPTP, thereby preventing the damaging fluxes of ions and loss of cytochrome c which lead to mitochondrial and cellular swelling and subsequent cell death by both apoptosis and necrosis [19]. Indeed, magnesium has been previously shown to improve the survival of neurons following both stroke and TBI [90–92].

Magnesium: Clinical Studies

Following on from the success of experimental studies, there have been a number of clinical studies carried out to determine the efficacy of magnesium as a novel therapeutic for the treatment of TBI. Studies of preeclampsia/eclampsia [93], myocardial infarction [94], subarachnoid hemorrhage [95], and ischemic stroke [96] had previously confirmed the safety profile and tolerability of magnesium salts in the clinical setting. Moreover, in dose optimization studies in stroke, there were no observed effects on heart rate, blood pressure, or blood glucose levels [88]. In light of this, magnesium was considered a desirable potential therapeutic agent as it was low cost, widely available, and had a predictable pharmacological profile [88].

Despite the promising experimental findings, clinical trials of magnesium in head injury to date have been mixed and have generally failed to reproduce the excellent preclinical results. Indeed, there is a marked lack of concordance between the experimental and clinical literature in terms of the efficacy following TBI. Notably, clinical trials of magnesium in stroke have also yielded disappointing results [97]. Presumably, there are major discrepancies between the design and findings of the experimental and clinical studies that may have influenced outcome, some of which are readily identifiable. Firstly, one of the clinical trials [98] had a standard patient care requirement to restore blood magnesium levels to normal. This was independent of placebo or treatment group allocation and meant that all patients received basic magnesium therapy. Having restored blood magnesium levels back to normal, those patients randomized to the magnesium treatment group received an additional infusion of magnesium. Obviously, such a protocol would have markedly influenced the results, greatly affecting comparisons between the placebo and treatment groups. Reports of no or a deleterious effect of magnesium therapy under these conditions is not surprising given the experimental data demonstrating that high dose magnesium therapy is deleterious to outcome following TBI [80]; the doses used in the clinical trial were considerably higher than those used successfully in experimental studies. In contrast, a smaller TBI trial where control patients did not receive any magnesium therapy, and which used much lower doses of magnesium, demonstrated a reduction in intraoperative brain swelling and more favorable outcome at 3 months in magnesium-treated patients compared to controls [99]. Clinical trials of magnesium in stroke have been met with similar problems, including different treatment windows in clinical compared to experimental studies. Despite some experimental studies demonstrating a therapeutic window of up to 24 h following TBI, the window of 12 h after stroke symptom onset may have been too long to enroll patients [97]. Indeed, other experimental stroke studies have shown a therapeutic window of just 6 h following permanent MCAO [100]. In light of such disappointing clinical findings, in 2005, the FAST-MAG trial began enrolling patients to receive magnesium treatment within 2 h of stroke symptom onset [96]. As of mid-2011, the trial had achieved approximately 75% of the enrolment goal, with the final trial findings highly anticipated.

It has been suggested that a further reason for the lack of efficacy in clinical trials is the low brain bioavailability of magnesium after peripheral administration [101]. Accordingly, several studies have investigated the bioavailability of magnesium in clinical patients. It has been reported that there is no significant change in CSF ionized magnesium following a 20-mmol bolus of magnesium sulfate [102]. Furthermore, hypomagnesemia was found to produce only a small increase in total and ionized CSF magnesium concentration, suggesting that the regulation of magnesium in CSF is largely maintained in acute brain injury, thereby limiting the brain bioavailability of magnesium. Efforts to increase CNS penetration of magnesium after acute CNS injury are ongoing [103]. Such observations certainly emphasize that further refinements are required in terms of our understanding of the optimal timing and dosing, as well as brain bioavailability of magnesium sufficient to produce a neuroprotective effect. In addition, the efficacy of magnesium therapy when combined with other neuroprotective approaches, such as hypothermia, may be beneficial. Indeed, experimental findings in this area are encouraging [104].

Conclusions

Magnesium continues to be of interest in the TBI field, despite the conflicting clinical findings. It is now widely accepted that intracellular magnesium decline is a ubiquitous feature of acute brain injury, with the deleterious effects on injury pathways and outcome well documented. It is also well accepted that if magnesium levels in injured tissue are restored to normal physiological levels, neuroprotection is evident. The challenge now is to understand the mechanisms of magnesium homeostasis in normal and injured tissue and particularly the role of magnesium transporters, such that a strategy to increase centrally active magnesium can be successfully developed.

References

1. Finfer SR, Cohen J. Severe traumatic brain injury. Resuscitation. 2001;48(1):77–90.
2. Hsiang J, Marshall LF. Head injury. In: Swash M, editor. Outcomes in neurological and neurosurgical disorders. New York: Cambridge University Press; 1998. p. 157–80.
3. McAllister TW. Neurobiological consequences of traumatic brain injury. Dialogues Clin Neurosci. 2011;13(3):287–300.
4. Finnie JW, Blumbergs PC. Traumatic brain injury. Vet Pathol. 2002;39(6):679–89.
5. Smith DH, Meaney DF, Shull WH. Diffuse axonal injury in head trauma. J Head Trauma Rehabil. 2003;18(4):307–16.
6. De Girolami U, Frosch MP, Anthony DC. The central nervous system. In: Cotran CR, Kumar V, Collins T, editors. Robbins pathological basis of disease. Philadelphia: W. B. Saunders Company; 1999. p. 1293–357.
7. McIntosh TK, Smith DH, Garde E. Therapeutic approaches for the prevention of secondary brain injury. Eur J Anaesthesiol. 1996;13(3):291–309.
8. Bullock R. Injury and cell function. In: Reilly P, Bullock R, editors. Head injury: pathophysiology and management of severe closed injury. London: Chapman and Hall; 1997. p. 121–41.
9. Povlishock JT, Christman CW. The pathophysiology of traumatic brain injury. In: Salzman SK, Faden AI, editors. The neurobiology of central nervous system trauma. New York: Oxford University Press; 1994. p. 109–20.
10. Birch NJ. Magnesium and the cell. London: Academic; 1993.
11. Ebel H, Gunther T. Magnesium metabolism: a review. J Clin Chem Clin Biochem. 1980;18(5):257–70.
12. Terasaki M, Rubin H. Evidence that intracellular magnesium is present in cells at a regulatory concentration for protein synthesis. Proc Natl Acad Sci U S A. 1985;82(21):7324–6.
13. Bara M, Guiet-Bara A. Potassium, magnesium and membranes. Review of present status and new findings. Magnesium. 1984;3(4–6):215–25.
14. Marcoux J, McArthur DA, Miller C, et al. Persistent metabolic crisis as measured by elevated cerebral microdialysis lactate-pyruvate ratio predicts chronic frontal lobe brain atrophy after traumatic brain injury. Crit Care Med. 2008;36(10):2871–7.
15. Vink R, McIntosh TK, Yamakami I, Faden AI. 31P NMR characterization of graded traumatic brain injury in rats. Magn Reson Med. 1988;6(1):37–48.
16. Tavazzi B, Signoretti S, Lazzarino G, et al. Cerebral oxidative stress and depression of energy metabolism correlate with severity of diffuse brain injury in rats. Neurosurgery. 2005;56(3):582–9.
17. Verweij BH, Muizelaar JP, Vinas FC, Peterson PL, Xiong Y, Lee CP. Improvement in mitochondrial dysfunction as a new surrogate efficiency measure for preclinical trials: dose–response and time -window profiles for administration of the calcium channel blocker Ziconotide in experimental brain injury. J Neurosurg. 2000;93:829–34.

18. Vink R, Cernak I. Regulation of intracellular free magnesium in central nervous system injury. Front Biosci. 2000;5:D656–65.
19. Lemasters JJ, Theruvath TP, Zhong Z, Nieminen AL. Mitochondrial calcium and the permeability transition in cell death. Biochim Biophys Acta. 2009;1787(11):1395–401.
20. Sherwood L. Human physiology: from cells to systems. Belmont: Wadsworth Publishing Company; 1997.
21. Hovda DA, Becker DP, Katayama Y. Secondary injury and acidosis. J Neurotrauma. 1992;9 Suppl 1:S47–60.
22. McIntosh TK, Vink R, Soares H, Hayes R, Simon R. Effect of noncompetitive blockade of N-methyl-D-aspartate receptors on the neurochemical sequelae of experimental brain injury. J Neurochem. 1990;55(4):1170–9.
23. Mendelow AD, Crawford PJ. Primary and secondary brain injury. In: Reilly P, Bullock R, editors. Head injury: pathophysiology and management of severe closed injury. London: Chapman and Hall; 1997. p. 72–88.
24. Scheff SW, Sullivan PG. Cyclosporin A significantly ameliorates cortical damage following experimental traumatic brain injury in rodents. J Neurotrauma. 1999;16(9):783–92.
25. Iseri LT, French JH. Magnesium: nature's physiologic calcium blocker. Am Heart J. 1984;108(1):188–93.
26. Hallak M, Berman RF, Irtenkauf SM, Janusz CA, Cotton DB. Magnesium sulfate treatment decreases N-methyl-D-aspartate receptor binding in the rat brain: an autoradiographic study. J Soc Gynecol Investig. 1994;1(1):25–30.
27. Brocard JB. Glutamate induced increases in intracellular free Mg(2+) in cultured cortical neurons. Neuron. 1993;11(4):751–7.
28. Choi DW. Excitotoxic cell death. J Neurobiol. 1992;23(9):1261–76.
29. Mayer ML, Westbrook GL, Guthrie PB. Voltage-dependent block by Mg2+ of NMDA responses in spinal cord neurones. Nature. 1984;309(5965):261–3.
30. Ascher P. Measuring and controlling the extracellular glycine concentration at the NMDA receptor level. Adv Exp Med Biol. 1990;268:13–6.
31. Faden AI, Demediuk P, Panter SS, Vink R. The role of excitatory amino acids and NMDA receptors in traumatic brain injury. Science. 1989;244(4906):798–800.
32. Zhang L, Rzigalinski BA, Ellis EF, Satin LS. Reduction of voltage-dependent Mg^{2+} blockade of NMDA current in mechanically injured neurons. Science. 1996;274(5294):1921–3.
33. Choi DW. Calcium-mediated neurotoxicity: relationship to specific channel types and role in ischemic damage. TINS. 1988;11(10):465–9.
34. Chan PH. Reactive oxygen radicals in signaling and damage in the ischemic brain. J Cereb Blood Flow Metab. 2001;21(1):2–14.
35. Globus MY, Alonso O, Dietrich WD, Busto R, Ginsberg MD. Glutamate release and free radical production following brain injury: effects of posttraumatic hypothermia. J Neurochem. 1995;65(4):1704–11.
36. Awasthi D, Church DF, Torbati D, Carey ME, Pryor WA. Oxidative stress following traumatic brain injury in rats. Surg Neurol. 1997;47(6):575–81.
37. Lewen A, Matz P, Chan PH. Free radical pathways in CNS injury. J Neurotrauma. 2000;17(10):871–90.
38. Mark RJ, Lovell MA, Markesbery WR, Uchida K, Mattson MP. A role for 4-hydroxynonenal, an aldehydic product of lipid peroxidation, in disruption of ion homeostasis and neuronal death induced by amyloid beta-peptide. J Neurochem. 1997;68(1):255–64.
39. Mark RJ, Pang Z, Geddes JW, Uchida K, Mattson MP. Amyloid beta-peptide impairs glucose transport in hippocampal and cortical neurons: involvement of membrane lipid peroxidation. J Neurosci. 1997;17(3):1046–54.
40. Weglicki WB, Phillips TM. Pathobiology of magnesium deficiency: a cytokine/neurogenic inflammation hypothesis. Am J Physiol. 1992;263(3 Pt 2):R734–7.
41. Nimmo AJ, Cernak I, Heath DL, Hu X, Bennett CJ, Vink R. Neurogenic inflammation is associated with development of edema and functional deficits following traumatic brain injury in rats. Neuropeptides. 2004;38(1):40–7.
42. Black PH. Stress and the inflammatory response: a review of neurogenic inflammation. Brain Behav Immun. 2002;16(6):622–53.
43. Severini C, Improta G, Falconieri-Erspamer G, Salvadori S, Erspamer V. The tachykinin peptide family. Pharmacol Rev. 2002;54(2):285–322.
44. Donkin JJ, Nimmo AJ, Cernak I, Blumbergs PC, Vink R. Substance P is associated with the development of brain edema and functional deficits after traumatic brain injury. J Cereb Blood Flow Metab. 2009;29(8):1388–98.
45. Turner RJ, Blumbergs PC, Sims NR, Helps SC, Rodgers KM, Vink R. Increased substance P immunoreactivity and edema formation following reversible ischemic stroke. Acta Neurochir Suppl. 2006;96:263–6.
46. Turner RJ, Helps SC, Thornton E, Vink R. A substance P antagonist improves outcome when administered 4 h after onset of ischaemic stroke. Brain Res. 2011;1393:84–90.
47. Vink R, Young A, Bennett CJ, et al. Neuropeptide release influences brain edema formation after diffuse traumatic brain injury. Acta Neurochir Suppl. 2003;86:257–60.
48. Donkin JJ, Turner RJ, Hassan I, Vink R. Substance P in traumatic brain injury. Prog Brain Res. 2007;161:97–109.
49. Malpuech-Brugere C, Nowacki W, Daveau M, et al. Inflammatory response following acute magnesium deficiency in the rat. Biochim Biophys Acta. 2000;1501(2–3):91–8.

50. Shogi T, Oono H, Nakagawa M, Miyamoto A, Ishiguro S, Nishio A. Effects of a low extracellular magnesium concentration and endotoxin on IL-1beta and TNF-alpha release from, and mRNA levels in, isolated rat alveolar macrophages. Magnes Res. 2002;15(3–4):147–52.
51. Hutchinson PJ, O'Connell MT, Rothwell NJ, et al. Inflammation in human brain injury: intracerebral concentrations of IL-1alpha, IL-1beta, and their endogenous inhibitor IL-1ra. J Neurotrauma. 2007;24(10):1545–57.
52. Morganti-Kossman MC, Lenzlinger PM, Hans V, et al. Production of cytokines following brain injury: beneficial and deleterious for the damaged tissue. Mol Psychiatry. 1997;2(2):133–6.
53. Taupin V, Toulmond S, Serrano A, Benavides J, Zavala F. Increase in IL-6, IL-1 and TNF levels in rat brain following traumatic lesion. Influence of pre- and post-traumatic treatment with Ro5 4864, a peripheral-type (p site) benzodiazepine ligand. J Neuroimmunol. 1993;42(2):177–85.
54. Zhu HD, Martin R, Meloni B, et al. Magnesium sulfate fails to reduce infarct volume following transient focal cerebral ischemia in rats. Neurosci Res. 2004;49(3):347–53.
55. Molina-Holgado E, Ortiz S, Molina-Holgado F, Guaza C. Induction of COX-2 and PGE(2) biosynthesis by IL-1beta is mediated by PKC and mitogen-activated protein kinases in murine astrocytes. Br J Pharmacol. 2000;131(1):152–9.
56. Rothwell N. Interleukin-1 and neuronal injury: mechanisms, modification, and therapeutic potential. Brain Behav Immun. 2003;17(3):152–7.
57. Shohami E, Ginis I, Hallenbeck JM. Dual role of tumor necrosis factor alpha in brain injury. Cytokine Growth Factor Rev. 1999;10(2):119–30.
58. Kristian T, Siesjo BK. Calcium in ischemic cell death. Stroke. 1998;29(3):705–18.
59. Hicks RR, Smith DH, Lowenstein DH, Saint Marie R, McIntosh TK. Mild experimental brain injury in the rat induces cognitive deficits associated with regional neuronal loss in the hippocampus. J Neurotrauma. 1993;10(4):405–14.
60. Murasato Y, Harada Y, Ikeda M, Nakashima Y, Hayashida Y. Effect of magnesium deficiency on autonomic circulatory regulation in conscious rats. Hypertension. 1999;34(2):247–52.
61. Vink R, Nimmo AJ, Cernak I. An overview of new and novel pharmacotherapies for use in traumatic brain injury. Clin Exp Pharmacol Physiol. 2001;28(11):919–21.
62. Ravishankar S, Ashraf QM, Fritz K, Mishra OP, Delivoria-Papadopoulos M. Expression of Bax and Bcl-2 proteins during hypoxia in cerebral cortical neuronal nuclei of newborn piglets: effect of administration of magnesium sulfate. Brain Res. 2001;901(1–2):23–9.
63. Kowaltowski AJ, Netto LE, Vercesi AE. The thiol-specific antioxidant enzyme prevents mitochondrial permeability transition. Evidence for the participation of reactive oxygen species in this mechanism. J Biol Chem. 1998;273(21):12766–9.
64. Muir JK, Raghupathi R, Emery DL, Bareyre FM, McIntosh TK. Postinjury magnesium treatment attenuates traumatic brain injury-induced cortical induction of p53 mRNA in rats. Exp Neurol. 1999;159(2):584–93.
65. Raghupathi R, Graham DI, McIntosh TK. Apoptosis after traumatic brain injury. J Neurotrauma. 2000;17(10):927–38.
66. Dong GX, Singh DK, Dendle P, Prasad RM. Regional expression of Bcl-2 mRNA and mitochondrial cytochrome c release after experimental brain injury in the rat. Brain Res. 2001;903(1–2):45–52.
67. Vink R, McIntosh TK, Demediuk P, Faden AI. Decrease in total and free magnesium concentration following traumatic brain injury in rats. Biochem Biophys Res Commun. 1987;149(2):594–9.
68. Vink R, McIntosh TK, Demediuk P, Weiner MW, Faden AI. Decline in intracellular free Mg^{2+} is associated with irreversible tissue injury after brain trauma. J Biol Chem. 1988;263(2):757–61.
69. Vink R, Nechifor M, editors. Magnesium in the CNS. Adelaide: Adelaide University Press; 2011.
70. Cernak I, Savic VJ, Kotur J, Prokic V, Veljovic M, Grbovic D. Characterization of plasma magnesium concentration and oxidative stress following graded traumatic brain injury in humans. J Neurotrauma. 2000;17(1):53–68.
71. Heath DL, Vink R. Traumatic brain axonal injury produces sustained decline in intracellular free magnesium concentration. Brain Res. 1996;738(1):150–3.
72. Smith DH, Cecil KM, Meaney DF, et al. Magnetic resonance spectroscopy of diffuse brain trauma in the pig. J Neurotrauma. 1998;15(9):665–74.
73. Suzuki M, Nishina M, Endo M, et al. Decrease in cerebral free magnesium concentration following closed head injury and effects of VA-045 in rats. Gen Pharmacol. 1997;28(1):119–21.
74. McIntosh TK, Faden AI, Yamakami I, Vink R. Magnesium deficiency exacerbates and pretreatment improves outcome following traumatic brain injury in rats: 31P magnetic resonance spectroscopy and behavioral studies. J Neurotrauma. 1988;5(1):17–31.
75. Vink R, Heath DL, McIntosh TK. Acute and prolonged alterations in brain free magnesium following fluid percussion-induced brain trauma in rats. J Neurochem. 1996;66(6):2477–83.
76. Cook NL, Van Den Heuvel C, Vink R. Are the transient receptor potential melastatin (TRPM) channels important in magnesium homeostasis following traumatic brain injury? Magnes Res. 2009;22(4):225–34.

77. Heath DL, Vink R. Neuroprotective effects of MgSO4 and MgCl2 in closed head injury: a comparative phosphorus NMR study. J Neurotrauma. 1998;15(3):183–9.
78. Heath DL, Vink R. Improved motor outcome in response to magnesium therapy received up to 24 hours after traumatic diffuse axonal brain injury in rats. J Neurosurg. 1999;90(3):504–9.
79. Heath DL, Vink R. Concentration of brain free magnesium following severe brain injury correlates with neurologic motor outcome. J Clin Neurosci. 1999;6(6):505–9.
80. Heath DL, Vink R. Optimization of magnesium therapy after severe diffuse axonal brain injury in rats. J Pharmacol Exp Ther. 1999;288(3):1311–6.
81. Hoane MR, Barbay S, Barth TM. Large cortical lesions produce enduring forelimb placing deficits in un-treated rats and treatment with NMDA antagonists or anti-oxidant drugs induces behavioral recovery. Brain Res Bull. 2000;53(2):175–86.
82. McDonald JW, Silverstein FS, Johnston MV. Magnesium reduces N-methyl-D-aspartate (NMDA)-mediated brain injury in perinatal rats. Neurosci Lett. 1990;109(1–2):234–8.
83. McIntosh TK, Vink R, Yamakami I, Faden I. Magnesium protects against neurological deficit after brain injury. Brain Res. 1989;482:252–60.
84. Smith DH, Okiyama K, Gennarelli TA, McIntosh TK. Magnesium and ketamine attenuate cognitive dysfunction following experimental brain injury. Neurosci Lett. 1993;157(2):211–4.
85. Heath DL, Vink R. Magnesium sulphate improves neurologic outcome following severe closed head injury in rats. Neurosci Lett. 1997;228(3):175–8.
86. Hoane MR, Barth TM. The behavioral and anatomical effects of MgCl2 therapy in an electrolytic lesion model of cortical injury in the rat. Magnes Res. 2001;14(1–2):51–63.
87. Muir KW, Lees KR. A randomized, double-blind, placebo-controlled pilot trial of intravenous magnesium sulfate in acute stroke. Stroke. 1995;26(7):1183–8.
88. Muir KW, Lees KR. Dose optimization of intravenous magnesium sulfate after acute stroke. Stroke. 1998;29(5):918–23.
89. Saatman KE, Bareyre FM, Grady MS, McIntosh TK. Acute cytoskeletal alterations and cell death induced by experimental brain injury are attenuated by magnesium treatment and exacerbated by magnesium deficiency. J Neuropathol Exp Neurol. 2001;60(2):183–94.
90. Marinov MB, Harbaugh KS, Hoopes PJ, Pikus HJ, Harbaugh RE. Neuroprotective effects of preischemia intraarterial magnesium sulfate in reversible focal cerebral ischemia. J Neurosurg. 1996;85(1):117–24.
91. Schanne FA, Gupta RK, Stanton PK. 31P-NMR study of transient ischemia in rat hippocampal slices in vitro. Biochim Biophys Acta. 1993;1158(3):257–63.
92. Sirin BH, Coskun E, Yilik L, Ortac R, Sirin H, Tetik C. Neuroprotective effects of preischemia subcutaneous magnesium sulfate in transient cerebral ischemia. Eur J Cardiothorac Surg. 1998;14(1):82–8.
93. Duley L. Pre-eclampsia, eclampsia, and hypertension. Clin Evid (Online). 2011;2011.
94. Flather M, Pipilis A, Collins R, et al. Randomized controlled trial of oral captopril, of oral isosorbide mononitrate and of intravenous magnesium sulphate started early in acute myocardial infarction: safety and haemodynamic effects. ISIS-4 (Fourth International Study of Infarct Survival) Pilot Study Investigators. Eur Heart J. 1994;15(5):608–19.
95. Ma L, Liu WG, Zhang JM, Chen G, Fan J, Sheng HS. Magnesium sulphate in the management of patients with aneurysmal subarachnoid haemorrhage: a meta-analysis of prospective controlled trials. Brain Inj. 2010;24(5):730–5.
96. Gorelick PB, Ruland S. IMAGES and FAST-MAG: magnesium for acute ischaemic stroke. Lancet Neurol. 2004;3(6):330.
97. Muir KW, Lees KR, Ford I, Davis S. Magnesium for acute stroke (intravenous magnesium efficacy in Stroke trial): randomised controlled trial. Lancet. 2004;363(9407):439–45.
98. Temkin NR, Anderson GD, Winn HR, et al. Magnesium sulfate for neuroprotection after traumatic brain injury: a randomised controlled trial. Lancet Neurol. 2007;6:29–38.
99. Dhandapani SS, Gupta A, Vivekanandhan S, Sharma BS, Mahapatra AK. Randomized controlled trial of magnesium sulphate in severe closed traumatic brain injury. Ind J Neurotrauma. 2008;5:27–33.
100. Yang Y, Qiu L, Fayyaz A, Shuaib A. Survival and histological evaluation of therapeutic window of post-ischemia treatment with magnesium sulfate in embolic stroke model of rat. Neurosci Lett. 2000;285:119–22.
101. McKee JA, Brewer RP, Macy GE, et al. Analysis of the brain bioavailability of peripherally administered magnesium sulfate: a study in humans with acute brain injury undergoing prolonged induced hypermagnesemia. Crit Care Med. 2005;33(3):661–6.
102. Brewer RP, Parra A, Borel CO, Hopkins MB, Reynolds JD. Intravenous magnesium sulfate does not increase ventricular CSF ionized magnesium concentration of patients with intracranial hypertension. Clin Neuropharmacol. 2001;24(6):341–5.
103. Lee JH, Roy J, Sohn HM, et al. Magnesium in a polyethylene glycol formulation provides neuroprotection after unilateral cervical spinal cord injury. Spine. 2010;35(23):2041–8.
104. Meloni BP, Zhu H, Knuckey NW. Is magnesium neuroprotective following global and focal cerebral ischaemia? A review of published studies. Magnes Res. 2006;19(2):123–37.

Chapter 19
Magnesium in Subarachnoid Hemorrhage: From Bench to Bedside

Jack Hou and John H. Zhang

Key Points

- Delayed cerebral vasospasm after aneurysmal subarachnoid hemorrhage was traditionally regarded as the main pathophysiological process leading to delayed brain injury.
- Recent studies suggest that early onset injurious processes, such as reactive oxygen species formation, cellular metabolism disruption, neuroinflammation, microthrombosis, and spreading ischemia, may play a greater role in causing delayed brain injury than previously assumed.
- Magnesium was shown to alleviate early brain injury by blocking voltage-gated calcium channels and NMDA receptor, blocking free radical formation in the mitochondria, inhibiting ischemia-induced glutamate release, relaxing vascular smooth muscle, preventing platelet aggregation, and suppressing cortical spreading depolarization.
- Clinical trials of magnesium in subarachnoid hemorrhage gave mixed results. Whether earlier and higher dosage topical administration of magnesium would yield greater benefit remains to be answered.

Keywords Subarachnoid hemorrhage • Cerebral aneurysm • Early brain injury • Vasospasm • Delayed ischemic neurological deficit • Clinical trial • Magnesium

Introduction

Aneurysmal subarachnoid hemorrhage (aSAH) is a devastating disease responsible for 5% of stroke cases and affects approximately 28,000 North Americans annually. Even though its incidence is dwarfed by other subtypes of stroke, the relative youth of the affected patients with a mean age of 50 years [1, 2] and that fewer than 40% of them so stricken return to functional life despite modern treatment mean that this disease accounts for around 25% of loss of productive life years, a magnitude comparable to ischemic stroke, the most frequent subtype of stroke [3–5].

The most common cause of SAH is head trauma [6, 7], although 75–80% of nontraumatic, spontaneous cases arise from rupturing of an intracranial aneurysm [8]. Autopsy studies estimate that 5% of the adult population harbor at least one saccular intracranial aneurysm [9], where 25–50% of these will become symptomatic during a person's lifetime with an overall annual risk of approximately 3%.

J. Hou, M.D. (✉) • J.H. Zhang, M.D., Ph.D.
Zhang Neuroscience Research Laboratories, Loma Linda University Medical Center, 11234 Anderson Street, Room 2562B, Loma Linda, CA 92354, USA
e-mail: Jackofknives@hotmail.com; johnzhang3910@yahoo.com

R.R. Watson et al. (eds.), *Magnesium in Human Health and Disease*, Nutrition and Health, DOI 10.1007/978-1-62703-044-1_19, © Springer Science+Business Media New York 2013

Like most diseases, the clinical syndrome of aSAH comes in different shapes and sizes, ranges from completely asymptomatic state to sudden death. Nevertheless, a classical scenario would be a 50-year-old female presenting with a new and sudden onset of severe headache, often referred to as the "worst headache of my life," accompanied by nausea and vomiting, photophobia, neck pain (meningismus), and possibly syncope. The signs of aSAH include reactive hypertension, nuchal rigidity, subhyaloid ocular hemorrhages, and, depending on the hemorrhaged volume and location of the hematoma, various degrees of focal neurological deficits and decreased level of consciousness. Efforts are made to promptly admit the patients under clinical suspicion of aSAH to the neurosurgical service, where the initial diagnostic test of choice is computer tomography (CT) without contrast enhancement before treatment is instigated. CT in the first 12 h after aSAH detects hyperdensities (white) representing blood in the subarachnoid basal cisterns with a sensitivity and specificity of 98% and 100%, respectively [10, 11].

Patients often require frequent and vigilant invasive monitoring of physiological and biochemical parameters in the intensive care unit, including maintaining the blood pressure, cerebral perfusion pressure, serum sodium level, and more. Two major concerns during the initial stabilization of patients are aneurysmal rebleeding and a syndrome termed delayed ischemic neurological deficit (DIND), discussed later. Rebleeding is prevented medically by prompt administration of tranexamic acid, an antifibrinolytic agent, and surgically by early aneurysmal treatment with either clipping or coiling, while DIND is a medical complication and the central focus of this chapter. The current DIND prophylaxis is nimodipine, an L-type voltage-gated calcium channel inhibitor. However, in the recent Cochrane review on nimodipine involving 3,361 patients [12], the authors concluded that the evidence for nimodipine is "not beyond all doubt." The contemporary management once DIND has occurred is the so-called triple-H therapy to medically induce hypertension, hemodilution, and hypervolemia, which only provides short-lasting clinical benefits and may not justify the risks associated with the procedure. The last resorts and experimental procedures to alleviate DIND include intra-arterial administration of vasodilators such as papaverine hydrochloride [13], transluminal balloon angioplasty [14], and cisternal irrigation with thrombolytics in an attempt to remove the blood clot [15, 16], although none has been shown to definitively improve outcome. Even with the treatment options available today, prognosis remains poor. The overall mortality ranges from 32% to 67% [2] where the majority of patients pass away within the first 2–3 weeks. Of the survivors, 30–40% have moderate to severe disability [5].

This chapter covers the neuroprotective effects of magnesium therapy implicated in aSAH. To formulate a rational treatment strategy for aSAH, it is, however, necessary to first understand its natural history in terms of the various pathophysiology processes in aSAH that are not directly manageable by surgical means and how they in concert orchestrate a vicious cascade and ultimately cause DIND. We then transition to a walkthrough on the dozen of studies on the clinical applicability of magnesium in quest for ameliorating DIND and improve long-term outcome. In doing that, we delve into the levels of magnesium in various body compartments in the acute stage of the disease, the pharmacokinetics of intravenous magnesium therapy and other alternative routes of administration, and finally unveiling the results of both pilot and multicenter clinical trials.

Overview of Pathophysiology

Aneurysmal subarachnoid hemorrhage (aSAH) arises from rupturing of an abnormally widened or ballooned arterial vessel due to weakness in the vessel wall, most commonly at the bifurcations of anterior and posterior communicating arteries. The exact pathogenesis of intracranial aneurysm is unknown; however, many risk factors, such as hypertension, substance abuse, pregnancy [17], autosomal dominant polycystic kidney disease [18], and a number of other congenital and acquired disorders, have been linked to its formation.

After the rupture of an aneurysm, arterial blood is rapidly ejected into the subarachnoid basal cisterns, a space restricted by the cranial vault, causing acute increase in intracranial pressure (ICP). The intensity of ICP elevation is linked to the hemorrhaged volume. Despite a reactive increase in systolic arterial pressure (a component of the Cushing's response), the cerebral perfusion pressure (CPP), derived from mean arterial pressure (MAP) minus ICP, decreases. The decreased CPP is followed by an acute vasoconstriction [19, 20] and the consequent decrease in cerebral blood flow (CBF), which can ultimately result in the global arrest of intracranial circulation [21]. Diminished blood flow promotes hemostasis but, if it continues, may manifest as loss of consciousness and death [22–24].

The clinical course in aSAH is biphasic. While the primary brain injury is devastating, approximately 16% of patients develop a clinically manifested secondary brain injury [25], coined as delayed ischemic neurological deficit (see Box 19.1), between day 3 and 12 with a peak incidence on day 7 following the disease onset. Using modern and current medical and surgical technology to treat aSAH, DIND is the leading cause of mortality and morbidity [26], killing 7%, and leads to severe deficit in

Box 19.1

The following definitions are the commonly used primary outcome measures in animal studies, observational clinical studies, and clinical trials:

Delayed ischemic neurological deficit (DIND) (incidence 20–30%) [10, 39]: It was defined clinically as a "acute or sub-acute new focal neurological deficit (motor or speech deficit) that had developed after aSAH, a decrease on Glasgow coma score (GCS) of ≥2 points lasting >6 h, not related to treatment (coiling or clipping) complications, rebleed, progressive hydrocephalus, electrolyte or metabolic disturbance, or infection" [40]. DIND is therefore a diagnosis of exclusion. This is a subjective diagnosis and can be difficult to assess in poor-grade, comatose patients, where variations in examination may be subtle or imperceptible [41].

Delayed cerebral infarction (DCI) (incidence 28–35%) [42]: No consensus exists for the exact inclusive timeline. A generally accepted definition is "the presence of cerebral infarction on CT or MRI of the brain within 6 weeks after aSAH, or on the latest CT or MRI made before death within 6 weeks, or proven at autopsy, not present on the CT or MRI between 24 and 48 h after early aneurysm occlusion, and not attributable to other causes such as surgical clipping or endovascular treatment. Hypodensities on CT imaging resulting from ventricular catheter or intraparenchymal hematoma should not be regarded as cerebral infarctions from DCI" [43].

Delayed cerebral vasospasm (incidence 45–70% depending on diagnostic modality): No consensus exists for the precise definition of cerebral vasospasm in clinical trials. In a nutshell, vasospasm suggests reduced arterial diameter due to contraction of myocytes within the vessel wall. Angiographic visualization of proximal arterial vasoconstriction was the first laboratory finding to be associated with DIND reported in 1951 by Ecker and Riemenschneider [44]. Digital subtraction angiography remains the gold standard and confirmatory diagnostic test for delayed cerebral vasospasm [45] but, due to its invasiveness and complications associated with the procedure, has been replaced by transcranial Doppler ultrasound (TCD) measurement of flow velocity for screening of patients [46, 47] in which mean absolute middle cerebral artery (MCA) velocity of >120 cm/s suggests vasospasm [47]. Differential diagnosis of vasospasm from hyperemia, characterized by increased blood flow velocities in both the MCA and the internal carotid artery (ICA), is facilitated by employing the ratio of these velocities, the so-called Lindegaard ratio. MCA/ICA ratio >3 suggests vasospasm [47]. Angiography and TCD are poor indicators of circulatory impairment at the microvascular level in the absence of proximal arterial vasospasm.

another 7% [27] of the total aSAH population, accounting for one-third of all patients with poor outcome [10, 28]. Despite improvement in therapeutic modalities, the incidence of DIND and functional clinical outcome have remained static [12, 29], and improvement of such parameters has been the Holy Grail in aSAH research for over 60 years.

The exact pathogenesis of DIND is incompletely understood. Until recently, delayed onset vasospasm of cerebral arteries was regarded as the single and most important contributor to DIND, as 40–70% of patients between day 3 and 14 after aSAH develop proximal arterial vasospasm identified with digital subtraction angiography (DSA) [30] or transcranial Doppler ultrasound (TCD) [31], coincides with the period of DIND. Previous studies found strong correlations between vasospasm, delayed cerebral infarction (DCI), DIND, and poor functional outcome [32–35]. However, inconsistency exists, as 21% of patients who developed DIND had no signs of vasospasm and the majority of (70–80%) of those who developed moderate to severe angiographic vasospasm were asymptomatic [33, 36]. The endothelin A receptor antagonist, Clazosentan, shown to significantly reduce moderate to severe angiographic vasospasm from 66% to 23% [37], has not, in phase III trial, had an effect on DCI and clinical outcome [38], suggesting that the correlation does not necessarily equal causal relationship. The role of proximal arterial vasospasm in the pathogenesis of DCI and DIND is probably smaller than previously assumed and may not capture the full pathophysiological spectrum of delayed brain injury in aSAH.

Primary Brain Injury

Generally, for any injury inflicted to the brain from external insult, whether vascular or traumatic, the initial primary brain injury is nonsalvageable and often permanent. Nevertheless, delayed effects of acute SAH-induced brain injury, also known as secondary brain injury, can potentially be minimized or prevented. The pathophysiology of secondary brain injury, but not the primary brain injury, is therefore the most important target of new therapeutic measures.

Direct Physical Trauma and Cerebral Hypoperfusion

The first brain injury incurred in aSAH is the traumatic compression of the brain from acute elevation in ICP reaching up to 162 mmHg [48]. Hematoma can have a mass effect, stretches the subarachnoid space, and compresses major arteries of the skull base and smaller conducting vessels, which then narrows the vessel lumen. Depending on the amount of blood ejected into the subarachnoid cisterns and the presence or absence of cerebral spinal fluid (CSF) outflow obstruction, two patterns of ICP elevation were described [48]. In most patients, with small hemorrhaged volume, ICP peaks and approaches diastolic arterial pressure, then falls to a trough marginally above baseline [21]. For the remaining patients, enlarging hematoma and acute hydrocephalus due to CSF outflow obstruction may cause persistent ICP elevation [48, 49]. The degree of ICP elevation is directly linked to neuroinflammation, malfunction of cell metabolism, and DIND [50].

Despite the massive catecholamine release at the time of hemorrhage leading to increased mean arterial pressure (MAP), with the severely elevated ICP, the cerebral perfusion pressure (CPP) decreases in the initial aSAH ictus [21] (MAP (↑) – ICP (↑↑) =>CPP (↓)). This effect is amplified by impaired autoregulation of cerebral blood flow (CBF) most prominent in the first 3 days of aSAH [51, 52], which renders the brain unable to maintain CBF in the presence of reduced CPP. The maintenance of an adequate CPP has therefore been the cornerstone of modern aSAH therapy in the intensive care unit.

In addition, cerebral vasoconstriction occurs as a result of aSAH [53–56]. In experimental studies, spasm of proximal conductive and small resistance (<100 μm) arteries is apparent immediately after aneurysmal rupture [57, 58]. Taken together, the fall of CPP and primary vasoconstriction leads to temporary cerebral hypoperfusion or global circulatory arrest, causing transient ischemia followed by reperfusion.

Secondary Brain Injury (Early Brain Injury)

Brain tissue, especially neurons, are metabolically very active, and their function and survival are dependent on an optimal local milieu and continued nutritional supply of glucose and oxygen. The damage inflicted by primary brain injury not only disrupts the delicate hemostasis but also has delayed and profound destructive effects. The pathological processes occurring in the initial 72 h are also known as "early brain injury." A number of critical, interrelated pathways have been implicated in early brain injury, including oxidative stress, disrupted cellular metabolism and apoptosis, neuroinflammation, proximal artery and microcirculatory vasospasm, alternations of nitric oxide (NO) levels, microthrombosis, spreading depolarization, and spreading ischemia, which may lead to neuronal cell death and cerebral edema.

Oxidative Stress

Free radical formation is a major mediator of injury occurring in ischemia and subsequent reperfusion, seen in aSAH [59]. Experimental and clinical studies have shown that free radicals generated in the early phase of aSAH overwhelm the endogenous antioxidant pathways [60–62].

Reperfusion injury is one way of forming reactive oxygen species (ROS) and is believed to result from two critical chemical reactions [63]. In ischemia, the breakdown of cellular storage of ATP to hypoxanthine occurs. During reperfusion, the sudden availability of oxygen acts as cofactor to allow xanthine oxidase to convert hypoxanthine accumulated during ischemia to uric acid, leading to the subsequent formation of superoxide and hydrogen peroxide as byproducts [63]. In the presence of the enormous amount of free iron from the breakdown of hemoglobin released from extravasated red blood cells, superoxide and hydrogen peroxide can participate in the Fenton reaction to produce the highly potent oxidant, hydroxyl radicals. Furthermore, inflammatory reaction in the early phase of disease, most prominently NADPH oxidase activity in the CSF, an enzyme of the respiratory burst pathway in neutrophils and a "professional" producer of ROS, was reported as an independent predictor of DIND [64].

$$O_2^- + Fe^{3+} \rightarrow O_2 + Fe^{2+}$$

$$Fe^{2+} + H_2O_2 \rightarrow Fe^{3+} + \cdot OH + OH^-$$

The Fenton reaction is where superoxide (O_2^-) and hydrogen peroxide (H_2O_2) in the presence of ferric iron (Fe^{3+}) produce the highly reactive hydroxyl radicals $\cdot OH$ [63].

ROS production is detrimental to brain tissue. ROS was demonstrated in aSAH to produce endothelial damage and proliferation of the intima lining and thereby amplifies delayed vasospasm [59]. ROS also causes lipid peroxidation of cell membranes leading to loss of membrane integrity and disruption of blood-brain barrier (BBB). The brain is especially susceptible to lipid peroxidation as the neuronal membrane is enriched with polyunsaturated fatty acids high in reactive hydrogen [65]. Disruption of neurocytic cell membranes was shown to inhibit cation pumps, such as Na+/K+ATPase and Ca^{2+} ATPase, and thus disrupts ionic homeostasis. The increased intracellular Ca^{2+} concentration as a result of decreased Ca^{2+} efflux activates phospholipase A2 with the subsequent release of arachidonic acids, including 6-ketoprostaglandin F1 alpha, leukotriene C4, and prostaglandin E2 [60], all of which enhance inflammation.

Previous studies support the notion that magnesium deficiency enhances ROS-induced cell injury. Dickens et al. in an in vitro study reported that low level of magnesium enhances ROS-induced intracellular oxidative stress in endothelial cells [66], whereas Rayssiguier et al. found that dietary Mg deficiency increases the susceptibility of lipoproteins to peroxidation in rats [67]. Furthermore, magnesium supplementation was shown to reduce ROS formation in the mitochondria via inhibition of NADPH oxidase [68]. The mechanisms of the antioxidant effect of magnesium is incompletely understood but is probably mediated, among others, through binding to phospholipids in the cell membrane and thus decreases phospholipid mobility [69] and alters membrane permeability and receptor function [70].

Cellular Metabolism

Low magnesium level in the CSF and serum is a frequent occurrence in brain injury, particularly in aSAH [71, 72]. In view of the vital roles played by magnesium in cellular homeostasis and metabolism, where it participates as cofactor for all kinases in the mitochondria, seven glycolytic enzymes, four TCA cycle enzymes [72], and more, magnesium deficiency can have profound implications on the energy balance of the injured neurocytic cells.

Mitochondria depend on Mg^{2+} for the production of ATP in the oxidative phosphorylation of the tricarboxylic acid (TCA) cycle. Phosphorylation by ATP in which free energy is harnessed to perform work is enzymatically catalyzed by kinases. Since all kinases require Mg^{2+} as a cofactor, ATP can only be utilized when it is coupled to Mg^{2+}. Mitochondria malfunction and disrupted energy balance were therefore reported in aSAH [73–76], which induces low cellular pH, high intracellular Ca^{2+} level, and oxidative injury, whereas magnesium therapy was demonstrated to maintain mitochondrial function in experimental traumatic brain injury [77] and middle cerebral artery occlusion rat models [78].

Adequate ATP supply is necessary for calcium homeostasis with the Ca^{2+}-ATPase pump. Apart from sustaining ATP synthesis, magnesium controls calcium level via blocking of the NMDA receptor and voltage-gated Ca^{2+} channels on the cell membrane and also has intrinsic antagonistic effects on Ca^{2+}-mediated muscle tonus, neurotransmitter release, and the propagation of action potentials [72].

Apoptosis is an important mediator of brain injury in aSAH [79, 80]. Calcium overload, caused by metabolic failure associated with ischemic injury and increased cellular membrane permeability from ROS, excessively stimulates apoptotic cascades and ultimately leads to the assembly of mitochondrial permeability transition pore (MPTP) and the consequent mitochondrial dysfunction and apoptosis [81–87]. Magnesium deficit amplifies apoptosis [88], while magnesium therapy has a preventive role in neuronal apoptosis [89] by both decreasing the expression of apoptosis-inducing p53-related factors [90] and decreasing caspase-3 expression [91]; however, this effect on apoptosis is reduced in the presence of ATP depletion [89, 91]. On the other hand, attenuation of apoptosis through extracellular events such as the ability of extracellular magnesium to block the ischemia-induced opening of the glutamate-NMDA channel [92, 93], thereby blunting calcium influx and the subsequent apoptotic cascade, is independent of intracellular ATP, exerting neuroprotective effect irrespective of energy status.

Neuroinflammation

There is mounting evidence, both clinically and by experimentation, that early activation of inflammatory cascade and the interaction between leukocytes and endothelia are implicated in DIND. This paradigm was first explored 30 years ago, when Rousseaux et al. discovered a characteristic fever curve that plateaus at 39°C beginning on the fourth day and lasting an average of 9 days that is highly associated (88%) with angiographic vasospasm compared to patients without vasospasm (18%) [94]. Since then, higher serum leukocyte counts and circulating as well as CSF immune complexes and cell adhesion molecule level have also been reported to directly correlate with the risk of DIND [95–104].

Cell adhesion molecules such as intercellular adhesion molecule-1 (ICAM-1), vascular cell adhesion molecule-1 (VCAM-1), and E-selectin facilitate leukocyte migration. Increased level of adhesion molecule expression in endothelial cells and increased serum levels occur within 24 h [105, 106]. Migration of neutrophils at first, then macrophages from the systemic circulation through endothelial cells to the blood clot soon follows. Leukocyte activity is related to the release of ROS, which causes endothelial injury and intima proliferation [23].

In experimental studies, increase in blood-brain barrier (BBB) permeability is seen within 36 h after aSAH, which peaks at 48 h and normalizes by day 3 [107], and this is believed to be secondary to the intense inflammatory response. Disrupted BBB causes vasogenic cerebral edema and secondary brain injury.

Another important sequelae of the inflammatory reaction is the production of endothelin by leukocytes within 5 days after aSAH [108–110], which, according to previous in vitro studies, may be secondary to the hemolysis process. Endothelin is a powerful vasoconstrictor causing decreased cerebral blood flow and ischemia, of which increased concentrations in aSAH were correlated to endothelial damage, DCI, and poor outcome [111].

Magnesium appears to reduce leukocyte activity in vitro and inhibit the inflammatory response, whereas magnesium deficiency stimulates inflammation by increasing macrophage synthesis of proinflammatory cytokines, among others IL-1β and TNF-α, probably through a calcium-mediated pathway [112]. Of clinical relevance is the potential of magnesium on reducing BBB permeability and cerebral edema from the attenuated inflammation [113].

Vasospasm and Microcirculation

Constriction of cerebral vessels undisputedly occurs in response to aSAH [53–56]. In animal studies, narrowing of proximal conductive and small resistance (<100 μm) arteries is apparent immediately after aneurysmal rupture [57, 58]. In humans, studies have shown that spasm of proximal arteries occurs with a delay of 3–7 days after aSAH [114–117]. Moreover, Uhl et al. affirmed the findings in animal studies that vasospasm also occurs in small vessels in humans within the first 72 h after aSAH [56]. The origin of cerebral vasoconstriction is multifactorial and is related to the combination of structural changes and increased vascular tone caused by the increased level of endothelin and hemoglobin derivatives and decreased availability of endothelium-derived nitric oxide (NO).

Histopathological studies of small vessels in animals and deceased aSAH patients who passed away within the first 72 h showed intima thickening with endothelial swelling and vacuolization, opening of interendothelial tight junctions, media muscle necrosis, and corrugation of elastica [54, 118–120]. As alluded to earlier, these changes are at least partially attributable to the production of ROS and the intense inflammatory reaction [121, 122], which augments the narrowing of vessel lumen. Magnesium prevents endothelial injury by protecting endothelium from ROS-induced injury and dampens inflammation.

There is growing evidence that oxyhemoglobin, released by the red blood cells during hemolysis of subarachnoid blood clot in the subacute stage of aSAH [123], is also an initiating factor of vasospasm. This notion was inspired by previous findings which indicated that the risk of DIND is directly associated with the size of the hematoma in the initial CT scan [124] and the onset of DIND coincides with the period of peak hemolysis [125]. Indeed, hemoglobin, oxyhemoglobin, and deoxyhemoglobin induce severe vasospasm in experimental models, and their elevated concentration in human perivascular space and CSF is parallel to the occurrence of DIND [125, 126]. The mechanism for which oxyhemoglobin produces cerebral arterial contractions is either by direct stimulatory effect on vascular smooth muscle or by scavenging endothelial nitric oxide (NO) [126]. Magnesium was shown to counteract the vasospastic effect of oxyhemoglobin where it ameliorates vasospasm in experimental SAH [127–129] and thereby reduces vasospasm, improves CBF, and maintains cell metabolism [130–132].

Endothelin-1 (ET-1) is a potent constrictor of cerebral arteries [133] and mediates its action mostly via endothelin A receptor (ETA) [134] present on the vascular smooth muscle cells [133, 135]. In experimental models of aSAH, ET-1 is upregulated and released from glial and white blood cells in response to inflammation and ischemic injury [109, 136], and oxyhemoglobin stimulates the synthesis and secretion of ET-1 in endothelial cell culture [137]. In human, plasma ET-1 surges immediately after aSAH, and ETA receptor expression increases 24 h later [138–140]. Consequently, the reactivity of cerebral blood vessels to ET-1 increases significantly after SAH [141–144]. ET-1 also has mitogenic and metabotropic effects where it induces histopathological changes of cerebral blood vessels similar to that seen in aSAH described earlier [145]. Several receptor signal transduction pathways were suggested to be involved in ET-1-induced vasoconstriction and include (a) increase in cytosolic free calcium concentration by facilitation of Ca^{2+} influx and mobilization of intracellular Ca^{2+}; (b) G-protein-mediated activation of phospholipase C (PLC), leading to phosphorylation of phosphatidyl inositol hydrolysis and rapid formation of inositol 1,4,5-triphosphate (IP3) and sustained diacylglycerol (DAG) accumulation; (c) activation of PI3K pathway leading to Rho kinase phosphorylation; and (d) activation of Src and its downstream MAPK pathway – all of which may ultimately lead to myosin light chain (MLC) phosphorylation and the subsequent vasoconstriction [133].

Magnesium appears to exert its vasodilatory effect by counteracting the above mechanisms. It was shown to regulate nuclear and cytoplasmic Ca^{2+} level in vascular myocytes via inhibition of voltage-gated L-type Ca^{2+} channels, thus reducing Ca^{2+} influx [146], so that it reduces the production and completely attenuates the vasoconstrictive effect of ET-1 [147, 148]. Magnesium also possesses intrinsic antagonistic effect on intracellular Ca^{2+}, where it competitively inhibits Ca^{2+} binding to calmodulin and thus prevents its stimulatory effect on myosin light-chain kinase (MLCK) and decreases vascular smooth muscle contractility [149, 150]. Furthermore, magnesium stimulates the synthesis and release of prostacyclin, a potent vasodilator, which counteracts the overproduction of vasoconstrictors in aSAH [151]. Consequently, magnesium sulfate showed reversal of delayed cerebral vasospasm in experimental SAH studied in the 3- to 7-day timeline, where topical therapy exerted greater vasodilatory effect than systemic administration on the cerebral arteries [127, 152] and resulted in improvement of CBF [153].

Nitric Oxide (NO)

A time-dependent change in NO availability occurs during the first 24 h of aSAH. In experimental studies, three stages are recognized. First, a decrease of NO is seen 10 min after aSAH, then returns to baseline at 3 h [154], and finally an increase above baseline at 24 h after aSAH [155]. Human studies have shown continuous decline in extracellular NO concentrations [156] followed by increased NO 24 h after aSAH [157]. The initial fall of NO is due to scavenging of NO by oxyhemoglobin to generate nitrate and ferric heme [154, 158], while the later rise in NO is due to induction of inducible NO synthase (NOS).

The production of NO is mediated by NOS. NOS exists in three isoforms, namely, endothelial (eNOS), neuronal (nNOS), and inducible (iNOS). nNOS and eNOS are Ca^{2+}-dependent enzymes. Elevated intracellular Ca^{2+} levels activate calmodulin, which in turn stimulates NO formation from L-arginine by NOS. iNOS, on the other hand, is Ca^{2+} independent and is inducible by stimulus such as stress, inflammation, and infection. nNOS and eNOS constitutively produce minute amount of NO in the nM range in contrast to iNOS, following an inducible latent period it produces NO in the uM range for an extended period of time [159].

Constitutive and inducible isotypes of NOS suggest distinctive functions. Constitutive NOS is responsible for the production of baseline NO to keep regional cells in an inhibitory state – and the activation of these cells is mediated by disinhibition. It thus lends reason to believe that the physiological level of NO produced by endothelia exerts neuroprotective effects but the supraphysiological level derived from nNOS and iNOS is neurotoxic [160–162].

The constitutive availability of NO is critical to a number of biological processes. NO inhibits platelet aggregation and leukocyte adhesion to endothelial cells and regulates CBF by cGMP-mediated vasodilation. In the hyperacute phase of aSAH, with the fall of NO, platelet aggregation occurs, leukocytes adhere to endothelium, and cerebral vasoconstriction causes decline in CBF [20, 21, 57, 154, 163, 164]. In the later phase, NO elevates abnormally above baseline, which may also have detrimental effects. NO is converted to peroxynitrite, a potent ROS, which causes lipid peroxidation of cell membrane and thereby damages mitochondria, endothelium, and smooth muscle [165] and ultimately causes cell death [166]. Thus, both low and high levels of NO may be pathological, and the pattern of cerebral NO fluctuation was associated with DIND and poor outcome after aSAH [157, 167, 168].

Magnesium deficiency in aSAH demonstrated reduced eNOS activity and consequently causes vasoconstriction [169, 170]. Additionally, magnesium therapy inhibits nNOS activity of cortical neurons in several experimental ischemia models [171, 172], probably by inhibition of NMDA-receptor-mediated Ca^{2+} influx, thus maintains physiological level of NO and reduces NO-mediated injury.

Microthromboembolism

Platelet activation is increasingly recognized as a major culprit for delayed ischemia in aSAH, in which circulatory impairment occurs in the absence of proximal arterial vasospasm [173, 174]. Supportive evidence includes reduction of platelet count in jugular venous blood and changes in platelet shape consistent with platelet activation seen within 5 min in experimental aSAH and within 48 h in patients [175, 176]. Platelet clumping in small blood vessels was also demonstrated within 10 min after animal studies [163] and on day 2 in human autopsy studies [177].

One of the postulated stimulus of platelet activation was ischemia-induced NMDA receptor activation of platelets which causes Ca^{2+}-dependent phospholipase A2 activation [178, 179] and the consequent release of arachidonic acid derivatives, among others prostaglandins and thromboxanes. Thromboxane A2 (TXA2) causes vasoconstriction and promotes platelet aggregation, and its level was associated with the development of DIND [180–182]. When platelets aggregate into a clot, they disgorge additional vasoconstrictive substances such as serotonin, ADP, and platelet-derived growth factor (PDGF) [183, 184]. In the later stage, clot dissolution within small vessels causes disruption of basement membrane due to degradation of collagen type IV via matrix metalloproteinases-9, which exposes subendothelial matrix and further enhances platelet adhesion, exacerbates vasogenic cerebral edema, and allows platelet access into the brain parenchyma [120] where they cause more inflammation.

Magnesium was demonstrated to have both intrinsic and extrinsic inhibitory effects on platelets. Not only does magnesium inhibits Ca^{2+}-dependent phospholipase A2 activation [185], it also antagonizes many platelet activators, such as TXA2 and β-thromboglobulin, where these effects are mediated by inhibition of intracellular Ca^{2+} mobilization [170, 186, 187]. Furthermore, the intrinsic membrane stabilizing properties of magnesium prevents platelet shape change [70]. Collectively, magnesium lowers the risk of platelet-dependent thrombosis in an independent and additive manner to aspirin [188–190] and thereby prevents microcirculatory cerebral blood flow restriction.

Spreading Ischemia, a Vicious Cycle to Brain Injury

A current paradigm shift in aSAH research has evolved in recent years to the field of electrophysiology and the uncoupling of neurovascular units, where the focus is shifted from the survival of individual cells, discussed earlier, to the spread of injurious impulses on a global scale. A plethora of noxious stimuli, among others, mechanical injury, potassium from hemolyzed blood clot, glutamate excitotoxicity, acute hyperexcitability, and hypoxia [191, 192], all of which evident in aSAH, may bring about a highly destructive phenomenon, termed spreading depolarization [191] (Box 19.2).

Cortical spreading depression was first described in rabbit cerebral cortex by Leao in 1944 [193]. Emerging evidence, both in animals and humans, suggests that cortical spreading depolarization (CSD) and spreading ischemia occur abundantly in acute and subacute stages of aSAH. Cat experiments with intracortical electrodes have provided the first evidence that CSDs occur during the initial

Box 19.2

Cortical spreading depolarization (CSD) describes waves of depolarization of cortical neurons that propagate through gray matter at 2–6 mm/min with an amplitude of approximately 60-mV extracellular direct current (DC) potential [220], much like a "brain tsunami."

Slow potential change (SPC) is a characteristic of spreading depolarization, where synchronous sustained depolarization of neurons is associated with slow return of baseline potential [215]. The low-frequency voltage change and high amplitude in spreading depolarization separate it from the epileptic activity seen in electrocorticography.

Nonspreading depression is a stationary zone of suppressed brain electrical activity in the area of the brain exposed to severe noxious stimulus. It foreshadows (precedes) spreading depolarization by minutes [215].

Spreading depression is migratory suppression of brain electrical activity secondary to spreading depolarization. Neurons cannot fire action potentials due to sustained depolarization above threshold at which voltage-gated sodium channels are inactivated [200]. There may be additional unknown mechanisms at play. It is regarded as an epiphenomenon of spreading depolarization.

Spreading hyperemia is a normal physiological response of neurovascular coupling to markedly increase the regional cerebral blood flow (rCBF) in order to maintain cellular ionic homeostasis as compensation for the increased energy demand seen in CSDs.

Spreading ischemia is a pathological response of inverse neurovascular coupling where spreading depolarization is followed by cerebral hypoperfusion (ischemia) due to severe microvascular spasm rather than the physiologic vasodilation.

hours of aSAH [194, 195]. Such finding was later replicated in a rat model where mock aSAH CSF replacement with hemolysis products, high potassium, and low glucose triggered CSDs and spreading ischemia, leading to cortical infarctions [196]. This pathological finding is in congruence with autopsy studies of aSAH patients, in whom 75% of fatal cases demonstrated widely scattered triangular or laminar ischemic cortical lesions, 12-fold more frequent than infarctions in the territories of large arteries [197, 198]. In humans, a prospective multicenter study (COSBID) [199] via subdural electrode strips placement on cerebral cortex showed electrocorticography-detected CSDs in 72% ($n=18$) of patients. In four patients, DCI developed in the recording area. DCI was preceded by nonspreading cortical depression and the subsequent CSDs of >60 min. A later study using opto-electrode PtiO2 sensor combined with laser Doppler flowmetry for CBF and electrocorticography demonstrated two patterns of CSD in aSAH patients [200]. They are isolated and recurrent CSDs. While isolated CSDs were observed in 12 out of 13 patients and were associated with either spreading hyperemia or ischemia, clustered CSDs were invariably followed by spreading ischemia of longer durations (up to 144 min). This finding in aSAH patients was later confirmed by Bosche et al. [201]. Collectively, recurrent CSDs were temporally associated with development of DCI in the recording areas, suggesting that energy depletion as a result of CSDs may contribute to the pathogenesis of DIND.

The core process of spreading depolarization is passive. High extracellular potassium levels from hemolyzed blood and depleted energy due to hypoxia trigger passive cation influx across the cell membrane that exceeds the ATP-dependent efflux of Na^+ and Ca^{2+} [200, 202]. Postulated candidates for neuronal cation influx include NMDA-receptor-controlled channels and slowly inactivating sodium and calcium-sensitive nonspecific cation channels [200]. As a result, the net dendritic cation current turns inward causing near-complete sustained depolarization. The spread of electrical impulses may occur via synaptic transmission of glutamate-induced calcium influx in the postsynaptic neurons, which removes magnesium block of N-methyl-D-aspartate (NMDA) receptors, and may lead to additional calcium influx [203]. Transcellular gap junctions may provide another means of neuron-to-neuron propagation of CSDs [204]. A CSD spreads along the gradient from high to low concentration of noxious stimulus [205–207] thereby invades the ischemic penumbra, the still viable cerebral tissue, and causes DCI. Repolarization is energy dependent [208], which may be severely deficient in aSAH, leading to prolonged slow potential change (SPC). Consequently, the extreme shunting of cation across the neuronal membrane induces cytopathological changes characterized by cytotoxic edema of neurons [202, 209], distortion of dendritic spines [202], and intracellular calcium overload, arguably the principle mediator of apoptosis from ischemia [210, 211]. Moreover, ROS is substantially increased in sustained depolarization [212], which may contribute to depolarization-induced damage.

The physiological response to increased energy demand in CSD is spreading vasodilation, mediated by the effect of increased production of NO and arachidonic acid metabolites in the vascular smooth muscle cells [213]. In aSAH patients, however, under the condition of increased vasoconstrictors and decreased vasodilators, an inverse neurovascular response of severe vasoconstriction that propagates in the tissue together with CSD was demonstrated, termed spreading ischemia [214]. The transient increase in O_2 demand exceeding vascular O_2 supply depletes energy and reduces cation pump activity, while glutamate transporters show reversed transport, together with failure of repolarization sustaining neuronal depolarization toward the vicious cycle of cell death [215]. This has led to the hypothesis that spreading ischemia could be involved in DIND after aSAH.

Indeed, CSDs and spreading ischemia may be a common final pathway of all the pathophysiological processes discussed earlier and that amelioration of this phenomenon may decrease the incidence of CSDs. In addition to the neuroprotective properties mentioned, magnesium blocks NMDA receptor and thereby inhibits excitatory neurotransmission [216, 217]. Consistently, in experimental aSAH rat model, prophylaxis administration of MgSO4 reduces the frequency and duration of CSDs and spreading ischemia leading to decreased DCI [218, 219].

Hypomagnesemia After aSAH

In serum and CSF, magnesium ions (Mg^{2+}) present in three forms [221]. The ultrafilterable fractions are able to penetrate the blood-brain barrier consisting of the 70% ionized, free form and the 10% complexed form to small anion chelators, such as citrate and lactate. The protein-bound state is nonultrafilterable and makes up the remaining 30%. While clinical laboratories generally measure total magnesium concentrations, only the ionized, free fraction of magnesium in the serum is physiologically active [222]. Nevertheless, hypomagnesemia is generally defined as serum total magnesium below 1.6 mg/dL [223].

Bergh et al. demonstrated that hypomagnesemia was evident in 38% of patients admitted within 72 h after aSAH [71, 224]. The exact timeline for the onset of hypomagnesemia in aSAH and whether it is already present in the hyperacute phase is currently unknown. However, since a higher proportion of hypomagnesemia (54%) was seen in patients examined within 12 h after the disease onset [71], the magnitude and prevalence of reduction in serum magnesium level are probably greater in the hyperacute phase and may return to normal level hours after aSAH.

The exact cause of hypomagnesemia after aSAH remains elusive. Renal elimination of magnesium ion is improbable since hypomagnesemia is already apparent in the hyperacute phase of aSAH. It is probably more likely related to the transient intracellular shift of magnesium ions. Using phosphorus-31 MRS (13P-MRS) in acute ischemic stroke patients, intracellular free magnesium elevation of 45% compared to control subjects was demonstrated [225]. The elevation persisted for 72 h and returned to baseline level thereafter. This phenomenon was associated with the acidotic phase of brain injury and the depletion of ATP in congruence with the notion that magnesium expulsion from cells is an active process and that the electrochemical gradient of magnesium favors influx of ions. Whether or not increased intracellular magnesium level embodies an intrinsic neuroprotective response is unclear. Nevertheless, Bergh et al. in an observational clinical study of 107 patients admitted within 2 days after aSAH have shown that hypomagnesemia at the time of admission, i.e., reduced extracellular magnesium level probably as a consequence of intracellular shift, was a predictor of the occurrence of DCI with adjusted hazard ratio being 1.9 (95% CI 0.7–4.7) [71].

Adverse Effects

Normal serum total magnesium (TMg) reference value lies in the range of 1.7–2.2 mg/dL in humans [223]. Generally, magnesium sulfate is well tolerated, of which some adverse effects, such as dizziness, headache, and flushing, may occur and usually do not warrant termination of treatment. Curare-like side effects such as muscle weakness, hyporeflexia, and cardiovascular suppression presented as bradycardia and hypotension [226] begin to appear as serum levels exceed 3–4 mg/dL [227]. When the serum levels reach 11–13 mg/dL potential fatal adverse effects such as respiratory depression, electrocardiogram abnormalities, and asystole have been reported. Intravenous bolus of calcium gluconate is an antidote that can quickly antagonize the effect of magnesium.

These adverse effects may be detected early or mitigated through mechanical ventilation and extensive cardiopulmonary and other system support monitoring in the intensive care unit, often warranted for high-grade aSAH patients irrespectively. Furthermore, the cardiosuppressive side effect may be taken advantage of to counteract the initial reactive hypertension secondary to aSAH in keeping the systolic blood pressure in the desired range of 120–150 mmHg.

Pharmacokinetics and Blood-Brain Barrier Penetration

For the neuroprotective mechanisms of magnesium therapy to take effect, except for the processes for which the sites of action are in the intravascular space, is dependent on magnesium penetration of the blood-brain barrier (BBB) to enter the CNS.

Several studies, in both animals and humans, have shown that magnesium crosses the blood-brain barrier. Previous studies demonstrated that increasing plasma total magnesium level from baseline by 65% in craniotomy patients [228], by 100% in aSAH patients [229], and by 350% in normal mongrel dogs [230] gave an elevation of CSF total magnesium by a modest but significant 15%, 16%, and 21%, respectively. Equilibrium was reached at around 1.5 h in craniotomy patients [228] and 3–4 h in normal dogs [230] after intravenous administration.

It is interesting to note that the concentration of total magnesium in the CSF is approximately 40% higher than serum total magnesium level given that systemic total magnesium was within normal limits [228, 230, 231]. The mechanism that maintains the normal CSF/plasma Mg ratio and prevents rapid equilibration with plasma probably involves active transport of this ion reinforced by the findings that normally a significantly higher concentration of Mg is maintained in CSF as compared to plasma and there is a tendency to maintain a fixed CSF concentration in face of changing plasma levels [230]. Additionally, with abnormally elevated plasma total magnesium level, a new steady state ratio between plasma and CSF is reached and limits the amount of the Mg entering the CSF, even in conditions, such as aSAH, traumatic brain injury, and stroke, where the BBB is expected to be disrupted [228, 230, 232].

Even though the rise in CSF magnesium concentration is modest as a whole, magnesium level is selectively increased in regions of pathology, such as seen in ischemic stroke [233] and seizures [234] from induced hypermagnesemia. To date, no study has examined the spatial distribution of magnesium in aSAH patients.

Since previous in vitro studies on magnesium neuroprotection, such as the antispasmodic properties, require a CSF Mg^{2+} concentration of approximately 10 mg/dL to take effect [127, 131, 235, 236] and that the low BBB penetration of systemically infused magnesium sulfate is unable to achieve the desired CSF concentration without clinical toxicity, scientists have been propelled to explore topical routes of administration. Mori et al. led the way by applying intracisternal route of administration and demonstrated reversal of angiographic vasospasm and improvement of cerebral blood flow measured with autoradiographic technique. Such effect required CSF Mg^{2+} concentration of greater than 3 mg/dL in rat [153] and canine [237] models of aSAH. In a recent clinical study, similar cerebrovascular results were produced by the intracisternal route achieving CSF Mg^{2+} level of 6.0 mg/dL [238].

Neuraxial administration of magnesium has long been explored in the realm of anesthesiology for perioperative analgesia, where it appears to be safe. A number of animal studies reported no injury associated with neuraxial isotonic magnesium sulfate administration [239–245]. Derived from a canine study [239], the extrapolated theoretical toxic intrathecal dose is over 500 mg in humans [246]. Of two case reports of intrathecal dose of over 500 mg in human parturients [247, 264], neither was associated with any permanent sequelae. Although in the case of inadvertent subarachnoid cistern infusion of 1,000-mg magnesium sulfate, the patient developed lower extremity weakness for 5 h followed by complete and uneventful recovery [264].

Clinical Trials

Encouraged by the overwhelmingly positive evidence in preclinical studies accompanied by the lack of major adverse effects and the readily availability, magnesium was "the face that launched a thousand ships." In the past 10 years, a dozen of pilot studies followed by four randomized controlled clinical trials which met a Jadad score of 3 or above were conducted [248]. For an overview, see Table 19.1.

Table 19.1 Clinical trials of magnesium in subarachnoid hemorrhage

Trial	Design	n	Start time	Duration	Bolus dose and target range	Results	Adverse effects
Pilot study							
Boet and Mee [249]	Monocenter, uncontrolled	10	<5 days	10 days	Bolus: 20 mM/20 min Target: 2x baseline or 2.0–2.5 mmol/L	TCD vasospasm (n=5) DIND (n=3) Long-term morbidity (n=2)	No serious side effects
Chia [250]	Monocenter, historical controlled	23	<5 days	Until discharge from ICU	Bolus: none Target: 1.0–1.5 mM/L	Significant less angio vasospasm No change in neurological outcome	No serious side effects
van den Bergh [251]	Dose finding study	14	<2 days	14 days	3 groups for dosing finding. Highest dose group: bolus 16 mM	Dose finding	No serious side effects
van Norden [252]	Monocenter, randomized, double-blind, placebo controlled	94	<4 days	18 days	Bolus: none Target: 1.0–2.0 mM/L	Safety study	Termination (n=6) due to nausea, headache, and renal failure with hypermagnesemia >2.5 (n=3); nimodipine-induced hypotension (n=3)
Yahia [253]	Monocenter, uncontrolled	19	<3 days	10 days	Bolus: none Target: 1.5–4.0 mM/L	DIND (n=2) TCD vasospasm (n=5) Angio vasospasm (n=9) GOS 4–5 (n=18)	No serious side effects
Prevedello [254]	Monocenter, controlled	72	<1 day in ICU	Until discharge from ICU	Bolus: 20 mM/20 min Target: 2x baseline, 1.8–2.6 mM/L	Shorter admission time if developed TCD vasospasm	No serious side effects
Schmid-Elsaesser [255]	Monocenter, randomized, magnesium versus nimodipine controlled	104	<4 days	7 days	Bolus: 10 mg/kg for 30 min Target: not controlled	No difference vs. nimodipine	No serious side effects
Stippler [256]	Monocenter, historical controlled	76	<2 days	12 days	Bolus: none Target: not controlled	Significantly reduced DIND Trend to improved outcome (mRS) No change in mortality and GOS	No serious side effects

Veyna [257]	Monocenter, prospective, randomized, single-blind, placebo controlled	40	<3 days	10 days	Bolus: 20 mM/30 min Target: 4–5.5 mg/dL	Trend to better GOS 4–5	No serious side effects
Muroi [258]	Monocenter, prospective, randomized, single-blind, placebo controlled	58	<3 days	12 days	Bolus: 16 mM/15 min Target: 2x baseline, <2.0 mM/L	Trend to fewer TCD vasospasm and DCI Significant better GOS by 3 months	Dose-dependent hypocalcemia, hypotension (n = 1), bradycardia, and atrial fibrillation (n = 1); hypermagnesemia from renal failure (n = 3)
Randomized controlled clinical trials, Jadad criteria 3							
Van den Bergh (MASH trial) [259]	Multicenter, randomized, double-blind, placebo controlled	283	<4 days	14 days	Bolus: none Target: 1.0–2.0 mM/L	Significant increase excellent outcome (mRS 0). Trends to reduced DIND by 34%, poor outcome (mRS ≤4) by 23% by 3 months	No serious side effects
Wong [260]	Monocenter, prospective, randomized, double-blind, placebo controlled	60	<2 days	14 days	Bolus: 20 mM/20 min Target: 2x baseline, <2.5 mM/L	No change in functional outcome (BI) Trend to reduced incidence and duration of TCD vasospasm	No serious side effects
Wong et al. (IMASH III) [261]	Multicenter phase III, randomized, double-blind, placebo controlled	387	<2 days	14 days	Bolus: 20 mM/30 min Target: 2x baseline, <2.5 mM/L	No difference vs. control in outcome, DCI and DIND	Severe limb weakness (n = 1) Severe hypocalcemia (n = 1)
Westermaier [262]	Monocenter, randomized, placebo controlled	110	NA	10 days	Bolus: 16 mM/30 min Target: 2.0–2.5 mM/L	Reduced DIND and DCI	No serious side effects

TCD, transcranial Doppler; *DIND*, delayed ischemic neurological deficit; *DCI*, delayed cerebral infarction; *Angio*, angiography; *GOS*, Glasgow outcome score; *mRS*, modified Rankin score; *BI*, Barthel index; *ICU*, intensive care unit; *IV*, intravenous

All of the mentioned studies recruited patients with aSAH and initiated intravenous magnesium sulfate infusion in the 48–120-h window and maintained an elevated serum magnesium level in the 1.0–2.5-mM/L range for 7–18 days. "Hard" primary clinical endpoints measured using Glasgow outcome score (GOS) or modified Rankin scale (mRS) were reported in six studies at 3 months [248, 257, 259] or 6 months [260–262]. Reports of pilot clinical trials suggest that magnesium trends toward having beneficial effects on clinical outcomes and some of the often-measured parameters for assessing the well-being of aSAH patients, such as the decreased incidence of TCD-verified vasospasm. However, caution should be exercised on the interpretation of these results, as there may be a publication bias toward reporting of positive trends so that positive pilot studies outweigh the negatives.

The heterogeneity of trial designs raises further concern. Across the different studies, regimens and the time points of initiating magnesium therapy varied widely; the clinical relevance of some of the measured primary paraclinical parameters is uncertain, and clinical outcome definitions were not in accord until after consensus reached in 2010 [43]. These shortcomings make pooling of data statistically difficult.

Given the results of promising pilot trials, two landmark clinical trials, the MASH II and IMASH III trials, were conducted. The Asian-Australian Intravenous Magnesium Sulfate for Aneurysmal Subarachnoid Hemorrhage (IMASH) III trial was a multicenter, randomized, double-blinded, saline-controlled phase III clinical trial [261]. A total of 387 patients were randomized within 48 h, for whom magnesium sulfate or saline infusion was initiated after confirmation of aSAH by CT or digital subtraction angiography (DSA). The time from ictus to the start of drug infusion was 31.7 h with a standard deviation of 15.5 h. Twenty-mM intravenous bolus of MgSO4 was administrated over 30 min followed by a sustained plasma total magnesium concentration of twice the baseline level up to 2.5 mM/L. In the magnesium-treated group, two patients terminated treatment from magnesium-related side effects, where one developed severe limb weakness and the other severe hypocalcemia. The difference in the incidence of hypotensive episodes (systolic blood pressure <90 mmHg) was statistically insignificant between the MgSO4- (15%) and saline-placebo-treated groups (13%). The percentages of patients with a favorable outcome (6-month GOSE 5–8), excellent outcome (mRS 0–1), and good outcome (mRS 0–2) were almost identical. The proportions of patients with DIND- and TCD-diagnosed vasospasm were also similar.

The Magnesium and Acetylsalicylic acid in Subarachnoid Hemorrhage (MASH) II trial conducted by van den Bergh et al. was a randomized, double-blinded, saline-controlled multicenter phase II trial, which included 283 patients [259]. The median time for the start of treatment was 28 h, where no bolus infusion was given and targeted serum magnesium level was in the 1.0–2.0-mM/L range. No major side effects were observed. Magnesium sulfate infusion led to statistically significant increase in excellent outcome (mRS 0), while trends toward reduction in DIND by 34%, risk reduction of poor outcome (mRS ≤4) by 23% at 3 months, and relative risk for excellent outcome by 3.4 at 3 months were reported. The lack of statistical significance of the latter finding may be the result of small sample size of the study. We await the results of MASH phase III trial in 2014 with a sample size of 1,200 patients and the primary clinical outcome being mRS at 3 months.

Nevertheless, in the recent meta-analysis performed by Wong et al. [248] consisting of four high-quality clinical trials (Jadad score ≥3), the authors conclude that current data does not lend support to a beneficial effect of intravenous magnesium sulfate infusion in leading to better clinical outcome and reducing DIND.

Summary and Future Perspectives

While our understanding of the pathophysiology of DIND has progressed significantly, this knowledge has, however, not been translated into clinically effective therapy. Possible causes of this mismatch include the multifactorial nature and complex pathophysiology of the disease. Indeed, even though aSAH is an entity in itself, it shares similarities to many other acute brain disorders, such as the mechanical forces at the time of aneurysmal rupture resulting in direct mechanical damage to the brain through shearing, tearing, and stretching seen in traumatic brain injury; ischemia and reperfusion evident in ischemic stroke; and blood clot formation present in intracerebral hemorrhage.

In this chapter, we explored aSAH-induced primary brain injury and the sequence of cellular, neurochemical, metabolic, and electrophysiological alterations initiated by aSAH that continue to develop over time leading to early brain injury. Manifested pathophysiology partially related to the hemolysis process includes reactive oxygen species formation, disrupted cellular metabolism, neuroinflammation, secondary vasoconstriction or delayed vasospasm, alternation of NO levels, and microthrombosis. All of which may be implicated in cortical spreading depolarization, excitotoxicity, and spreading ischemia and ultimately cause cell death. Magnesium may exert beneficial effects on aSAH-induced brain injury via multimodal mechanisms, where most of them relies upon its calcium antagonistic and membrane stabilizing prosperities, such that it noncompetitively blocks NMDA receptor and voltage-gated calcium channels, blocks free radical production in the mitochondria, inhibits ischemia-induced glutamate release, relaxes vascular smooth muscles, prevents platelet aggregation, and suppresses cortical spreading depolarizations.

In search of the Holy Grail to improved outcome, 50 or so hopeful neuroprotective therapies have been tested, where most aimed to reverse delayed proximal vasospasm, and yet none was unequivocally beneficial in humans. Given the failure of current therapeutic focus and the numerous identified molecular events in the ischemic cascade, it is clear that alleviation of a single pathological process may not provide sufficient protection, and an agent such as magnesium that inhibits multiple injurious processes may be needed.

Nevertheless, past clinical trials have been disappointing. There are three potential design defects in previous studies: (1) failure to achieve therapeutic dose in the CSF because of limiting systemic adverse effects, (2) failure to treat patients early enough after aSAH onset, and (3) failure to employ sample size large enough to detect modest benefits of magnesium therapy.

Previous clinical trials have mostly kept serum magnesium level in the range from 1.0 to 2.5 mM/L. Although it remains possible that higher target magnesium levels such as 3.0 mM/L can lead to improvement in functional outcome, increased dosage is limited by cardiac suppression and hypotensive episodes. Given that the CNS bioavailability of systemically infused magnesium in patients with acute brain injury is limited, direct intracisternal infusion should be considered to circumvent the complication of systemic toxicity and to achieve a CSF level adequate in promoting neuroprotective effects [238].

A new line of thought in treating aSAH approaches the adage held in the management of ischemic stroke, where "time is brain." As alluded to earlier, many pathological processes are activated in the initial hours of aneurysmal rupture and may evolve with time, which emphasizes on the effort at preventing the progression of "early brain injury" promptly. The MASH II and IMASH III trials initiated therapy with a median time of 28 and 32 h from aSAH ictus, respectively, with the primary objective at treating DIND rather than preventing the hyperacute pathological processes from occurring. A possible explanation of the lack of effect could therefore be the time delay. The easily recognizable clinical presentation of aSAH and the well-established safety profile of magnesium make it an appealing agent for infield therapy before definitive diagnosis is made and could be accomplished within 1–2 h or less from aSAH ictus. In the recently published pilot study of paramedic-initiated intravenous magnesium sulfate for the FAST-MAG trial (Field Administration of Stroke Therapy) for acute ischemic stroke patients, the treatment was shown to be safe and feasible [263]. Whether earlier administration of magnesium can improve clinical outcome remains to be answered.

References

1. Nieuwkamp DJ, de Gans K, Algra A, et al. Timing of aneurysm surgery in subarachnoid haemorrhage–an observational study in The Netherlands. Acta Neurochir. 2005;147(8):815–21.
2. Hop JW, Rinkel GJ, Algra A, van Gijn J. Case-fatality rates and functional outcome after subarachnoid hemorrhage: a systematic review. Stroke. 1997;28(3):660–4.

3. Sudlow CL, Warlow CP. Comparable studies of the incidence of stroke and its pathological types: results from an international collaboration. International Stroke Incidence Collaboration. Stroke. 1997;28(3):491–9.
4. Huang CY, Chan FL, Yu YL, Woo E, Chin D. Cerebrovascular disease in Hong Kong Chinese. Stroke. 1990;21(2):230–5.
5. Johnston SC, Selvin S, Gress DR. The burden, trends, and demographics of mortality from subarachnoid hemorrhage. Neurology. 1998;50(5):1413–8.
6. Hall JR. Impact of traumatic subarachnoid hemorrhage on outcome in nonpenetrating head injury. Part II: relationship to clinical course and outcome variables during acute hospitalization. J Trauma. 1997;42(6):1196–7.
7. Taneda M, Kataoka K, Akai F, Asai T, Sakata I. Traumatic subarachnoid hemorrhage as a predictable indicator of delayed ischemic symptoms. J Neurosurg. 1996;84(5):762–8.
8. Wirth FP. Surgical treatment of incidental intracranial aneurysms. Clin Neurosurg. 1986;33:125–35.
9. McCormick WF, Acosta-Rua GJ. The size of intracranial saccular aneurysms. An autopsy study. J Neurosurg. 1970;33(4):422–7.
10. Haley Jr EC, Kassell NF, Torner JC. The International Cooperative Study on the Timing of Aneurysm Surgery. The North American experience. Stroke. 1992;23(2):205–14.
11. Adams Jr HP, Kassell NF, Torner JC. Usefulness of computed tomography in predicting outcome after aneurysmal subarachnoid hemorrhage: a preliminary report of the Cooperative Aneurysm Study. Neurology. 1985;35(9):1263–7.
12. Dorhout Mees SM, Rinkel GJ, Feigin VL, et al. Calcium antagonists for aneurysmal subarachnoid haemorrhage. Cochrane Database Syst Rev. 2007;18(3):CD000277.
13. Kassell NF, Helm G, Simmons N, Phillips CD, Cail WS. Treatment of cerebral vasospasm with intra-arterial papaverine. J Neurosurg. 1992;77(6):848–52.
14. Newell DW, Eskridge JM, Aaslid R. Current indications and results of cerebral angioplasty. Acta Neurochir Suppl. 2001;77:181–3.
15. Findlay JM, Kassell NF, Weird BK, et al. A randomized trial of intraoperative, intracisternal tissue plasminogen activator for the prevention of vasospasm. Neurosurgery. 1995;37(1):168–76.
16. Kodama N, Sasaki T, Kawakami M, Sato M, Asari J. Cisternal irrigation therapy with urokinase and ascorbic acid for prevention of vasospasm after aneurysmal subarachnoid hemorrhage. Outcome in 217 patients. Surg Neurol. 2000;53(2):110–7.
17. Vega-Basulto SD, Lafontaine-Terry E, Gutie Rrez-Muñoz FG, Roura-Carrasco J, Pardo-Camacho G. Intracranial hemorrhage due to aneurysms and arteriovenous malformations during pregnancy and puerperium. Neurocirugia. 2008;19(1):25–34.
18. Ruggieri PM, Poulos N, Masaryk TJ, et al. Occult intracranial aneurysms in polycystic kidney disease: screening with MR angiography. Radiology. 1994;191(1):33–9.
19. Alkan T, Tureyen K, Ulutas M, et al. Acute and delayed vasoconstriction after subarachnoid hemorrhage: local cerebral blood flow, histopathology, and morphology in the rat basilar artery. Arch Physiol Biochem. 2001;109(2):145–53.
20. Bederson JB, Levy AL, Ding WH, et al. Acute vasoconstriction after subarachnoid hemorrhage. Neurosurgery. 1998;42(2):352–60.
21. Bederson JB, Germano IM, Guarino L. Cortical blood flow and cerebral perfusion pressure in a new noncraniotomy model of subarachnoid hemorrhage in the rat. Stroke. 1995;26(6):1086–91.
22. Grote E, Hassler W. The critical first minutes after subarachnoid hemorrhage. Neurosurgery. 1988;22(4):654–61.
23. Jackowski A, Crockard A, Burnstock G, Russell RR, Kristek F. The time course of intracranial pathophysiological changes following experimental subarachnoid haemorrhage in the rat. J Cereb Blood Flow Metab. 1990;10(6):835–49.
24. Prunell GF, Mathiesen T, Diemer NH, Svendgaard NA. Experimental subarachnoid hemorrhage: subarachnoid blood volume, mortality rate, neuronal death, cerebral blood flow, and perfusion pressure in three different rat models. Neurosurgery. 2003;52(1):165–75.
25. Frontera JA, Fernandez A, Schmidt JM, et al. Defining vasospasm after subarachnoid hemorrhage: what is the most clinically relevant definition? Stroke. 2009;40(6):1963–8.
26. Dorsch NW. Therapeutic approaches to vasospasm in subarachnoid hemorrhage. Curr Opin Crit Care. 2002;8(2):128–33.
27. Kassell NF, Sasaki T, Colohan AR, Nazar G. Cerebral vasospasm following aneurysmal subarachnoid hemorrhage. Stroke. 1985;16(4):562–72.
28. Tettenborn D, Dycka J. Prevention and treatment of delayed ischemic dysfunction in patients with aneurysmal subarachnoid hemorrhage. Stroke. 1990;21(12 Suppl):IV85–9.
29. Etminan N, Vergouwen MD, Ilodigwe D, Macdonald RL. Effect of pharmaceutical treatment on vasospasm, delayed cerebral ischemia, and clinical outcome in patients with aneurysmal subarachnoid hemorrhage: a systematic review and meta-analysis. J Cereb Blood Flow Metab. 2011;31(6):1443–51.

30. Roos YB, de Haan RJ, Beenen LF, et al. Complications and outcome in patients with aneurysmal subarachnoid haemorrhage: a prospective hospital based cohort study in the Netherlands. J Neurol Neurosurg Psychiatry. 2000;68(3):337–41.
31. Vora YY, Suarez-Almazor M, Steinke DE, Martin ML, Findlay JM. Role of transcranial Doppler monitoring in the diagnosis of cerebral vasospasm after subarachnoid hemorrhage. Neurosurgery. 1999;44(6):1237–47.
32. Fisher CM, Roberson GH, Ojemann RG. Cerebral vasospasm with ruptured saccular aneurysm–the clinical manifestations. Neurosurgery. 1977;1(3):245–8.
33. Vergouwen MD, Ilodigwe D, Macdonald RL. Cerebral infarction after subarachnoid hemorrhage contributes to poor outcome by vasospasm-dependent and -independent effects. Stroke. 2011;42(4):924–9.
34. Rabinstein AA, Friedman JA, Weigand SD, et al. Predictors of cerebral infarction in aneurysmal subarachnoid hemorrhage. Stroke. 2004;35(8):1862–6.
35. Fergusen S, Macdonald RL. Predictors of cerebral infarction in patients with aneurysmal subarachnoid hemorrhage. Neurosurgery. 2007;60(4):658–67.
36. Alaraj A, Charbel FT, Amin-Hanjani S. Peri-operative measures for treatment and prevention of cerebral vasospasm following subarachnoid hemorrhage. Neurol Res. 2009;31(6):651–9.
37. Macdonald RL, Kassell NF, Mayer S, et al. Clazosentan to overcome neurological ischemia and infarction occurring after subarachnoid hemorrhage (CONSCIOUS-1): randomized, double-blind, placebo-controlled phase 2 dose-finding trial. Stroke. 2008;39(11):3015–21.
38. Macdonald RL, Higashida RT, Keller E, et al. Clazosentan, an endothelin receptor antagonist, in patients with aneurysmal subarachnoid haemorrhage undergoing surgical clipping: a randomised, double-blind, placebo-controlled phase 3 trial (CONSCIOUS-2). Lancet Neurol. 2011;10(7):618–25.
39. Bederson JB, Connolly ES, Batjer HH, et al. Guidelines for the management of aneurysmal subarachnoid hemorrhage: a statement for healthcare professionals from a special writing group of the Stroke Council, American Heart Association. Stroke. 2009;40(3):994–1025.
40. Wong GK, Poon WS, Chan MT, et al. Plasma magnesium concentrations and clinical outcomes in aneurysmal subarachnoid hemorrhage patients: post hoc analysis of intravenous magnesium sulphate for aneurysmal subarachnoid hemorrhage trial. Stroke. 2010;41(8):1841–4.
41. Schmidt JM, Wartenberg KE, Fernandez A, et al. Frequency and clinical impact of asymptomatic cerebral infarction due to vasospasm after subarachnoid hemorrhage. J Neurosurg. 2008;109(6):1052–9.
42. Vergouwen MD, Etminan N, Ilodigwe D, Macdonald RL. Lower incidence of cerebral infarction correlates with improved functional outcome after aneurysmal subarachnoid hemorrhage. J Cereb Blood Flow Metab. 2011;31(7):1545–53.
43. Vergouwen MD, Vermeulen M, van Gijn J, et al. Definition of delayed cerebral ischemia after aneurysmal subarachnoid hemorrhage as an outcome event in clinical trials and observational studies: proposal of a multidisciplinary research group. Stroke. 2010;41(10):2391–5.
44. Ecker A, Riemenschneider PA. Arteriographic demonstration of spasm of the intracranial arteries, with special reference to saccular arterial aneurysms. J Neurosurg. 1951;8(6):660–7.
45. Okada Y, Shima T, Nishida M, et al. Comparison of transcranial Doppler investigation of aneurysmal vasospasm with digital subtraction angiographic and clinical findings. Neurosurgery. 1999;45(3):443–9.
46. Newell DW, Grady MS, Eskridge JM, Winn HR. Distribution of angiographic vasospasm after subarachnoid hemorrhage: implications for diagnosis by transcranial Doppler ultrasonography. Neurosurgery. 1990;27(4):574–7.
47. Lindegaard KF, Nornes H, Bakke SJ, Sorteberg W, Nakstad P. Cerebral vasospasm diagnosis by means of angiography and blood velocity measurements. Acta Neurochir. 1989;100(1–2):12–24.
48. Nornes H, Magnaes B. Intracranial pressure in patients with ruptured saccular aneurysm. J Neurosurg. 1972;36(5):537–47.
49. Asano T, Sano K. Pathogenetic role of no-reflow phenomenon in experimental subarachnoid hemorrhage in dogs. J Neurosurg. 1977;46(4):454–66.
50. Hayashi T, Suzuki A, Hatazawa J, et al. Cerebral circulation and metabolism in the acute stage of subarachnoid hemorrhage. J Neurosurg. 2000;93(6):1014–8.
51. Jakubowski J, Bell BA, Symon L, Zawirski MB, Francis DM. A primate model of subarachnoid hemorrhage: change in regional cerebral blood flow, autoregulation carbon dioxide reactivity, and central conduction time. Stroke. 1982;13(5):601–11.
52. Schmieder K, Möller F, Engelhardt M, et al. Dynamic cerebral autoregulation in patients with ruptured and unruptured aneurysms after induction of general anesthesia. Zentralbl Neurochir. 2006;67(2):81–7.
53. Pennings FA, Bouma GJ, Ince C. Direct observation of the human cerebral microcirculation during aneurysm surgery reveals increased arteriolar contractility. Stroke. 2004;35(6):1284–8.
54. Hatake K, Wakabayashi I, Kakishita E, Hishida S. Impairment of endothelium-dependent relaxation in human basilar artery after subarachnoid hemorrhage. Stroke. 1992;23(8):1111–6.

55. Sehba FA, Flores R, Muller A, et al. Adenosine A(2A) receptors in early ischemic vascular injury after subarachnoid hemorrhage. Laboratory investigation. J Neurosurg. 2010;113(4):826–34.
56. Uhl E, Lehmberg J, Steiger HJ, Messmer K. Intraoperative detection of early microvasospasm in patients with subarachnoid hemorrhage by using orthogonal polarization spectral imaging. Neurosurgery. 2003; 52(6):1307–15.
57. Sehba FA, Ding WH, Chereshnev I, Bederson JB. Effects of S-nitrosoglutathione on acute vasoconstriction and glutamate release after subarachnoid hemorrhage. Stroke. 1999;30(9):1955–61.
58. Sehba FA, Friedrich V, Makonnen G, Bederson JB. Acute cerebral vascular injury after subarachnoid hemorrhage and its prevention by administration of a nitric oxide donor. J Neurosurg. 2007;106(2):321–9.
59. Chrissobolis S, Miller AA, Drummond GR, Kemp-Harper BK, Sobey CG. Oxidative stress and endothelial dysfunction in cerebrovascular disease. Front Biosci. 2011;16:1733–45.
60. Gaetani P, Marzatico F, Rodriguez y Baena R, et al. Arachidonic acid metabolism and pathophysiologic aspects of subarachnoid hemorrhage in rats. Stroke. 1990;21(2):328–32.
61. Marzatico F, Gaetani P, Cafè C, Spanu G, Rodriguez y Baena R. Antioxidant enzymatic activities after experimental subarachnoid hemorrhage in rats. Acta Neurol Scand. 1993;87(1):62–6.
62. Lin CL, Hsu YT, Lin TK, et al. Increased levels of F2-isoprostanes following aneurysmal subarachnoid hemorrhage in humans. Free Radic Biol Med. 2006;40(8):1466–73.
63. Kennedy TP, Rao NV, Hopkins C, et al. Role of reactive oxygen species in reperfusion injury of the rabbit lung. J Clin Invest. 1989;83(4):1326–35.
64. Provencio JJ, Fu X, Siu A, et al. CSF neutrophils are implicated in the development of vasospasm in subarachnoid hemorrhage. Neurocrit Care. 2010;12(2):244–51.
65. Zhang W, Li P, Hu X, et al. Omega-3 polyunsaturated fatty acids in the brain: metabolism and neuroprotection. Front Biosci. 2011;17:2653–70.
66. Dickens BF, Weglicki WB, Li YS, Mak IT. Magnesium deficiency in vitro enhances free radical-induced intracellular oxidation and cytotoxicity in endothelial cells. FEBS Lett. 1992;311(3):187–91.
67. Rayssiguier Y, Gueux E, Bussière L, Durlach J, Mazur A. Dietary magnesium affects susceptibility of lipoproteins and tissues to peroxidation in rats. J Am Coll Nutr. 1993;12(2):133–7.
68. Garcia LA, Dejong SC, Martin SM, et al. Magnesium reduces free radicals in an in vivo coronary occlusion-reperfusion model. J Am Coll Cardiol. 1998;32(2):536–9.
69. Reinhart RA. Magnesium metabolism. A review with special reference to the relationship between intracellular content and serum levels. Arch Intern Med. 1988;148(11):2415–20.
70. Träuble H, Eibl H. Electrostatic effects on lipid phase transitions: membrane structure and ionic environment. Proc Natl Acad Sci USA. 1974;71(1):214–9.
71. van den Bergh WM, Algra A, van der Sprenkel JW, Tulleken CA, Rinkel GJ. Hypomagnesemia after aneurysmal subarachnoid hemorrhage. Neurosurgery. 2003;52(2):276–82.
72. Altura BM. Introduction: importance of Mg in physiology and medicine and the need for ion selective electrodes. Scan J Clin Lab Invest Suppl. 1994;217:5–9.
73. Marzatico F, Gaetani P, Silvani V, et al. Experimental isobaric subarachnoid hemorrhage: regional mitochondrial function during the acute and late phase. Surg Neurol. 1990;34(5):294–300.
74. Marzatico F, Gaetani P, Rodriguez y Baena R, et al. Bioenergetics of different brain areas after experimental subarachnoid hemorrhage in rats. Stroke. 1988;19(3):378–84.
75. Rodriguez y Baena R, Gaetani P, Silvani V, Spanu G, Marzatico F. Effect of nimodipine on mitochondrial respiration in different rat brain areas after subarachnoid haemorrhage. Acta Neurochir Suppl. 1988;43:177–81.
76. d'Avella D, Cicciarello R, Zuccarello M, et al. Brain energy metabolism in the acute stage of experimental subarachnoid haemorrhage: local changes in cerebral glucose utilization. Acta Neurochir. 1996;138(6):737–43.
77. Xu M, Dai W, Deng X. Effects of magnesium sulfate on brain mitochondrial respiratory function in rats after experimental traumatic brain injury. Chin J Traumatol. 2002;5(6):361–4.
78. Lin JY, Chung SY, Lin MC, Cheng FC. Effects of magnesium sulfate on energy metabolites and glutamate in the cortex during focal cerebral ischemia and reperfusion in the gerbil monitored by a dual-probe microdialysis technique. Life Sci. 2002;71(7):803–11.
79. Aoki K, Zubkov AY, Ross IB, Zhang JH. Therapeutic effect of caspase inhibitors in the prevention of apoptosis and reversal of chronic cerebral vasospasm. J Clin Neurosci. 2002;9(6):672–7.
80. Nau R, Haase S, Bunkowski S, Brück W. Neuronal apoptosis in the dentate gyrus in humans with subarachnoid hemorrhage and cerebral hypoxia. Brain Pathol. 2002;12(3):329–36.
81. Zoratti M, Szabo I. The mitochondrial permeability transition. Biochim Biophys Acta. 1995;1241(2):139–76.
82. Schinder AF, Olson EC, Spitzer NC, Montal M. Mitochondrial dysfunction is a primary event in glutamate neurotoxicity. J Neurosci. 1996;16(19):6125–33.
83. Zamzami N, Hirsch T, Dallaporta B, Petit PX, Kroemer G. Mitochondrial implication in accidental and programmed cell death: apoptosis and necrosis. J Bioenerg Biomembr. 1997;29(2):185–93.

84. Zamzami N, Susin SA, Marchetti P, et al. Mitochondrial control of nuclear apoptosis. J Exp Med. 1996;183(4):1533–44.
85. Gottlieb RA. Mitochondria and apoptosis. Biol Signals Recept. 2001;10(3–4):147–61.
86. Kroemer G, Dallaporta B, Resche-Rigon M. The mitochondrial death/life regulator in apoptosis and necrosis. Annu Rev Physiol. 1998;60:619–42.
87. Mignotte B, Vayssiere JL. Mitochondria and apoptosis. Eur J Biochem. 1998;252(1):1–15.
88. Martin H, Richert L, Berthelot A. Magnesium deficiency induces apoptosis in primary cultures of rat hepatocytes. J Nutr. 2003;133(8):2505–11.
89. Türkyilmaz C, Türkyilmaz Z, Atalay Y, Söylemezoglu F, Celasun B. Magnesium pre-treatment reduces neuronal apoptosis in newborn rats in hypoxia-ischemia. Brain Res. 2002;955(1–2):133–7.
90. Lee JS, Han YM, Yoo DS, et al. A molecular basis for the efficacy of magnesium treatment following traumatic brain injury in rats. J Neurotrauma. 2004;21(5):549–61.
91. Tang YN, Zhao FL, Ye HM. Expression of caspase-3 mRNA in the hippocampus of seven-day-old hypoxic-ischemic rats and the mechanism of neural protection with magnesium sulfate. Zhonghua Er Ke Za Zhi. 2003;41(3):212–4.
92. Dubinsky JM, Brustovetsky N, Pinelis V, Kristal BS, Herman C, Li X. The mitochondrial permeability transition: the brain's point of view. Biochem Soc Symp. 1999;66:75–84.
93. Okiyama K, Smith DH, Gennarelli TA, et al. The sodium channel blocker and glutamate release inhibitor BW1003C87 and magnesium attenuate regional cerebral edema following experimental brain injury in the rat. J Neuroschem. 1995;64(2):802–9.
94. Rousseaux P, Scherpereel B, Bernard MH, Graftieaux JP, Guyot JF. Fever and cerebral vasospasm in ruptured intracranial aneurysms. Surg Neurol. 1980;14(6):459–65.
95. Neil-Dwyer G, Cruickshank J. The blood leucocyte count and its prognostic significance in subarachnoid haemorrhage. Brain. 1974;97(1):79–86.
96. Spallone A, Acqui M, Pastore FS, Guidetti B. Relationship between leukocytosis and ischemic complications following aneurysmal subarachnoid hemorrhage. Surg Neurol. 1987;27(3):253–8.
97. Maiuri F, Gallicchio B, Donati P, Carandente M. The blood leukocyte count and its prognostic significance in subarachnoid hemorrhage. J Neurosurg Sci. 1987;31(2):45–8.
98. Pellettieri L, Nilsson B, Carlsson CA, Nilsson U. Serum immunocomplexes in patients with subarachnoid hemorrhage. Neurosurgery. 1986;19(5):767–71.
99. Ostergaard JR, Kristensen BO, Svehag SE, Teisner B, Miletic T. Immune complexes and complement activation following rupture of intracranial saccular aneurysms. J Neurosurg. 1987;66(6):891–7.
100. Kasuya H, Shimizu T. Activated complement components C3a and C4a in cerebrospinal fluid and plasma following subarachnoid hemorrhage. J Neurosurg. 1989;71(5 Pt 1):741–6.
101. Polin RS, Bavbek M, Shaffrey ME, et al. Detection of soluble E-selectin, ICAM-1, VCAM-1, and L-selectin in the cerebrospinal fluid of patients after subarachnoid hemorrhage. J Neurosurg. 1998;89(4):559–67.
102. Mack WJ, Mocco J, Hoh DJ, et al. Outcome prediction with serum intercellular adhesion molecule-1 levels after aneurysmal subarachnoid hemorrhage. J Neurosurg. 2002;96(1):71–5.
103. Mocco J, Mack WJ, Kim GH, et al. Rise in serum soluble intercellular adhesion molecule-1 levels with vasospasm following aneurysmal subarachnoid hemorrhage. J Neurosurg. 2002;97(3):537–41.
104. Aihara Y, Kasuya H, Onda H, Hori T, Takeda J. Quantitative analysis of gene expressions related to inflammation in canine spastic artery after subarachnoid hemorrhage. Stroke. 2001;32(1):212–7.
105. Handa Y, Kubota T, Kaneko M, et al. Expression of intercellular adhesion molecule 1 (ICAM-1) on the cerebral artery following subarachnoid haemorrhage in rats. Acta Neurochir. 1995;132(1–3):92–7.
106. Lin C-L, Dumont AS, Calisaneller T, et al. Monoclonal antibody against E selectin attenuates subarachnoid hemorrhage-induced cerebral vasospasm. Surg Neurol. 2005;64(3):201–5.
107. Germanò A, d'Avella D, Imperatore C, Caruso G, Tomasello F. Time-course of blood-brain barrier permeability changes after experimental subarachnoid haemorrhage. Acta Neurochir. 2000;142(5):575–80.
108. Bertsch T, Kuehl S, Muehlhauser F, et al. Source of endothelin-1 in subarachnoid hemorrhage. Clin Chem Lab Med. 2001;39(4):341–5.
109. Fassbender K, Hodapp B, Rossol S, et al. Endothelin-1 in subarachnoid hemorrhage: an acute-phase reactant produced by cerebrospinal fluid leukocytes. Stroke. 2000;31(12):2971–5.
110. Fassbender K, Hodapp B, Rossol S, et al. Inflammatory cytokines in subarachnoid haemorrhage: association with abnormal blood flow velocities in basal cerebral arteries. J Neurol Neurosurg Psychiatry. 2001;70(4):534–7.
111. Juvela S. Plasma endothelin and big endothelin concentrations and serum endothelin-converting enzyme activity following aneurysmal subarachnoid hemorrhage. J Neurosurg. 2002;97(6):1287–93.
112. Shogi T, Oono H, Nakagawa M, et al. Effects of a low extracellular magnesium concentration and endotoxin on IL-1beta and TNF-alpha release from, and mRNA levels in, isolated rat alveolar macrophages. Magnes Res. 2002;15(3–4):147–52.

113. Kaya M, Küçük M, Kalayci RB, et al. Magnesium sulfate attenuates increased blood-brain barrier permeability during insulin-induced hypoglycemia in rats. Can J Physiol Pharmacol. 2001;79(9):793–8.
114. Kagstrom EGT, Hanson J, Galera R. Changes in cerebral blood flow after subarachnoid haemorrhage. Excerpta Medica Int Cong Series. 1966;110:629–33.
115. Bergvall U, Steiner L, Forster DM. Early pattern of cerebral circulatory disturbances following subarachnoid haemorrhage. Neuroradiology. 1973;5(1):24–32.
116. Bergvall U, Galera R. Time relationship between subarachnoid haemorrhage, arterial spasm, changes in cerebral circulation and posthaemorrhagic hydrocephalus. Acta Radiol Diagn. 1969;9:229–37.
117. Weir B, Grace M, Hansen J, Rothberg C. Time course of vasospasm in man. J Neurosurg. 1978;48(2):173–8.
118. Sasaki T, Kassell NF, Zuccarello M, et al. Barrier disruption in the major cerebral arteries during the acute stage after experimental subarachnoid hemorrhage. Neurosurgery. 1986;19(2):177–84.
119. Zubkov AY, Tibbs RE, Clower B, et al. Morphological changes of cerebral arteries in a canine double hemorrhage model. Neurosci Lett. 2002;326(2):137–41.
120. Friedrich V, Flores R, Muller A, Sehba FA. Escape of intraluminal platelets into brain parenchyma after subarachnoid hemorrhage. Neuroscience. 2010;165(3):968–75.
121. Park KW, Metais C, Dai HB, Comunale ME, Sellke FW. Microvascular endothelial dysfunction and its mechanism in a rat model of subarachnoid hemorrhage. Anesth Analg. 2001;92(4):990–6.
122. Bevan JA, Bevan RD, Walters CL, Wellman T. Functional changes in human pial arteries (300 to 900 micrometer ID] within 48 hours of aneurysmal subarachnoid hemorrhage. Stroke. 1998;29(12):2575–9.
123. Macdonald RL, Weir BK. A review of hemoglobin and the pathogenesis of cerebral vasospasm. Stroke. 1991;22(8):971–82.
124. Brouwers PJ, Dippel DW, Vermeulen M, et al. Amount of blood on computed tomography as an independent predictor after aneurysm rupture. Stroke. 1993;24(6):809–14.
125. Pluta RM, Afshar JK, Boock RJ, Oldfield EH. Temporal changes in perivascular concentrations of oxyhemoglobin, deoxyhemoglobin, and methemoglobin after subarachnoid hemorrhage. J Neurosurg. 1998;88(3):557–61.
126. Cook DA, Vollrath B. Free radicals and intracellular events associated with cerebrovascular spasm. Cardiovasc Res. 1995;30(4):493–500.
127. Ram Z, Sadeh M, Shacked I, Sahar A, Hadani M. Magnesium sulfate reverses experimental delayed cerebral vasospasm after subarachnoid hemorrhage in rats. Stroke. 1991;22(7):922–7.
128. Pyne GJ, Cadoux-Hudson TA, Clark JF. Magnesium protection against in vitro cerebral vasospasm after subarachnoid haemorrhage. Br J Neurosurg. 2001;15(5):409–15.
129. Miura K. Changes in Mg++ concentration of CSF after subarachnoid hemorrhage and Mg++–effects on the contractions of bovine cerebral artery. No Shinkei Geka. 1988;16(11):1251–9.
130. Altura BT, Altura BM. Interactions of Mg and K on cerebral vessels–aspects in view of stroke. Review of present status and new findings. Magnesium. 1984;3(4–6):195–211.
131. Seelig JM, Wei EP, Kontos HA, Choi SC, Becker DP. Effect of changes in magnesium ion concentration on cat cerebral arterioles. Am J Physiol. 1983;245(1):H22–6.
132. Torregrosa G, Perales AJ, Salom JB, et al. Different effects of Mg^{2+} on endothelin-1- and 5-hydroxytryptamine-elicited responses in goat cerebrovascular bed. J Cardiovasc Pharmacol. 1994;23(6):1004–10.
133. Rubanyi GM, Polokoff MA. Endothelins: molecular biology, biochemistry, pharmacology, physiology, and pathophysiology. Pharmacol Rev. 1994;46(3):325–415.
134. Zubkov AY, Rollins KS, Parent AD, Zhang J, Bryan RM. Mechanism of endothelin-1-induced contraction in rabbit basilar artery. Stroke. 2000;31(2):526–33.
135. Dawson DA, Sugano H, McCarron RM, Hallenbeck JM, Spatz M. Endothelin receptor antagonist preserves microvascular perfusion and reduces ischemic brain damage following permanent focal ischemia. Neurochem Res. 1999;24(12):1499–505.
136. Pluta RM, Boock RJ, Afshar JK, et al. Source and cause of endothelin-1 release into cerebrospinal fluid after subarachnoid hemorrhage. J Neurosurg. 1997;87(2):287–93.
137. Ohlstein EH, Storer BL. Oxyhemoglobin stimulation of endothelin production in cultured endothelial cells. J Neurosurg. 1992;77(2):274–8.
138. Zimmermann M, Seifert V. Endothelin and subarachnoid hemorrhage: an overview. Neurosurgery. 1998;43(4):863–75.
139. Jośko J, Hendryk S, Jedrzejowska-Szypuła H, et al. Influence endothelin ETA receptor antagonist–BQ-123–on changes of endothelin-1 level in plasma of rats with acute vasospasm following subarachnoid hemorrhage. J Physiol Pharmacol. 1998;49(3):367–75.
140. Vikman P, Beg S, Khurana TS, et al. Gene expression and molecular changes in cerebral arteries following subarachnoid hemorrhage in the rat. J Neurosurg. 2006;105(3):438–44.
141. Nakagomi T, Ide K, Yamakawa K, et al. Pharmacological effect of endothelin, an endothelium-derived vasoconstrictive peptide, on canine basilar arteries. Neurol Med Chir. 1989;29(11):967–74.

142. Alafaci C, Jansen I, Arbab MA, et al. Enhanced vasoconstrictor effect of endothelin in cerebral arteries from rats with subarachnoid haemorrhage. Acta Physiol Scand. 1990;138(3):317–9.
143. Alafaci C, Salpietro FM, Iacopino DG, Edvinsson L, Tomasello F. Endothelin: an endothelium-derived vasoactive peptide and its possible role in the pathogenesis of cerebral vasospasm. Ital J Neurol Sci. 1991;12(3 Suppl 11):55–8.
144. Papadopoulos SM, Gilbert LL, Webb RC, D'Amato CJ. Characterization of contractile responses to endothelin in human cerebral arteries: implications for cerebral vasospasm. Neurosurgery. 1990;26(5):810–5.
145. Kasuya H, Weir BK, White DM, Stefansson K. Mechanism of oxyhemoglobin-induced release of endothelin-1 from cultured vascular endothelial cells and smooth-muscle cells. J Neurosurg. 1993;79(6):892–8.
146. Altura BM, Zhang A, Cheng TP, Altura BT. Extracellular magnesium regulates nuclear and perinuclear free ionized calcium in cerebral vascular smooth muscle cells: possible relation to alcohol and central nervous system injury. Alcohol. 2001;23(2):83–90.
147. Berthon N, Laurant P, Fellmann D, Berthelot A. Effect of magnesium on mRNA expression and production of endothelin-1 in DOCA-salt hypertensive rats. J Cardiovasc Pharmacol. 2003;42(1):24–31.
148. Kemp PA, Gardiner SM, March JE, Rubin PC, Bennett T. Assessment of the effects of endothelin-1 and magnesium sulphate on regional blood flows in conscious rats, by the coloured microsphere reference technique. Br J Pharmacol. 1999;126(3):621–6.
149. Alborch E, Salom JB, Perales AJ, et al. Comparison of the anticonstrictor action of dihydropyridines (nimodipine and nicardipine) and Mg^{2+} in isolated human cerebral arteries. Eur J Pharmacol. 1992;229(1):83–9.
150. Perales AJ, Torregrosa G, Salom JB, et al. In vivo and in vitro effects of magnesium sulfate in the cerebrovascular bed of the goat. Am J Obstet Gynecol. 1991;165(5 Pt 1):1534–8.
151. Nadler JL, Goodson S, Rude RK. Evidence that prostacyclin mediates the vascular action of magnesium in humans. Hypertension. 1987;9(4):379–83.
152. Mori K, Miyazaki M, Hara Y, et al. Novel vasodilatory effect of intracisternal injection of magnesium sulfate solution on spastic cerebral arteries in the canine two-hemorrhage model of subarachnoid hemorrhage. J Neurosurg. 2009;110(1):73–8.
153. Mori K, Miyazaki M, Iwata J, Yamamoto T, Nakao Y. Intracisternal infusion of magnesium sulfate solution improved reduced cerebral blood flow induced by experimental subarachnoid hemorrhage in the rat. Neurosurg Rev. 2008;31(2):197–203.
154. Sehba FA, Schwartz AY, Chereshnev I, Bederson JB. Acute decrease in cerebral nitric oxide levels after subarachnoid hemorrhage. J Cereb Blood Flow Metab. 2000;20(3):604–11.
155. Yatsushige H, Calvert JW, Cahill J, Zhang JH. Limited role of inducible nitric oxide synthase in blood-brain barrier function after experimental subarachnoid hemorrhage. J Neurotrauma. 2006;23(12):1874–82.
156. Sakowitz OW, Wolfrum S, Sarrafzadeh AS, et al. Relation of cerebral energy metabolism and extracellular nitrite and nitrate concentrations in patients after aneurysmal subarachnoid hemorrhage. J Cereb Blood Flow Metab. 2001;21(9):1067–76.
157. Ng WH, Moochhala S, Yeo TT, Ong PL, Ng PY. Nitric oxide and subarachnoid hemorrhage: elevated level in cerebrospinal fluid and their implications. Neurosurgery. 2001;49(3):622–6.
158. Cooper CE. Nitric oxide and iron proteins. Biochim Biophys Acta. 1999;1411(2–3):290–309.
159. Félétou M, Köhler R, Vanhoutte PM. Endothelium-derived vasoactive factors and hypertension: possible roles in pathogenesis and as treatment targets. Curr Hypertens Rep. 2010;12(4):267–75.
160. Samdani AF, Dawson TM, Dawson VL. Nitric oxide synthase in models of focal ischemia. Stroke. 1997;28(6):1283–8.
161. Gahm C, Holmin S, Mathiesen T. Nitric oxide synthase expression after human brain contusion. Neurosurgery. 2002;50(6):1319–26.
162. Gahm C, Holmin S, Mathiesen T. Temporal profiles and cellular sources of three nitric oxide synthase isoforms in the brain after experimental contusion. Neurosurgery. 2000;46(1):169–77.
163. Sehba FA, Mostafa G, Friedrich V, Bederson JB. Acute microvascular platelet aggregation after subarachnoid hemorrhage. J Neurosurg. 2005;102(6):1094–100.
164. Kim P, Schini VB, Sundt TM, Vanhoutte PM. Reduced production of cGMP underlies the loss of endothelium-dependent relaxations in the canine basilar artery after subarachnoid hemorrhage. Circ Res. 1992;70(2):248–56.
165. Moro MA, Almeida A, Bolaños JP, Lizasoain I. Mitochondrial respiratory chain and free radical generation in stroke. Free Radic Biol Med. 2005;39(10):1291–304.
166. Szabó C, Dawson VL. Role of poly(ADP-ribose) synthetase in inflammation and ischaemia-reperfusion. Trends Pharmacol Sci. 1998;19(7):287–98.
167. Pluta RM, Oldfield EH, Boock RJ. Reversal and prevention of cerebral vasospasm by intracarotid infusions of nitric oxide donors in a primate model of subarachnoid hemorrhage. J Neurosurg. 1997;87(5):746–51.
168. Pluta RM, Dejam A, Grimes G, Gladwin MT, Oldfield EH. Nitrite infusions to prevent delayed cerebral vasospasm in a primate model of subarachnoid hemorrhage. JAMA. 2005;293(12):1477–84.

169. Pearson PJ, Evora PR, Seccombe JF, Schaff HV. Hypomagnesemia inhibits nitric oxide release from coronary endothelium: protective role of magnesium infusion after cardiac operations. Ann Thorac Surg. 1998;65(4):967–72.
170. Shechter M. The role of magnesium as antithrombotic therapy. Wien Med Wochenschr. 2000;150(15–16):343–7.
171. Sun X, Mei Y, Tong E. Effect of magnesium on nitric oxide synthase of neurons in cortex during early period of cerebral ischemia. J Tongji Med Univ. 2000;20(1):13–5. 42.
172. Garnier Y, Middelanis J, Jensen A, Berger R. Neuroprotective effects of magnesium on metabolic disturbances in fetal hippocampal slices after oxygen-glucose deprivation: mediation by nitric oxide system. J Soc Gynecol Investig. 2002;9(2):86–92.
173. Vergouwen MD, Vermeulen M, Coert BA, Stroes ES, Roos YB. Microthrombosis after aneurysmal subarachnoid hemorrhage: an additional explanation for delayed cerebral ischemia. J Cereb Blood Flow Metab. 2008;28(11):1761–70.
174. Dóczi TP. Impact of cerebral microcirculatory changes on cerebral blood flow during cerebral vasospasm after aneurysmal subarachnoid hemorrhage. Stroke. 2001;32(3):817.
175. Hirashima Y, Hamada H, Kurimoto M, Origasa H, Endo S. Decrease in platelet count as an independent risk factor for symptomatic vasospasm following aneurysmal subarachnoid hemorrhage. J Neurosurg. 2005;102(5):882–7.
176. Denton IC, Robertson JT, Dugdale M. An assessment of early platelet activity in experimental subarachnoid hemorrhage and middle cerebral artery thrombosis in the cat. Stroke. 1971;2(3):268–72.
177. Stein SC, Browne KD, Chen X-H, Smith DH, Graham DI. Thromboembolism and delayed cerebral ischemia after subarachnoid hemorrhage: an autopsy study. Neurosurgery. 2006;59(4):781–7.
178. Dumuis A, Pin JP, Oomagari K, Sebben M, Bockaert J. Arachidonic acid released from striatal neurons by joint stimulation of ionotropic and metabotropic quisqualate receptors. Nature. 1990;347(6289):182–4.
179. Dumuis A, Sebben M, Haynes L, Pin JP, Bockaert J. NMDA receptors activate the arachidonic acid cascade system in striatal neurons. Nature. 1988;336(6194):68–70.
180. Juvela S, Ohman J, Servo A, Heiskanen O, Kaste M. Angiographic vasospasm and release of platelet thromboxane after subarachnoid hemorrhage. Stroke. 1991;22(4):451–5.
181. Juvela S, Hillbom M, Kaste M. Platelet thromboxane release and delayed cerebral ischemia in patients with subarachnoid hemorrhage. J Neurosurg. 1991;74(3):386–92.
182. Juvela S. Cerebral infarction and release of platelet thromboxane after subarachnoid hemorrhage. Neurosurgery. 1990;27(6):929–35.
183. Friedrich V, Flores R, Muller A, Sehba FA. Luminal platelet aggregates in functional deficits in parenchymal vessels after subarachnoid hemorrhage. Brain Res. 2010;1354:179–87.
184. Okada Y, Copeland BR, Mori E, et al. P-selectin and intercellular adhesion molecule-1 expression after focal brain ischemia and reperfusion. Stroke. 1994;25(1):202–11.
185. Bara M, Guiet-Bara A, Durlach J. Analysis of magnesium membraneous effects: binding and screening. Magnes Res. 1988;1(1–2):29–34.
186. Ravn HB, Vissinger H, Kristensen SD, Husted SE. Magnesium inhibits platelet activity–an in vitro study. Thromb Haemost. 1996;76(1):88–93.
187. Ravn HB, Vissinger H, Kristensen SD, et al. Magnesium inhibits platelet activity – an infusion study in healthy volunteers. Thromb Haemost. 1996;75(6):939–44. Available at: http://www.ncbi.nlm.nih.gov/pubmed/8822590
188. Ravn HB, Kristensen SD, Vissinger H, Husted SE. Magnesium inhibits human platelets. Blood Coagul Fibrinolysis. 1996;7(2):241–4.
189. Adams JH, Mitchell JR. The effect of agents which modify platelet behaviour and of magnesium ions on thrombus formation in vivo. Thromb Haemost. 1979;42(2):603–10.
190. Hughes A, Tonks RS. Magnesium, adenosine diphosphate and blood platelets. Nature. 1966;210(5031):106–7.
191. Leao AA. Further observations on the spreading depression of activity in the cerebral cortex. J Neurophysiol. 1947;10(6):409–14.
192. Hossmann KA. Viability thresholds and the penumbra of focal ischemia. Ann Neurol. 1994;36(4):557–65.
193. Leao AA. Spreading depression of activity in the cerebral cortex. Fed Proc. 1944;3:28.
194. Hubschmann OR, Kornhauser D. Cortical cellular response in acute subarachnoid hemorrhage. J Neurosurg. 1980;52(4):456–62.
195. Hubschmann OR, Kornhauser D. Effect of subarachnoid hemorrhage on the extracellular microenvironment. J Neurosurg. 1982;56(2):216–21.
196. Dreier JP, Ebert N, Priller J, et al. Products of hemolysis in the subarachnoid space inducing spreading ischemia in the cortex and focal necrosis in rats: a model for delayed ischemic neurological deficits after subarachnoid hemorrhage? J Neurosurg. 2000;93(4):658–66.
197. Birse SH, Tom MI. Incidence of cerebral infarction associated with ruptured intracranial aneurysms. A study of 8 unoperated cases of anterior cerebral aneurysm. Neurology. 1960;10:101–6.
198. Neil-Dwyer G, Lang DA, Doshi B, Gerber CJ, Smith PW. Delayed cerebral ischaemia: the pathological substrate. Acta Neurochir. 1994;131(1–2):137–45.

199. Dreier JP, Woitzik J, Fabricius M, et al. Delayed ischaemic neurological deficits after subarachnoid haemorrhage are associated with clusters of spreading depolarizations. Brain. 2006;129(Pt 12):3224–37.
200. Kager H, Wadman WJ, Somjen GG. Conditions for the triggering of spreading depression studied with computer simulations. J Neurophysiol. 2002;88(5):2700–12.
201. Bosche B, Graf R, Ernestus RI, et al. Recurrent spreading depolarizations after subarachnoid hemorrhage decreases oxygen availability in human cerebral cortex. Ann Neurol. 2010;67(5):607–17.
202. Kraig RP, Nicholson C. Extracellular ionic variations during spreading depression. Neuroscience. 1978;3(11):1045–59.
203. Mayer ML, Westbrook GL, Guthrie PB. Voltage-dependent block by Mg^{2+} of NMDA responses in spinal cord neurones. Nature. 1984;309(5965):261–3.
204. Largo C, Cuevas P, Somjen GG, Martín del Río R, Herreras O. The effect of depressing glial function in rat brain in situ on ion homeostasis, synaptic transmission, and neuron survival. J Neurosci. 1996;16(3):1219–29.
205. Czéh G, Aitken PG, Somjen GG. Membrane currents in CA1 pyramidal cells during spreading depression (SD) and SD-like hypoxic depolarization. Brain Res. 1993;632(1–2):195–208.
206. Jing J, Aitken PG, Somjen GG. Interstitial volume changes during spreading depression (SD) and SD-like hypoxic depolarization in hippocampal tissue slices. J Neurophysiol. 1994;71(6):2548–51.
207. Aitken PG, Tombaugh GC, Turner DA, Somjen GG. Similar propagation of SD and hypoxic SD-like depolarization in rat hippocampus recorded optically and electrically. J Neurophysiol. 1998;80(3):1514–21.
208. LaManna JC, Rosenthal M. Effect of ouabain and phenobarbital on oxidative metabolic activity associated with spreading cortical depression in cats. Brain Res. 1975;88(1):145–9.
209. Takano T, Tian G-F, Peng W, et al. Cortical spreading depression causes and coincides with tissue hypoxia. Nat Neurosci. 2007;10(6):754–62.
210. Dietz RM, Weiss JH, Shuttleworth CW. Zn^{2+} influx is critical for some forms of spreading depression in brain slices. J Neurosci. 2008;28(32):8014–24.
211. Gwag BJ, Canzoniero LM, Sensi SL, et al. Calcium ionophores can induce either apoptosis or necrosis in cultured cortical neurons. Neuroscience. 1999;90(4):1339–48.
212. Dreier JP, Körner K, Ebert N, et al. Nitric oxide scavenging by hemoglobin or nitric oxide synthase inhibition by N-nitro-L-arginine induces cortical spreading ischemia when K^+ is increased in the subarachnoid space. J Cereb Blood Flow Metab. 1998;18(9):978–90.
213. Koehler RC, Roman RJ, Harder DR. Astrocytes and the regulation of cerebral blood flow. Trends Neurosci. 2009;32(3):160–9.
214. Dreier JP, Major S, Manning A, et al. Cortical spreading ischaemia is a novel process involved in ischaemic damage in patients with aneurysmal subarachnoid haemorrhage. Brain. 2009;132(Pt 7):1866–81.
215. Dreier JP. The role of spreading depression, spreading depolarization and spreading ischemia in neurological disease. Nat Med. 2011;17(4):439–47.
216. Rothman SM. Synaptic activity mediates death of hypoxic neurons. Science. 1983;220(4596):536–7.
217. Kass IS, Cottrell JE, Chambers G. Magnesium and cobalt, not nimodipine, protect neurons against anoxic damage in the rat hippocampal slice. Anesthesiology. 1988;69(5):710–5.
218. van den Bergh WM, Zuur JK, Kamerling NA, et al. Role of magnesium in the reduction of ischemic depolarization and lesion volume after experimental subarachnoid hemorrhage. J Neurosurg. 2002;97(2):416–22.
219. van der Hel WS, van den Bergh WM, Nicolay K, Tulleken KA, Dijkhuizen RM. Suppression of cortical spreading depressions after magnesium treatment in the rat. Neuroreport. 1998;9(10):2179–82.
220. Olsen TS. Pathophysiology of the migraine aura: the spreading depression theory. Brain. 1995;118(Pt 1):307–8.
221. Speich M, Bousquet B, Nicolas G. Reference values for ionized, complexed, and protein-bound plasma magnesium in men and women. Clin Chem. 1981;27(2):246–8.
222. Musso CG. Magnesium metabolism in health and disease. Int Urol Nephrol. 2009;41(2):357–62.
223. Fong J, Gurewitsch ED, Volpe L, et al. Baseline serum and cerebrospinal fluid magnesium levels in normal pregnancy and preeclampsia. Obstet Gynecol. 1995;85(3):444–8.
224. van den Bergh WM, Algra A, Rinkel GJ. Electrocardiographic abnormalities and serum magnesium in patients with subarachnoid hemorrhage. Stroke. 2004;35(3):644–8.
225. Helpern JA, Vande Linde AM, Welch KM, et al. Acute elevation and recovery of intracellular [Mg^{2+}] following human focal cerebral ischemia. Neurology. 1993;43(8):1577–81.
226. Ryu JH, Sohn IS, Do SH. Controlled hypotension for middle ear surgery: a comparison between remifentanil and magnesium sulphate. Br J Anaesth. 2009;103(4):490–5.
227. Crozier TA, Radke J, Weyland A, et al. Haemodynamic and endocrine effects of deliberate hypotension with magnesium sulphate for cerebral-aneurysm surgery. Euro J Anaesthesiol. 1991;8(2):115–21.
228. Fuchs-Buder T, Tramer MR, Tassonyl E. Cerebrospinal fluid passage of intravenous magnesium sulfate in neurosurgical patients. J Neurosurg Anesthesiol. 1997;9(4):324–8.
229. Wong GK, Yeung DK, Ahuja AT, et al. Intracellular free magnesium of brain and cerebral phosphorus-containing metabolites after subarachnoid hemorrhage and hypermagnesemic treatment: a ^{31}P–magnetic resonance spectroscopy study. Clinical article. J Neurosurg. 2010;113(4):763–9.

230. Oppelt WW, MacIntyre I, Rall DP. Magnesium exchange between blood and cerebrospinal fluid. Am J Physiol. 1963;205(5):959–62.
231. Altura BT, Altura BM. The role of magnesium in etiology of strokes and cerebrovasospasm. Magnesium. 1982;1:277–91.
232. Wong GK, Lam CW, Chan MT, Gin T, Poon WS. The effect of hypermagnesemic treatment on cerebrospinal fluid magnesium level in patients with aneurysmal subarachnoid hemorrhage. Magnes Res. 2009;22(2):60–5.
233. Sjostrom LG, Wester P. Accumulation of magnesium in rat brain after intravenously induced hypermagnesemia. Cerebrovasc Dis. 1995;4:241.
234. Hallak M, Berman RF, Irtenkauf SM, Evans MI, Cotton DB. Peripheral magnesium sulfate enters the brain and increases the threshold for hippocampal seizures in rats. Am J Obstet Gynecol. 1992;167(6):1605–10.
235. Altura BT, Altura BM. Withdrawal of magnesium causes vasospasm while elevated magnesium produces relaxation of tone in cerebral arteries. Neurosci Lett. 1980;20(3):323–7.
236. Turlapaty PD, Altura BM. Magnesium deficiency produces spasms of coronary arteries: relationship to etiology of sudden death ischemic heart disease. Science. 1980;208(4440):198–200.
237. Mori K, Yamamoto T, Miyazaki M, et al. Optimal cerebrospinal fluid magnesium ion concentration for vasodilatory effect and duration after intracisternal injection of magnesium sulfate solution in a canine subarachnoid hemorrhage model. J Neurosurg. 2011;114(4):1168–75.
238. Mori K, Yamamoto T, Nakao Y, et al. Initial clinical experience of vasodilatory effect of intra-cisternal infusion of magnesium sulfate for the treatment of cerebral vasospasm after aneurysmal subarachnoid hemorrhage. Neurol Med Chir. 2009;49(4):139–44.
239. Simpson JI, Eide TR, Schiff GA, et al. Intrathecal magnesium sulfate protects the spinal cord from ischemic injury during thoracic aortic cross-clamping. Anesthesiology. 1994;81(6):1493–9.
240. Bahar M, Chanimov M, Grinspun E, Koifman I, Cohen ML. Spinal anaesthesia induced by intrathecal magnesium sulphate. Anaesthesia. 1996;51(7):627–33.
241. Chanimov M, Cohen ML, Grinspun Y, et al. Neurotoxicity after spinal anaesthesia induced by serial intrathecal injections of magnesium sulphate. An experimental study in a rat model. Anaesth. 1997;52(3):223–8. Available at: http://www.ncbi.nlm.nih.gov/pubmed/9124662
242. Jellish WS, Zhang X, Langen KE, et al. Intrathecal magnesium sulfate administration at the time of experimental ischemia improves neurological functioning by reducing acute and delayed loss of motor neurons in the spinal cord. Anesthesiology. 2008;108(1):78–86.
243. Karasawa S, Ishizaki K, Goto F. The effect of intrathecal administration of magnesium sulphate in rats. Anaesthesia. 1998;53(9):879–86.
244. Takano Y, Sato E, Kaneko T, Sato I. Antihyperalgesic effects of intrathecally administered magnesium sulfate in rats. Pain. 2000;84(2–3):175–9.
245. Tsai SK, Huang SW, Lee TY. Neuromuscular interactions between suxamethonium and magnesium sulphate in the cat. Br J Anaesth. 1994;72(6):674–8.
246. Mao J, Price DD, Mayer DJ. Mechanisms of hyperalgesia and morphine tolerance: a current view of their possible interactions. Pain. 1995;62(3):259–74.
247. Goodman EJ, Haas AJ, Kantor GS. Inadvertent administration of magnesium sulfate through the epidural catheter: report and analysis of a drug error. Int J Obstet Anesth. 2006;15(1):63–7.
248. Wong GK, Boet R, Poon WS, et al. Intravenous magnesium sulphate for aneurysmal subarachnoid hemorrhage: an updated systemic review and meta-analysis. Crit Care. 2011;15(1):R52.
249. Boet R, Mee E. Magnesium sulfate in the management of patients with Fisher Grade 3 subarachnoid hemorrhage: a pilot study. Neurosurgery. 2000;47(3):602–6.
250. Chia RY, Hughes RS, Morgan MK. Magnesium: a useful adjunct in the prevention of cerebral vasospasm following aneurysmal subarachnoid haemorrhage. J Clin Neurosci. 2002;9(3):279–81.
251. van den Bergh WM, Albrecht KW, Berkelbach van der Sprenkel JW, Rinkel GJ. Magnesium therapy after aneurysmal subarachnoid haemorrhage a dose-finding study for long term treatment. Acta Neurochir. 2003;145(3):195–9.
252. van Norden AG, van den Bergh WM, Rinkel GJ. Dose evaluation for long-term magnesium treatment in aneurysmal subarachnoid haemorrhage. J Clin Pharm Ther. 2005;30(5):439–42.
253. Yahia AM, Kirmani JF, Qureshi AI, Guterman LR, Hopkins LN. The safety and feasibility of continuous intravenous magnesium sulfate for prevention of cerebral vasospasm in aneurysmal subarachnoid hemorrhage. Neurocrit Care. 2005;3(1):16–23.
254. Prevedello DM, Cordeiro JG, de Morais AL, et al. Magnesium sulfate: role as possible attenuating factor in vasospasm morbidity. Surg Neurol. 2006;65(14–1):20.
255. Schmid-Elsaesser R, Kunz M, Zausinger S, et al. Intravenous magnesium versus nimodipine in the treatment of patients with aneurysmal subarachnoid hemorrhage: a randomized study. Neurosurgery. 2006;58(6):1054–65.
256. Stippler M, Crago E, Levy EI, et al. Magnesium infusion for vasospasm prophylaxis after subarachnoid hemorrhage. J Neurosurg. 2006;105(5):723–9.

257. Veyna RS, Seyfried D, Burke DG, et al. Magnesium sulfate therapy after aneurysmal subarachnoid hemorrhage. J Neurosurg. 2002;96(3):510–4.
258. Muroi C, Terzic A, Fortunati M, Yonekawa Y, Keller E. Magnesium sulfate in the management of patients with aneurysmal subarachnoid hemorrhage: a randomized, placebo-controlled, dose-adapted trial. Surg Neurol. 2008;69(1):33–9.
259. van den Bergh WM, Algra A, van Kooten F, et al. Magnesium sulfate in aneurysmal subarachnoid hemorrhage: a randomized controlled trial. Stroke. 2005;36(5):1011–5.
260. Wong GK, Chan MT, Boet R, Poon WS, Gin T. Intravenous magnesium sulfate after aneurysmal subarachnoid hemorrhage: a prospective randomized pilot study. J Neurosurg Anesthesiol. 2006;18(2):142–8.
261. Wong GK, Poon WS, Chan MT, et al. Intravenous magnesium sulphate for aneurysmal subarachnoid hemorrhage (IMASH): a randomized, double-blinded, placebo-controlled, multicenter phase III trial. Stroke. 2010;41(5):921–6.
262. Westermaier T, Stetter C, Vince GH, et al. Prophylactic intravenous magnesium sulfate for treatment of aneurysmal subarachnoid hemorrhage: a randomized, placebo-controlled, clinical study. Crit Care Med. 2010;38(5):1284–90.
263. Saver JL, Kidwell C, Eckstein M, Starkman S. Prehospital neuroprotective therapy for acute stroke: results of the Field Administration of Stroke Therapy-Magnesium (FAST-MAG) pilot trial. Stroke. 2004;35(5):e106–8.
264. Lejuste MJ. Inadvertant intrathecal administration of magnesium sulfate. S Afr Med J. 1985;68(6):367–8.

Chapter 20
Magnesium and Alcohol

Teresa Kokot, Ewa Nowakowska-Zajdel, Małgorzata Muc-Wierzgoń, and Elżbieta Grochowska-Niedworok

Key Points

- Excessive alcohol consumption causes tissue magnesium loss (brain, liver, heart, and skeletal muscles) which leads to a decrease of serum Mg concentration. Its excretion increases by two- to threefold.
- Liver is the organ most exposed to alcohol effects, since it is the main location of ethanol metabolism.
- Magnesium deficiency, as a side effect of alcohol consumption, may be the predisponent to most disorders (e.g., fatty hepatitis, cirrhosis, *ischemic heart disease*, dilated cardiomyopathy, muscle dystrophy).
- Most of neurological disorders (starting from Korsakoff's syndrome and ending up with brain stroke) are caused by the effect of alcohol on the vessels, leading to vein and arteriole contraction, followed by uvea damage on microcapillary level.
- Neuromuscular hyperactivity caused by magnesium deficiency may be included in the common symptoms of chronic alcoholism and deficiency of this element.
- Supplementation of magnesium has a beneficial effect on the body, since it decreases intensity of clinical symptoms related to excessive alcohol consumption.

Keywords Alcohol consumption • Magnesium deficiency • Liver ethanol metabolism • Hepatic disorders • Cardiovascular disorders • Psychoneuromuscular diseases • Neuromuscular hyperactivity • Supplementation of Magnesium

Introduction

Alcohol is a socially acceptable drug, and its excessive consumption is still an up-to-date social and health problem. According to the World Health Organization, alcohol is the third risk factor for population health [1].

T. Kokot, M.D., Ph.D. (✉) • E. Nowakowska-Zajdel, M.D., Ph.D. • M. Muc-Wierzgoń, Prof.
• E. Grochowska-Niedworok, Ph.D.
Department of Internal Medicine and Department of Human Nutrition, Medical University of Silesia, Żeromskiego 7, Bytom 41-902, Poland
e-mail: teresa.kokot@gmail.com; ewanz@onet.eu; mwierzgon@sum.edu.pl; travel1@poczta.onet.pl

R.R. Watson et al. (eds.), *Magnesium in Human Health and Disease*, Nutrition and Health,
DOI 10.1007/978-1-62703-044-1_20, © Springer Science+Business Media New York 2013

Approximately 75% of the world population consumes alcohol, and in 10% of this group, alcohol consumption causes serious health problems. It is described that men are five times more susceptible to alcoholism compared to women [1].

Alcoholism causes failure of numerous organs and systems. There is evidence for being the arterial hypertension factor; moreover, excessive alcohol consumptions increases the risk of brain stroke and heart attack [2, 3]. In the course of alcoholic disease, degeneration of cardiac muscle fibers, symptoms of steatosis, and dilated cardiomyopathy may occur, as well as vitamin B1 deficiency, with harmful effect to the heart [3]. Adverse effects of excessive alcohol consumption affect also the hematologic, neurologic, and gastric systems. As a result of secondary vitamin B12 deficiency, which may be observed in alcoholic disease, hematopoiesis disorders with secondary anemia, leucopenia, and thrombocytopenia occur.

Immunity disorders as well as the degeneration of antibody production result in the increased occurrence of infectious diseases such as tuberculosis [1]. Many studies show that excessive alcohol consumption is responsible for cirrhosis, oral and oropharyngeal cancer, and malicious tumors of the gastrointestinal tract and breasts [1]. It also causes social pathologies, suicides, and traffic accidents [1].

Numerous studies show that Mg deficiency may have a key impact on excessive alcohol consumption, with simultaneous magnesium homeostasis disorders caused by alcohol. Alcoholism decreases Mg level in serum and tissues which leads to secondary organ alterations. However, the experimental trials show that administration of an excessive dose of Mg may decrease the ethanol demand (observations on rats addicted to alcohol). In this "vicious circle," it is difficult to distinguish the actual effect of Mg maintenance disorders and the alcohol effect on the toxicomania establishment [4]. This data confirm that Mg affects the tendency to alcohol addiction and plays a significant role in its therapy [4].

Magnesium and Its Importance in Human Body

Magnesium (Mg^2) is a microelement indispensable for life. It is one of the four most important cations in human body (sodium, potassium, calcium, and magnesium) and the second, after potassium, intracellular element [5–10]. A body of a grown man contains about 24 g (1,000 mmol/l) of magnesium, 60% of which is located in bones, 39% in soft tissues (brain, heart, liver, kidneys, and skeletal muscles), and only 1% in intercellular fluids including plasma. About 16% of systemic magnesium is exchanged. Normal magnesium concentration amounts to 0.8–1.0 mmol/l which is only 1% of systemic resources. Total magnesium concentration in serum concerns three fractions: bound with proteins (mostly albumins), 19%; with anions (citrate, lactate, and bicarbonate) and with phospholipids, 14%; and with ionized calcium, 67% [2, 11].

In the cell, magnesium may be found mostly inside the mitochondria, endoplasmic reticulum, nucleus, and cytoplasm [10, 12]. In the cytoplasm, it forms a complex with ATP and ADP; in the nucleus, it is bound with DNA, RNA, and free nucleotides [5]. Magnesium bound with DNA plays a significant role in the secondary and tertiary DNA structure stabilization. However, the effect of this element on gene regulation remains unexplained [Wolf and Cittadini 2003]. Mg is of a great significance in cell cycle, creating a complex with ATP. This complex activates a phosphorylation cascade in the cell with protein kinases. Moreover, it modulates the histone phosphorylation [13, 14]. It has also an impact on functions of enzymes participating in DNA (endonuclease) repair, replication (topoisomerases II, polymerases I), and transcription (ribonucleases H) [13]. Magnesium shows high biochemical activity due to its relatively small radius compared to the size of nucleus (0.86 and 1.14 A for Mg^2 and Ca^2, respectively) [13]. Concentration of free magnesium varies between 0.5 and 0.6 mmol/l, of which 2–3% of Mg is free.

Cell and serum magnesium levels are affected by hormonal and metabolic disorders, including hyperaldosteronism, diabetes, arterial hypertension, and alcoholism [2]. Along with the decrease of cell and serum magnesium levels, decrease of potassium level, increase of sodium, and frequently an increase of Ca may be observed, which increases the systemic homeostasis disorders and affects tissue and organ disorders [10, 15].

Magnesium Deficiency Caused by Alcohol Consumption

Primary magnesium deficiency results most often from its insufficient supply in diet. In men, it usually has chronic, however mild course. Secondary deficiency is most often caused by absorption and excretion disorders, occurs in course of numerous endocrinopathies, during excessive tissue accumulation, in bone diseases, and as a consequence of alcohol consumption (acute intoxication or chronic alcohol disease) [4].

Acute Alcohol Intoxication

Excessive alcohol consumption causes a significant (5–10%) decrease (outflow) of Mg in hepatocytes, in dose-dependent time and manner, resulting from temporary decrease of the cellular ATP level [12].

Magnesium deficiency causes the formation of its first metabolism product, acetaldehyde, and, subsequently – from acetaldehyde and brain biogenic amines – "Tetrahydroisochinolines", true pseudo-opioids determining the habit by bonding with endorphin receptors.

The changes in PKC translocation and accumulation of magnesium may be observed after an excessive consumption (within 8 min) of a single dose of alcohol. It takes 75 min to normalize the processes [5].

Excessive alcohol consumption causes tissue magnesium loss which leads to a decrease of serum Mg concentration. At the same time, its excretion increases by two- to threefold. The decrease of serum Mg concentration affects the brain, liver, heart, and skeletal muscles variably. Its deficiency in those tissues leads to cell cycle, energy production, and protein synthesis disorders and, as a result, cell, tissue, and organ function disorders. Changes in tissue Mg levels are accompanied by changes in other ion levels: K, Ca, and Na (particularly decrease of K level and increase of Na and Ca levels) [5].

The observed hypermagnesuria lasts for a short period of time, and it relates mostly to acute alcohol intoxication. The surplus magnesium in urine is also caused by other disorders such as coexisting liver insufficiency, stress, or iatrogenic factors.

Chronic Alcohol Intoxication

Based on the previous scientific research, it has been found that the secondary magnesium deficiency in the course of chronic alcoholism causes many clinical disorders. The mechanisms causing alcohol-related magnesium deficiency depend on various clinical conditions. One of the basic factors predisposing to magnesium deficiency is its insufficient supply in diet, which may result from lack of appetite and secondary nutrition disorders.

In liver diseases, especially chronic or accompanied by cirrhosis and secondary insufficiency, tissue Mg deficiency is a common complication. It is related to the impaired appetite as well as coexistent disorders in absorption and digestion (damaged gastrointestinal mucous or loss of the element caused by fatty diarrhea or increased ammonium production) as well as metabolic disorders (hypoalbuminemia, dyselectrolytemia). Magnesium deficiency results also from treatment with diuretic medications and glucocorticoids [4].

In pancreas diseases, in the course of excessive alcohol consumption, an abnormal organ distribution of magnesium occurs apart from coexistent intestine function disorders [4]. Magnesium loss occurs apart from coexistent intestine function disorders in excessive sweating (delirium tremens), above-mentioned side effects of medications, or loss of the element with urine.

Magnesium deficiency, as a side effect of alcohol consumption, may be the predisponent to most of the previously described disorders, such as brain stroke, sarcopenia, cardiomyopathy, fatty hepatitis, cirrhosis, and others. Therefore, it is important to explain the pathomechanism of the effect of magnesium transport disorders caused by alcohol consumption on the functions of brain and its vessels, skeletal muscles, cardiac muscle, and liver cells.

The Effect of Alcohol on Magnesium Balance in the Body as well as Tissue and Organ Function Disorders

Alcohol, Magnesium Deficiency, and Liver Cell Damage

Liver is the organ most exposed to alcohol effects, since it is the main location of ethanol metabolism. Liver cells are most commonly used as a model in research on the effect of ethanol on human body [5]. Acute and chronic alcohol consumption lead to changes in Mg homeostasis and transport in liver cells. In both cases, the concentration of free Mg^{2+} decreases (which has been experimentally confirmed), and, at the same time, the level of Na^{+} and Ca^{2+} increases [5].

Single doses of alcohol have a small impact on Mg concentration changes or even none. However, Mg homeostasis and transport disorders show a dependence on the time of use of the stimulant [5]. Acute alcohol consumption is associated with liver cell Mg displacement, which, to a large extent, depends on the temporary decrease of cellular ATP [6, 16, 17]. Displacement of Mg is possible due to the Na-dependent pump and as a result of ethanol metabolism by the alcohol dehydrogenase [5].

The use of alcohol also affects the liver cell response to catecholamine stimulation. The effect of catecholamine use will be also a sudden Mg displacement from the cell by means of Na/Mg ion exchange [5]. The use of even small doses of alcohol – 0.01% (1.5 mM) – is sufficient to inhibit the Mg accumulation in liver cells for 60 min [12]. This effect depends on the incorrect distribution of PKCε bound with the cell membrane.

Chronic use of alcohol in diet for 3 weeks causes loss of 22–25% of the total Mg accumulated in liver cells (in cytoplasm, mitochondria, and endoplasmic reticulum) [5]. This deficiency is associated with a decrease of the ATP level by 17% [5, 6]. Also, an incorrect PKC kinase translocation in the cell is observed [5]. Scientific research shows that 12-day period is necessary to regain the PKCσ translocation and return to the correct Mg accumulation in the cell.

Acute and chronic alcohol use cause faulty translocation of PKCσ to the cell membrane. Researches confirm the significance of PKC isoform in the course of Mg accumulation in liver cells. From among all PKC isoforms, only PKCε is particularly sensitive on the cellular level [12].

Monitoring of the metabolic efficiency of liver and determining its energetic reserve as well as regenerative potential are extremely important to assess the toxic effect of alcohol. One of the methods that may be used to assess the metabolic efficiency of liver is testing the course of ketogenesis by means of determining the concentration of ketones, i.e., acetoacetic acid and β-hydroxybutyrate, and the calculation of molar concentration proportion, i.e., KBR factor [6].

Decreased magnesium volume caused by ethanol may lead to cirrhosis (Mg loss in liver is associated with an increased accumulation of collagen).

Alcohol, Magnesium Deficiency, and Cardiovascular Disorders

Within the cardiac muscle, magnesium acts as calcium antagonist and decreases both the conductivity and excitability. It has a protective effect on the heart by counteracting hypoxia and ischemia. Magnesium protects blood vessel walls by combating excess calcium and changes in connective tissue as well as by the direct antispasmodic effect. Deficiency of this element may cause cardiac muscle diseases and changes within blood vessels [2]. Magnesium deficiency, because of its effect on morphotic blood elements (thrombocyte and erythrocyte stabilizer), may be defined as a vessel risk factor [1].

Excessive alcohol consumption affects the development of alcohol-related dilated cardiomyopathy [5, 18, 19]. It probably results from lack of alcohol dehydrogenase in the cytoplasm of cardiac muscle cells and the presence of alcohol induced by cyt P450-2E1 in sarcoplasmic reticulum of cardiomyocytes [5, 18]. Both these enzymes are responsible for the conversion of alcohol to acetaldehyde, which is considered to have an adverse effect that may be observed not only in cardiomyocytes but also in skeletal and liver muscles.

Alcohol, Magnesium Deficiency, and Nervous System Disorders

Magnesium is a neurotropic ion with a soothing effect. Its deficiency causes scattered neuromuscular hyperactivity, which spreads from the cerebral cortex to subcortical layer through the extradural neuron and neuromuscular bonds (curarizing action) through peripheral and autonomic innervations [4].

Acute and chronic alcohol consumption is the cause of many neurological disorders, starting from Korsakoff's syndrome [20] and ending up with brain stroke [21, 22]. Most of those disorders are caused by the effect of alcohol on the vessels, leading to vein and arteriole contraction, followed by uvea damage on microcapillary level [5, 23, 24].

Magnesium deficiency also facilitates vessel damage by affecting calcium phosphate, protein, carbohydrate, and glycoprotein metabolisms; aging changes in the connective tissue; and fatty protein changes which facilitate atherosclerotic lipid disorders in the blood [4].

Decrease of extracellular Mg concentration in alcoholism is responsible for small arteriole and blood vessel contraction. The increase of cell Ca level associated with decrease of Mg level causes (pulse) increase of brain vessel contraction, which, in effect, increases the incidence of hemorrhagic stroke. Research in alcohol addicts shows rapid Mg loss in brain cells, especially in astrocytes. Apart from astrocyte Mg content, changes may be noticed also in brain vessel endothelium. Glial cells are necessary to maintain metabolic Mg balance [5, 23]. Acute as well as chronic exposure of those cells to alcohol causes a decrease of cell Mg concentration, which leads to changes in function and concentration of protein kinase C (PKC) isoforms, tyrosine phosphorylation, and changes in oxidation processes. Ethanol decreases also the content of magnesium in brain cells, brain vessels, smooth muscle cells, as well as in the in vivo extracellular environment [5]. Decrease of extracellular Mg concentration affects the activity of these cells and results in the increase of c-fos and c-jun proto-oncogenes and NF-kB expression, which has a potential significance for the development of arterial hypertension, vessel diseases, or brain stroke [5].

In chronic alcohol disease, a decrease of Mg concentration in brain may be observed, particularly in the neuron myelin sheath [...]. Selective decrease of brain Mg level may be attributed to respiratory alkalosis which occurs after alcohol withdrawal [5]. Neuromuscular hyperactivity caused by Mg deficiency may be included in the common symptoms of chronic alcoholism and deficiency of this element. Secondarily, trembling, myoclonus, excitation states, over timidity, depression tendencies or, on the contrary, aggressive behavior occur. Moderate doses of alcohol help to decrease or eliminate neurological symptoms of magnesium deficiency. In alcoholics, the consumption of alcohol eases the symptoms from the nervous system, such as neuromuscular hyperactivity which occur as a result of deficiency of this element. Sudden decrease of soothing effect of alcohol which masks the neuromuscular hyperactivity caused by Mg deficiency cooperates with the occurrence of the withdrawal syndrome. This syndrome, in terms of clinical symptoms, paradoxically, originates from nervous cell Mg deficiency [4]. The occurrence of pre-delirium epilepsy or delirium tremens in course of alcohol withdrawal is associated with many aspects, including magnesium deficiency, respiratory alkalosis, pyridoxine deficiency, decrease of adrenal cortex inner secretion, or decrease of brain oxygen metabolism. At the same time, there is a particular correlation between the abnormal EEG record, alcohol withdrawal, and decrease of serum Mg concentration. The correlation between cerebrospinal fluid Mg concentration decrease and the decrease of serum Mg is maintained with simultaneous maintenance of gradient between those two concentrations.

Exclusive magnesium deficiency is also the cause of peripheral nervous system disorder. Lack of cooperation between magnesium and B group vitamins may be the cause of alcohol multinervous inflammations. Inflammation of the optic nerve is a frequent symptom in the course of alcoholism [4]. Multinervous inflammation of other than alcohol-related origin is accompanied by an increase of serum Mg concentration.

In alcohol addicts with serious psychoneuromuscular disorders, a decrease of serum Mg concentration and increase of PCK activity are the most specific symptoms [4].

Alcohol, Magnesium Deficiency, and Skeletal Muscle Disorders

Magnesium increases the muscle efficiency by participating in energy reserve accumulation and its utilization, as in case of myosin B creation. The effect of magnesium on skeletal muscles completes the muscle soothing effect by decreasing the neuromuscular striated fiber excitability (miorelaxing effect).

Many tests show that Mg deficiency in alcoholics leads to skeletal muscle dystrophy and impairment of contraction strength. These changes are the result of muscle protein synthesis decrease, protease activity increase, lack of nucleic acid stabilization, but mostly the generation of acetaldehyde from ethanol [24].

Research on the mechanisms leading to skeletal muscle dystrophy and decrease of contraction strength shows evidence for the occurrence of protein transcription and translation changes [5]. Those changes mostly include the decrease of type II protein synthesis compared to type I in muscle fibers (i.e., plantaris muscle or soleus muscle). In mixed muscle fibers, for example, in the calf, the synthesis of both types of fibers decreases to a comparable extent [...]. Research by other authors showed a simultaneous decrease of total mRNA and proteins in muscle bioptats in rats fed with ethanol for 2 weeks [5]. Those changes are associated with the muscle contraction release and the elongation of relaxation phase. In hypomagnesemia, in the course of alcoholism, a decrease of any isometric quadriceps muscle contraction may be observed, which is closely correlated with the decrease of serum Mg concentration and coupled with a decrease of the balance between creatine kinase, AD, and phosphocreatine concentration, as well as Mg concentration in muscles [5].

Summary

Until recently, the physiological role of magnesium has been ignored and magnesium metabolism disorders have not been diagnosed due to rare estimation of this electrolyte in the blood. Mechanisms responsible for Mg loss caused by excessive alcohol consumption as well as the protective effect of magnesium supplementation have not been researched in detail so far. Knowledge of magnesium biochemistry and pathophysiology allows to recognize the role of this element in the conditions of excessive exposition of the body to ethanol.

Magnesium ions, which are present in all cells of the body, participate actively in numerous pathophysiological effects. In the course of many idiopathic diseases, i.e., civilization diseases, alcoholism, diabetes, etc., or due to use of various therapeutic methods, its metabolism may be unbalanced. From the clinical point of view, supplementation of Mg^{+2} has a beneficial effect on the body, since it decreases the intensity of clinical symptoms related to the excessive alcohol consumption.

References

1. Szczeklik E, et al. Internal Medicine 2010. Carcow; Medical Publisher. 2010.
2. Touyz R. Magnesium in clinical medicine. Front Biosci. 2004;9:1278–93.
3. Lucas D, Brown R, Wassef M, Giles T. Alcohol and the cardiovascular system. JACC. 2005;45(12):1916–24.
4. Durlach J. Magnesium in Clinical Practice. London. John Libbey, 1988.
5. Romani AM. Magnesium homeostasis and alcohol consumption. Magnes Res. 2008;21(4):197–204.
6. Young A, Cefaratti C, Romani A. Chronic EtOH administration alters liver MG2+ homeostasis. Am J Physiol Gastrointest Liver Physiol. 2003;284:G57–67.
7. Gunther T. Functional compartmentation of intracellular magnesium. Magnesium. 1986;5:53–9.
8. Gunther T. Mechanisms and regulation of Mg $^{2+}$efflux and Mg^{2+}influx. Miner Electrolyte Metab. 1993;19:250–65.
9. Romani A, Scarpa A. Regulation of cell magnesium. Arch Biochem Biophys. 1992;298:1–12.
10. Romani A, Scarpa A. Regulation of cellular magnesium. Front Biosci. 2000;5:D720–34.
11. Kokot F, Hyla-Klekot L, Kokot S. Laboratory tests. Scope and interpretation of standards, Katowice. Medical Publisher; 2011.
12. Torres L, Konopnika B, Berti-Mattera L, Liedtke C, Romani A. Defective translocation of PKCϵ in EtOH-induced inhibition of Mg^2 accumulation in rat hepatocytes. Alcohol Clin Exp Res. 2010;34(9):1659–69.
13. Wolf FI, Trapani V. Cell (patho)physiology of magnesium. Clin Sci. 2008;114:27–35.
14. Pasternak K, Kocot J, Horecka A. Biochemistry of magnesium. J Elementol. 2010;15(3):601–16.
15. Sontia B, Touyz RM. Role of magnesium in hypertension. Arch Biochem Biophys. 2007;458:33–9.
16. Tessman P, Romani A. Acute ethanol administration affects Mg^{2+}homeostasis in liver cells: evidence for the activation of a Na^+/Mg^{2+}exchanger. Am J Physiol. 1998;275:G1106–16.
17. Young A, Berti-Mattera L, Romani A. Effect of repeated doses of ethanol on hepatic Mg^2. Homeostasis and mobilization. Alcohol Clin Exp Res. 2007;31(7):1240–51.
18. Lucas DL, Brown RA, Wassef M, Giles TD. Alcohol and the cardiovascular system: research challenges and opportunities. J Am Coll Cardiol. 2005;45:1916–24.
19. Brown RA, Crawford M, Natavio M, Petrowski P, Ren J. Dietary magnesium supplementation attenuates ethanol-induced myocardial dysfunction. Alcohol Clin Exp Res. 1998;22:2062–72.
20. Berman MO. Severe brain dysfunction: alcoholic Korsakoffs syndrome. Alcohol Health Res World. 1990;14:120–9.
21. Ashley MJ. Alcohol consumption, ischemic heart disease and cerebrovascular disease: an epidemiology perspective. J Stud Alcohol. 1982;43:869–87.
22. Donahue RP, Abbott RD, Reed DM, Yano K. Alcohol and hemorrhagic stroke: the Honolulu heart program. JAMA. 1986;225:2311–4.
23. Price TNC, Burke JF, Mayne LV. A novel human astrocyte cell line (A735) with astrocyte- specific neuro-transmitter function. In vitro Cell Dev Biol Animal. 1999;35:279–88.
24. Preedy VR, Reilly ME, Patel VB, Richardson PJ, Peters TJ. Protein metabolism in alcoholism: effects on specific tissues and the whole body. Nutrition. 1999;15(7/8):604–8.

Index

R.R. Watson et al. (eds.), *Magnesium in Human Health and Disease*, Nutrition and Health,
DOI 10.1007/978-1-62703-044-1, © Springer Science+Business Media New York 2013

MIX
Papier aus verantwortungsvollen Quellen
Paper from responsible sources
FSC® C105338

If you have any concerns about our products,
you can contact us on
ProductSafety@springernature.com

In case Publisher is established outside the EU,
the EU authorized representative is:
Springer Nature Customer Service Center GmbH
Europaplatz 3, 69115 Heidelberg, Germany

Printed by Libri Plureos GmbH
in Hamburg, Germany